Multimedia over Cognitive Radio Networks

Algorithms, Protocols, and Experiments

Multimedia over Cognitive Radio Networks

Algorithms, Protocols, and Experiments

Edited by
Fei Hu and Sunil Kumar

CRC Press
Taylor & Francis Group
Boca Raton   London   New York

CRC Press is an imprint of the
Taylor & Francis Group, an **Informa** business

CRC Press
Taylor & Francis Group
6000 Broken Sound Parkway NW, Suite 300
Boca Raton, FL 33487-2742

First issued in paperback 2016

© 2015 by Taylor & Francis Group, LLC
CRC Press is an imprint of Taylor & Francis Group, an Informa business

No claim to original U.S. Government works

Version Date: 20141014

ISBN 13: 978-1-138-03401-3 (pbk)
ISBN 13: 978-1-4822-1485-7 (hbk)

Visit the Taylor & Francis Web site at
http://www.taylorandfrancis.com

and the CRC Press Web site at
http://www.crcpress.com

To Fang Yang, Gloria Yang Hu, Edwin and Edward (twins)

Also dedicated to Prof. Sunil Kumar's family (Rajni, Paras, and Shubham)

Contents

Preface

Rapid development of multimedia applications (such as video streaming) over wireless networks requires wide channel bandwidth. Even when the available radio spectrum is allocated, large portions of it are not used efficiently. Lately, cognitive radio networks (CRNs), which use intelligent dynamic spectrum access techniques to use the unoccupied spectrum, have received much attention. In CRN, the unlicensed secondary users (SUs), also known as cognitive users, search intelligently the vacant spectral bands and access them to maximize SUs performance. An SU needs to hand off to other vacant channels if its current channel is needed by licensed primary users (PUs). The main issues in CRN design involve *spectrum analysis* and *spectrum decision*. The aim of *spectrum analysis* is to use the signal processing and machine learning algorithms to extract the patterns of the sensed idle spectrum (from the *spectrum-sensing* module). Examples of patterns include channel quality changes in a time window, channel holding time, and the probability of control channel saturation in the next time slot. These patterns are used by the *spectrum decision* module to generate accurate network control *actions* based on the reasoning of the system *state* in the current time slot. Examples of *actions* include the spectrum hand-off, allocation of spectral bands to SUs, adjustment of the video source rate, data allocation in the orthogonal frequency division multiplexing (OFDM) channel, and spectrum scanning rate.

Many challenging issues related to multimedia transmission over CRN still remain unresolved. Examples of these issues include QoS (quality of service) support for delivering a video stream over dynamic channels that change each time the PU takes the channel back, using queuing theory to model the delay/jitter parameters when an SU performs spectrum hand-off, QoE (quality of experience) support that reflects the users' satisfaction with a video clip's resolution, achieving higher-priority transmission for users, and integrating the video coding with CRN protocols in order to achieve a better QoS/QoE performance. This book covers the algorithms, protocols, and experiments for the delivery of multimedia traffic over CRNs.

Features of the Book: Compared to other books on CRNs, this book has the following two special features: (1) Emphasis on understanding video streaming in the dynamic spectrum access environment in CRNs: This book covers the important CRN protocol designs for transmitting video flows over CRNs. It has physical/MAC/routing layer protocols for a mobile and dynamic spectrum environment. In the physical layer, it considers different modulation/encoding schemes; in the MAC layer, it focuses on new algorithms for scheduling and communication among neighboring users in the context of spectrum hand-off; in the routing layer, it provides multichannel, multihop, spectrum-adaptive routing algorithms. (2) Explanation of both theoretical and experimental designs: Most CRN books provide just the protocol and algorithm details without linking them to the experimental design. This book explains how the universal software radio peripheral (USRP) boards could be used for real-time, high-resolution video transmission. It also discusses how a USRP board can sense the spectrum dynamics and how it can be controlled by GNU software. A separate chapter discusses how the ns-2 could be used to build a simulated CRN platform.

Target Audience: This book is suitable for the following types of readers:

(1) *College students*: This book can serve as a textbook or reference book for college courses on CRNs, which could be offered in computer science, electrical and computer engineering, and information technology and science departments. It can also be used as a part of a wireless network course.

(2) *Researchers*: Because the book explains multimedia over CRNs from protocols to experimental design, it will be very useful for researchers (including graduate students) who are interested in multimedia applications in a dynamic spectrum environment.

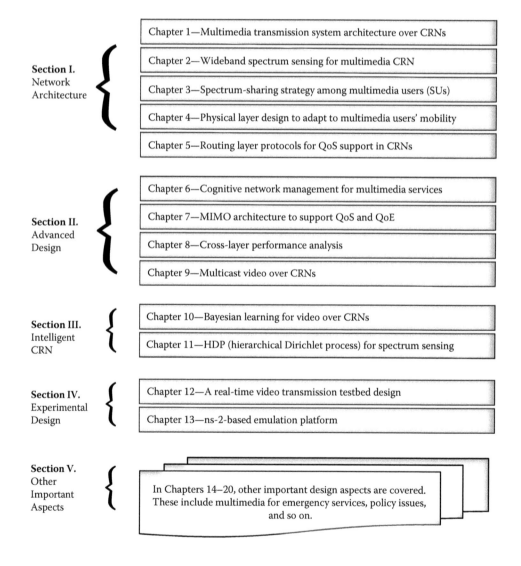

Section I.
Network Architecture

- Chapter 1—Multimedia transmission system architecture over CRNs
- Chapter 2—Wideband spectrum sensing for multimedia CRN
- Chapter 3—Spectrum-sharing strategy among multimedia users (SUs)
- Chapter 4—Physical layer design to adapt to multimedia users' mobility
- Chapter 5—Routing layer protocols for QoS support in CRNs

Section II.
Advanced Design

- Chapter 6—Cognitive network management for multimedia services
- Chapter 7—MIMO architecture to support QoS and QoE
- Chapter 8—Cross-layer performance analysis
- Chapter 9—Multicast video over CRNs

Section III.
Intelligent CRN

- Chapter 10—Bayesian learning for video over CRNs
- Chapter 11—HDP (hierarchical Dirichlet process) for spectrum sensing

Section IV.
Experimental Design

- Chapter 12—A real-time video transmission testbed design
- Chapter 13—ns-2-based emulation platform

Section V.
Other Important Aspects

In Chapters 14–20, other important design aspects are covered. These include multimedia for emergency services, policy issues, and so on.

(3) *Computer scientists*: Algorithms on video coding/transmission over CRNs as well as Python programming for USRP boards are provided in this book. Computer scientists could refer to these principles in their own design.

(4) *Engineers*: Many useful CRN design principles are explained in the book, which would be useful for engineers in industry in their product design work. The first 10 chapters provide concrete CRN protocol details.

Book Organization: This book consists of 20 chapters (five sections) that cover the most important aspects of multimedia over CRNs. These chapters include CRN architecture, protocols, algorithms, and experimental design.

Editors

Dr. Fei Hu is currently an associate professor in the Department of Electrical and Computer Engineering at the University of Alabama, Tuscaloosa, Alabama. He obtained his PhD degrees at Tongji University (Shanghai, China) in the field of signal processing (in 1999) and at Clarkson University (Potsdam, New York) in electrical and computer engineering (in 2002). He has published more than 160 journal/conference papers and books. Dr. Hu's research has been supported by U.S. National Science Foundation, Cisco, Sprint, and other sources. His research expertise can be summarized as *3S: Security, Signals, Sensors.*

Dr. Sunil Kumar is currently a professor and Thomas G. Pine Faculty Fellow in the Department of Electrical and Computer Engineering at San Diego State University (SDSU), San Diego, California, USA. He received his PhD in electrical and electronics engineering from the Birla Institute of Technology and Science (BITS), Pilani, India, in 1997. From 1997 to 2002, Dr. Kumar was a postdoctoral researcher and adjunct faculty at the University of Southern California, Los Angeles. He also worked as a consultant in industry on JPEG2000- and MPEG-4-related projects, and was a member of the US delegation in JPEG2000 standardization activities. Prior to joining SDSU, Dr. Kumar was an assistant professor at Clarkson University, Potsdam, New York (2002–2006). He was an ASEE Summer Faculty Fellow at the Air Force Research Lab in Rome, New York, during the summer of 2007 and 2008, where he conducted research in Airborne Wireless Networks. Dr. Kumar is a senior member of IEEE and has published more than 125 research articles in international journals and conferences, including three books/book chapters. His research has been supported by grants/awards from the National Science Foundation, U.S. Air Force Research Lab, Department of Energy, California Energy Commission, and other agencies. His research areas include wireless networks, cross-layer and QoS-aware wireless protocols, and error-resilient video compression.

Contributors

Bashar I. Ahmad
Department of Engineering
University of Cambridge
Cambridge, United Kingdom

Gianmarco Baldini
European Commission
Joint Research Centre
Institute for the Protection and Security
  of the Citizen
Ispra, Italy

Ke Bao
Department of Electrical and Computer
  Engineering
University of Alabama
Tuscaloosa, Alabama

Tamer Başar
Coordinated Science Laboratory
University of Illinois at Urbana-Champaign
Champaign, Illinois

Abdur Rahim Biswas
CREATE-NET (Center for REsearch
  And Telecommunication
  Experimentation for NETworked
  communities)
Trento, Italy

Faouzi Bouali
Department of Signal Theory
  and Communications
Universitat Politécnica de Catalunya
Barcelona, Spain

Athina Bourdena
Department of Informatics Engineering
Technological Educational Institute of Crete
Heraklion, Greece

Emmanuel Casseau
University of Rennes 1, IRISA, INRIA
Rennes, France

Abdelaali Chaoub
Laboratory of Electronic and Communication
Mohammadia School of Engineers
Mohammed V-Agdal University
Rabat, Morocco

Chihkai Chen
Department of Electrical Engineering
University of California, Los Angeles
Los Angeles, California

Xiang Chen
Aerospace Center
Tsinghua University
Beijing, China

Nan Cheng
Department of Electrical and Computer
Engineering
University of Waterloo
Waterloo, Ontario, Canada

Markus Fiedler
Department of Communication Systems
Faculty of Computing
Blekinge Institute of Technology
Karlskrona, Sweden

Matthieu Gautier
University of Rennes 1, IRISA, INRIA
Rennes, France

Fei Hu
Department of Electrical and Computer
Engineering
University of Alabama
Tuscaloosa, Alabama

Xin-Lin Huang
School of Electronics and Information
Tongji University
Shanghai, China

Elhassane Ibn-Elhaj
Department of Telecommunication
National Institute of Posts
and Telecommunications
Rabat, Morocco

Reema Imran
Department of Electrical Engineering
University of Jordan
Amman, Jordan

Muhammad Awais Javed
School of Electrical Engineering
and Computer Science
The University of Newcastle
Callaghan, Australia

Jamil Yusuf Khan
School of Electrical Engineering
and Computer Science
The University of Newcastle
Callaghan, Australia

Sunil Kumar
Department of Electrical and Computer
Engineering
San Diego State University
San Diego, California

Cong Ling
Department of Electrical and Electronic
Engineering
Imperial College London
London, United Kingdom

Ning Lu
Department of Electrical and Computer
Engineering
University of Waterloo
Waterloo, Ontario, Canada

Jon W. Mark
Department of Electrical and Computer
Engineering
University of Waterloo
Waterloo, Ontario, Canada

George Mastorakis
Department of Informatics Engineering
Technological Educational Institute of Crete
Heraklion, Greece

Constandinos X. Mavromoustakis
Department of Computer Science
University of Nicosia
Nicosia, Cyprus

Yakim Y. Mihov
Faculty of Telecommunications
Technical University of Sofia
Sofia, Bulgaria

Arumugam Nallanathan
Department of Informatics
King's College London
London, United Kingdom

Ricardo Neisse
European Commission
Joint Research Centre
Institute for the Protection and Security
 of the Citizen
Ispra, Italy

Duy Trong Ngo
School of Electrical Engineering
 and Computer Science
The University of Newcastle
Callaghan, Australia

Homayoun Nikookar
Faculty of Electrical Engineering
Mathematics and Computer Science
Delft University of Technology
Delft, The Netherlands

Maha Odeh
Interdisciplinary Centre for Security
Reliability and Trust
University of Luxembourg
Luxembourg City, Luxembourg

Ganda Stéphane Ouedraogo
University of Rennes 1, IRISA, INRIA
Rennes, France

Evangelos Pallis
Department of Informatics Engineering
Technological Educational Institute of Crete
Heraklion, Greece

Jordi Pérez-Romero
Department of Signal Theory
 and Communications
Universitat Politécnica de Catalunya
Barcelona, Spain

Adrian P. Popescu
Department of Communication Systems
Faculty of Computing
Blekinge Institute of Technology
Karlskrona, Sweden

Alexandru O. Popescu
Department of Communication Systems
Faculty of Computing
Blekinge Institute of Technology
Karlskrona, Sweden

Barbaros Preveze
Department of Electrical and Electronics
 Engineering
Cankaya University
Ankara, Turkey

Mickaël Raulet
INSA of Rennes, IETR
Rennes, France

Mubashir Husain Rehmani
Department of Electrical Engineering
COMSATS Institute of Information
 Technology
Wah Cantt, Pakistan

Walid Saad
Wireless@VT
Bradley Department of Electrical
 and Computer Engineering
Virginia Tech
Blacksburg, Virginia

Yasir Saleem
Department of Computer Science
 and Networked System
Sunway University
Selangor, Malaysia

Oriol Sallent
Department of Signal Theory
 and Communications
Universitat Politécnica de Catalunya
Barcelona, Spain

Olivier Sentieys
University of Rennes 1, IRISA, INRIA
Rennes, France

Xuemin (Sherman) Shen
Department of Electrical and Computer
  Engineering
University of Waterloo
Waterloo, Ontario, Canada

Hongjian Sun
School of Engineering and Computing
  Sciences
Durham University
Durham, United Kingdom

Alberto Trombetta
DiSTA
University of Insubria
Varese, Italy

Christos Verikoukis
Telecommunications Technological Centre
  of Catalonia (CTTC)
Barcelona, Spain

Jing Wang
Research Institute of Information
  Technology
Tsinghua National Laboratory for Information
  Science and Technology
Tsinghua University
Beijing, China

Jun Wu
School of Electronics and Information
Tongji University
Shanghai, China

Yang Yan
Research Institute of Information
  Technology
Tsinghua National Laboratory for Information
  Science and Technology
Tsinghua University
Beijing, China

Kung Yao
Department of Electrical Engineering
University of California, Los Angeles
Los Angeles, California

Yong Yao
Department of Communication Systems
Faculty of Computing
Blekinge Institute of Technology
Karlskrona, Sweden

Hervé Yviquel
INSA of Rennes, IETR
Rennes, France

Ning Zhang
Department of Electrical and Computer
  Engineering
University of Waterloo
Waterloo, Ontario, Canada

Ming Zhao
Research Institute of Information Technology
Tsinghua National Laboratory for Information
  Science and Technology
Tsinghua University
Beijing, China

Xiaofeng Zhong
Research Institute of Information Technology
Tsinghua National Laboratory for Information
  Science and Technology
Beijing, China

and

Department of Electronic Engineering
Tsinghua University
Beijing, China

Haibo Zhou
Department of Electrical Engineering
Shanghai Jiao Tong University
Shanghai, China

Nizar Zorba
Department of Electrical Engineering
Qatar University
Doha, Qatar

NETWORK ARCHITECTURE TO SUPPORT MULTIMEDIA OVER CRN

I

Chapter 1

A Management Architecture for Multimedia Communication in Cognitive Radio Networks

Alexandru O. Popescu, Yong Yao,
Markus Fiedler, and Adrian P. Popescu

*Department of Communications and Computer Systems, School of Computing,
Blekinge Institute of Technology, Karlskrona, Sweden*

Contents

1.1 Introduction

As mobility and computing become ever more pervasive in society and business, new mobile services are emerging for multimedia communication that are creating new challenges for telecommunication operators. New advanced infrastructure needs to be created to coexist with legacy infrastructures, which are expected to progressively fade out over a longer time period. Studies by Ericsson (November 2011) indicate, for instance, that mobile data traffic is expected to grow tenfold by 2016 (Release, Ericsson Press 2011). In particular, Internet traffic, especially video, is expected to drive the increase in mobile traffic by nearly 60% per year.

In addition, considering the rate at which the spectrum share of mobile devices is increasing and the customer base is expanding in densely populated areas, overcrowding in the operating bands is unavoidable, particularly for multimedia communication. Today, there are several possible solutions to solve the problem of spectrum crunch.

Long-Term Evolution (LTE) and LTE-Advanced (LTE-A) systems have been designed to support high data rates and a large number of users. However, even though the design objectives of LTE-A and WiMAX are to support high data rates and a large number of users, high-mobility devices located at the edge of a cell may experience degraded service levels [1]. The main reasons for this are limited possibilities of reconfiguring terminals and networks depending on spectrum availability, ineffective spectrum usage, and nonoptimal use of radio resources as well as insufficiently flexible deployment of base stations (BSs). Accordingly, these limitations of 4G systems, often referred to as 4G bottlenecks, are expected to become a major issue for quality of service (QoS) over the next couple of years.

Combining LTE and Wi-Fi in a single service is an interesting solution, although the slow adoption of femtocells by mobile industry has left the market open for low-cost (and often free) Wi-Fi service. The fact that femtocell technology still does not work properly has led mobile operators to promote the use of Wi-Fi to offload traffic from their networks. As of 2009, the US-based operator Sprint actively mandated that all smartphones used within its network should also be Wi-Fi capable, helping to relieve network load. In addition, this is a worldwide trend, where, for instance, KDDI Japan is already offloading as much as 50% of its wireless traffic to its 220,000 public hotspots, whereas AT&T has surpassed 2.7 billion Wi-Fi hotspot connections. Even developing countries in Africa and Southeast Asia are heavily investing in public access Wi-Fi.

Subsequently, Wi-Fi systems began to be used in combination with ad hoc mode, where aggregated Wi-Fi networks continue to expand provided that smartphones have the capability of directly connecting to other Wi-Fi enabled smartphones in ad hoc mode, bypassing the access point. This creates a small ad hoc network that can be extended to span several hops by including more Wi-Fi-enabled phones in each path. Finally, ad hoc networks spanning entire cities can be envisioned that would allow the mobile device to communicate without using any cellular infrastructure. Naturally, this would pose a serious threat to the revenues of current mobile operators.

Accordingly, new management architectures and business models must be developed to merge the interests of individual users, who want to make free or very cheap calls, with the operator objectives of retaining control and sources of revenues.

The basic idea of cognitive radio networks (CRNs) is to allow unlicensed users use underutilized spectrum in highly dynamic radio environments. Key features include, among others, pervasive wireless computing and communications, cognitive radio (CR) technology, IPv6, wearable devices with artificial intelligence (AI) facilities, and a unified global standard [2,3]. Nevertheless, the typical decentralized approach to CR functionality has dictated that adaptations to the operational parameters of cognitive radio devices (CRDs) have mainly been based on local observations made by individual CR users.

However, peer discovery without network support is typically time and energy consuming, requiring beacon signals with sophisticated scanning and security procedures [1]. By using LTE network assistance in terms of synchronization, identity and security management, beacon signal configuration, and reserving peer discovery resources, the discovery process of devices can be made more user-friendly and energy efficient. This is true regardless of the subsequent device-to-device communications taking place in the cellular spectrum or the use of noncellular technologies. Accordingly, it is important to mention that 3GPP has already done studies on the advantages of network-controlled discovery and communications for LTE Release-12 [1].

Moreover, CR functionality for multimedia communication requires the development of flexible terminals and a rethinking of the classical wireless network architecture [4,5]. For purposes such as flow control in spectrum sensing, selection, and sharing of radio environments, cooperation among protocols in different layers is necessary [4,6]. CRNs push the CR concept a step further by looking at the network as a association of CRDs. Typically, CRN architectures cover the whole TCP/IP protocol stack and do not conform to the strictly enforced traditional layered approach. To support video and audio streaming, there is a need to adopt a cross-layer design for information exchange at different layers in the protocol stack. New architectural models must be developed for communication management in CRNs [7]. Subsequently, a novel management architecture designed and developed at the application layer is advanced for horizontal spectrum sharing, also known as infrastructure-based CRN (currently under standardization by IEEE P.1900). For the objective of integration with low-power BSs in 4G networks, a centralized entity called the support node (SN) provides necessary functionality in terms of specific hardware and software [7].

The architectural solution enables expansion of the technological limits in 4G mobile systems to include CR facilities, and makes it possible to increase spectrum utilization within particular geographical areas. The theoretical spectral efficiency improvement of 4G systems has been shown to suffer from performance constraints in practical scenarios [1]. To increase the throughput, higher bandwidths together with transmissions over shorter distances are necessary. Integration with multihop CRNs can in this case yield higher throughput (i.e., more bandwidth through dynamic spectrum access) and also minimize the transmission distances (because direct communication with the BS is no longer necessary).

The CRN management architecture suggested in this chapter incorporates sensing and prediction, addressing and routing, and middleware and decision making. Accordingly, to coordinate spectrum access and exchange of signaling information, a common control channel (CCC) is necessary to enable cooperation among CR users [4]. Dimensioning of a CCC is considered in this chapter with regard to a specific CRN scenario. Particular focus is placed on the overhead required to maintain a network topology, which enables end-to-end (e2e) optimization of routing paths according to user and environmental constraints.

The rest of this chapter is organized as follows. Section 1.2 considers related work in the field. Section 1.3 describes the operations performed by SN and network members. Section 1.4 presents a basic CRN management scenario. Section 1.5 introduces the architecture for the suggested management framework. Section 1.6 is about CCC dimensioning aspects as well as implementation details for the network model used to communicate among network hosts. Further, with regard to secondary user (SU) service completion, Section 1.7 investigates the performance of a basic communication scenario through numerical analysis and simulation. Finally, Section 1.8 concludes the chapter.

1.2 Related Work

Recent research in the CR area has focused on new solutions for energy and spectral efficiency in wireless communication (green operation), spectrum assignment for opportunistic spectrum access, routing and handover mechanisms, and decision making and prediction algorithms to minimize the impact of SUs on primary users (PUs) [8–10].

For instance, Peng et al. [11] present a general model and utility functions for optimizing utilization in spectrum allocation. Given the spectrum heterogeneity, a framework for solving the spectrum access problem is suggested where the problem is reduced to a global optimization scheme based on graph theory. It is further pointed out that, because the global optimization problem is Non-deterministic Polynomial-time-hard, a heuristic approach to vertex labeling is required. Accordingly, a set of approximation algorithms is described for both centralized and distributed approaches to spectrum allocation, each with its specific advantages and drawbacks. Nevertheless, this study only considers a static network environment and focuses on optimizing a snapshot of the network at a particular time moment. For a dynamic network environment, the spectrum allocation problem becomes more complex, given the need for recomputations as the topology changes.

On the contrary, Won-Yeol and Akyldiz [12] advance a spectrum decision framework to allocate a set of spectra to CR users by considering the application requirements together with the dynamic nature of the spectrum bands. Two categories of applications are considered. These are real-time and best-effort applications where the suggested spectrum decisions are classified into minimum-variance-based spectrum decision and maximum-capacity-based spectrum decision. The aim is to minimize the capacity variance of the selected spectrum bands for real-time applications and maximize the total network capacity for best-effort applications. In addition, a dynamic resource management scheme is developed to adaptively coordinate the spectrum decisions dependent on the time-varying CRN capacity. This is achieved through a centralized resource manager located at the BS, which gathers information from CR users and accordingly performs spectrum decisions. Basically, CR users carry out two main tasks, which are spectrum sensing and quality monitoring. Subsequently, this enables a so-called event detection that allows the CRN to reconfigure its resource allocation as needed in order to maintain the service quality. Simply put, the advanced spectrum decision framework provides a hierarchical QoS guaranteeing scheme that consists of spectrum sharing (to assign channel and transmission power for short-term service qualities) and spectrum decision (to determine the best spectrum for sustaining the service quality over the long term).

In Ref. [13], the authors describe a spectrum-aware mobility management scheme for CR cellular networks. A novel network architecture is developed to support spectrum mobility management, user mobility management, and intercell resource allocation. Essentially, infrastructure-based CRNs are considered, which consist of multiple cells where each cell has a single BS with corresponding SUs in its coverage area. The cell coverage of each BS is furthermore regarded as the PU activity region. Hence, in order to avoid interference with the PU operation, each BS determines appropriate actions in support of an upper-level control node through observations and also through reporting of CR users located within the cell coverage area. This is possible because CR users are assumed to have a single wideband radio frequency (RF) transceiver that can simultaneously sense multiple contiguous spectrum bands without RF reconfiguration. The network determines a suitable spectrum band and also the target cell according to both the current spectrum utilization and the stochastic connectivity model. Increased cell capacity is, in other words, achieved through bandwidth aggregation by sharing spectrum owned by other cellular operators, or by opportunistically utilizing unused spectrum bands otherwise licensed to PUs. In addition, a number of four different hand-off schemes are defined for CRNs with regard to the mobility management of users and spectrum. Ultimately, a cost-based hand-off decision mechanism is suggested to minimize QoS degradation caused by user mobility.

Moreover, in Ref. [14] the authors focus on providing a systematic overview of CR communication and networking from the standpoint of the lower layers. In particular, the study investigates how the physical, medium access control, and network layers are involved in enabling multihop or relay communication in CRNs. It is concluded that the design of routing and control mechanisms in a multihop CR scenario requires cooperative CR users to optimize the overall detection capability of the CR network. In other words, spectrum-aware routing requires cooperative spectrum sensing that can take advantage of the dynamically available spectrum holes and adapt the path computations to the changing environment. Naturally, spectrum-aware routing, also called opportunistic spectrum routing, further requires support from intermediate (relay) nodes.

1.3 Network Model

A cognitive network is partitioned into a number of so-called CRNs. Every CRN is served by an SN and contains a number of SUs referred to as CRDs in the remainder of this chapter. The SN is responsible for populating the available spectrum opportunities (SOPs) within its geographical coverage area and also for keeping track of the current network members and operational conditions. A knowledge database is used to maintain all relevant CRN information. A CRD wishing to join a CRN must first contact the responsible SN by sending a join message to the SN requesting an operational zone (called a device spectrum opportunity [DSOP] in the continuation [15]) within an available spectrum opportunity in the particular CRN. SOPs can be partitioned in three categories with reference to the activity and holding times of licensed users: static, dynamic, and opportunistic [4,5,16]. However, in our CRN communication framework, only static/dynamic SOPs are considered because the holding time of an e2e path must be long enough to allow it to be traversed.

The functions provided by an SN for basic support of multimedia communication in a CRN are as follows:

1. Collection of information regarding the specific CRN, where data are collected from all network elements. This is used to identify and represent the available SOPs and to build a statistics database for decision making.
2. Support for two CRN routing mechanisms, greedy and optimized [7]. The routing mechanisms must also accommodate support for intra- and inter-CRN routing decisions, that is, computation of e2e routing paths among CRDs within the same or different CRNs.

3. Provision of bootstrapping and adaptation support for SUs. While SOPs are available in the CRN coverage area, an SN continues to allow SUs to join the CRN and to provide adaptations to network members for maintained service.
4. Spectrum sensing for the geographic coverage area of the CRN.

The functions provided by a CRD are as follows:

1. Spectrum sensing and monitoring regarding the used resources (i.e., channel utilization/occupancy). The collected information is used to detect possible conflicts in the CRN and update the deserving SN.
2. Routing and support for crossing routes, based on e2e routing decisions received from the particular SN. That is, network hosts are required to reserve resources for crossing communication paths to meet the e2e goals of different sources and destinations.
3. End-to-end decision making for opportunistic routing purposes. If no more static/dynamic SOPs are available in the particular geographical area, arriving CRDs must make their own decision regarding communication, which entails per-packet dynamic routing over opportunistically available channels with limited or no QoS guarantees. This means that SUs may be required to handle elastic traffic (i.e., share the same channel), which further degrades the existing QoS owing to resource sharing among a varying number of SUs generating different traffic.

A CRN is designed to cover an area in the range of a microcell network, that is, a radius of around 1000 meters. The goal is to provide added spectrum utilization and improved capacity at cells in crowded metropolitan areas such as office centers, airports, and malls. Compared to typical microcells where the spatial multiplexing interference created by the subdivision of cells is managed with the help of optimized power controls, a CR approach makes it possible to use the entire frequency spectrum, not just the fixed spectrum chunks allocated for operators [17]. This allows for optimization in both transmit power and the frequency domain, thus ensuring a considerable increase in the available communication capacity.

1.4 Basic Scenario

Four fundamental operations need to be controlled in the management of a CRN [4]: spectrum sensing, spectrum decision, spectrum sharing, and spectrum mobility. Another CRD-associated operation is mobility. Hence, ad hoc algorithms can be combined with spectrum mobility algorithms to provide an e2e solution for communication. Collaboration between CRDs and the SN is therefore vital and provides us with the necessary operations to communicate within CRNs. The joining of unlicensed users is done in a first-come, first-serve order through a scheduler to fairly partition the available DSOPs. If no static/dynamic DSOPs are available or predicted to be available in the CRN coverage area, the SN rejects the joining of newly arriving CRDs. Otherwise, after checking the available resources in its locally stored database, the SN computes a reply to the requesting CRD containing all the necessary information to allow it to join the CRN. A reply message typically contains parameters such as geographic point, frequency assignment, maximum power output, DSOP expected duration time, and a list of neighboring CRDs (already bootstrapped network hosts). Moreover, the SN constructs a database with information collected from all network elements. The collected data are used to identify and represent available DSOPs and to thus build a statistics database for future decision making [15].

The SN performs global spectrum sensing of the entire CRN coverage area to detect SOPs available for SU operation. Moreover, to identify local conflicts within the CRN, network members are expected to perform their own spectrum sensing for their respective channel assignments (depicted here as a DSOP, provided that an unknown modulation is used in the CRN, where every SU needs a portion of an available SOP to operate in). Any detected inconsistency must be communicated back to the SN for centralized decision making. This implies that a conflict in the DSOP of a network member may require an adaptation of its operational parameters, which is decided by the SN [15]. In addition, a CRD has the facility to communicate with another CRD in the CRN by conveying a communication request to the local SN. Depending on the type of request, unoptimized or optimized, the reply is either the virtual identifier (VID) of the destination or a complete optimized e2e path to reach it. That is, all intermediate hops and channel assignments along the path are optimized according to the requirements of the source.

1.5 Communication Architecture

The architecture demands an intelligent wireless communication system able to

- Monitor the radio environment continuously.
- Learn from the radio environment.
- Facilitate communication among multiple users.
- Adapt the e2e performance to statistical variations of the radio environment as well as user preferences and behavior.
- Facilitate self-adapting methods for communication.
- Solve diverse conflicts among users.
- Provide self-awareness.

The so-called cognitive engine, defined as a set of algorithms that together perform sensing, learning, optimization, and adaptation control of the CRN, is used [18]. Hence, our framework can be seen as a software suite implemented at BSs and CRDs with specific radio capabilities. The main components of the CRN management architecture (depicted in Figures 1.1 and 1.2) are middleware, software-defined radio (SDR), and overlays.

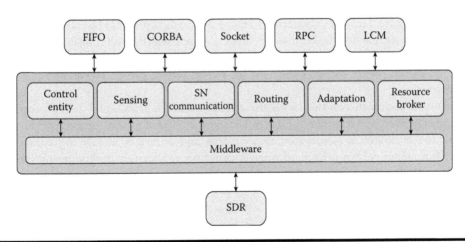

Figure 1.1   Cognitive radio device.

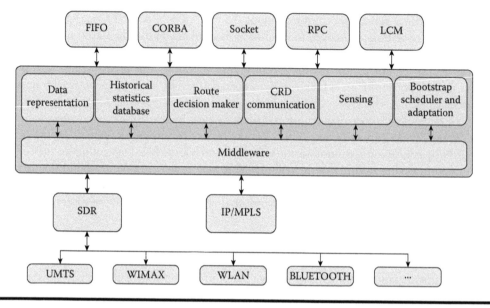

Figure 1.2 Support node.

The CRN architecture has been partially developed and implemented at Blekinge Institute of Technology. A short description of the components is as follows.

1.5.1 *Middleware*

Operator and user expectations in the CR area have governed the research for solving the complexity of management, security, and scalability provisioning. New BS designs need to be developed that can incorporate CR functionality and increase spectrum utilization through the use of dynamic spectrum access [15]. Because CRNs use low-layer (link layer and physical layer) information to improve radio communication performance, it is necessary to point out that full control of the lower layers is often limited owing to hidden information in the form of proprietary hardware and software. However, cooperation among protocols at different layers is necessary in a CRN for purposes such as flow and control over spectrum sensing (RF and transmit power output observations), spectrum selection, and spectrum sharing [4,6]. This type of direct communication between nonadjacent layers indicates the need for a cross-layer approach to ensure a functional implementation.

Because this approach conflicts with the strictly enforced traditional layered approach, a specific middleware, as presented in Figures 1.1 and 1.2, is used to address some of the stated problems. The middleware is software that bridges and abstracts underlying components of similar functionality, exposing it through a common application programming interface (API) originally developed for seamless handover purposes [19]. This offers the advantage of flexibility in present and future development of new services and applications. Furthermore, by using a middleware-based architecture with various overlays and underlays, the convergence of different technologies is simplified. The goal in this case is to develop a framework for communicating in CRNs, which requires minimal changes to the applications using the platform.

1.5.2 Software-Defined Radio

SDR is a class of reconfigurable radio where the physical layer behavior can be significantly changed as a consequence of changes in software; that is, the same hardware entity can perform different functions at different times. Simply put, an SDR is a device that provides RF and intermediate frequency functionality, including waveform synthesis in the digital domain [20]. This enables a CRD to relay messages to other CRDs and SNs over different channels (Figure 1.3).

1.5.3 Support Node

The SN is a centralized entity, implemented in the form of a software suite, that provides the following functionalities (in the form of overlays).

Data representation: A knowledge database for the data collected in the particular multihop CRN. In our case, this is based on the geometric structure of a datacube composed from several two-dimensional $[0,1] \times [0,1]$ Cartesian coordinate spaces (CCSs) [21]. Each CCS represents a different CR-dimension, where all dimensions are functions of time. Thus, the operational parameters of each network host can be described in every CR-dimension, for example, frequency (depicted as channels, depending on the available SOPs and used modulation [15]), power (transmit output power), and geographic point mapped to a set of virtual space coordinates (x,y) referred to as VID Φ. Because the selection is on the basis of the actual geographical location of the hosts in the physical CRN coverage area, visualization can be used to determine if additional CRDs can join the particular CRN or not. Every mapped device is represented in the data cube with a geometric portion of the space dimension CCS called zone, which surrounds VID Φ [7,21]. In addition, such devices have information about adjacent zones and the network member devices responsible for them. Because network hosts with geographical coordinates located close together access resources (such as SOPs) in the same area of the CRN, their VIDs are also mapped close

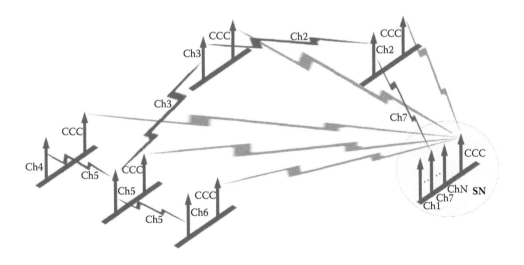

Figure 1.3 Message relaying between hosts in the CRN.

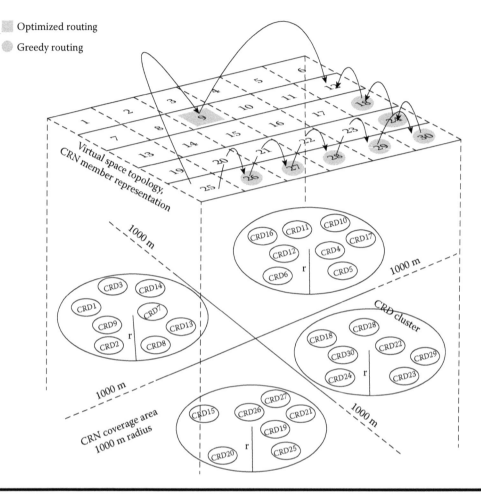

Figure 1.4 CRN overview depicting CRD clusters, optimized and greedy routing (From A. O. Popescu, et al., Communication mechanisms for cognitive radio networks, *11th IEEE Pervasive Computing and Communication*, San Diego, CA, March 2013.).

to each other in the virtual space dimension representation. Simply put, the SN imposes a particular geometric structure on its CRN coverage area to create a structured CRN topology (Figure 1.4).

Historical-statistics database (HSDB): The collected CRN data are subject to constant changes, where moving devices means that PUs may come and go, altering communication relationships in the CRN. Therefore, e2e path computations should not be based solely on current available operational states, but also on associated statistics computed and measured over long time periods for the particular CRN coverage area. Prediction models can be used to determine the holding times of DSOPs, which an optimized e2e path is set to traverse. Thus, having historical statistical data recorded in the HSDB overlay enables the decision maker algorithms to learn from experience and to more accurately predict future changes through methods such as the ones suggested in Refs. [22,23].

Decision maker: This is used for e2e route computation. The operational data collected by the SN and network members, together with HSDB-retained information, allow the decision maker overlay to compute an e2e route according to user preferences. The e2e path is subject to centralized decision making with reference to a particular temporal framing and various QoS/QoE parameters, for example, cost, throughput, delay, service availability, security, and privacy levels. Different mechanisms can be used for decision making, for example, context-aware, fuzzy logic, analytic hierarchy processing [22,24]. The computed e2e path is propagated back to the requesting CRD.

CRD communication: This function handles all communication between the SN and the CRD terminal. Specific messages are exchanged over a CCC, which is conceptualized to handle all necessary data exchange with the SN, for example, bootstrap request, adaptation request, update request, communication request, and resource reply. This overlay collects all required information from the necessary overlays and properly formats the CRD messages before they are transmitted out on the CCC.

Additional functionalities: The SN provides other functionalities, for example, bootstrapping procedures, for joining unlicensed CR users, keeping the data representation up to date through global spectrum sensing of the CRN coverage area and updates received from the network members, maintaining a member list of all current network hosts and associated operational parameters in different CR-dimensions, and handling inter-domain communication hand-off to other SNs. Because a device is bootstrapped one SN at a time, handover schemes for devices moving from one CRN to another must also be developed.

1.5.4 Cognitive Radio Device

The following overlays are used in the CRD architecture (software suite).

Control entity: It handles the actions related to user control, contextual-awareness, and security facilities. This refers to informing the SN about the user context and service-defined preferences, that is, the type of service as well as other relevant information for the e2e route computation. This overlay offers the end user the possibility of defining preferences for best decisions to be taken with the help of generic models, distributed measurements, and data exchange.

Sensing: This handles actions related to multiband operation and fast frequency scanning in collaboration with the SDR underlay. Gathering spectrum usage information enables forecasting for licensed and unlicensed users as well as for the user itself (e.g., user activity, available resources, movement prediction). This requires solving a number of challenging research questions such as the hidden PU problem, detecting spread spectrum PUs, sensing duration and frequency, cooperative sensing, and security.

SN-communication entity: This handles signaling communication between the terminal itself and the SN. This is done over a CCC and through specific message types formatted to handle all necessary data exchange, for example, bootstrap reply, adaptation reply, update reply, communication reply, and resource request.

Routing: This function controls the routing so that the destination is reached. Depending on the type of requested communication, that is, unoptimized or optimized, the SN replies either with a message containing the VID of the destination or with a complete optimized e2e path. The e2e path is computed according to user preferences, and the reply message from the SN has all the information necessary to reach the destination, that is, intermediate CRDs (all hops) together with channel assignments along the path that are optimized according to, for example, cost, throughput, delay, and security level [7].

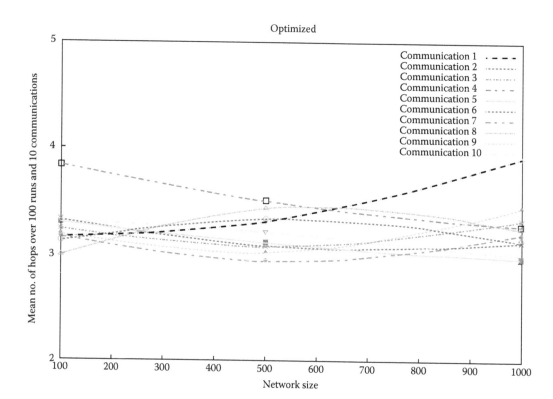

Figure 1.5   CRN cluster-optimized routing performance.

Unoptimized routing employs a distributed hop-by-hop routing model (greedy routing, Figure 1.5) toward the VID of the destination without a comprehensive network overview. That means we have in this case limited or no QoS guarantees. In other words, a source only requires the VID of the destination from the SN and routes toward it according to a recti-linear distance geometry along the *x* and *y* dimensions. The VIDs of neighbors in the virtual space topology are cross-checked with the VID of the destination, and the neighbor closest to the destination is selected as next hop [7]. Two network members are considered neighbors if the coordinate spans of their geometric zones overlap along one dimension and abut along the other dimension. On the contrary, an SN-computed and -optimized routing path allows for the e2e goals of the source to be achieved at the cost of larger overhead and computa-tional latency. The SN imposes different constraints on the routing path, denoting ad hoc packet routing with aggregated throughput obtained by employing suitable multiobjective, multiconstrained optimization algorithms [7,15]. The destination can thus be reached accord-ing to specific QoS framings that depend on the particular domain, networking conditions, and user requirements. If an intermediate device (a CRN network host) between a particular source and destination fails during an ongoing communication, a recomputation of the e2e path may be required. The routing overlay at the affected device (i.e., for which its next hop is no longer available) must contact the SN to request an updated route to the destination, if one is still available. This may be necessary if a PU takes over the DSOP of an SU that is part of the CRN and the particular e2e path. However, if the SN fails or no other network host with a suitable DSOP holding time is available, the routing overlay at individual devices must make its own decision regarding communication. This entails a fully distributed per-packet dynamic routing over opportunistically available channels, that is, limited or no QoS guarantees as in the unoptimized routing case.

Adaptation: While sensing the operational environment, a network host may discover that adaptations are needed in one or more CR-dimensions to maintain service continuity [15]. This can occur owing to node mobility, where a moving device is no longer able to access its assigned resources (such as a DSOP) and needs to update them according to its new geographical location. The particular device sends an adaptation request to the local SN and waits for a reply. When a reply message is received, the actual changes to the operational parameters are executed by the "Adaptation" overlay of the device. The changes may require adaptations to be done in different CR-dimensions, for example, frequency, power, and space, and entail collaboration with the SDR underlay, which allows for the physical layer behavior to be modified as a consequence of these changes.

Resource broker: It monitors the available resources of the particular CRD. Every network host is required to keep track of its vital operational parameters, for example, remaining processing capacity, remaining battery capacity, and available bandwidth of its assigned data channel (DSOP). It is necessary to point out that even if all data channels have the same initial capacity (assuming a uniform modulation), the available bandwidth may vary depending on the geographical positions of the network hosts (subject to interference from other devices and obstacles affecting some DSOPs more than others), ongoing communications (current channel load due to, for example, elastic traffic [25]), and other environmental factors. Hence, the actual channel capacity available to a network host at a specific time must be determined either through periodic or on-demand updates among the network members.

1.6 Common Control Channel

The CCC is intended for CRN control purposes, which means that it must meet the signaling needs between SNs and CRDs in the CRN framework. Toward this goal, we use a CCC that is available at all CRDs (global coverage) in the CRN all the time [4,26]. To avoid interference with the licensed bands, CCC is placed in out-of-band frequencies such as the unused parts of the UHF band recently released by the discontinuation of analog TV. Unlike in-band CCC, where channels used for data transmissions are also used to convey control messages, an out-of-band CCC separates control data from the user data transmission. To access the CCC, network members are thus required to have a dual-mode radio, that is, an extra narrow-band low-bit-rate transceiver to exchange control messages with the SN. Even though out-of-band CCC frequencies are used, it is assumed that they may change with respect to geographical domains, because no worldwide harmonization of the CCC exists today, and different frequencies may be available in different countries. Three different possibilities are available for CCC detection by SUs: out-of-band, in-band, and combined solution [27]. The suggested CRN framework is based on a centralized cognitive framework and aimed for BS implementation. Typically, the cognitive framework can be implemented in different ways, such as centralized, distributed, or mixed (a hybrid between centralized and distributed), each with specific advantages and drawbacks [28]. An in-band CCC detection method is considered most suitable for our purposes [4].

In addition, information about the modulation (radio access technology) used for the particular out-of-band CCC must be conveyed via the initial in-band signaling as well. This allows a CRD that wishes to join a CRN to directly tune into the out-of-band CCC, thus removing the need to sense the entire frequency spectrum to discover it. A typical scenario occurs when users wish to join the CRN because of service starvation owing, for example, to overcrowding in the regular operating bands of a particular operator. Because CRN management requires an exchange of data among network members, the information carried over the CCC consists of on-demand compiled messages; that is, information is only communicated over the CCC by a terminal request.

1.6.1 CCC Model

The exchanged signaling messages among network members are mainly generated owing to node mobility and the complexity of the particular CRN scenario, thus creating the need to update the data (topology) maintained at the local SN. The frequency of these updates is related to the complexity of the particular scenario; that is, it depends on the CRN coverage area characteristics and population size. Hence, different CRNs may need to use the CCC differently. Some CRNs may require a more frequent use of the CCC, whereas others may require less frequent use. This indicates that the bandwidth needed for a CCC can vary with different geographical domains. Given that message types of various sizes are exchanged over the CCC (between the SN and network members), the amount of data necessary to represent the different parameters (in bits) is computed as follows.

We assume the following information is available:

1. The device id D_{id} identifies and differentiates between network hosts (CRDs). To keep the identifier size small, the id is the phone number of a user, because the aim is to implement CR functionality at BSs with CR-available capabilities. A 67-bit number is used to represent the (phone) numbers, 34 bits of which are used for the actual number (allowing for 10-digit numbers), and the remaining bits are reserved for prefixes if a host in the local CRN wishes to communicate with non-CR users located in other geographical domains.

2. The frequency band interval between F_{min} and F_{max}, which specifies the DSOP where the available communication channel for a particular host is located. For simplicity, frequency modulation is assumed, and the CRN operating frequencies are in the range of 0 Hz to 10 GHz [27]. Thus, two variables of 34 bits each can be used to define the minimum and maximum frequencies.

3. Geographic point information, G_{gps}, given by the global positioning system (GPS) latitude and longitude coordinates. The latitude is defined as XX° YY′ ZZ″ together with a north/south pointer G_{NS} that can be represented by 1 bit. The range of latitude degrees XX° is from 0° to 90°, and it can be encoded by 7 bits. One degree of latitude is 60 arc minutes (approximately 100 km), implying a range of 0 to 59 arc minutes and seconds, respectively. This requires 6 bits each. An additional 7 bits are used to represent the hundredths of a second (from 0 to 99) for increased positioning precision. Meridians of longitude are similar, though ranging from 0° to 180° instead, thus adding another bit to the set of coordinates. Moreover, a west/east pointer G_{WE} represented by 1 bit is used to identify on which side of the prime meridian (the zero line, which passes through Greenwich, UK) the longitude coordinate is located in. Altogether, this results in a total of 55 bits to represent a complete G_{gps} set of coordinates with a positioning precision of approximately 0.3 m.

4. The maximum transmit power is used by the SN to set the output level for network members accessing specific DSOPs in the CRN coverage area. The power control makes it possible for network hosts to operate in the CRN with a minimum of interference to other devices. However, not all devices usually have the same maximum output power capabilities, and the required power classes (maximum output power level) typically vary with the used bands and modulations. Considering power class 1 mobile devices, a maximum transmit output power of 2000 mW is feasible, which requires 11 bits to define the output power variable P_{max}.

5. The expectation time, which is the time period during which a particular SOP is expected to be available for SU operation. This information can, for example, be obtained based on long-term traffic measurements and analysis of a particular CRN coverage area as well as a statistical prediction method. A SOP duration foresight of maximum 1 day, or 86,400 seconds, is assumed, which requires two variables to be communicated, t_{start} and t_{end}. t_{start} depicts (in seconds) the time moment when a particular SOP becomes available, and t_{end} depicts (in seconds) the time moment when a particular SOP's availability expires. Thus, 17 bits are needed to represent each of the two variables t_{start} and t_{end}, that is, 34 bits in total. To ensure that unpredicted SOPs expire the next day, the t_{end} variable of the reply contains a maximum value of 86,400. Assuming no unexpected events (such as PUs unexpectedly taking over the used SOP), a stationary host accessing a DSOP (within a static SOP) may continue to use the same DSOP even after the first expectation time expires. However, an updated expectation time must first be requested from the SN, thus implying that every CRN member must contact the SN at least once a day to exchange control information.

6. The starting t_{min} and ending t_{max} time moments define the time interval when a network host requires an optimized e2e path. This is different from the DSOP expectation time interval described earlier. For simplicity, a maximum waiting time of 100 seconds is assumed, which requires 7 bits to represent each of the two variables t_{min} and t_{max} (i.e., 14 bits in total, where $t_{min} < t_{max}$).

7. Virtual space coordinates are imposed by the SN to create a structured CRN topology. Identifiers of current network members in the CRN geographical coverage area are mapped to a set of virtual space coordinates (x,y) (in the CCS representation of the CRN space dimension) referred to as VID Φ. The Φ virtual coordinates are represented by 8 bits per CCS axis, x and y; that is, a total of 16 bits are required to represent a complete set of CRN virtual space coordinates $VID_{x,y}$. This is enough to represent the virtual space coordinates for a network population of thousands of devices, and it can easily be expanded for larger network populations.

8. The virtual cluster radius length is referred to as R_{length}. Because several hosts may be available throughout different geographical areas of a CRN (for a microcell coverage area), virtual clusters are created by the SN, which groups together network members depending on location and accessed resources (such as SOPs) in the particular CRN geographic area. That is, the CRN coverage area is subdivided into clusters, where each cluster comprises a number of member devices. The radius of the clusters is not a fixed number, and it may vary from one CRN to another. A device is considered to belong to a particular cluster if its geographic coordinates are within the radius of that particular cluster. The virtual cluster radius is represented by 17 bits, for a maximum cluster radius of 700 m represented in centimeters. Naturally, the most basic cluster scenario is when the entire CRN area (a microcell of 1000 m radius) is covered by only two clusters (Figure 1.6).

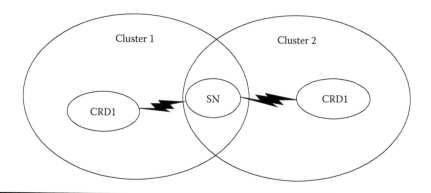

Figure 1.6 Basic scenario with two clusters.

Table 1.1 CRD Message Types

Update request	$D_{id}(i)$, $D_{id}(n)$, A
Adaptation request	$D_{id}(s)$, $G_{gps}(s)$, A
Bootstrap request	$D_{id}(s)$, $G_{gps}(s)$, A
Communication request	$D_{id}(s)$, $D_{id}(d)$, A, t_{min}, $t_{max}(s)$, $Cv(s)$
Resource reply	$D_{id}(s)$, $Rv(s)$, A

Devices in the same cluster are represented as neighbors in the CRN virtual space topology (illustrated in Figure 1.4). Considering the geographical proximity between adjacent clusters, devices therein are assumed to be able to communicate directly without any intermediate relays. However, this requires an initial dimensioning of the cluster sizes in CRNs with regard to the specific environmental constraints to ensure limited signal degradation and maximum throughput for the available channels, during inter- and intracluster communication among devices. The number of clusters used in a specific CRN depends on the cluster sizes, and we consider factors such as local, geographical, or regulatory aspects; required uplink/downlink data rates by devices in the particular geographical area; rated power output of devices in the CRN; and fading characteristics of the channels in the coverage area caused by obstacles [7,15].

Furthermore, because devices in the CRN are not always stationary, this process can potentially be very disruptive given the need to update the network topology and the used resources [15]. A communicating device moving away from the location where it accesses an available DSOP may have insufficient signal strength to maintain the connection. This can cause signal-to-noise ratio (SNR) and bit-error-rate (BER) problems. Thus, the SN must provide topology updates, initiated on demand by requests from network hosts. In other words, network hosts are required to keep track of their own movements and cluster boundary crossings, which further implies that every device must be aware of its own current location and the coverage area of the cluster it resides in. This can be done through bootstrap and adaptation messages where the center coordinates and cluster radius are provided together with other necessary information, as shown in Table 1.1. In such a case, a simple comparison of the current geographical coordinates in relation to the center coordinates and radius of the cluster allows a particular network host to be aware of when the cluster boundaries are crossed. The geographical coordinates can be any type of positioning coordinates, although for simplicity purposes we only consider GPS coordinates.

The parameters described earlier are sufficient to compute most of the control messages used in the CRN management architecture. Table 1.1 depicts the CRD-generated messages *bootstrap request, adaptation request, update request, communication request*, and *resource reply*. Each message type is used by network hosts to pass a number of relevant parameters to the SN, where the variable A can take values that satisfy $0 \leq A \leq 4$, and it is used to inform the SN about the type of CRD message received:

Update request: The value $A = 0$ informs the SN that this is an update request message from intermediate device i with ID $D_{id}(i)$. This message type is used if a network host detects that one of its neighbors is no longer available. As such, this requires an update of the particular host's neighbor list and the network topology maintained at the SN. For instance, if an intermediate device along an e2e path fails in its attempts to forward packets to one of its neighbors, the neighbor is considered void. An update neighbor list request is send to the deserving SN, which also allows it to update the network topology. $D_{id}(n)$ is the id of neighbor n (in the neighbor list of intermediate device i) that is no longer reachable.

Adaptation request: The value $A = 2$ informs the SN that this is an adaptation request message from source device s with ID $D_{id}(s)$, where $G_{gps}(s)$ contains its current geographical position. This message type is used if a network host requires adaptations to its operational parameters as a result of mobility or a PU taking over its DSOP. Provided the particular device has moved to a new geographical location (located in a different cluster), its entire neighbor list must be updated as well.

Bootstrap request: The value $A = 2$ informs the SN that this is a bootstrap request message from source device s with ID $D_{id}(s)$, where $G_{gps}(s)$ contains its current geographical position. This request is similar to the adaptation request, although with a lower priority assigned to it. This is because preserving service continuity for already bootstrapped network members is considered to be more important than bootstrapping new arrivals. That is, network hosts have priority in accessing available resources in the CRN, whereas newly arriving CRDs are only bootstrapped if resources are available and not needed for any pending adaptation requests in the particular area of the CRN.

Communication request: In the greedy routing approach, no path optimization mechanisms are employed. This means that no e2e goals are defined in the context vector $Cv(s)$ of source device s. Subsequently, only the source $D_{id}(s)$ and the destination $D_{id}(d)$ IDs are included in the communication request message to the deserving SN. However, in the case of an optimized routing request, the user context parameters (such as service-defined preferences, i.e., type of service) for the path must also be conveyed. QoS requirements are in such a case enclosed in the context vector $Cv(s)$ of source device s and include the following parameters: minimum required throughput Th_{min} for the service, the starting t_{min} and ending t_{max} times for the time interval when the resources required by source device s must become available (this is different from the DSOP availability time interval t_{start} and t_{end}), the maximum delay Δ_r tolerated for the route, the maximum allowed cost Co_{max} for the particular service, and the minimum expected security level Se_{min} [15]. The value $A = 3$ informs the SN that this is a communication request from a source device with ID $D_{id}(s)$ to a destination device with ID $D_{id}(d)$.

Resource reply: This is used to reply to an SN-initiated resource request. Every network host included in the initial request must relay its currently available resources to the SN. Thus, the parameters relayed in resource vector $Rv(s)$ by the source device s comprise the device's remaining processing capacity, battery capacity, security capabilities, and the available channel capacity of its assigned data channel. The data channel characteristics of a network host are typically described by several metrics, for example, throughput, delay, and security level. However, the security level over a channel depends on the capabilities of the particular transmitter and receiver, and it can be determined by the SN from a consideration of their commonly supported protocols. The value $A = 4$ informs the SN that this is a resource reply message from the source device with ID $D_{id}(s)$.

Table 1.2 shows the SN-generated messages (depending on the received requests), which are *update reply, adaptation reply, bootstrap reply, communication reply,* and *resource request*. Each message type is used by the SN to pass a number of relevant parameters to network hosts, where the variable B can take values satisfying $0 \le B \le 4$, and it is used to inform the hosts of the type of message received.

Update reply: $D_{id}(s)$ is the ID of source device s that the response is aimed at. $D_{id}(n)$ is the ID of the new neighbor n of source device s. $Ch_{Fmin,Fmax}(n)$ comprises the channel assignment of neighbor n (given frequency modulation as mentioned earlier). The value $B = 0$ indicates that the received message is an update message.

Table 1.2   SN Message Types

Adaptation Reply
$D_{id}(s)$, $Ch_{Fmin, Fmax}(s)$, $P_{max}(s)$
$VID_{x,y}(s)$, $t_{start}(s)$, $t_{end}(s)$, $D_{id}(n \mid n = 1, 2,...x)$
$Ch_{Fmin, Fmax}(n \mid n = 1, 2,...x)$
$VID_{x,y}(n \mid n = 1, 2,...x)$, $G_{gps}(c)$, $R_{length}(c)$, B
Bootstrap Reply
$D_{id}(s)$, $Ch_{Fmin, Fmax}(s)$, $P_{max}(s)$
$VID_{x,y}(s)$, $t_{start}(s)$, $t_{end}(s)$, $R_{length}(c)$, B
$D_{id}(n \mid n = 1,2,...x)$, $G_{gps}(c)$
$Ch_{Fmin, Fmax}(n \mid n = 1, 2,...x)$, $VID_{x,y}(n \mid n = 1, 2,...x)$
Communication Reply
$D_{id}(s)$, $D_{id}(d)$, $D_{id}(i \mid i = 1, 2,...x)$
$Ch_{Fmin, Fmax}(i \mid i = 1, 2,...x)$, B
Update Reply: $D_{id}(s)$, $D_{id}(n)$, $Ch_{Fmin, Fmax}(n)$, B
Resource Request: $D_{id}(s)$, B

Adaptation reply and bootstrap reply: These messages are identical, although the computational priority at the SN is higher for the adaptation message. $D_{id}(s)$ is the ID of source device s that the response targets. $P_{max}(s)$ is the maximum allowed transmit power output for the requesting device, whereas $Ch_{Fmin,Fmax}(s)$ is its new channel (i.e., DSOP) assignment [20]. $t_{start}(s)$ and $t_{end}(s)$ define the time interval during which the assigned DSOP is predicted to be available for SU operation, which in this particular case is source device s. $VID_{x,y}(s)$ is the VID of the requesting source device. $D_{id}(n \mid n = 1, 2,...x)$ is the neighbor list of the source device, provided there are x neighbors. $Ch_{Fmin,Fmax}(n \mid n = 1, 2,...x)$ denotes the channel assignments for the neighbors, and $VID_{x,y}(n \mid n = 1, 2,...x)$ denotes the VIDs of the neighbors. $G_{gps}(c)$ denotes the center GPS coordinates for virtual cluster c, and $R_{length}(c)$ is the radius in centimeters for virtual cluster c, within which the particular source device is located. The value $B = 1$ indicates that the received message is an adaptation message, whereas the value $B = 2$ indicates a received bootstrap message (differentiation is needed as the two messages have different priorities).

Communication reply: $D_{id}(s)$ is the ID of source device s that the response is aimed at. $D_{id}(d)$ is the ID of the destination device d in the requested route. $D_{id}(i \mid i = 1, 2,...x)$ denotes the IDs of all intermediate nodes i along the path to the destination, and $Ch_{Fmin,Fmax}(i \mid i = 1, 2,...x)$ encloses the channel assignment for each respective intermediate hop. The variable $B = 3$ indicates that the received message is a communication message. The routing path is compiled by the SN according to e2e goals by using the optimization algorithms in Ref. [15].

Resource request: This is used by SN to request the present state of available resources at a particular source device s with ID $D_{id}(s)$. This message is typically transmitted to network hosts located throughout regions that an optimized e2e path must traverse. The value $B = 4$ indicates that the received message is a resource request message.

A computational example of the size of an adaptation reply message (ARM) compiled by the SN is as follows:

$$ARM = (D_{id}(s)) + (Ch_{Fmin,Fmax}(s)) + (P_{max}(s)) + (D_{id}(n \mid n = 1, 2,...x)) + (Ch_{Fmin,Fmax}(n \mid n = 1, 2,...x))$$
$$+ (t_{start}, t_{end}(s)) + (VID_{x,y}(s)) + (VID_{x,y}(n \mid n = 1, 2,...x)) + G_{gps}(c) + R_{length}(c) + (B)$$

Assuming that a network host has four neighbors in its new geographical location, the size (in bits) of ARM can be calculated as follows:

$$ARM(4) = 80 + (44 + 44) + 14 + 4 \times 80 + 4 \times (44 + 44) + (20 + 20) + 16 + 4 \times 16 + 55 + 17 + 3 = 1049$$

Likewise, a bootstrap reply message can be computed with similar information and size, whereas other control messages may contain more or less data. For instance, in the case of a communication reply message (CRM) where an e2e path optimized by the SN is replied back to the requesting host, the distance between the source and the destination can vary for different requests. A variable number of intermediate hops along the path implies routing overheads of different sizes. Typically, the overhead contained in an optimized CRM needs to accommodate device IDs (for source, destination, and intermediate hops), channel assignments (for every intermediate hop to the destination), and message type identifier. Channel assignments are necessary for optimized routes given that network hosts only have such information for adjacent neighbors. In our case, an optimized path entails bypassing the adjacent neighbors and communicating directly with network hosts located in a neighboring cluster, naturally within the limits of a device's capabilities (such as, transmit power output). Hence, an optimized routing path typically reaches the destination through less intermediate hops as compared to a greedy routing approach (Figure 1.4), even for distant spacing between the source and destination. Put differently, the overhead enclosed in the CRM that describes an optimized path is lightweight, thanks to an optimization mechanism that reduces the number of intermediate hops from the source to the destination [7].

1.6.2 Implementation Details

Assuming equally sized clusters with a radius of not less than 150 m, the maximum number of viable virtual clusters in the CRN coverage area described earlier (a microcell with 1000 m radius) is 36 [7]. Depending on the particular CRN environment, clusters with both smaller and larger radii may be necessary. It is important to mention that, in the case of a smaller cluster radius, the signaling requirements may also increase owing to node mobility and the exchange of ARMs. Hence, the CCC must be dimensioned to handle signaling loads in accordance with the particular scenario, that is, the used number of clusters.

To estimate the amount of overhead required for optimizing communications with a particular 36-virtual-cluster solution (i.e., a medium-cluster-size scenario), we have implemented a topology construction algorithm for the two-dimensional [0,1] × [0,1] CCS representation of the CRN space dimension [7]. The virtual space topology construction algorithm has been implemented with the C++ object-oriented programming language, and it is used to simulate the mean number of hops (path lengths) required for 10 random communications over 100 runs and different network populations (n) of 100, 500, and 1000 simulated devices. The communications per simulated run and network population are done for randomly selected network hosts, where each communication is optimized according to the shortest path and least cost constraints. In Figure 1.5, it can be observed that, for each network population, the mean path length per simulated communication over 100 runs is between 3 and 4 hops, regardless of the network population. In other words, provided the CRN coverage area is subdivided into virtual clusters where each cluster is responsible for the hosts within its area, the mean path lengths converge according to a uniform distribution, regardless of the network population. That is, the routing tables no longer grow with the network population [7,15].

Thus, the average path length (regardless of the network population) that can be expected for a random communication in a CRN optimized according to a 36-virtual-cluster subdivision of its coverage area, is computed with

$$\overline{H}_n = \frac{1}{30} \sum_{c=1}^{30} \overline{h}_c \tag{1.1}$$

$\overline{H}_n$ is the average path length, computed from the mean path lengths of all 10 simulated communications. $\overline{h}_c$ indicates the mean path length for each respective communication taken over all runs per network population. Assuming 10 simulated communications and network populations (n) of 100, 500, and 1000 simulated devices, a total of $10 \times 3 = 30$ mean path lengths is obtained. Accordingly, from the mean path lengths presented in Table 1.3, the grand average path length for this particular case (computed with Equation 1.1) is 3.25. In addition, over the 100 simulated runs, 95% of the communication path lengths for each network population (n) of 100, 500, and 1000 simulated devices are within ±3.1%, ±3.0%, and ±3.2%, respectively, of the grand average path length. The 95% confidence interval (CI) is computed for each network population:

$$CI = \overline{H}_n \pm 1.96 \frac{\sigma}{\sqrt{n}} \tag{1.2}$$

Table 1.3   Mean Path Length per Communication over All Runs and Different Network Populations

Network Population	Com1	Com2	Com3	Com4	Com5	Com6	Com7	Com8	Com9	Com10
100	3.17	3.33	3.24	3.84	3.30	3.13	3.17	3.00	3.20	3.32
500	3.32	3.09	3.09	3.51	3.12	3.34	2.95	3.44	3.03	3.22
1000	3.92	3.13	3.33	3.28	2.97	3.15	3.20	3.26	3.46	2.99

$\overline{H}_n$ is the average path length, computed from the mean path lengths of all 10 simulated communications. The critical value for the 95% CI is 1.96, with a standard normal (0,1) z-distribution. σ is the standard deviation, given by $\sigma = \sqrt{Var(x)}$, where $Var(x)$ is the variance of the grand average path length for each network population, taken over all communications and runs. n is the total number of communications performed over all runs; that is, given that 10 communications are simulated for each run, the total number of communications over 100 runs per network population is $100 \times 10 = 1000$. Subsequently, the amount of CRM overhead for this particular case can be computed as follows:

$$CRM = (D_{id}(s)) + \left(\overline{x} \times (D_{id}(i \mid i = 1,2,...x)) + (Ch_{F_{min}, F_{max}}(n \mid n = 1,2,...x))\right) + (B)$$

$D_{id}(s)$ is the source device id, and $D_{id}(i \mid i = 1, 2,...x)$ is the id of the intermediate hops, given a total number of x intermediate hops (including the destination, which is the last hop). $\overline{x}$ is the average path length for a random optimized communication according to the suggested virtual cluster solution. $Ch_{F_{min}, F_{max}}(i \mid i = 1, 2,...x)$ denotes the operating frequencies (channel assignments) of all intermediate hops along the path, including the destination. We assume a frequency modulation where each network member occupies a channel in the frequency range F_{min} to F_{max}. The value $B = 3$ indicates that the received message is of type CRM. Given an average path length of 3.25 hops, the amount of overhead (in bits) required to optimize communications according to the particular 36-virtual-cluster solution is

$$80 + 3.25 \times (80 + (44 + 44)) + 3 = 629$$

It is important to mention that for a scenario with differently dimensioned cluster sizes, the grand average path length may also vary, which requires a different optimization overhead.

1.7 System Performance

Depending on several factors, which are presented in Section 1.6, the cluster sizes in the CRN coverage are dimensioned to ensure limited signal degradation and maximum throughput for the available channels during inter- and intracluster communication among devices. As such, different-sized clusters in various numbers can be available throughout different CRNs. However, because all inter-CRN communication (regardless of where in the CRN the source is located) must go through the BS, only the network hosts located within the clusters in the immediate vicinity of the BS can transmit data directly to it. Subsequently, in this case we present the system model for a basic scenario where only two clusters are located near the BS, and each cluster accommodates only one transmitting host at a time within its area. With this particular model, only one hop is necessary to achieve communication among the network hosts and the BS. This enables us to study the direct impact PU activity has on SU service completion, which further allows an optimization problem to be formulated and addressed.

1.7.1 Queuing Modeling

As indicated in Figure 1.6, two clusters denoted as *CL*-1 and *CL*-2 are assumed to share a single radio channel that is licensed to PUs and is denoted as *c*. The activity of PUs in this channel is assumed to be spatially invariant within the geographical area of the two clusters. In other words, at a particular time moment, PU's activity (i.e., being either present or absent in channel *c*) is observed to be the same at every SU located within the two clusters. To avoid radio interference between the two clusters during SU communication, we assume that channel *c* is divided into two identical subchannels denoted as c_1 and c_2. In our case, for simplicity, this is done by frequency modulation techniques such as frequency division multiple access (FDMA) [15]. Subsequently, c_1 and c_2 are allocated to *CL*-1 and *CL*-2, respectively, for use by SUs.

Typically, channel *c* is said to become available for SUs when PUs are absent. Thus, in such a case, SUs in *CL*-1 and *CL*-2 can opportunistically access c_1 and c_2 in each respective cluster. Moreover, it is important to mention that channel *c* can only be used by one PU at a time. This means that if channel *c* is already used by one PU, a newly arrived PU is blocked from transmitting data. Accordingly, if channel *c* is available for SUs, each of the two subchannels c_1 and c_2 can only be used by one SU at a time. Similarly, if the available subchannel in each respective cluster is used by an SU, data transmission for newly arrived SUs within each particular cluster is blocked by the SN. Naturally, when PUs return to channel *c*, SUs using either c_1 or c_2 are dropped by the SN. Because our current system model only assumes one subchannel per cluster (i.e., no adaptation is possible for SUs to retain service), the data transmissions of dropped SUs are forced to be terminated.

We assume that PU arrivals to channel *c* follow the Poisson stream with mean rate λ_p. The time periods of PUs transmitting data are assumed to be exponentially distributed with mean value $1/\mu_p$. For each cluster *CL*-1 and *CL*-2, the arrivals of SUs at subchannels c_1 and c_2 are assumed to independently follow Poisson streams with mean rates γ_1 and γ_2. Accordingly, the time periods of SUs transmitting data via the two respective subchannels are assumed to independently follow exponential distributions with mean rates δ_1 and δ_2.

With regard to the above assumptions, a loss-system-based continuous time Markov chain (CTMC) queuing model is constructed. We let a set of three integers (i_1, i_2, j) denote a system state such that i_1 SUs are using subchannel c_1, i_2 SUs are using subchannel c_2, and j PUs are using channel *c*. Therefore, the system state space is obtained as $S = \{(i_1, i_2, j)\}$, where the three integers are constrained by $i_1 \in \{0,1\}$, $i_2 \in \{0,1\}$, $j \in \{0,1\}$, $(i_1 + j) \in \{0,1\}$, and $(i_2 + j) \in \{0,1\}$. The system state diagram is shown in Figure 1.7, whereas the system state transitions are described as follows:

- Both $(0,0,0) \leftrightarrow (1,0,0)$ and $(0,1,0) \leftrightarrow (1,1,0)$ mean that an SU in *CL*-1 accesses and releases c_1 with rates γ_1 and δ_1, respectively.
- Both $(0,0,0) \leftrightarrow (0,1,0)$ and $(1,0,0) \leftrightarrow (1,1,0)$ means that an SU in *CL*-2 accesses and releases c_2 with rates γ_2 and δ_2, respectively.

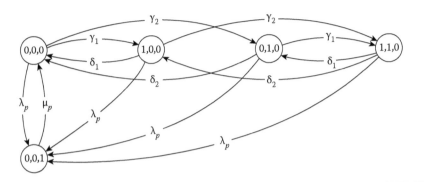

Figure 1.7 State diagram of the modeled system.

■ $(0,0,0) \leftrightarrow (0,0,1)$ means that a PU accesses and releases channel c with mean rates λ_p and μ_p, respectively.
■ Transitions from $(1,0,0)$, $(0,1,0)$, and $(1,1,0)$ to $(0,0,1)$ mean that a PU occupies channel c with rate λ_p, so that the SUs using either c_1 or c_2 are forced to be terminated.

Let $\pi_{i_1,i_2,j}$ denote the steady-state probability of state (i_1,i_2,j). We have the balance equations:

$$\pi_{0,0,0} = \frac{\pi_{1,0,0}\delta_1 + \pi_{0,1,0}\delta_2 + \pi_{0,0,1}\mu_p}{(\gamma_1 + \gamma_2 + \lambda_p)} \tag{1.3}$$

$$\pi_{1,0,0} = \frac{\pi_{0,0,0}\gamma_1 + \pi_{1,1,0}\delta_2}{(\gamma_2 + \delta_1 + \lambda_p)} \tag{1.4}$$

$$\pi_{0,1,0} = \frac{\pi_{0,0,0}\gamma_2 + \pi_{1,1,0}\delta_1}{(\gamma_1 + \delta_2 + \lambda_p)} \tag{1.5}$$

$$\pi_{1,1,0} = \frac{\pi_{1,0,0}\gamma_2 + \pi_{0,1,0}\gamma_1}{(\delta_1 + \delta_2 + \lambda_p)} \tag{1.6}$$

$$\pi_{0,0,1} = \frac{(\pi_{0,0,0} + \pi_{1,0,0} + \pi_{0,1,0} + \pi_{1,1,0})\lambda_p}{\mu_p} \tag{1.7}$$

where all steady-state probabilities are subject to the constraint

$$\pi_{0,0,0} + \pi_{1,0,0} + \pi_{0,1,0} + \pi_{1,1,0} + \pi_{0,0,1} = 1 \tag{1.8}$$

With Equations 1.3 to 1.8, a set of linear equations can be constructed, and the steady-state probabilities of all states can be computed.

1.7.2 Performance Metrics for SUs

With the foregoing queuing model described, we study four performance metrics: blocking probabilities of SUs in clusters *CL*-1 and *CL*-2, dropping probability of SUs, and service-completion throughput of SUs.

Given the system state at (i_1, i_2, j), we know that the SUs in both clusters *CL*-1 and *CL*-2 are blocked when channel c is occupied by PUs; that is, $j = 1$. If $j = 0$, blocking of newly arrived SUs in *CL*-1 and *CL*-2 occurs for $i_1 = 1$ and $i_2 = 1$, respectively, that is, when subchannels c_1 and c_2 are already in use by other SUs. Let $P_{bl,1}$ and $P_{bl,2}$ denote the probabilities of blocking SUs in *CL*-1 and *CL*-2, respectively:

$$P_{bl,1} = \pi_{1,0,0} + \pi_{1,1,0} + \pi_{0,0,1} \tag{1.9}$$

$$P_{bl,2} = \pi_{0,1,0} + \pi_{1,1,0} + \pi_{0,0,1} \tag{1.10}$$

For a system state (i_1, i_2, j), the dropping of SUs occurs when we have three mutually considered conditions: (1) $j = 0$, (2) a PU occupies channel c, and (3) $(i_1 + i_2) > 0$. In other words, the case of dropping SUs refers to three particular system states: (1,0,0), (0,1,0), and (1,1,0). In each of the three states, the rate of dropping SUs is equal to λ_p, as shown in Figure 1.7. Thus, let P_{dr} denote the dropping probability of the SUs. To compute P_{dr}, we divide the total rate of dropping SUs by the actual arrival rate of SUs into the system. For the system at the three states (1,0,0), (0,1,0), and (1,1,0), the numbers of dropped SUs due to channel occupancy by PUs are equal to 1, 1, and 2, respectively. Hence, the total rate of dropping SUs equals $\lambda_p(\pi_{1,0,0} + \pi_{0,1,0} + 2\pi_{1,1,0})$. Given that the actual arrival rate is $(\gamma_1(1 - P_{bl,1}) + \gamma_2(1 - P_{bl,2}))$, we have

$$P_{dr} = \frac{\lambda_p(\pi_{1,0,0} + \pi_{0,1,0} + 2\pi_{1,1,0})}{\gamma_1(1 - P_{bl,1}) + \gamma_2(1 - P_{bl,2})} \tag{1.11}$$

We define the service-completion throughput of SUs as the average rate of SUs completing the transmission using subchannels. Thus, let R denote the service-completion throughput of SUs. According to the expectation definition $\sum_{i=1}^{\infty} x_i \Pr(x_i)$, R is computed as follows:

$$\sum_{\forall i_1, i_2, j}^{(i_1, i_2, j) \in S} (i_1 \delta_1 + i_2 \delta_2) \pi_{i_1, i_2, j} \tag{1.12}$$

1.7.3 Performance Evaluation

Both numerical and simulation results are reported; the simulation experiments were conducted to validate the numerical analysis.

1.7.3.1 Parameter Settings

We consider the parameters of the PU arrival rate as $\lambda_p \in \{0.03, 0.06, 0.09, 0.12\}$ s^{-1} and the service rate as $\mu_p = 0.16$ s^{-1} [23]. Naturally, for SUs, the two clusters cover particular geographical areas, where the arrival rates of SUs are considered equal: $\gamma_s = 0.68$ s^{-1}; that is, $\gamma_1 + \gamma_2 = \gamma_s = 0.68$ s^{-1}. This furthermore entails that the two clusters may cover differently sized geographical areas, which means that the arrival rates of SUs to them may be different. Therefore, we introduce a selection probability $\alpha \in \{0.1, 0.3, 0.5\}$, by which γ_1 and γ_2 are constrained as $\gamma_1 = \alpha\gamma_s$ and $\gamma_2 = (1-\alpha)\gamma_s$, respectively [23]. Assuming that the SU service rates over two identical subchannels c_1 and c_2 are the same, they are considered equal to 0.82 s^{-1}; that is, $\delta_1 = \delta_2 = 0.82$ s^{-1}. For each parameter setting,

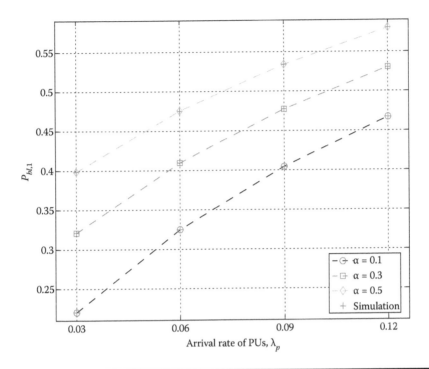

Figure 1.8 **Blocking probability of SUs in *CL*-1 versus λ_p.**

the simulator runs for a simulation time of 10^7 s. The numerical results of $P_{bl,1}$, $P_{bl,2}$, P_{dr}, and R are shown in Figures 1.8, 1.9, 1.10, and 1.11, respectively. The simulation results are indicated with the marker "+," and we observe that they closely match the numerical results. Further, we report in Table 1.4 the numerical results of the actual arrival rate of SUs into the system. A more detailed discussion on the obtained results follows.

1.7.3.2 Given Identical Selection Probability

Figures 1.8, 1.9, and 1.10 indicate that $P_{bl,1}$, $P_{bl,2}$, and P_{dr} increase with λ_p, respectively. This is because more PUs are requesting channel resources when λ_p is increasing. Hence, the availability of the two subchannels for SUs is reduced. As a result, the service-completion throughput of SUs, R, is decreased with λ_p, as observed in Figure 1.11.

1.7.3.3 Given Identical Arrival Rates

Figure 1.8 shows that $P_{bl,1}$ increases with α. Figure 1.9 shows on the contrary that $P_{bl,2}$ decreases with α. The reason for this is that when α is increasing, the competition among SUs becomes higher for subchannel $c1$ and lower for subchannel $c2$. Put differently, given $\alpha < 0.5$, the competition among SUs in *CL*-1 is always lower than in *CL*-2.

However, when α is close to 0.5, the arrivals of SUs are more evenly distributed between the two clusters. This means that the available resources are better used; the total blocking probability of SUs decreases with increasing α, and it is smallest when $\alpha = 0.5$. This can be observed in Table 1.4, where the actual arrival rate of SUs into the system increases with α.

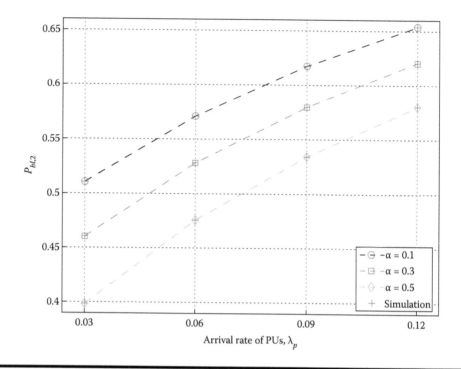

Figure 1.9 Blocking probability of SUs in *CL*-2 versus λ_p.

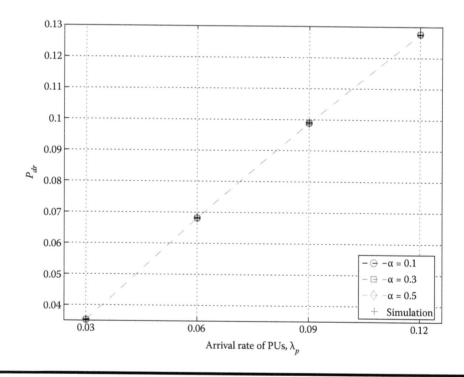

Figure 1.10 Dropping probability of SUs versus λ_p.

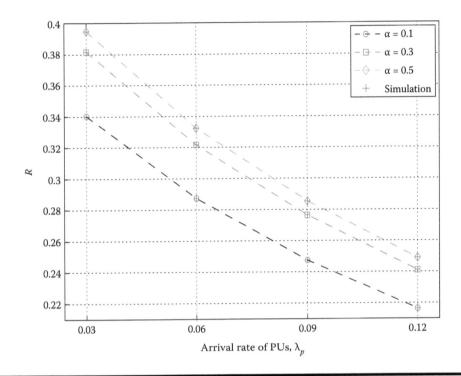

Figure 1.11 Service-completion throughput of SUs versus λ_p.

Table 1.4 Numerical Results of SUs Actual Arrival Rate into the System

α	$\gamma_1(1 - P_{bl,1}) + \gamma_2(1 - P_{bl,2})$			
	$\lambda_p = 0.03$	$\lambda_p = 0.06$	$\lambda_p = 0.09$	$\lambda_p = 0.12$
0.1	0.3527	0.3084	0.2747	0.2480
0.3	0.3955	0.3451	0.3067	0.2763
0.5	0.4090	0.3567	0.3168	0.2854

In addition, in Figure 1.11 it can be observed that the total service-completion throughput of SUs, R, increases with α. Moreover, Figure 1.10 shows that P_{dr} is the same for all three values of $\alpha \in \{0.1, 0.3, 0.5\}$. This is because the total subchannel availability for SUs is the same regardless of how the arrival rates are distributed between the two clusters. That is, for a particular α value, if more SUs end up being dropped from *CL*-1 or *CL*-2, it makes no difference to the total P_{dr}.

1.7.3.4 Performance Optimization

The aforementioned performance evaluation addresses an optimization problem regarding optimizing the service-completion throughput of SUs. For our particular model, the optimization can be done by adjusting the geographical sizes of two different clusters. The goal is to balance the arrival rates of SUs between the two clusters, that is, achieve an α value that is as close as possible to 0.5. Consequently, this allows us to minimize the total blocking probability, which furthermore provides the best service-completion throughput, as can be seen in Figure 1.11. Naturally, this means that the arrival rates of SUs is an important part of the cluster dimensioning and needs to be considered in addition to the aforementioned factors.

1.8 Conclusions

A novel architectural solution has been advanced for multimedia communication in CRNs. The architecture is based on the use of a middleware with a common set of APIs and a number of overlay and underlay entities. This allows communication among network members according to e2e goals by centrally optimizing the computed routing paths. The architecture is described with an emphasis on the associated research challenges. More specifically, two of the most important elements of this architecture are the solutions adopted for data representation and routing. Following an investigation of the system performance with regard to the particular optimization solution used, we conclude, through simulation and numerical analysis, that the initial cluster dimensioning plays a vital role in securing satisfactory service-completion throughput for SUs. Balancing the arrivals of SUs between the clusters enables the blocking probability of SUs to be minimized and thus leads to a better utilization of the available resources. An important aspect for future work is further development of the system model to better emulate the described multihop scenario used in the suggested architecture, together with an updated numerical and simulation model for performance evaluation.

References

1. D. Astely, E. Dahlman, G. Fodor, S. Parkvall, and J. Sachs, LTE release 12 and beyond, *IEEE Communications Magazine*, vol. 51, No. 7, pp. 154–160, 2013.
2. S. Akhtar, 2G-5G networks: Evolution of technologies, standards and deployment, *Encyclopedia of Multimedia Technology and Networking*, 2009, http://123seminarsonly.com/Seminar-Reports/012/64344596-2G-5G-Networks-Encyclopedia-Paper.pdf
3. X. Li, A. Gani, R. Salleh, and O. Zakaria, The future of mobile wireless communication networks, *International Conference on Communication Software and Networks*, pp. 554–557, Macau, China, 2009.
4. I. A. Akyildiz, L. Won-Yeol, and K. Chowdhury, Spectrum management in cognitive radio ad hoc networks, *IEEE Network*, vol. 23, No. 4, pp. 6–12, 2009.
5. I. A. Akyildiz, L. Won-Yeol, and K. Chowdhury, CRAHNs: Cognitive radio ad hoc networks, *Ad Hoc Networks*, vol. 7, No. 5, pp. 810–836, 2009.
6. L. Berlemann, S. Mangold, G. R. Hiertz, and B. Walke, Spectrum load smoothing: Distributed quality-of-service support for cognitive radios in open spectrum, *European Transactions on Telecommunications*, vol. 17, No. 3, pp. 395–406, 2006.
7. A. O. Popescu, Y. Yao, and M. Fiedler, Communication mechanisms for cognitive radio networks, *11th IEEE Pervasive Computing and Communication*, San Diego, CA, March 2013.
8. Z. Hasan, H. Boostanimehr, and V. K. Bhargava, Green cellular networks: A survey, some research issues and challenges, *IEEE Communications Surveys Tutorials*, vol. 13, No. 4, pp. 524–540, 2011.
9. S. Murugesan, Harnessing green IT: Principles and practices, *IT Professional*, vol. 10, No. 1, pp. 24–33, 2008.
10. C. Yan, Z. Shunqing, X. Shugong, and G. Y. Li, Fundamental trade-offs on green wireless networks, *IEEE Communications Magazine*, vol. 49, No. 6, pp. 30–37, 2011.

11. C. Peng, H. Zheng, and B. Y. Zhao, Utilization and fairness in spectrum assignment for opportunistic spectrum access, *Mobile Networks and Applications*, vol. 11, No. 4, pp. 555–576, 2006.
12. L. Won-Yeol and I. F. Akyldiz, A spectrum decision framework for cognitive radio networks, *IEEE Transactions on Mobile Computing*, vol. 10, No. 2, pp. 161–174, 2011.
13. L. Won-Yeol and I. F. Akyldiz, Spectrum-aware mobility management in cognitive radio cellular networks, *IEEE Transactions on Mobile Computing*, vol. 11, No. 4, pp. 529–542, 2012.
14. L. Ying-Chang, C. Kwang-Cheng, G. Y. Li, and P. Mahonen, Cognitive radio networking and communications: An overview, *IEEE Transactions on Vehicular Technology*, vol. 60, No. 7, pp. 3386–3407, 2011.
15. A. O. Popescu, D. Erman, M. Fiedler, and A. P. Popescu, On routing in cognitive radio networks, *9th IEEE International Conference on Communications*, Bucharest, Romania, June 2012.
16. H. Khalife, N. Malouch, and S. Fdida, Multihop cognitive radio networks: To route or not to route, *IEEE Network*, vol. 23, No. 4, pp. 20–25, 2009.
17. S. Mangold, L. Berlemann, and S. S. Nandagopalan, Spectrum sharing with value-orientation for cognitive radio, *European Transactions on Telecommunications*, vol. 17, No. 3, pp. 383–394, 2006.
18. B. Le, T. W. Rondeau, and C. W. Bostian, Cognitive radio realities, *Wireless Communications and Mobile Computing*, vol. 7, No. 9, pp. 1037–1048, 2007.
19. A. P. Popescu, D. Erman, K. D. Vogeleer, A. O. Popescu, and M. Fiedler, ROMA: A middleware framework for seamless handover, *Network Performance Engineering*, vol. 5233, pp. 784–794, 2010.
20. B. Ackland, D. Raychaudhuri, M. Bushnell, C. Rose, I. Seskar, T. Sizer, D. Samardzija, et al., *High Performance Cognitive Radio Platform with Integrated Physical and Network Layer Capabilities*, Tech. Report NSF Grant CNS-0435370, 2005, http://nsf.gov/awardsearch/showAward.do?AwardNumber = 0435370
21. A. O. Popescu and M. Fiedler, On data representation in cognitive radio networks, *8th Swedish National Computer Networking Workshop*, Stockholm, Sweden, June 2012.
22. Y. Yao, S. R. Ngoga, D. Erman, and A. P. Popescu, Competition-based channel selection for cognitive radio networks, *IEEE Wireless Communications and Networking Conference*, pp. 1432–1437, Paris, France, April 2012.
23. Y. Yao, S. R. Ngoga, D. Erman, and A. P. Popescu, Performance of cognitive radio spectrum access with intra- and inter-handoff, *IEEE International Conference on Communications*, pp. 1539–1544, Ottawa, Canada, June 2012.
24. M. Kassar, B. Kervella, and G. Pujolle, An overview of vertical handover decision strategies in heterogeneous wireless networks, *Computer and Communications*, vol. 31, No. 10, pp. 2607–2620, 2008.
25. B. Martinez, A. O. Popescu, V. Pla, and A. P. Popescu, Cognitive radio networks with elastic traffic, *8th IEEE Euro-NF Conference on Next Generation Internet*, Karlskrona, Sweden, June 2012.
26. D. Raychaudhuri and J. Xiangpeng, A spectrum etiquette protocol for efficient coordination of radio devices in unlicensed bands, *14th IEEE Personal, Indoor and Mobile Radio Communications*, vol. 1, pp. 172–176, Beijing, China, September 2003.
27. J. Perez-Romero, O. Salient, R. Agusti, and L. Giupponi, A novel on-demand cognitive pilot channel enabling dynamic spectrum allocation, *2nd IEEE International Symposium on New Frontiers in Dynamic Spectrum Access Networks*, pp. 46–54, Dublin, Ireland, April 2007.
28. R. W. Thomas, L. A. DaSilva, and A. B. MacKenzie, Cognitive networks, *IEEE Symposium on New Frontiers in Dynamic Spectrum Access Networks*, pp. 352–360, Baltimore, MD, November 2005.

Chapter 2

Paving a Wider Way for Multimedia over Cognitive Radios: An Overview of Wideband Spectrum Sensing Algorithms

Bashar I. Ahmad,[1] Hongjian Sun,[2] Cong Ling,[3] and Arumugam Nallanathan[4]

[1]*Engineering Department, University of Cambridge, Trumpington Street, Cambridge*

[2]*School of Engineering and Computing Sciences, Durham University, South Road, Durham*

[3]*Department of Electrical and Electronic Engineering, Imperial College London, South Kensington Campus, London*

[4]*Department of Informatics, King's College London, Strand, London*

Contents

2.1 An Introduction to Spectrum Sensing for Cognitive Radio

With conventional static spectrum allocation policies, a licensee, that is, a primary user (PU), is permitted to use a particular spectrum band over relatively long periods of time. Such inflexible allocation regimes have resulted in a remarkable spectrum underutilization in space or time as reported in several empirical studies conducted in densely populated urban environments [1,2]. By enabling an unlicensed transmitter, that is, a secondary user (SU), to opportunistically access or share these fully or partially unused licensed spectrum gaps, the cognitive radio (CR) paradigm (a term attributed to Mitola [3]) has emerged as a prominent solution to the persistent spectrum scarcity problem [4]. It fundamentally relies on a dependable spectrum awareness routine to identify vacant spectrum bands and limit any introduced interference, possibly below a level agreed a priori with the network PUs. Therefore, spectrum sensing, which involves scanning the radio frequency (RF) spectrum in search of a spectrum opportunity, is considered to be one of the most critical components of a CR [3–8].

There are several approaches to spectrum sharing and PU/SU coexistence in CR networks [see 6–9 for an overview]. In this chapter, we predominantly focus on the interweaving systems, where an SU is not permitted to access a spectrum band when a PU transmission is present, thus ensuring minimal interference. With alternative methods, that is, underlay and overlay systems, the PU and SUs can simultaneously access a spectral subband. They use multiaccess techniques such as spread spectrum and/or assume the availability of a considerable amount of information on the network PUs (e.g., codebooks, operation patterns, propagation channel information) to reduce the resulting coexistence interference. These two paradigms still, however, rely to an extent on spectrum awareness to establish the spectrum status. In essence, the spectrum sensing problem entails the SUs reliably determining whether a particular spectral subband, for example, $\mathcal{B}_l$, is vacant or the PU is present. This corresponds to the classical binary hypothesis testing problem that aims to distinguish between the two hypotheses:

$$\mathcal{H}_0 : \mathbf{y} = \mathbf{w}$$
$$\mathcal{H}_1 : \mathbf{y} = \mathbf{x} + \mathbf{w}$$

(2.1)

such that $\mathcal{H}_0$ and $\mathcal{H}_1$ signify the absence and presence of a transmission in $\mathcal{B}_l$, that is, $\mathbf{x} = [x(t_1), x(t_2), \ldots, x(t_M)]^T$, respectively. Whereas, $\mathbf{y} = [y(t_1), y(t_2), \ldots, y(t_M)]^T$ is the vector encompassing the M collected samples of the received signal at the SU. In classical digital signal processing (DSP), the sampling instants $\{t_m\}_{m=1}^{M}$ are uniformly distributed, where $t_m = mT_{US}$ and $f_{US} = 1/T_{US}$ is the data acquisition rate. To shorten the notation, let $x[m] = x(mT_{US})$ and $y[m] = y(mT_{US})$. For simplicity, $\mathbf{w} \sim \mathcal{N}\left(0, \sigma_w^2\right)$ is zero-mean additive white Gaussian noise (AWGN) with covariance σ_w^2; more general noise models can be considered. The probability of successfully detecting the PU presence is $P_D = Pr\{\mathcal{H}_1 | \mathcal{H}_1\}$, and the probability of a false alarm, that is, missing a spectral opportunity, is $P_{FA} = Pr\{\mathcal{H}_1 | \mathcal{H}_0\}$. To decide between $\mathcal{H}_0$ and $\mathcal{H}_1$ is a classical detection problem with long-established solutions [10,11]. This is customarily accomplished by comparing the outcome of a test statistic $\mathfrak{T}(\mathbf{y})$ with a predetermined threshold value γ according to

$$\mathfrak{T}(\mathbf{y}) \underset{\mathcal{H}_1}{\overset{\mathcal{H}_0}{\lessgtr}} \gamma.$$

(2.2)

Thus, the sensing problem boils down to formulating an effective test statistic and appropriately setting the comparison threshold value to determine the available spectral opportunities without introducing harmful interference. As noted in [8], whether the detection problem is tackled within a deterministic or Bayesian statistical framework, the resultant $\mathfrak{T}(\mathbf{y})$ is a form of the likelihood ratio $Pr\left(\mathbf{y}|\mathcal{H}_1\right)/Pr\left(\mathbf{y}|\mathcal{H}_0\right)$ [10,11]. Next, we consider a number of traditional spectrum sensing techniques aimed at a single monitored RF band $\mathcal{B}_l$, that is, narrowband spectrum sensing (NBSS).

2.1.1 Conventional Narrowband Techniques

Coherent, energy, and feature detectors are among the most commonly used NBSS algorithms in CR networks [4–6,8]. They have different operational and implementation requirements as discussed below.

1. *Coherent detector*: Otherwise known as a matched filter, it utilizes the optimal test statistic that maximizes the signal-to-noise ratio (SNR) in the presence of additive noise [10,11]. Its test statistic simply involves correlating the suspected PU transmission x with the received signal according to

$$\mathfrak{T}(\mathbf{y}) \triangleq \mathbf{x}^H \mathbf{y} \underset{\mathcal{H}_1}{\overset{\mathcal{H}_0}{\lessgtr}} \gamma$$

(2.3)

   where $(.)^H$ is the conjugate transpose operation. The PU signal structure is assumed to be perfectly known at the receiver, for example, signal variance, modulation type, packet format, channel coefficients, and so on. One of the main advantages of the coherent detector is that it requires only a small number of data measurements to achieve predefined probabilities of detection P_D and false alarm P_{FA}, where $\mathcal{O}(1/\text{SNR})$ samples suffice even in low-SNR regions (i.e., when SNR $\ll$ 1). The SNR is defined as SNR $= P_{S,l}/\sigma_{w,l}^2$ such that $P_{S,l}$ and $\sigma_{w,l}^2$ are the powers of a transmission present in $\mathcal{B}_l$ and AWGN variance, respectively. However, in low-SNR conditions, the detector's performance drastically degrades owing to the difficulty in maintaining synchronization between the transmitter and the receiver. Accurate synchronization is a fundamental requirement of the coherent detector. In addition, the complexity of the match filter increases with the diversity of potential PUs because a distinct detector per signal structure is imperative. The coherent detector is inflexible and can be unsuitable for CR networks, which often include several PUs using different transmission technologies and dynamically adapting their transmission characteristics.

2. *Energy detector*: The energy detector, also known as the radiometer, is a noncoherent detector widely regarded as one of the simplest approaches for deciding between $\mathcal{H}_0$ and $\mathcal{H}_1$. Its test statistic is given by

$$\mathcal{T}(\mathbf{y}) \triangleq \sum_{m=1}^{M} |y[m]|^2 \underset{\mathcal{H}_1}{\overset{\mathcal{H}_0}{\gtrless}} \gamma. \tag{2.4}$$

This detector does not assume any knowledge of the PU signal structure or synchronization with the transmitter. It demands $\mathcal{O}(1/\mathrm{SNR})$ signal measurements in high-SNR cases (i.e., when $\mathrm{SNR} \gg 1$) and $\mathcal{O}(1/\mathrm{SNR}^2)$ samples in low-SNR regions to deliver the desired P_{FA} and P_D [5]. Note that if the present AWGN noise power/variance is known a priori, the energy detector is the optimal detector according to the Neyman–Pearson criterion [10]. Whereas in Equation 2.4 the time domain samples can be used to determine the energy level in the monitored frequency band $\mathcal{B}_l$, the energy detector can be implemented in the frequency domain by taking the fast Fourier transform (FFT) of $\{y[m]\}_{m=1}^{M}$ and summing the squared magnitude of the resultant FFT bins that belong to $\mathcal{B}_l$, that is, $\mathcal{T}(\mathbf{y}) \triangleq \sum_{f_k \in \mathcal{B}_l} |Y(f_k)|^2$. The FFT is an optimized version of the discrete Fourier transform (DFT) given by $Y(f_k) = \sum_{m=1}^{M} y[m]e^{-i2\pi km/M}$ such that $k = 0,1,\ldots,M-1, f_k = kf_{US}/M$, and f_{US} is the uniform sampling rate.

3. *Feature detection*: Communication signals inherently incorporate distinct features such as symbol periods, training sequences, and cyclic prefixes to facilitate their detection at the intended receiver. Feature detectors exploit such unique structures to determine the presence of a transmission in $\mathcal{B}_l$ by formulating its test statistics as a function of the incoming signal second-order statistics

$$2\mathcal{T}(\mathbf{y}) \triangleq \mathfrak{F}\left\{\mathbb{E}\left[\mathbf{y}\mathbf{y}^H\right]\right\} \underset{\mathcal{H}_1}{\overset{\mathcal{H}_0}{\gtrless}} \gamma \tag{2.5}$$

where $\mathbb{E}[\cdot]$ is the expectation operator, and $\mathfrak{F}[\cdot]$ is a generic function. A detailed discussion of such detectors is presented in [8] with the relevant references. Here, we focus on a particular feature detector known as the cyclostationary feature detector.

Most transmissions are modulated sinusoidal carriers with particular symbol periods. Their means and autocorrelation functions exhibit periodicity; that is, they are wide sense cyclostationary (WSCS) signals. The cyclostationary detector capitalizes on these built-in periodicities and uses the cyclic spectral density (CSD) function of the incoming WSCS signal [12,13]. Let T_p be the underlying cyclostationarity period. The sampled transmission $\{x[m]\}_{m=1}^{M}$ cyclic autocorrelation function (CAF) is defined by

$R_x^{\tilde{c}}[m] \triangleq \mathbb{E}\left[x[n]x*[n+m]e^{-2\pi\tilde{c}n} \right]$, where $R_x^{\tilde{c}}[m] \neq 0$ if $\tilde{c} = i/T_p$ (i is a nonzero integer) and $R_x^{\tilde{c}}[m] = 0$ if $\tilde{c} \neq i/T_p$. The cyclic frequency is $\tilde{c} \neq 0$. The CSD is the discrete-time Fourier transform (DTFT) of the CAF, that is, $S_x(\tilde{c}, f_k) = \sum_{m=-\infty}^{+\infty} R_x^{\tilde{c}}[m]e^{-j2\pi f_k m}$, and an FFT-type implementation can be used. Unlike a PU transmission, the present noise is not WSCS. It is typically assumed to be wide sense stationary (WSS), and the DTFT of $R_w^{\tilde{c}}[m]$ is $S_w(\tilde{c}, f_k) = 0$ for $\tilde{c} \neq 0$. The cyclic detector test statistics can be expressed by

$$\mathfrak{T}(\mathbf{y}) = \sum_{\tilde{c}} \sum_{f_k} \hat{S}_y(\tilde{c}, f_k)\left[S_x(\tilde{c}, f_k) \right]^* \underset{\mathcal{H}_1}{\overset{\mathcal{H}_0}{\gtrless}} \gamma$$

assuming a known transmission $S_x(\tilde{c}, f_k)$ with multiple periods; x^* is the conjugate of x [5]. The estimated CSD of the received signal is denoted by $\hat{S}_y(\tilde{c}, f_k)$, and it is obtained from $\mathbf{y}$. Although the cyclostationary detector can reliably differentiate between various PU modulated signals and the present noise, its complexity and computationally cost are relatively high as the calculation of the 2D CSD function is involved. Its performance and sampling requirements in terms of delivering desired P_{FA} and P_D values are generally intractable [5].

An alternative feature detection technique uses the properties of the covariance matrix in Equation 2.5 to identify the presence of a PU signal, namely, the fact that the signal and noise covariance matrices are distinguishable. An example is the covariance detectors in [14], where the test statistic is expressed in terms of a sample covariance matrix maximum and minimum eigenvalues, that is, $\mathfrak{T}(\mathbf{y}) = \upsilon_{max}/\upsilon_{min}$, and no information on the transmitter signal is required. Other methods promote particular structures of the PU signal covariance matrix and usually demand knowledge of certain signal characteristics, for example [15].

The aforementioned NBSS algorithms are compared in Table 2.1, outlining their advantages and disadvantages. The main features such as the amount of prior information the CR needs to detect the presence of the PUs are outlined. If a single PU is utilizing $\mathcal{B}_l$ and its transmission structure is fully known to the SU, then the coherent detector is the best candidate for sensing

Table 2.1 Comparison between Common Narrowband Spectrum Sensing Techniques

Detector	Prior Knowledge	Advantages	Limitations
Coherent	PU full signal structure	Optimal performance	PU signal dependent
		Low computational complexity	Demands synchronization
Energy	Noise power	No signal knowledge	Does not distinguish between users
		Low computational complexity	Limited by noise power estimation
Feature	Partial knowledge of the PU structure	Distinguishes PUs from SUs	High computational complexity
		Robust to noise and interference	Long sensing times

the spectrum with the highest performance, lowest complexity, and shortest sensing time. This presumes accurate synchronization between the PU and the SU. Because this scenario rarely occurs in CR networks, where other transmitters can opportunistically access a vacant B_l, other detectors become more viable candidates. For example, feature detectors could be deployed when partial knowledge of the PU transmissions is available, for example, cyclic prefixes, modulation scheme, preambles, and so on. They are robust against noise uncertainties and interference and can distinguish between different types of signals. Nevertheless, the feature detectors, for example, the cyclostationary detector, require more complex processing, sensing time, and power resources. On the other hand, the energy detector is a simple and low-complexity option, which does not require any prior knowledge of the PU signal or synchronization. Its sensing time is also notably low for relatively high-SNR regions. However, the radiometer does not differentiate between PUs, SUs, and potential interferers. Its performance is highly dependent on accurate estimation of the present noise power/variance, which determines the threshold values to restrain P_{FA}. Inaccurate estimation of the noise power can limit the attained detection quality when the SNR is lower than a particular level, that is, the SNR wall reported in [16].

2.1.2 Cooperative Sensing

One of the key challenges of realizing a spectrum sensing routine is the well-known hidden terminal problem faced in wireless communications. It pertains to the scenario where the PU transmission is undermined by channel shadowing or multipath fading, and an SU is located in the PU deep fading region. This can lead to the SU arriving at a decision that the sensed spectrum band is vacant; any subsequent utilization can cause severe interference with the PU. To enhance the CR network sensitivity, the network can merge the sensing results of a few of its spatially distributed CRs to exploit their inherent spatial diversity. Each of these CRs experiences different channel conditions, and their cooperation can alleviate the hidden terminal problem [4–6,8,17,18]. For example, in Figure 2.1, three SUs are overseeing the spectral subband used by the PU transmitter. The two SUs, SU 2 and SU 3, are located in the PU deep fading region. While SU 2 cooperates with SU 1 to establish the presence of a PU transmission, SU 2 fails to detect the PU and subsequently transmits over the monitored spectral subband, causing detrimental interference. This simple example illustrates the basic idea behind cooperative spectrum sensing.

A key aspect of collaborative sensing is efficient cooperation schemes that substantially improve the network reliability. They should minimize the bandwidth and power requirements that are associated with the control channel over which information is exchanged among the network SUs. Below, we briefly address the three common information fusing schemes to combine the sensing results of I collaborating SUs in a CR network.

1. *Hard decision fusion*: With hard combining, each of the SUs makes a decision on the presence of the PU and shares a single bit to represent its binary decision $d_i = \{0,1\}$; that is, "0" and "1" signify $\mathcal{H}_0$ and $\mathcal{H}_1$, respectively. The final decision on the spectrum band status is based on a voting metric that can be expressed by

$$\mathcal{V}_{HF} = \sum_{i \in I} d_i \underset{\mathcal{H}_1}{\overset{\mathcal{H}_0}{\lessgtr}} \nu \qquad (2.6)$$

where ν is the voting threshold. The final decision based on Equation 2.6 is simply a combining logic that takes the following forms: (1) AND logic where $\nu = I$ and all the collaborating

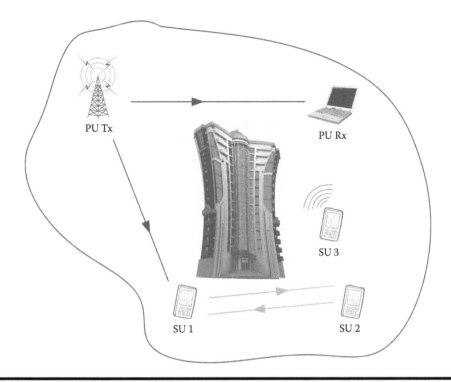

Figure 2.1 Cooperation among SUs to circumvent SU 2 interfering with the PU; SU 2 and SU 3 are in the PU deep fading region.

SUs should decide that the subband is not in use to deem it vacant, (2) OR logic where $\nu = 1$ and a PU transmission is considered to be present if one of the SUs arrived at the decision $d_i = 1$, and (3) a majority vote where $\nu = \lceil I / 2 \rceil$, ensuring that at least half the SUs detected an active PU before deciding $\mathcal{H}_1$. The ceiling function $\lceil x \rceil$ yields the smallest integer greater than or equal to x. Each of the above voting strategies reflects a different view on opportunistic access; for example, the OR logic guarantees minimum network interference at the expense of missed spectral opportunities. On the other hand, the AND logic prioritizes increasing the opportunistic throughput without restraining the possible interference. Note that the overheads of the hard combining in terms of the information exchange are minimal as the 1-bit is shared by each SU.

2. *Soft decision fusion*: In this approach, the I SUs share their sensing statistics; that is, $\mathfrak{T}_i(\mathbf{y})$, $i = 1, 2, \ldots, I$. A weighted sum of the sensing statistics is used as the decision metric according to

$$\mathcal{V}_{SF} = \sum_{i \in I} \varpi_i \mathfrak{T}_i(\mathbf{y}) \tag{2.7}$$

where ϖ_i is the weight allocated to the ith SU. A simple choice of weights is a uniform prior without considering the quality of the channel between the SU and the PU; that is, $\varpi_i = 1$. The weights $\{\varpi_i\}_{i=1}^{I}$ can be proportional to the ith link quality, for example, the SNR of the channel between the PU and the ith participating SU [19]. Although soft combining in Equation 2.7 necessitates the exchange of large quantities of data compared with hard fusion, it can lead to optimal cooperative spectrum sensing [8].

3. *Hybrid decision fusion*: This approach combines both soft and hard combining techniques, seeking to harness hard fusion's low transmission overhead and soft fusion's superior performance. Generally, sharing more statistical information among the SUs results in a better fusion outcome and vice versa. An example is the hybrid technique proposed in [20], where each SU sends two bits of information related to the monitored subband, that is, softened hard combining, to enhance the network sensing dependability.

In practice, several other cooperative sensing design challenges should be taken into account, such as feasibility issues of the control channel, optimizing the overheads associated with information exchange, collaborative network implementation or clustering (e.g., centralized or distributed or ad hoc fusion center), cooperative sequential detection, censoring or sleeping, and so on. Although the objective of this subsection is to briefly introduce cooperative spectrum sensing in CR networks, several comprehensive overviews on this topic are available, and the reader is referred to [4–8,17,18] with extensive references lists therein. Most importantly, cooperative sensing is typically implemented at a network level higher than the considered physical layer. We are predominantly interested in determining the status of a given monitored subband (or a number of them) at a single CR, and the cooperative sensing concept can leverage the obtained statistic at each of these CRs. Nonetheless, in Sections 2.2 and 2.3, we address certain wideband spectrum sensing techniques that are particularly amenable to collaborative multiband detection.

2.1.3 Wideband Spectrum Sensing and Nyquist Sampling

In CR networks supporting multimedia applications over wireless links, high opportunistic throughput is sought to fulfill stringent QoS requirements. SUs also ought to strive to minimize interruptions to the data transmission/exchange with the targeted receiver. This often arises when the PU returns to using the opportunistically accessed spectrum band. Wideband spectrum sensing enables a CR to meet the aforementioned demands by achieving spectrum awareness over wide frequency ranges typically consisting of a number of spectrum bands with different licensed users. If a PU reappears, the availability of several other possible vacant subbands facilitates the seamless hand-off from one spectral channel to another.

A natural approach to a wideband spectrum sensing system model is to divide the total monitored bandwidth into L nonoverlapping subbands/channels because the licensed RF spectrum inherently possesses such a structure. For simplicity, and without the loss of generality, we assume that these subbands are contiguous and are of equal width denoted by B_C. Hence, the overseen wide bandwidth has a total width of LB_C and is given by $\mathcal{B} = \left[f_{min}, f_{min} + LB_C \right]$; normally, the initial frequency point is $f_{min} = 0$. This model, which supports heterogeneous wireless devices that may adopt different wireless technologies for their transmissions, is depicted in Figure 2.2. At any point in time or geographic location, the maximum number of concurrently active channels is expected to be L_A and the received signal at the SU is given by $y(t) = \sum_{k=1}^{K} y_k(t)$ such that $K \leq L_A$. In accordance with the low spectrum utilization premise that motivated the CR paradigm from the outset, we reasonably assume $L_A \ll L$ and that the single-sided joint bandwidth of the active channels does not exceed $B_A = L_A B_A$.

The primary objective of the CR is to scan $\mathcal{B} = \bigcup_{l=1}^{L} \mathcal{B}_l$ and determine which of the monitored subbands are vacant; this is also known as multiband spectrum sensing (MSS). Assuming that

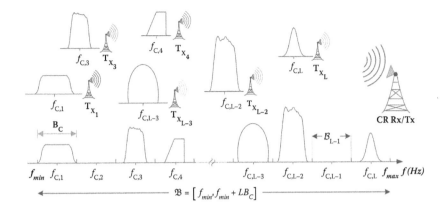

Figure 2.2 Wide monitored bandwidth divided into L nonoverlapping contiguous subbands used by various transmitters. The cognitive radio scans $\mathscr{B}$ in search of a spectral opportunity.

the network transmissions are uncorrelated, the MSS problem reduces to the following binary hypothesis testing for each subband:

$$\mathfrak{T}(\mathbf{y}_l) \underset{\mathcal{H}_{1,l}}{\overset{\mathcal{H}_{0,l}}{\lessgtr}} \gamma_l, \ l = 1, 2, \ldots, L \tag{2.8}$$

The aim is to discriminate between

$$\mathcal{H}_{0,l}: \mathbf{y}_l = \mathbf{w}_l$$
$$\mathcal{H}_{1,l}: \mathbf{y}_l = \mathbf{x}_l + \mathbf{w}_l \tag{2.9}$$

where $\mathcal{H}_{0,l}$ and $\mathcal{H}_{1,l}$ indicate that the lth channel is vacant and occupied, respectively.

Digitally performing the simultaneous sensing of the L spectral channels requires a wide RF front-end and sampling the incoming signal $y(t)$ at rates exceeding the Nyquist rate $f_{Nyq} = 2B$ (no prior knowledge of the activity of the system subbands is presumed). This prevents the adverse effects of the aliasing phenomenon that can hinder accomplishing virtually any DSP task in accordance with the Nyquist sampling criterion [21]. For considerably wide bandwidths (e.g., several gigahertz), f_{Nyq} can be prohibitively high (tens of gigahertz), demanding specialized data acquisition hardware and high-speed processing modules with high memory and power consumption requirements. Such solutions can be very challenging and infeasible for portable devices supporting multimedia communications. It is noteworthy that whilst newly designed wideband compact selective antennas are continuously emerging, for example [22], the development of analog-to-digital converters (ADCs) with high resolution and reasonable power consumption is relatively lagging behind [23,24]. Therefore, the sampling rate can be the bottleneck in realizing efficient wideband spectrum sensing routines. This triggered an immense interest in novel sampling techniques that can mitigate the Nyquist criterion and permit sampling at remarkably low rates without compromising the sensing quality, that is, sub-Nyquist data acquisition. Instead of concurrently processing the L subbands, the SU can sweep across $\mathscr{B}$

and filter out data relating to each individual system subband. A narrowband detector is then utilized to confirm the status of the filtered spectral channel. In this case, $f_{Nyq} = 2B_C \ll 2B$. This approach is dubbed sequential wideband Nyquist spectrum sensing and is discussed in Section 2.2.

2.1.4 Sensing Performance and Trade-Offs

Before introducing several MSS algorithms, here we highlight the performance measures frequently adopted to assess the sensing quality along with a few of the associated trade-offs. As will be apparent from the following text, there is no unified performance metric for wideband spectrum sensing, and the selected measure is dependent on the specific scenario and its parameters.

2.1.4.1 Probabilities of Detection and False Alarm

The receiver operating characteristic (ROC) is one of the most commonly used detection performance metrics. For a particular spectral subband, for example, $\mathcal{B}_l$, it captures the relationship between the probability of false alarm and detection, given by

$$P_{FA,l} = Pr\left\{ \mathfrak{T}(\mathbf{y}_l) \geq \gamma_l \,|\, \mathcal{H}_{0,l} \right\} \quad \text{and} \quad P_{D,l} = Pr\left\{ \mathfrak{T}(\mathbf{y}_l) \geq \gamma_l \,|\, \mathcal{H}_{1,l} \right\} \tag{2.10}$$

respectively. They are interrelated via the detection threshold γ_l, whose value trades $P_{D,l}$ for the probabilities of detecting a spectrum opportunity $1 - P_{FA,l}$ and vice versa. In certain instances, the probability of missed detection $P_{M,l} = 1 - P_{D,l}$ is examined instead of $P_{D,l}$, and the ROC probabilities are plotted against the SNR. The reliability of a spectrum sensing routine can be reflected in its ability to fulfill certain probabilities of detection; that is, $P_{D,l} \geq \eta_l$, and flase alarm; that is, $P_{FA,l} \leq \rho_l$. To illustrate the relationship between $P_{FA,l}$, $P_{D,l}$, and the number of collected transmission measurements M, we next consider the narrowband match filter and energy detectors.

From the coherent detector in Equation 2.3, we have $\mathfrak{T}(\mathbf{y}_l) \sim \mathcal{N}\left(0, MP_{S,l}\sigma_{w,l}^2 \,|\, \mathcal{H}_{0,l}\right)$ and $\mathfrak{T}(\mathbf{y}_l) \sim \mathcal{N}\left(MP_{S,l}, MP_{S,l}\sigma_{w,l}^2 \,|\, \mathcal{H}_{1,l}\right)$, where the a priori known $\mathbf{x}_l$ is deterministic [5]. We recall that $P_{S,l}$ and $\sigma_{w,l}^2$ are the signal power and AWGN variance, respectively. It follows that

$$P_{D,l} = Q\left(Q^{-1}\left(P_{FA,l}\right) + MP_{s,l} \,/\, \sigma_{w,l}\, P_{S,l}\right)$$

$$\hat{M} = \left[Q^{-1}\left(P_{FA,l}\right) - Q^{-1}\left(P_{D,l}\right)\right]^2 \mathrm{SNR}^{-1} \tag{2.11}$$

The number of data samples required to achieve the desired operating point $(P_{FA,l}, P_{D,l})$ is denoted by $\hat{M}$; $Q(x)$ is the tail probability of a zero-mean Gaussian random variable. For the energy detector in Equation 2.4, we have $\mathfrak{T}(\mathbf{y}_l) \sim \mathcal{N}\left(M\sigma_{w,l}^2, 2M\sigma_{w,l}^2 \,|\, \mathcal{H}_{0,l}\right)$ and $\mathfrak{T}(\mathbf{y}_l) \sim \mathcal{N}\left(M\sigma_{w,l}^2 + MP_{S,l}, 2M\sigma_{w,l}^2 + 4M\sigma_{w,l}^2 P_{S,l} \,|\, \mathcal{H}_{1,l}\right)$. Note that the central limit theorem (CLT) is employed to approximate the chi-squared distribution of the energy detector by a normal distribution $\mathcal{N}\left(\bar{m},\sigma^2\right)$ with mean and variance equal to $\bar{m}$ and σ^2, respectively. Subsequently, we can write

$$P_{D,l} = Q\left(\left[\sigma_{w,l}\sqrt{2M}Q^{-1}\left(P_{FA,l}\right) - M\sigma_{w,l}P_{s,l}\right]\Big/\left[\sigma_{w,l}\sqrt{2M\sigma_{w,l}^2 + 4MP_{s,l}}\right]\right)$$

$$\hat{M} = 2\left[Q^{-1}\left(P_{FA,l}\right) - \sqrt{1+2\text{SNR}}\,Q^{-1}\left(P_{D,l}\right)\right]^2 \text{SNR}^{-2}$$

(2.12)

It can be noticed from Equations 2.11 and 2.12 that the number of data samples $\hat{M}$ is a design parameter that can be manipulated to achieve the required $P_{D,l} \geq \eta_l$ and $P_{FA,l} \leq \rho_l$ at the expense of increasing the sensing time, because classically $T_{ST} = M/f_{US}$ and $f_{US} \geq f_{Nyq}$ is the uniform sampling rate. This has implications for the delivered opportunistic throughput discussed below. Deciding a subband's status intrinsically relies on the test statistics threshold, that is, γ_l in Equation 2.8. It determines the detector operational point/region, and the complete ROC plot is generated by testing all the possible threshold values. The explicit dependence of $P_{FA,l}$ and $P_{D,l}$ on γ_l is discarded to simplify the notation.

For the studied wideband spectrum sensing problem, the ROC can be defined by the two vectors $\mathbf{P}_{FA} = [P_{FA,1}, P_{FA,2}, \ldots, P_{FA,L}]^T$ and $\mathbf{P}_D = [P_{D,1}, P_{D,2}, \ldots, P_{D,L}]^T$ encompassing the probabilities of the L system subbands. The aforestated reliability measure can be extended to the multiband environment via

$$\mathbf{P}_{FA} \preceq \rho \quad \text{and} \quad \mathbf{P}_D \succeq \eta$$

(2.13)

where $\rho = [\rho_1, \rho_2, \ldots, \rho_L]^T$ and $\eta = [\eta_1, \eta_2, \ldots, \eta_L]^T$. Each of the symbols $\succeq$ and $\preceq$ refers to an element-by-element vector comparison. The monitored channels can have different sensing requirements, and the available design parameters, for example, M, are selected such that Equation 2.13 is satisfied for all the system subbands. Alternatively, the SU can combine the probabilities of the spectrum bands via

$$\bar{P}_{FA} = \sum_{l=1}^{L} a_l P_{FA,l} \quad \text{and} \quad \bar{P}_D = \sum_{l=1}^{L} b_l P_{D,l}$$

(2.14)

where $\{a_l\}_{l=1}^{L}$ and $\{b_l\}_{l=1}^{L}$ are the weighting parameters. They reflect the importance, interference provisions, and the confidence level of the test statistics per channel. The sensing reliability can take the form of $\bar{P}_{FA} \leq \bar{\rho}_l$, and $\bar{P}_D \geq \bar{\eta}_l$. Using $b_l = 1$ leads to a simple averaging approach, $\bar{P}_D = \dfrac{1}{L}\sum_{l=1}^{L} P_{D,l}$. This creates the risk of a low detection rate for one particular subband (e.g., owing to low PU transmission power), drastically affecting the SU overall multiband detection across $\mathfrak{B}$. On the other hand, if a particular subband, for example, $\mathcal{B}_l$, has low interference constraints, a marginal b_l value can be assigned.

Recalling that the objective of the wideband spectrum sensing is to detect a sufficient amount of spectrum opportunities at an SU without causing harmful interference to the PUs, the probability of missing a spectral opportunity can be defined as

$$\tilde{P}_{MSO} = Pr\left\{\cap_{l=1}^{L}\mathcal{H}_{1,l}\Big|\cup_{l=1}^{L}\mathcal{H}_{0,l}\right\}.$$

(2.15)

Entries $\cap_{l=1}^{L}\mathcal{H}_{1,l}$ and $\cup_{l=1}^{L}\mathcal{H}_{0,l}$ stipulate that the hypothesis testing outcome for all the surveyed subbands are "1" and at least one channel hypothesis testing results in "0," respectively. This implies that a missed spectrum opportunity occurred because at least one of the monitored spectral channels was vacant. Whereas in Equation 2.15 one unoccupied subband is sought by the CR, a more generic formulation can incorporate multiple opportunities [25]. In [26], multiband detection performance measures that are independent of the sensing algorithm are proposed. Let N_{SO} be the number of correctly identified inactive channels, N_{DSO} the pursued number of vacant subbands (i.e., spectral opportunities), N_I the number of occupied channels declared vacant, and N_d the maximum permitted number of falsely identified vacant subbands (i.e., the interference limit). The sensing quality can be empirically examined in terms of the *probability of insufficient spectrum opportunities* $P_{ISO}(N_{SO}) = Pr\{N_{SO} < N_{DSO}\}$ and the *probability of excessive interference* $P_{EI}(N_I) = Pr\{N_I > N_{PI}\}$.

2.1.4.2 Opportunistic Throughput and Sensing Time

The main advantage of wideband spectrum sensing is its ability to provide superior opportunistic throughput R_O to meet onerous QoS requirements for the network SUs. The R_O values are the sums of the possible achieved data transmission rates leveraged by exploiting the network vacant spectral subbands in accordance with

$$R_O \triangleq \sum_{l=1}^{L} r_{0,l}\left(1 - P_{FA,l}\right) + r_{I,l}\left(1 - P_{D,l}\right) \tag{2.16}$$

we have $r_{0,l} = B_C \log_2\left(1 - P_{S,l}^{SU} / \sigma_{w,l}^2\right)$ is furnished correctly unveiling a spectral opportunity, that is, $\{\mathcal{H}_{0,l} \mid \mathcal{H}_{0,l}\}$, and $r_{I,l} = B_C \log_2\left(1 - P_{S,l}^{SU} / \left\{P_{S,l}^{PU} + \sigma_{w,l}^2\right\}\right)$ is obtained when inadvertently interfering with an active PU transmission, that is, $\{\mathcal{H}_{0,l} \mid \mathcal{H}_{1,l}\}$. $P_{S,l}^{SU}$ and $P_{S,l}^{PU}$ are the transmissions power over $\mathcal{B}_l$ pertaining to an SU and a PU, respectively. For an interference-free network, $r_{I,l} = 0$. It is clear from Equation 2.16 that an effective wideband spectrum sensing routine can substantially enhance R_O. This is depicted in Figure 2.3, which displays the opportunistic throughput R_O for a varying number of subbands, $P_{FA,l}$, and SU transmission power.

The opportunistic spectrum access operation at a CR involves spectrum sensing followed by transmitting over the identified vacant system subbands. Let $T_{Total} = T_{ST} + T_{OT}$ be the total access time consisting of the sensing functionality slot T_{ST} and the opportunistic transmission time slot T_{OT}. Note that throughout this chapter, the sensing time T_{ST} is assumed to incorporate the associated processing time affected by the detector complexity, computational cost, and the available processing resources at the SU. Thus, the total leveraged throughput according to Equation 2.16 is $R_T(T_{ST}) = \dfrac{T_{Total} - T_{ST}}{T_{Total}} R_O$, and the optimization of T_{ST} can be formulated as

$$\hat{T}_{ST} = \underset{0 < T_{ST} \leq T_{Total}}{\arg\max} \; R_T(T_{ST}) \tag{2.17}$$

$$\text{s.t. } \mathbf{P}_{FA} \preceq \rho \text{ and } \mathbf{P}_D \succeq \eta.$$

Variations of Equation 2.17 with different constraints can be adopted, for example, optimizing the number of captured samples M that maximizes R_T instead of T_{ST} as in [27]. Moreover, network medium access control (MAC) and frame structuring techniques other than the sequential sensing and transmission regime can be applied. This includes a range of T_{ST} or T_{Total} strategies

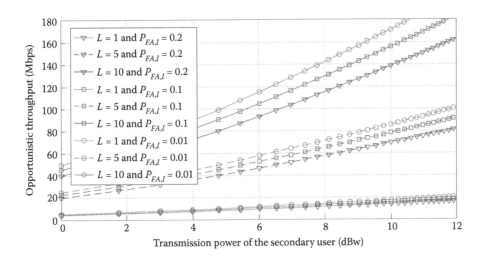

Figure 2.3 Opportunistic throughput for a varying number of surveyed subbands (5 MHz each) and SU transmitted power such that $P_{D,l} = 1$ and SNR $\geqslant$ 0dB.

involving partitioning into subslots [28] and administering concurrent sensing-transmission during T_{Total} via parallel decoding and sensing [29]. A number of alternative opportunistic throughput boosting techniques are proposed in the literature, for example, using a dual radio architecture with parallel sensing and transmission modules [30], detection with adaptive sensing time algorithms that take the SNR value into account [31], and many others.

Finally, there are several practical network considerations that should be weighed when performing wideband spectrum sensing. For example, transmission power control and interference trade-offs; vacant, single, or multiple subband allocation; number of collaborating SUs simultaneously overseeing $\mathfrak{B}$; and so on. They are typically decided based on the required detection quality and the gained opportunistic throughput; see [4,6–8] for further details.

2.2 Nyquist Multiband Spectrum Sensing

In this section, we described two wide spectrum sensing approaches that use the classical Nyquist data acquisition paradigm where $y[m] = y(t_m) = y(m/f_{US})$ and $f_{US} \geq f_{Nyq}$ is the uniform sampling rate. They are (1) sequential multiband Nyquist spectrum sensing (SMNSS) and (2) parallel multiband Nyquist spectrum sensing (PMNSS).

2.2.1 Sequential MSS Using Narrowband Techniques

In SMNSS, a narrowband detector (see Section 2.1.1) is applied to one system subband at a time. This circumvents the need to digitize the wide monitored frequency range $\mathfrak{B}$, and instead processes each channel $\mathcal{B}_l$ separately, where $f_{Nyq} = 2B_C$. Below, two common SMNSS methods are outlined, and their block diagrams are depicted in Figure 2.4.

1. *Demodulation:* A local oscillator (LO) is used at the SU. It down-converts the signal in each subband to the origin (or any intermediate frequency) by multiplying $y(t)$ by the channel's carrier frequency, followed by filtering and low-rate sampling. This is a widely used

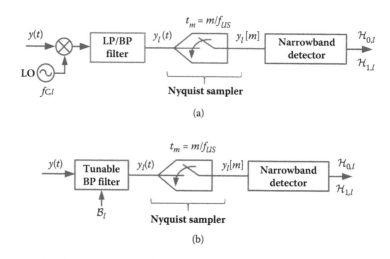

Figure 2.4 Two common approaches to sequential Nyquist multiband spectrum sensing.

technique in wireless communications (i.e., superheterodyne receiver architecture) and intrinsically relies on prior knowledge of the location/carrier frequency of the present transmission. When the positions of the active subbands and their carrier frequencies are not known, as in MSS, the standard demodulation technique cannot be implemented efficiently. Furthermore, the accurate generation of the carrier frequencies can demand bulky, energy-hungry phased locked loop circuits.

2. *Tunable bandpass filter:* A tunable analog bandpass filter is used to filter out the data belonging to each of the monitored subbands prior to sampling. Implementing a tunable power-efficient analog bandpass filter with a sharp cut-off frequency and high out-of-band attenuation poses serious design challenges, especially for portable devices.

A critical limitation of the demodulation and tunable bandpass filtering methods is the delay introduced by sweeping the spectrum when one subband is inspected at a time. This severely increases the aggregate sensing time T_{ST} necessary to scan the system's L channels, hinders fast processing, and degrades the network opportunistic throughput. Sequential techniques are also inflexible, requiring fine tuning of analog components to a particular channels' layout. The concept of initially performing coarse wideband sensing, that is, low-quality detection, to minimize the number of subbands that ought to be searched is proposed in [32]. It is a two-stage sequential spectrum sensing where a robust narrowband detector, for example, a cyclostationary detector, can be employed in the second stage for high-quality results. Other sequential methods, for example, sequential probability ratio tests, exist, and a good overview is given in [8].

2.2.2 Parallel Multiband Detection at Nyquist Rates

Few detection algorithms that simultaneously sense all the monitored subbands are discussed here. A trivial solution to the parallel sensing problem is to use a bank of L sequential multiband sensing modules, for example, those in Figure 2.4. Each uses a sampling rate of $f_{US} \geqslant 2B_C$ and is dedicated to a particular system channel. Different narrowband detectors can be assigned to scan $\mathfrak{B}$ based on the subbands' requirements, that is, heterogeneous architectures. This analog-based solution requires a bulky, inflexible, power-hungry analog front-end filter bank with high complexity, especially if different detectors are used. On the contrary, we are predominantly interested in digitally implementing the wideband sensing task, that is, a "software-based" solution with minimal analog front-end infrastructure and proven flexibility. This can be realized by estimating the spectrum of the incoming multiband signal from its samples collected at sufficiently high rates, that is, $f_{US} \geqslant f_{Nyq}$ and $f_{Nyq} = 2LB_C$. Next, three spectrum-estimation-based PMNSS techniques are addressed.

2.2.2.1 Multiband Energy Detector

The multiband energy detector (MBED) is an extension of the classical narrowband energy detector; its block diagram is shown in Figure 2.5. It is one of the most widely used multiband detection methods and relies on estimating the energy in each subband using the simplest power spectral density (PSD) estimator, that is, periodogram. The periodogram can be viewed as a simple estimate of the PSD formed using a digital filter bank of bandpass filters, and it involves the scaled squared magnitude of the signal's DFT/FFT [33]. The test statistic is given by

$$\mathfrak{T}(\mathbf{y}) \triangleq \sum_{f_n \in \mathcal{B}_l} | \hat{X}_W(f_n) |^2 \underset{\mathcal{H}_{1,l}}{\overset{\mathcal{H}_{0,l}}{\lessgtr}} \gamma_l, \quad l = 1,2,\ldots,L \tag{2.18}$$

where $\hat{X}_W(f_n)$ is the windowed DFT/FFT of the received signal and only the frequency bins that fall in $\mathcal{B}_l$ are considered. A windowing function $w_i(t)$ can be introduced, $\hat{X}(f_n) = \sum_{m=1}^{M} y[m] w_i[m] e^{-j2\pi mn/M}$, to minimize the experienced spectral leakage. The windowing function is defined within a signal time analysis window $T_i = [\tau_i, \tau_i + T_W]$ starting at the initial time instant τ_i and has a width $T_W = M/f_{US}$ such that $w_i(t) = w(t)$ if $t \in T_i$ and $w_i(t) = 0$ if $t \notin T_i$. The fixed tapering template $w(t)$ is chosen from a wide variety of available windowing functions, each with distinct characteristics. An extensive seminal overview of windowing/tapering functions is given in [34]. Clearly, if no tapering is applied, then $w_i(t) = 1$ if $t \in T_i$ and zero otherwise. A number of estimates over overlapping or nonoverlapping time windows are often averaged to improve the periodogram PSD estimation accuracy, for example, Bartlett and Welch periodograms [33]. This results in adding an averaging block to Figure 2.5 before the thresholding operation, where $T_{ST} = \left| \underset{j=1}{\overset{J}{\cup}} T_j \right|$, and J is the number of averaged spectrum estimates.

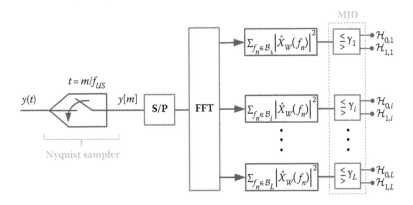

Figure 2.5   Block diagram of the multiband energy detector.

It can be observed from Equation 2.18 and Figure 2.5 that the test statistic for each sub-band has its own threshold value. Quan et al. [5] proposed jointly choosing the threshold values across all the system subbands, that is $\gamma = [\gamma_1, \gamma_2, \ldots, \gamma_L]^T$, to optimize the network opportunistic throughput constrained by satisfying Equation 2.13 and keeping the overall network interference below a certain level. This is known as multiple joint detection (MJD), which is a benchmark MSS algorithm. Several extensions of MJD have emerged; for example, MJD with a dynamically changing sensing time is proposed in [31]. A weighted version of Equation 2.18, that is,

$$\mathfrak{T}(\mathbf{y}_l) \triangleq \sum_{f_n \in B_l} a_n \, | \, \hat{X}_W(f_n)|^2,$$ is investigated in [35] to reflect correlated concurrent transmissions over the system subbands. It brings notable improvements over the original MJD when the level of correlation among the present transmissions is known in advance. The earlier joint detection in [36] utilized a bank of feature detectors.

2.2.2.2 Multitaper Spectrum Estimation

The multitaper PSD estimator (MT-PSDE) achieves superior estimation results by using carefully designed tapering functions, unlike the periodogram, where a fixed windowing function $w(t)$ is deployed [7,37,38]. It utilizes multiple orthogonal prototype filters with Slepian sequences or discrete prolate spherical wave functions as coefficients to improve the variance of the estimated spectrum without compromising the level of the incurred spectral leakage. The general discrete-time multitaper PSD estimator is expressed by

$$\hat{X}_{MT}(f_n) = \frac{1}{\overline{\lambda}} \sum_{k=1}^{N_{MT}} \lambda_k \left| X_k^{ES}(f_n) \right|^2 \tag{2.19}$$

such that the eigenspectrum $X_k^{ES}(f_n)$, which deploys the kth discrete orthogonal taper function $\mathbf{v}_k = [\mathbf{v}_k[1], \mathbf{v}_k[2], \ldots, \mathbf{v}_k[M]]^T$, is defined by

$$X_k^{ES}(f_n) = \sum_{m=1}^{M} y[m] \mathbf{v}_k[m] e^{-j2\pi mf_n}, \quad k = 1, 2, \ldots, N_T. \tag{2.20}$$

The average of the eigenspectra scaling factors $\{\lambda_k\}_{k=1}^{N_{MT}}$ is given by $\overline{\lambda} = \sum_{k=1}^{N_{MT}} \lambda_k$, where λ_k is the eigenvalue of the eigenvector $\mathbf{v}_k$ such that $\mathbf{M}\mathbf{v}_k = \lambda_k \mathbf{v}_k$, and the $(k,m)^{th}$ entry of the $M \times M$ matrix $\mathbf{M}$ is $\sin(2\pi B_C(k - m))/\pi(k - m)$. Frequently, $\lambda_k = 1$ is assumed because the dominant eigenvalues are typically close to unity, and/or $\overline{\lambda}$ is discarded because it does not influence the detector success rate. In [38], an adaptive multitaper estimator is introduced where $\{\lambda_k\}_{k=1}^{N_{MT}}$ scaling entries are replaced with weights that are optimized for a particular processed signal and its PSD characteristics. Note that the number of employed tapers can be bound by $N_{MT} \leqslant \lfloor MB_C \rfloor$ representing the number of degrees of freedom available to control the estimation variance; for example, $N_{MT} \in \{1, 2, \ldots, 16\}$ is recommended in [7]. After obtaining the PSD estimate $\hat{X}_{MT}(f_n)$, the energy of the signal in each of the system channels is measured similar to the classical MBED. The resultant values per spectral channel are compared to predetermined threshold values to decide between $\mathcal{H}_{0,l}$ and $\mathcal{H}_{1,l}$. Haykin [4,7] proposed a multitaper-singular-value-decomposition-based cooperative scheme among CRs to measure the level of present interference and improve the quality of spectrum sensing; it is a soft-combining approach. It was shown in [37] that the multitaper technique delivers premium sensing quality compared with the classical energy detector for a single CR and multiple collaborating SUs.

Interestingly, a number of digital filter bank sensing approaches are addressed in [39], where it was shown that the theory of multitaper spectrum estimation can be formulated within the filter bank framework. Filter bank techniques have a long history in the DSP field with established solutions.

Although the MT-PSDE has nearly optimal performance and is robust against noise inaccuracies, it has notably high computational and implementation complexities compared to the periodogram. Filter-bank-based multicarrier communication techniques, for example, Orthogonal frequency-division multiplexing (OFDM), are widely viewed as the modulation schemes of choice for CRs [6,8]. This is due to their ability to flexibly adapt the spectrum shape of the transmitted signal based on the available spectral opportunities. They have built-in FFT/IFFT processors that can be utilized to estimate the spectrum over wide frequency ranges, making periodogram-type estimators a prime candidate. It was shown in [39] that such filter bank estimators adequately adapted to a multicarrier communication technique can produce accurate PSD estimates similar to the MT-PSDE. The use of a filter-bank-type detector has the added advantage that the filter bank can be used for the CR sensing and transmission functionalities.

2.2.2.3 Wavelet-Based Sensing

The underlying assumption in the adopted system model is that the SUs survey spectral subbands with known locations and widths (see Section 2.1.3). However, the present heterogeneous transmissions can have different bandwidths and occupy parts of the predefined subbands. In [40] a wavelet-based MSS approach was proposed where the present transmissions have arbitrary positions and boundaries within the wide overseen frequency range $\mathfrak{B}$. The wavelet-based detector models the incoming signal as a train of transmissions, each band-limited or confined to a spectral subband with an unknown position or width as demonstrated in Figure 2.6. The signal PSD within each of the active channels is assumed to be smooth but exhibits discontinuities or singularities at the subbands' boundaries or edges. The continuous wavelet transform (CWT) of the PSD of the incoming WSS signal facilitates the detection of these singularities, which reveal the location and width of the present transmissions.

In the frequency domain, the CWT can be expressed by

$$\mathcal{W}_{\vartheta}^{X}\left(f\right)=\mathcal{P}_{Y}\left(f\right)*\varphi_{\vartheta}\left(f\right)\tag{2.21}$$

where $\mathcal{P}_{Y}\left(f\right)$ is the PSD of the received wideband signal $y(t)$, $*$ is the convolution operator,

$$\varphi_{\vartheta}\left(f\right)=\frac{1}{\vartheta}\varphi\left(\frac{f}{\vartheta}\right),\tag{2.22}$$

$\varphi(f)$ is the frequency response of the wavelet smoothing function, and ϑ is the dilation factor. The latter dyadic scale can take values that are powers of 2. To make the pursued discontinuities

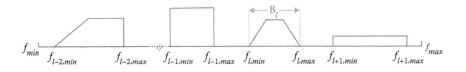

Figure 2.6　The PSD of arbitrarily placed transmissions of varying widths in $\mathfrak{B}=\left[f_{min},f_{max}\right]$.

more pronounced and their characterization easier, derivatives of the $\mathcal{W}_\vartheta^X(f)$ are used [40]. Such approaches are known as wavelet-modules-maxima. For example, the local maxima of the first-order derivative of $\mathcal{W}_\vartheta^X(f)$ and the zero crossings of the second-order derivative are employed to find the active subbands' boundaries and locations, respectively. Controlling $\varphi_\vartheta(f)$ provides additional flexibility, making wideband wavelet-based detection suitable for dynamic spectrum structures. They also possess the requisite properties to adaptively tune the time and frequency resolution, where a high-frequency resolution aids locating the subband edges.

To improve the edge detection procedure at the expense of higher complexity, the wavelet multiscale product $\tilde{\mathcal{W}}_{N_W}^X(f)$ or sum $\bar{\mathcal{W}}_{N_W}^X(f)$ can be applied where

$$\tilde{\mathcal{W}}_{N_W}^X(f) = \prod_{n=1}^{N_W} \frac{d\mathcal{W}_{2^n}^X(f)}{df} \quad \text{and} \quad \bar{\mathcal{W}}_{N_W}^X(f) = \sum_{n=1}^{N_W} \frac{d\mathcal{W}_{2^n}^X(f)}{df}. \tag{2.23}$$

The N_W summands/multiplicands in Equation 2.23 are the first-order derivatives of the CWT with the dyadic scales 2^n, $n = 1, 2 \ldots, N_W$. Higher-order derivatives and/or N_W values can be exploited to enhance the multiband detection sensitivity.

The wavelet-based multiband detector is not robust against interferers and noise. Their impact can be minimized by appropriately setting the detection threshold or increasing N_W in Equation 2.23 as suggested in [41]. A digital implementation of the wavelet-based multiband detector consists of the following four blocks: (1) uniform sampler $f_{US} \geqslant f_{Nyq}$, (2) discrete-time PSD estimator (e.g., a periodogram that involves the scaled squared magnitude of the FFT of the received signal), (3) wavelet transform of the estimated PSD, and (4) local maximum detector to extract the edges of the active subbands. Other advanced PSD estimation techniques, for example, the multitaper estimator, can be used, and the wavelet-based detector generally has a higher complexity compared with the MBED.

2.2.3 Comparison of Various Nyquist Multiband Detection Methods

All the foregoing Nyquist wideband spectrum sensing techniques are compared in Table 2.2. Sequential sensing methods demand an analog filtering module that permits processing one subband at a time. Although this facilitates sampling at relatively low rates $f_{US} \geqslant 2B_C$, it introduces severe delays and imposes stringent space as well as power consumption requirements. On the other hand, the parallel detectors simultaneously scan all the system subbands by digitally processing the entire overseen frequency range. They use excessively high sampling rates, especially for ultra-wide bandwidths. This inevitable trade-off motivated researchers to study novel sampling approaches to overcome the data acquisition bottleneck of digitally accomplishing the sensing task. Such algorithms are dubbed sub-Nyquist detectors and are discussed in the rest of this chapter. We divide them into two categories: compressive and noncompressive; their pros and cons are outlined in Section 2.5.

2.3 Compressive Sub-Nyquist Wideband Sensing

Compressed sampling (CS) or compressive sensing promotes the reconstruction of sparse signals from a small number of their measurements collected at significantly low sub-Nyquist rates. In [42,43], comprehensive overviews of CS and its various aspects are given with an extensive list of references. Noting the low spectrum utilization premise in CR networks, the processed wideband signal is inherently sparse in the frequency domain because only a few of the overseen subbands are concurrently

Table 2.2 Comparison between Nyquist MSS Techniques Where $f_{US} \geq f_{Nyq}$

Category	Detector	Advantages	Limitations
Sequential	Demodulation	Widely used and simple	Slow and complex inefficient hardware
	Tunable BPF	Simple and effective	Complex hardware and requires tuning
	Two-stage sensing	Faster high-quality sensing	Complex hardware and expensive
Parallel	Bank of SMNSS	High-quality sensing	Complex, bulky, inefficient, and expensive
	Multiband Energy	Simple and low complexity	Not robust against noise or interference
	Multitaper	Accurate and robust	Relatively high complexity
	Wavelet-based	Used for unknown subbands	Not robust against noise or interference

active $L_A \ll L$. CS facilitates wideband spectrum sensing with data acquisition rates $\alpha \ll f_{Nyq} = 2LB_C$. These rates are proportional to the joint bandwidth of the active subbands, that is, B_A, representing the information rate in lieu of the entire monitored spectrum $\mathcal{B}$. Owing to the current immense interest in CS, new compressive multiband detection algorithms are regularly emerging. In this section, a number of widely cited and state-of-the-art CS approaches are addressed.

In CS, the SU collects M sub-Nyquist samples of the signal of interest that encompasses the present transmissions, that is, $x(t)$, via

$$\mathbf{y} = \mathbf{\Phi}\mathbf{x} \tag{2.24}$$

where $\mathbf{y} \in \mathbb{C}^M$ is the samples vector, and $\mathbf{x} \in \mathbb{C}^N$ is the discrete-time representation of the transmissions captured at or above the Nyquist rate. The measurement matrix is $\mathbf{\Phi} \in \mathbb{C}^{M \times N}$ such that $M < N$, and noiseless observations are assumed in Equation 2.24. The signal is analyzed within the time window $T_j = \left[\tau_j, \tau_j + T_W \right]$, and the sensing time is $T_{ST} = \left| \cup_{j=1}^{J} T_j \right|$; usually $J = 1$, and $T_{ST} = T_W$. According to the Nyquist criterion, the number of Nyquist samples in T_j is given by $N = \lfloor T_W f_{US} \rfloor$, where $f_{US} \geq f_{Nyq}$. Because the CS average sampling frequency is defined by $\alpha = M/T_W$, the achieved reduction in the data acquisition rate is reflected in the compression ratio $C = f_{Nyq} / \alpha \approx N/M$.

For the DFT transform basis matrix $\mathbf{D} \in \mathbb{C}^{N \times N}$, we have $\mathbf{x} = \mathbf{\Psi}^{-1}\mathbf{f}$ such that $\mathbf{\Psi} = \mathbf{D}$, and $\mathbf{\Psi}^{-1}$ is the inverse DFT matrix. The sparse vector $\mathbf{f} \in \mathbb{C}^N$, which is the frequency representation of the present transmissions, is characterized by $\|\mathbf{f}\|_0 \leq K_S$, where K_S is the sparsity level and $K_S \ll N$. The ℓ_0 "norm" $\|\mathbf{f}\|_0$ is defined as the number of nonzero entries in $\mathbf{f}$. The relationship between the compressed samples and the signal spectrum can be expressed by

$$\mathbf{y} = \mathbf{\Upsilon}\mathbf{f} \tag{2.25}$$

and $\mathbf{\Upsilon} = \mathbf{\Phi}^{-1}\mathbf{\Psi}$ is the sensing matrix. With CS, we can exactly recover $\mathbf{f}$ from the $M < N$ noise-free linear measurements; for example, $M = \mathcal{O}\left(K_S \log(N / K_S)\right)$ suffices, furnishing substantial reductions in the sampling rate. This is facilitated by the sparsity constraint on **f**, which makes solving the underdetermined system of linear equations in Equation 2.25 feasible with closed-form performance guarantees. Such guarantees impose certain conditions on the measurement matrix **Φ** or more generally on the sensing matrix **Υ** [42,43]. Recovering **f** from **y** entails solving an optimization whose basic statement is given by

$$\hat{\mathbf{f}} = \arg\min_{\mathbf{f} \in \Sigma_{K_S}} \| \mathbf{f} \|_0 \quad \text{s.t.} \quad \mathbf{y} = \mathbf{\Upsilon} \tag{2.26}$$

where, $\Sigma_{K_S} = \left\{ \mathbf{f} \in \mathbb{C}^N : \| \mathbf{f} \|_0 \leqslant K_S \right\}$. Although tackling Equation 2.26 directly entails combinatorial computational complexity, a plethora of effective and efficient sparse recovery techniques have been developed, for example, convex relaxation, greedy, Bayesian, nonconvex, and brute-force algorithms; see [42,44,45] for an overview. Note that noise can be added to Equation 2.25 to represent noisy signal observations; that is, $\mathbf{y} = \mathbf{\Phi}\mathbf{x} + \varepsilon$ and ε is the additive measurements noise vector. Sparsifying basis/frames other than the DFT can be employed to promote the sparsity property. Motivated by wavelet-based detection to identify the edges of the active spectral channels, the earliest papers on CS-based MSS utilize $\mathbf{\Psi} = \mathbf{\Gamma}\mathbf{D}\mathbf{W}$, where $\mathbf{\Gamma}$ is an $N \times N$ differentiation matrix, $\mathbf{W}$ applies the wavelet-smoothing operation, and $\mathbf{x}$ is a realization/estimate of the discrete-time signal autocorrelation function [46,47].

In the majority of the theoretical treatments of the CS problem based on Equation 2.25, the measurement matrix **Φ** is assumed to be random and drawn from a sub-Gaussian distribution [42]. This implies the availability of signal measurements collected at or above the Nyquist rate as in [46], which defeats the purpose of sub-Nyquist sampling. On the contrary, the compressed samples in **y** should be collected directly from the received wideband analog signal $y(t)$ without the need for capturing the Nyquist samples first. This can be achieved by using the analog to information converter (AIC) shown in Figure 2.7 [48]; it is known as the random demodulator (RD). The incoming signal is multiplied by a pseudorandom chipping sequence switching at a rate of $f_p = 1/T_p \geqslant f_{Nyq}$ followed by an integrator and a low-rate (sub-Nyquist) uniform sampler with period T_{US}. Generating a fast chipping sequence can be easily achieved in practice, unlike sampling at excessively high rates with specialized ADCs. The RD was devised to process multitone signals made up of pure sinusoids, for example, located at multiples of an underlying resolution frequency Δ_f. The integrator can be implemented using a low-pass filter (LPF), f_{US} is proportional to $\mathcal{O}\left(K_S \log\left(2B / K_S\Delta_f \right) \right)$, and K_S is the number of pure tones present in the double-sided processed bandwidth $2B$. Several other lower bounds on $f_{US} \ll f_{Nyq}$ for

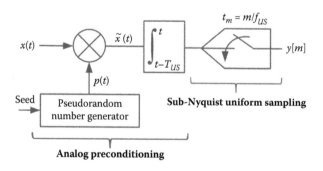

Figure 2.7   Block diagram of the random demodulator sub-Nyquist CS sampler. (Adapted From J. A. Tropp et al., *IEEE Transactions on Information Theory*, vol. 56, no. 1, pp. 520–544, 2010.)

$K_S \ll 2B/\Delta_f$ are derived in [48]. Although compressed model-based spectrum estimation algorithms are proposed in [49] to improve the RD performance, it remains unsuitable for multiband signals with each transmission occupying a particular system subband. The RD is also very sensitive to modeling mismatches during fine hardware–software calibration procedures [42,50]. An alternative AIC approach employs a slow uniform sampler that randomly skips samples. It is equivalent to using $\boldsymbol{\Phi}$ that randomly selects rows out of an $N \times N$ unity matrix [42]. This results in a nonuniform sampling (NUS) scheme, that is, random sampling on grid (RSG), as discussed in Section 2.4.

Assuming that $\mathbf{y}$ is available and that DFT is the sparsifying basis, multiband detection can be performed via

$$\mathfrak{T}(\mathbf{y}) \triangleq \sum_{f_n \in \mathcal{B}_l} |\hat{X}(f_n)|^2 \underset{\mathcal{H}_{1,l}}{\overset{\mathcal{H}_{0,l}}{\lessgtr}} \gamma_l, \quad l = 1, 2, \ldots, L \tag{2.27}$$

The estimated spectrum $\hat{X}(f_n)$ from the compressed samples in $\mathbf{y}$ refers to the entry in vector $\hat{\mathbf{f}}$ representing the frequency point f_n. We recall that $\hat{\mathbf{f}}$ is obtained by solving Equation 2.25 using a sparse approximation algorithm. Therefore, the detector in Equation 2.27 is a sub-Nyquist MBED whose sampling rate is $\alpha \ll f_{Nyq}$. The number of frequency points per subband depends on the spectral resolution dictated by the number of Nyquist samples N in the signal analysis time window where $N = \lfloor T_W f_{Nyq} \rfloor$. In addition, $\boldsymbol{\Psi}$ can be a frame to increase the spectral resolution, where $\mathbf{D} \in \mathbb{C}^{N \times \hat{N}}$ and $\hat{N} > N$. Figure 2.8 displays the block diagram of several sub-Nyquist wideband spectrum sensing techniques. For the test statistic $\mathfrak{T}(\mathbf{y})$ in Equation 2.27, we have $\mathfrak{F}\{\hat{X}(f_n), f_n \in \mathcal{B}_l\} = \sum_{f_n \in \mathcal{B}_l} |\hat{X}(f_n)|^2$. Two simple CS-based sensing techniques for a generic $\boldsymbol{\Phi}$ can be expressed as follows:

1. CS Method 1 (CS-1) [15,51]: The estimated spectral points that belong to each channel are grouped to calculate $\mathfrak{T}(\mathbf{y})$ and $N = \lfloor T_{ST} f_{Nyq} \rfloor$.
2. CS Method 2 (CS-2) [52]: Unlike CS-1, L DFT points are calculated, that is, $\mathbf{f} \in \mathbb{C}^L$, and one DFT point is recovered per monitored subband. The sensing time T_{ST} is divided into $\lfloor T_{ST} / T_W \rfloor$ subwindows, each of width $T_W = L/f_{Nyq}$. Within each of these partitions, an $\hat{\mathbf{f}}$ solution is determined where $\boldsymbol{\Phi} \in \mathbb{C}^{M \times L}$. To improve the estimation accuracy, J of the CS estimates are averaged over $T_{ST} = J T_W$. This emulates the scenario of J spatially distributed CRs collaboratively overseeing $\mathfrak{B}$.

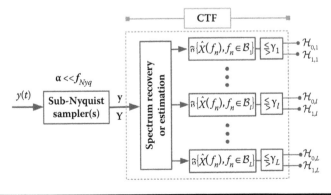

Figure 2.8   Block diagram of several sub-Nyquist MSS techniques. $\hat{X}(f)$ **is the recovered/estimated spectrum, CTF: Continuous-to-finite and** α **is the data acquisition rate such that** $\alpha \ll f_{Nyq}$.

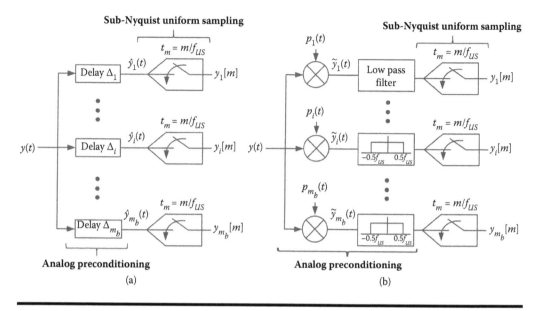

Figure 2.9 Block diagrams of two sub-Nyquist CS samplers: (a) multicoset sampling and (b) modulated wideband converter. (Part (a) from M. Mishali and Y. C. Eldar, *IEEE Transactions on Signal Processing*, vol. 57, no. 3, pp. 993–1009, 2009, and part (b) from M. Mishali and Y. C. Eldar, *IEEE Journal of Selected Topics in Signal Processing*, vol. 4, no. 2, pp. 375–391, 2010).

Compressive samplers that are particularly suitable for multiband signals include multicoset sampling (MCS) and the modulated wideband converter (MWC) (see Figure 2.9) [53,54]. They consist of a bank of m_b samplers collecting uniformly distributed measurements at sub-Nyquist rates. The resulting measurements vectors for all the data acquisition branches can be combined in the $M \times m_b$ matrix $\mathbf{Y} = [\mathbf{y}_1, \mathbf{y}_2, ..., \mathbf{y}_{mb}]$, which is termed the multiple measurements vector (MMV). Similarly, the targeted vector from each bank is stacked in $\mathbf{F} = [\mathbf{f}_1, \mathbf{f}_2, ..., \mathbf{f}_{mb}]$ and

$$\mathbf{Y} = \mathbf{Y}\mathbf{F} \tag{2.28}$$

where all columns $\mathbf{F}$ have the same sparsity pattern. Several sparse recovery algorithms can be applied to solve the MMV case, for example, minimum variance distortionless response and range of extended greedy techniques [42]. Next, we briefly describe MCS, the MWC, and the multirate asynchronous sub-Nyquist sampling (MASS) systems, highlighting their main features.

2.3.1 Multicoset Sampling and Blind Spectrum Sensing

The MCS, otherwise known as periodic nonuniform sampling, was proposed by Feng [55] as a sub-Nyquist data acquisition approach that promotes the accurate reconstruction of deterministic multiband signals. MCS selects a number of measurements from an underlying grid whose equidistant points are separated by a period less or equal to $T_{Nyq} = 1/f_{Nyq} = 1/2B$. The uniform grid is divided into M_b blocks of uniformly distributed samples. In each block, the fixed set $\mathcal{D}$ of length

$|\mathcal{D}| = m_b < M_b$ denotes the indices of the retained samples in the block; the remaining $M_b - m_b$ samples are discarded. The set

$$\mathcal{D} = \{\varrho_1, \varrho_2, \ldots, \varrho_{m_b}\}, \quad 0 \leqslant \varrho_1 < \varrho_2 < \ldots < \varrho_{m_b} \leqslant M_b - 1 \tag{2.29}$$

is referred to as the sampling pattern. The samples of the received wideband signal $y(t)$ in the i_{th} branch/coset are given by

$$y_i[m] = y\big((mM_b + \varrho_i)T_{Nyq}\big), \quad m \in \mathbb{Z} \tag{2.30}$$

and they are all shifted by the delay $\Delta_i = \varrho_i T_{Nyq}$ with respect to the origin. An example of an MCS sequences is shown in Figure 2.10 with $M_b = 12$ and $m_b = 3$ instants selected.

The multicoset sampling scheme can be implemented by a bank of m_b uniform samplers, each running at an acquisition rate of $f_{US} = f_{Nyq}/M_b$ and preceded by a delay as depicted in Figure 2.9a. The rate of the uniform samplers f_{US} in each of the system branches should satisfy

$$f_{US}^{MCS} \geqslant B_C, \tag{2.31}$$

and typically, $f_{US} = B_C$. The MCS average sampling rate is $\alpha_{MCS} = m_b f_{Nyq}/M_b$, which is lower than Nyquist for $m_b < M_b$. For signals with unknown spectral support (i.e., where locations of the active subbands in B are unknown), the MCS minimum permissible rate is

$$\alpha_{MCS} \geqslant 4L_A B_C \tag{2.32}$$

recalling that L_A is the maximum number of concurrently active spectral channels [53,55]. Accordingly, the minimum number of required multicoset sampling channels is $m_b^{MCS} \geq 4L_A$. The rate in Equation 2.32 is twice the Landau rate $f_{Landau} = 2L_A B_C$, which is the theoretical minimum sampling rate that permits the exact recovery of the continuous-time signal from its measurements according to the Landau theorem [56]; the maximum spectrum occupancy is assumed [42].

From Equation 2.30, it can be shown that the spectrum of the incoming signal, that is, $Y(f) = \int_{-\infty}^{+\infty} y(t)e^{-j2\pi ft}\,dt$, and the spectrum attained from the samples in the ith branch are related via

$$Y_i^d(f) = \sum_{m=-\infty}^{+\infty} y_i[m]e^{-j2\pi f((mM_b+\varrho_i)T_{Nyq})} = B_C \sum_{n=-L}^{L-1} Y\big(f + nB_C\big)e^{j2\pi \varrho_i n/M_b} \tag{2.33}$$

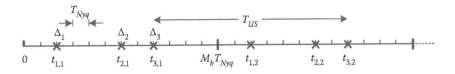

Figure 2.10   MCS with $M_b = 12$ and $m_b = 3$. Uniform samplers 1, 2, and 3 capture the measurements sets $\{y(t_{1,1}),y(t_{1,2}),\ldots\}$, $\{y(t_{2,1}),y(t_{2,2}),\ldots\}$, and $\{y(t_{3,1}),y(t_{3,2}),\ldots\}$, respectively.

such that $f \in \hat{\mathcal{B}}$, $\hat{\mathcal{B}} = [0, B_C]$, $y(t)$ is real, and $M_b = 2L$. It is noted that Equation 2.33 incorporates a frequency point per monitored spectral subband. Subsequently, we can write

$$\mathbf{y}(f) = \mathbf{A}\mathbf{x}(f), \quad f \in \hat{\mathcal{B}} \tag{2.34}$$

where $\mathbf{y}(f)$ is a vector of length m_b combining all $\{Y_i^d(f)\}_{i=1}^{m_b}$. The (i,k)th entry of the $m_b \times M_b$ matrix $\mathbf{A}$ is $B_C e^{j2\pi \varrho_i k / M_b}$, and $\mathbf{x}(f)$ contains the M_b required unknowns for each frequency f, that is, $\{Y(f + nB_C)\}_{n=-L}^{L-1}$. To fully recover the signal's infinite resolution spectrum, Equation 2.34 has to be solved for $f \in \hat{\mathcal{B}}$, where $\mathbf{x}(f)$ is sparse because only few of the subbands are simultaneously active. The resulting infinite number of equations in Equation 2.34 for $\{\mathbf{y}(f), f \in \hat{\mathcal{B}}\}$ is referred to as an infinite measurement vector (IMV). It is reasonable to assume that the set of all vectors $\{\mathbf{x}(f), f \in \hat{\mathcal{B}}\}$ has common support because the nonzero values for each f pertain to the same active subbands [42]. Sampling patterns that permit the exact recovery of $Y(f)$ and thereby the underlying continuous-time signal from its multicoset samples were studied in [57]; searching all possibilities is a combinatorial problem.

In [53], signal reconstruction from its multicoset samples was introduced within the CS framework. Let the support set defined by $\mathcal{K} = \{i : Y(f + iB_C) \neq 0, f \in \hat{\mathcal{B}}, i = -L, -L+1, \ldots, L-1\}$ be the indices of the $K_S \leqslant 2L_A$ active subbands and $|\mathcal{K}| = K_S$ (positive and negative frequencies are included). For the wideband spectrum sensing problem, unveiling the unknown $\mathcal{K}$ suffices. The continuous to finite (CTF) algorithm/block robustly detects $\mathcal{K}$ and reduces the IMV to an MMV of finite dimensions via

$$\mathbf{V} = \mathbf{A}\mathbf{U} \tag{2.35}$$

where frame $\mathbf{Q} = \mathbf{V}\mathbf{V}^H$ can be constructed using $\mathbf{Q} = \sum_{f \in \hat{\mathcal{B}}} \mathbf{y}(f)\mathbf{y}^H(f)$ from roughly $2K_S$ snapshots of $\mathbf{y}(f)$ [42,53]. The decomposition performed to attain $\mathbf{V}$ from $\mathbf{Q}$ minimizes the impact of the noise present in the received signal, where $y[m] = x[n] + w[n]$ and $x[n]$ represent the active transmissions. The received signal $y(t)$ in Equation 2.33 is presumed to consist of noiseless transmissions to simplify the notation. Assuming that the conditions in Equation 2.31 and 2.32 are satisfied, it is shown in [53] that the underdetermined system in Equation 2.35 has a unique solution matrix $\mathbf{U}_0$ with the minimal number of non-identically zeros rows. The indices of the latter rows coincide with the support set $\mathcal{K}$, which in turn reveals the occupied system subbands. A block diagram of the CTF module is depicted in Figure 2.11 [54]. Its input $\mathbf{y}(f)$ is the DTFT or FFT of the data samples produced in all the MCS branches. As well as accomplishing multiband detection, the CTF module plays a critical role in recovering the detected transmissions at the

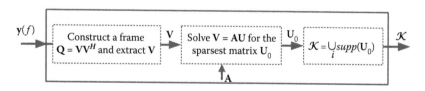

Figure 2.11 Block diagram of the CTF module used in the MCS and MWC systems. (From M. Mishali and Y. C. Eldar, *IEEE Journal of Selected Topics in Signal Processing*, vol. 4, no. 2, pp. 375–391, 2010.)

SU for postprocessing tasks, if required [42,53]. Other approaches to the IMV problem exist, for example, MUSIC-type algorithms [55]. An interesting discussion on the correlation and relationship between the initial work on MCS and CS is given in [58].

Therefore, MCS facilitates wideband spectrum sensing with substantially low sub-Nyquist sampling rates and $\alpha_{MCS} \ll f_{Nyq}$ in the low spectrum utilization regime, that is, $L_A \ll L$. The CTF algorithm can also significantly reduce the sub-Nyquist MSS computational complexity as the sensing matrix $\mathbf{A}$ in Equation 2.35 is of fixed dimension $m_b \times M_b$ for an infinite spectral resolution. On the contrary, directly applying CS with a DFT sparsifying basis/frame to the detection problem, for example, CS-1, yields sensing matrices whose dimensions grow proportional to the desired resolution. Nevertheless, implementing multicoset sampling involves accurate time interleaving among the m_b relatively slow uniform samplers that directly process the incoming wideband signal. This necessitates a high-bandwidth track and hold sampling device, which is difficult to build and might require specialized fine-tuned ADCs [54]. Maintaining accurate time shifts on the order of $1/f_{Nyq}$ in accordance with Equation 2.10 is challenging to realize in hardware, especially for f_{Nyq} in excess of several gigahertz. Inaccurate shifts can notably degrade the quality of the spectrum recovery.

2.3.2 Modulated Wideband Converter

The MWC data acquisition system depicted in Figure 2.9b aims to exploit advances in the CS field and circumvents the MCS drawbacks [54]. It comprises m_b bank of modulators and LPFs. In the ith branch, $i = 1,2,\ldots,m_b$, the received signal $y(t)$ is multiplied by a periodic chipping waveform $p_i(t)$ of period T_p. The modulated output $\tilde{y}_i(t) = y(t)p_i(t)$ is low-pass filtered and subsequently sampled at a sub-Nyquist sampling rate equal to $f_{US} = 1/T_{US}$; the analog filter cut-off frequency is $0.5f_{US}$. To be able to recover the spectrum of the sampled signal or identify the active subbands, a typical MWC configuration imposes

$$f_b = \frac{1}{T_p} \geqslant B_C, f_{US}^{MWC} \geqslant f_p, \alpha_{MWC} \geqslant 4L_A B_C \qquad (2.36)$$

as the frequency of the periodic waveform, uniform sampling rate per branch, and the MWC overall average sampling rate, respectively. This implies that

$$m_b^{MWC} \geqslant 4L_A \qquad (2.37)$$

sampling channels are required for $f_{US} = B_C$. With the MWC, the number of deployed modulators m_b can be reduced at the expense of increasing the uniform sampling rate per branch [54].

It can be shown that the DTFT of the samples in the ith branch is given by

$$Y_i^d(f) = \sum_{m=-\infty}^{+\infty} y_i[m]e^{-j2\pi fm T_{US}} = \sum_{n=-L_0}^{L_0} c_{i,n} Y\left(f - nf_p\right), \ f \in \tilde{B} \qquad (2.38)$$

where $\tilde{B} = \left[-0.5f_{US}, 0.5f_{US}\right]$ is the range dictated by the LPF. The coefficients $c_{i,n}$ in Equation 2.38 are the Fourier expansion coefficients of $p_i(t)$ such that $c_{i,n} = f_p \int_0^{T_p} p_i(t)e^{-j2\pi nt/T_p}dt$. To ensure that the $2L$ overseen subbands (including negative frequencies) are present in $Y_i^d(f)$, we have

$L_0 = \lceil 0.5 T_p (f_{Nyq} + f_{US}) \rceil - 1 = L$ for $f_p = B_C$. The mixing periodic function $p_i(t)$ should have a transition speed $f_{Tran} \lesssim f_{Nyq}$ within T_p; that is, T_p is divided into $M_{PS} \geqslant 2L + 1$ slots within which the chipping sequence can alter its values. Equation 2.38 leads to

$$\mathbf{y}(f) = \mathbf{B}\mathbf{x}(f), \quad f \in \tilde{\mathcal{B}} \tag{2.39}$$

where $\mathbf{y}(f)$ is a vector of length m_b with the ith element being $Y_i^d(f)$. The $(i,n)^{th}$ entry of $m_b \times 2L_0 + 1$ matrix $\mathbf{B}$ is the $c_{i,n}$ Fourier coefficient, and the entries of vector $\mathbf{x}(f)$ are the required $\left\{ Y(f - nf_p) \right\}_{n=-L_0}^{L_0}$ for $f \in \tilde{\mathcal{B}}$. It is noticed that the multicoset sampling formulation in Equation 2.34 and that of the MWC in Equation 2.39 are very similar. Thus, the CTF algorithm is also utilized in the MWC system, and similar performance guarantees are derived in [54]. We note that the CTF in the MWC is less computationally demanding compared with MCS. Constructing frame $\mathbf{Q}$ in MWC does not involve interpolating the slow sub-Nyquist data streams where $\mathbf{Q} = \int_{f \in \tilde{\mathcal{B}}} \mathbf{y}(f) \mathbf{y}^H(f) df = \sum_{m=-\infty}^{+\infty} \bar{\mathbf{y}}[m] \bar{\mathbf{y}}^T[m]$ and $\bar{\mathbf{y}}[m] = [y_1[m], y_2[m], \ldots, y_{m_b}[m]]^T$.

In principle, any periodic function, that is, $p_i(t) = p_i(t + T_p)$, with low mutual correlation and high-speed transitions exceeding f_{Nyq} is admissible. A popular choice is the sign-altering function with M_{PS} sign intervals within T_P; other sign patterns can be used [42]. This flexibility is crucial, and the high-speed chipping signals can be easily generated using a standard shift register. Synchronization of the m_b uniform samplers can be enforced by driving all samplers from a single master clock. The LPFs in MWC do not have to be ideal because mismatches, for example, rugged filter responses, can be compensated for in the digital domain [42]. They also limit the bandwidth of the digitized signal in each of the system branches to approximately $\pm B_C$; that is, off-the-shelf ADCs can be employed. In general, the MWC is robust against noise and model mismatches compared with the RD and MCS. Similar to the latter, the lower bound on the MWC sub-Nyquist sampling rate implies $\alpha_{MWC} \ll f_{Nyq}$ for low spectrum occupancy. Because the objective here is MSS, recovering the signal's spectral support using the CTF block is sufficient. Finally, note that the MWC consists of a bank of RDs. The relationship between the RD and the MWC is thoroughly treated in [59].

2.3.3 Multirate Asynchronous Sub-Nyquist Sampler

The MASS system in Figure 2.12 was proposed in [60] as a CS-based multiband detection approach. It utilizes a bank of q uniform samplers, each running at a distinct sub-Nyquist sampling rate $f_{US,i}$, $i = 1, 2, \ldots, q$. Most notably, MASS does not impose synchronization among its q channels, that is, they are asynchronous. Let T_{ST} be the width of the signal observation window (in seconds), and $\mathcal{M} = \{M_i\}_{i=1}^{q}$ be the set encompassing the number of captured uniform samples in all the system branches. Thus, the multirate approach average sampling rate is $\alpha_{MASS} = \sum_{i=1}^{q} f_{US,i} = M / T_{ST}$, and $M = \sum_{i=1}^{q} M_i$ is the total number of collected measurements. The aim is to achieve $M \ll N$ such that $N = \lfloor T_{ST} f_{Nyq} \rfloor$ denotes the number of Nyquist samples, and $f_{Nyq} = 2B$. In a typical MASS configuration, we have

$$M_i = \tilde{\rho}_i \sqrt{N}, \quad M_i \in \mathcal{M}, \quad \tilde{\rho}_i \in \mathbb{P}, \quad M_1 < M_2 \cdots < M_q \tag{2.40}$$

where $\mathbb{P}$ is the set of prime numbers, and the chosen measurements' numbers in $\mathcal{M}$ usually have consecutive values. Both $\mathcal{M}$ and q dictate the achieved compression ratio N/M.

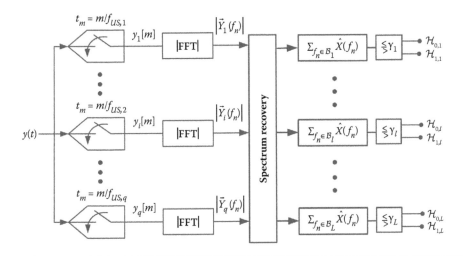

Figure 2.12 Block diagram of the MASS-based sub-Nyquist multiband detection system. (From H. Sun et al., *IEEE Transactions on Signal Processing*, vol. 60, no. 1, pp. 6068–6073, 2009.)

According to Figure 2.12, the DFT, or FFT, is applied in each sampling branch, and the absolute value of the resultant is taken. Let $Y_{Nyq}^d(f) = \sum_{n=0}^{N-1} y(nT_{Nyq})e^{-j2\pi f n T_{Nyq}}$ be the DTFT of the incoming wideband signal sampled at the Nyquist rate, and $T_{Nyq} = 1/f_{Nyq}$. It can be shown that the relationship between the DFT from the M_i measurements in the ith branch and $Y_{Nyq}^d(f)$ can be expressed by

$$Y_i^d[m] = \frac{M_i}{N} \sum_{n=-\lfloor 0.5N \rfloor}^{\lfloor 0.5N \rfloor} Y_{Nyq}^d[n] \sum_{l=-\infty}^{+\infty} \delta[n-(m+lM_i)], \quad m \in \left[-\lfloor 0.5M_i \rfloor, \lfloor 0.5M_i \rfloor\right] \quad (2.41)$$

such that $\delta[n]$ is a Kronecker delta. By expressing Equation 2.41 in a matrix format, we can write

$$\mathbf{y}_i = \mathbf{C}_i \mathbf{f} \quad (2.42)$$

where $\mathbf{y}_i \in \mathbb{C}^{M_i \times 1}$ is the output of the FFT block in the ith sampling channel, $\mathbf{C}_i \in \mathbb{R}^{M_i \times N}$ is the sensing matrix, and $\mathbf{f} \in \mathbb{C}^N$ is the desired signal spectrum. The (m,n)th entry of $\mathbf{C}_i$ is given by $\frac{M_i}{N} \sum_{l=-\infty}^{+\infty} \delta[n-(m+lM_i)]$. This implies that each column of the sensing matrix contains only one nonzero value equal to M_i/N, and in each row the maximum number of nonzero entries is $\lceil N/M_i \rceil$. Although f_{Nyq} guarantees that no aliasing is present in $Y_{Nyq}^d(f)$, the sub-Nyquist rates in each of the MASS branches wrap the wideband signal spectrum content onto itself in $Y_i^d[m]$. Nonetheless, the sparsity constraint $\|\mathbf{f}\|_0 \leq K_S$, and the operating conditions set in Equation 2.40 ensure that the probability of a spectral overlap is very small [26]. By aggregating the data from all the q system branches, we can write

$$\breve{\mathbf{y}} = \breve{\mathbf{C}}\mathbf{f} \quad (2.43)$$

where $\breve{\mathbf{y}} = \left[|\mathbf{y}_i|, |\mathbf{y}_2|, \ldots, |\mathbf{y}_q|\right]^T$ and $\breve{\mathbf{C}} = \left[\mathbf{C}_1, \mathbf{C}_2, \ldots, \mathbf{C}_q\right]^T$ are the concatenated absolute values of the FFT outputs and the associated disjoint sensing matrices, respectively. Note that $\breve{\mathbf{y}} \in \mathbb{R}^{M \times 1}$ and $\breve{\mathbf{C}} \in \mathbb{R}^{M \times N}$ can be of high dimensions because $M = \sum_{i=1}^{q} M_i$. The desired sparse real vector is $\check{\mathbf{f}} = |\mathbf{f}|$, and determining the spectrum magnitude at a resolution of $1/N$ accomplishes the MSS task. The spectrum recovery block in the MASS system entails solving Equation 2.43 using one of the standard sparse recovery algorithms from the CS literature. The energy per subband can be subsequently measured to establish its status. It is shown in [26] that $\check{\mathbf{f}}$ can be reliably reconstructed from Equation 2.43 provided that

$$q_{MASS} > 2K_S - 1 \tag{2.44}$$

sampling branches are employed. For low spectrum utilization, that is, $K_S \ll N$, the MASS detector can substantially reduce the sampling rate where $M \sim \mathcal{O}(K_S\sqrt{N})$.

Although MASS is asynchronous, its computational complexity can be high given the sizes of the handled matrices. Nevertheless, it circumvents the need for specialized analog preconditioning modules as in the RD, MCS, and MWC systems that can render the CS sampler inflexible and expensive. Most importantly, MASS is particularly suitable for implementation by spatially distributed CRs. Each radio can have one (or a few) sampling channels and transmit its M_i measurements to a fusion center, that is, soft combining. Because synchronization among different channels is not required, SUs do not need to share their sensing matrices. The latter aspect is a limiting factor for implementing other CS approaches across a network because each row of their sensing matrix is uniquely generated (e.g., chipping sequences), and synchronization among the collaborating radios is essential to ensure a reasonable detection/spectrum-recovery quality. Thus, MASS can effectively leverage spatial diversity in CR networks.

2.3.4 Remarks on CS Detectors and Other Techniques

The original MCS and MWC design objective is to be able to fully recover the processed multiband signal $x(t)$ from its sub-Nyquist samples. For the wideband spectrum sensing task, their average sampling rates are bounded by $\alpha \geq 4L_AB_C = 2f_{Landau}$ (the maximum expected spectrum occupancy is assumed). In addition, the other addressed CS detection techniques impose similar requirements by requesting $M \sim \mathcal{O}(\kappa K_S)$ measurements in T_j. The spectrum sparsity level denoted by $K_S \sim \mathcal{O}(2L_AB_C)$ depends on the spectral resolution, and the constant κ is typically significantly larger than 2, as in MASS.

If the final goal is multiband detection for CR, full signal reconstruction is not necessary. This can ease the data acquisition requirements and even alleviate the sparsity constraint on the overseen spectrum [61–63]. These two advantages can be realized by formulating the detection problem in terms of recovering the PSD of the received wideband signal, which is assumed to consist of $K \leq L$ WSS transmissions. The energy in each system subband can be subsequently measured to determine the channel's status; see Figure 2.8. It is shown in [63] that the MCS and MWC systems permit the exact recovery of the incoming signal PSD (not the underlying signal realization) when its average sampling rate satisfies

$$\alpha_{SC} \geq 2L_AB_C, \quad L_A \ll L \tag{2.45}$$

assuming a sparse spectrum. Hence, for low spectrum utilization, the sampling rate can be half of that imposed by the original MCS and MWC systems. In this case, the computationally efficient CTF algorithm can be used. Most remarkably, a sampling rate exceeding

$$\alpha_{NSC} \geq 0.5 f_{Nyq} \tag{2.46}$$

is sufficient for nonsparse signals. This implies that the sampling rates of the multiband detector can be as low as half of the Nyquist rate even for high spectrum occupancy; that is, the joint bandwidth of the simultaneously active system channels can be arbitrarily close to the total overseen bandwidth. Because the majority of the CS-based detectors are prone to interference and noise uncertainty, a CS-based feature detector is proposed in [64]. It is robust against such adverse effects and exploits the cyclostationarity feature of the incoming transmissions, assuming that their 2-D cyclic spectrum is sparse. Its main ideas are derived from the classical cyclostationary detector described in Section 2.1.1.

In summary, CS facilitates performing parallel wideband spectrum sensing at significantly low sub-Nyquist sampling rates; that is, it mitigates the encountered data acquisition rate limitation. This comes at the expense of more complex processing techniques, for example, sparse recovery algorithms that can be highly nonlinear, and specialized hardware to precondition the digitized signal. There are several remaining challenges that require further analysis, for example, devising flexible CS samplers, the effect of noise on CS-based detectors, model-based recovery techniques that take the communication signal structures into account, and efficient implementations, to name a few.

2.4 Alias-Free Sampling for Sub-Nyquist Wideband Sensing

The sampling process, which converts a continuous-time signal $x(t)$ into its discrete-time representation $x(t_m)$, is typically modeled by

$$x(t_m) = x(t)s(t), \quad m \in \mathbb{Z} \tag{2.47}$$

as depicted in Figure 2.13. The sampling signal $s(t) = \sum_{m \in \mathbb{Z}} \delta(t - t_m)$ comprises an infinite series of Dirac delta pulses positioned at the data acquisition time instants $\{t_m\}_{m \in \mathbb{Z}}$. In classical DSP, uniform sampling is utilized, and the captured measurements are equidistant; that is, $t_m = mT_{US} = m/f_{US}$. The multiplication in Equation 2.47 translates into a convolution in the frequency domain, and the PSD of the sampled waveform is given by

$$\mathcal{P}_X^d(f) = \mathcal{P}_X(f) * \mathcal{P}_S(f) \tag{2.48}$$

where $\mathcal{P}_X(f)$ is the PSD of the processed WSS signal, and $\mathcal{P}_S(f)$ is the spectrum of the sampling signal. For uniform sampling, it can be shown that $\mathcal{P}_S(f) = f_{US} \sum_{n \in \mathbb{Z}} \delta(f - nf_{US})$, and the spectrum of the discrete-time signal can be expressed by

$$\mathcal{P}_X^d(f) = f_{US} \sum_{n \in \mathbb{Z}} \mathcal{P}_X(f - nf_{US}) \tag{2.49}$$

It is made up of identical copies (that is, aliases) of the continuous-time signal spectrum $\mathcal{P}_X(f)$ shifted by multiples of the uniform sampling rate.

Assume that a transmission $x_l(t)$ occupies an unknown spectral band $\mathcal{B}_l$, $|\mathcal{B}_l| = B_C$, within the overseen frequency range $\mathfrak{B}$ of total width $B = LB_C$. If $f_{US} < f_{Nyq}$ ($f_{Nyq} = 2LB$), more than one

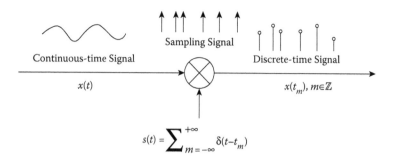

Figure 2.13   An ideal sampler and for uniform sampling $t_m = mT_{US}$.

spectral replica $\mathcal{P}_{X_l}(f - nf_{US})$, $n \in \mathbb{Z}$ of the transmission will be present in $\mathcal{P}_X^d(f)$. Without prior knowledge of the $\mathcal{B}_l$ position, we have no means of identifying which of the overseen system subbands is truly occupied by examining the spectrum of sampled signal, that is, $\mathcal{P}_X^d(f)$. The simultaneous presence of more than one transmission will lead to overlap among their replicas in $\mathfrak{B}$, rendering multiband detection infeasible when $f_{US} < f_{Nyq}$. This in ability to unambiguously identify the spectral components of the underlying continuous-time signal from the spectrum of the sampled data is referred to as spectrum aliasing, which can cause irresolvable processing problems. Sampling above the Nyquist rate eliminates the aliasing phenomenon and satisfies the Shannon sampling theorem requirements to fully recover $x(t)$ from $x(mT_{US})$, $m \in \mathbb{Z}$. If the position of the sole active subband is known in advance, bandpass sampling can be used, and $f_{US} \geq 2B_C$ suffices [21]; this is not the case in wideband spectrum sensing.

NUS is an alternative data acquisition approach that offers additional flexibility and new opportunities because of its potential to suppress spectrum aliasing. It intentionally uses non-uniformly distributed sampling instants, unlike scenarios where the irregularity of the collected measurements is viewed as a deficiency, for example, inaccessibility of the signals in certain periods, hardware imperfections, etc. Here, we consider randomized nonuniform sampling (RNUS), which can be regarded as an aliasing repression measure. It promotes performing wide multiband detection at remarkably low sub-Nyquist rates as illustrated in [65–68]. The utilization of randomized sampling in conjunction with appropriate processing algorithms, for example, adapted spectrum estimators, to eliminate/suppress the effect of aliasing is a methodology referred to as digital alias-free signal processing (DASP). A few monographs on the topic exist, for example [69–73].

In Figure 2.14, a CR is surveying the frequency range [0.5,1] GHz by estimating the spectrum of an incoming signal using the sub-Nyquist sampling rate $\alpha = 96$ MHz. A single PU transmission residing in $\mathcal{B}_l = [780,800]$MHz is present; its location is unknown to the CR. With uniform sampling, replicas of the single transmission are spread all over $\mathfrak{B}$ as shown in Figure 2.14a. They are indistinguishable from one another, and most of the overseen spectral subbands can be erroneously regarded as occupied. In Figure 2.14b, where RNUS is employed, the previously observed stiff-coherent aliasing is no longer present. It is significantly suppressed, and instead a broadband white-noise component is added. The latter is known as smeared or incoherent aliasing, and it does not hinder the correct identification of the active subbands, as is evident from Figure 2.14b. This demonstrates the aliasing-suppression capabilities of randomized sampling, which is leveraged here to devise an effective sub-Nyquist multiband detection routine. Next, we briefly discuss the notion of alias-free sampling, list a few RNUS schemes, and introduce DASP-based detection.

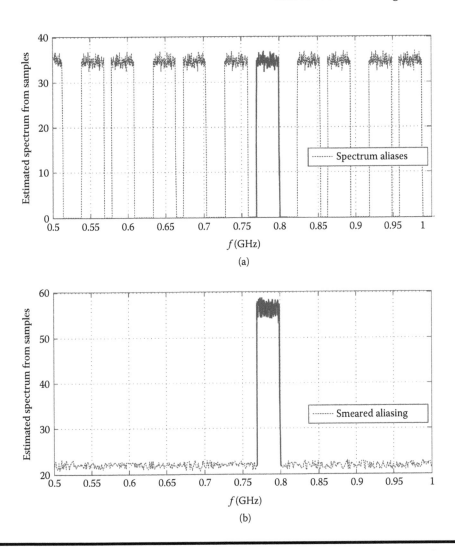

Figure 2.14 Estimated spectrum within the overseen wideband [0.5,1] GHz from $\{x(t_m)\}_{m=1}^{M}$ captured at α = 95 MHz. One transmission is present (solid line) at an unknown location. (a) Uniform sampling; (b) Random nonuniform sampling.

2.4.1 Alias-Free Sampling Notion

Alias-free behavior is typically related to the spectral analysis of a randomly sampled signal $x(t)$, for example, estimating the PSD $\mathcal{P}_X(f)$ from $\{x(t_m)\}_{m=1}^{M}$ rather than reconstructing it [69,70,72,73]. From Equation 2.48, the total elimination of spectral aliasing is achieved when

$$\mathcal{P}_S(f) = \delta(f) \tag{2.50}$$

Early papers on alias-free sampling, for example [74], showed that Equation 2.50 can be satisfied, and $\mathcal{P}_X(f)$ can be exactly estimated from arbitrarily slow nonuniformly distributed signal samples collected over infinitely long periods of time. Literal alias-free behavior is observed only in asymptotic regimes, that is, as M tends to infinity. In practice, the signal is analyzed for a limited duration of time $|T_j| = T_W$, and M is finite. Thus, eliminating spectrum aliasing is unattainable, and the benign smeared-aliasing component is typically sustained. As a result, several criteria have been proposed in the literature to affirm the alias-free nature of a RNUS scheme for a finite T_W [69,70,72]. For example, a scheme is alias free if it satisfies the following stationarity condition [72]

$$\sum_{m=1}^{M} p_m(t) = \alpha \tag{2.51}$$

such that $p_m(t)$ is the probability density function (PDF) of the sampling instant t_m. The average sampling rate is defined by $\alpha = M/T_W$ and $T_j = [\tau_i, \tau_i + T_W]$. Different alias-free criteria can lead to contradictory assessment results for the same scheme.

In the context of the studied wideband spectrum sensing problem, alias-free sampling and processing simply refer to the ability of the randomized sampling scheme and the deployed estimator to sufficiently attenuate spectrum aliasing within the overseen wide frequency range $\mathcal{B}$. As long as this suppression permits the reliable identification of the active system subbands, the sampling process is deemed to be suitable. We acknowledge that the term *alias free* can be misleading because $\mathcal{P}_X^d(f)$ in practice can never be completely free of aliasing.

2.4.2 Randomized Sampling Schemes

We now outline a number of data acquisition strategies that are typically used in DASP and are adequate for the MSS task. Each scheme has its own spectrum aliasing-suppression characteristics necessitating separate analysis, as noted in [70,72].

1. *Total random sampling (TRS):* Its concept is drawn from Monte Carlo integration over a finite integral. All the M sampling instants of a TRS sequence are independent identically distributed (IID) random variables. Their PDFs $\{p_m(t)\}_{m=1}^{M}$ have nonzero values only within the signal time analysis window T_j, and for a uniform prior case, they are given by

$$p_m(t) = \begin{cases} 1/T_W & t \in T_j \\ 0 & \text{elsewhere}, \quad m = 1, 2 \ldots, M \end{cases} \tag{2.52}$$

2. *Random sampling on grid (RSG):* It randomly selects M samples out of the total N_g possible sample positions that can be in general arbitrarily distributed within T_j. For simplicity, let the nominal time locations be $T_g = 1/f_g$ apart, that is, form an underlying uniform grid. Any of the grid points can be selected only once with equal probability, and $\binom{N_g}{M}$ possible distinct sampling sequences of length M exist. Typically, we set $f_g = f_{Nyq}$ and $N_g = N = \lfloor T_W f_{Nqy} \rfloor$. The RSG scheme accommodates the practical constraint of having a minimum distance between any two consecutive samples.

3. *Stratified random sampling (SRS):* It divides T_j into N_S disjointed subintervals, that is, $\{S_k\}_{k=1}^{N_S}$, where $T_j = \cup_{k=1}^{N_S} S_k$. Let M_k be the number of collected samples in stratum S_k and $M = \sum_{k=1}^{N_S} M_k$ be the total number of measurements. Various methods exist for choosing

the number of samples per subinterval. For simplicity, assume $M_m = 1$ and the PDF of the M random independent sampling instants is given by

$$p_m(t) = \begin{cases} 1/|\mathcal{S}_m| & t \in \mathcal{T}_j \\ 0 & \text{elsewhere,} \qquad m = 1, 2 \ldots, M \end{cases} \qquad (2.53)$$

Choosing $|\mathcal{S}_m|$, for example, to improve the spectrum estimation quality demands *a priori* knowledge of the processed signal. A practical approach is to assume equal subintervals $|\mathcal{S}_k| = T_W / M = 1/\alpha$, that is, stratified sampling with equal partitions (SSEP). Figure 2.15 depicts a realization of an SSEP sequence. Another popular SRS scheme is antithetical stratified sampling where $M_m = 2$, and these two samples are equidistant from the stratum center.

4. *Jittered random sampling (JRS):* It can be expressed as the intentional departure of the sampling instants from their nominal uniform sampling grid. It is modeled by $t_m = mT_{US} + \epsilon_m$, $m = 1, 2, \ldots, M$, where T_{US} is a sub-Nyquist uniform sampling period, and $\{\epsilon_m\}_{m=1}^M$ are zero-mean IID random variables with PDF $p_\epsilon(t)$. The PDF of the mth sample point is $p_m(t) = p_\epsilon(t - mT_{US})$. Uniform and Gaussian PDFs with varying widths/variances are among the common choices of $p_\epsilon(t)$.

5. *Additive random sampling (ARS):* Its sampling instants are described by $t_{m+1} = t_m + \epsilon_m$, $m = 1, 2, \ldots, M$, where $\{\epsilon_m\}_{m=1}^M$ are zero-mean IID random variables with PDF $p_\epsilon(t)$. ARS was one of the earliest alias-free schemes and was proposed in [74]. Its mth sampling instant is the sum of m IID random variables, and hence its PDF is given by

$$p_m = \overset{m}{\circledast} p_\epsilon(t), \quad m = 1, 2, \ldots, M \qquad (2.54)$$

where $\overset{m}{\circledast}$ is the m-fold convolution operation. Steps $\{\epsilon_m\}_{m=1}^M$ often involve a Gaussian or Poisson distribution. In [70], correlated $\{\epsilon_m\}_{m=1}^M$ is suggest as a means of improving the ARS aliasing-suppression impact, that is, correlated ARS.

There are many other DASP-oriented randomized and deterministic NUS schemes, such as data acquisition driven by the level of the processed signal (that is, zero-crossing and level-crossing sampling) and the previously discussed MCS scheme. The MCS aliasing-suppression characteristics are studied in [75].

2.4.3 Reliable Alias-Free Sampling-Based MSS

The proposed DASP-based sub-Nyquist wideband spectrum sensing approach relies on nonparametric spectral analysis similar to the majority of parallel sensing methods. It can be represented

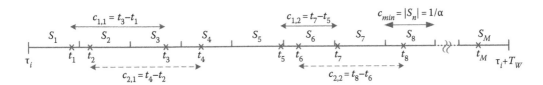

Figure 2.15 An SSEP sequence (crosses are the sampling instants). For two ADCs, samples collected by ADC 1 and 2 are $\{c_{1,1}, c_{1,2}, \ldots\}$ and $\{c_{2,1}, c_{2,2}, \ldots\}$ seconds apart, respectively.

by the block diagram in Figure 2.8 and involves the following three steps: (1) randomly sample the incoming signal at the rate $\alpha \ll f_{Nyq}$, (2) estimate the spectrum of the multiband signal at selected frequency points, and (3) compare the estimation outcome with pre-set thresholds. Establishing the status of the overseen L system subbands does not require determining the details of the spectral shape within $\mathcal{B}$. This premise is exploited here, and estimating a frequency representation that facilitates the multiband detection task is pursued (i.e. not necessarily the signal's exact PSD).

The DASP-based detector adopts the periodogram-type spectrum estimator given by

$$\hat{X}_{NUS}\left(f_n\right) = \sum_{j=1}^{J} \beta \left| \sum_{m=1}^{M} y(t_m) w_j(t_m) e^{-i2\pi f_n t_m} \right|^2 \qquad (2.55)$$

where β is a scalar dependent on the sampling scheme, and the processed signal is assumed to be WSS. The M nonuniformly distributed measurements are contaminated with AWGN, that is, $\left\{ y(t_m) = x(t_m) + n(t_m) \right\}_{m=1}^{M}$. They are collected within a time analysis window $T_j = \left[\tau_j, \tau_j + T_W \right]$. The total signal observation window or sensing time is given by $T_{ST} = \left| \cup_{j=1}^{J} T_j \right|$, and the average sampling rate is $\alpha_{NUS} = M/T_W$. The windowing function $w_j(t)$ is introduced to minimize spectral leakage, where $w_j(t) = w(t)$ for $t \in T_j$ and $w_j(t) = 0$ for $t \notin T_j$. Recalling that spectrum sensing does not require determining the signal exact signal PSD, $\hat{X}_{NUS}\left(f\right)$ is shown to yield a frequency representation that facilitates MSS regardless of the value of α [65–68,70]. It is noted that the statistical characteristics of Equation 2.55 is dependent on the randomized sampling scheme, that is, the PDFs of the sampling instants. To demonstrate the suitability of Equation 2.55 for the detection task, assume that $\left\{ t_m \right\}_{m=1}^{M}$ are generated according to the TRS scheme, and $J = 1$. It can be shown that

$$C(f) = \mathbb{E}\left[\hat{X}_{NUS}\left(f\right) \right] = \frac{M}{(M-1)\alpha}\left[P_S + \sigma_w^2 \right] + \frac{1}{E_W} \mathcal{P}_X(f) * \left| W(f) \right|^2 \qquad (2.56)$$

where $\beta = M/(M-1)E_W$, $E_W = \int_{\tau_j}^{\tau_j + T_W} w^2(t)\, dt$ is the energy of the employed windowing function $w(t)$, $W(f) = \int_{-\infty}^{+\infty} w(t) e^{-i2\pi ft}\, dt$, and $\mathcal{P}_X(f)$ is the PSD of the current transmissions. The powers of the processed multiband signal and measurements noise are denoted by P_S and σ_w^2, respectively. From Equation 2.56, $C(f)$ consists of a detectable feature given by the signals' windowed PSD, that is, $\mathcal{P}_X(f) * W(f)/E_W$, plus a component that represents the smeared-aliasing phenomenon. Unlike the stiff-coherent spectrum aliasing experienced in uniform sampling, $M\left[P_S + \sigma_w^2 \right]/(M-1)\alpha$ is a frequency-independent component and merely serves as an amplitude offset. It does not hamper the sensing operation (see Figure 2.14). Therefore, $\hat{X}_{NUS}\left(f\right)$ is an unbiased estimator of $C(f)$, a detectable frequency representation that allows detection of any activity within the monitored bandwidth. The width of T_j is chosen such that the distinguishable spectral features of the active subbands are reserved by $\mathcal{P}_X(f) * W(f)/E_W$, and the spectral leakage is kept below a certain level. Similar to the CS-2 method and to minimize the sensing routine computational complexity, one frequency point per system spectral channel can be inspected to decide between $\mathcal{H}_{0,l}$ and $\mathcal{H}_{1,l}$. To maintain relatively smooth spectrographs, $T_W \geq c/B_C$, $c > 1$ serves as a practical guideline [65].

The average sampling rate α and the number of averaged estimates J are the available design parameters that can restrain the level of estimation error in $\hat{X}_{NUS}\left(f\right)$; the variance expressions should be derived to evaluate the estimation accuracy. To ensure satisfying certain probabilities of detection and false alarm, that is, $P_{FA,l} \leqslant \rho_l$ and $P_{D,l} \geq \eta_l$, $l = 1,2,...,L$, prescriptive guidelines can be obtained for the DASP-based detection. They are based on a statistical analysis of the undertaken spectrum estimation. Closed-form formulas are presented in [65–68] for a number of RNUS schemes illustrating that the sensing time is a function of the sampling rate, SNR ratio, maximum spectrum occupancy B_A, and requested system probabilities $\mathbf{P}_D$ and $\mathbf{P}_{FA}$; thus

$$T_{ST} \geq \mathfrak{F}(\alpha, \mathbf{P}_D, \mathbf{P}_{FA}, B_A, \text{SNR}) \tag{2.57}$$

Such recommendations clearly depict the trade-off between the sensing time, sub-Nyquist sampling rate, and achievable detection performance. Assuming transmissions of equal power levels, system subbands of identical performance requirements ($P_{FA} \leqslant \rho$ and $P_D \geq \eta$), and nonoverlapping signal windows, the SSEP randomized scheme has

$$T_{ST} \geq \left\{ \frac{2B_A Q^{-1}(\rho)(1+\text{SNR}^{-1}) - Q^{-1}(\eta)\left[2B_A(0.5+\text{SNR}^{-1})+\alpha\right]}{(\alpha - B_A)/T_W} \right\}^2 \tag{2.58}$$

Most importantly, the provided MSS reliability guidelines, for example, Equation 2.58, affirm that the DASP-based detector sampling rate can be arbitrarily low at a predetermined additional sensing time and vice versa. Hence, there is no lower bound on the sampling rate, contrary to the compressive sensing counterpart. Different NUS schemes have different properties, and guidelines in the form of Equation 2.57 can be obtained. In Figure 2.16, we display the impact of the sub-Nyquist sampling rate α, requested ROC probabilities, and SNR on the sensing time T_{ST} when the overseen frequency range is of width $B = 100$ MHz and $B_A = 5$ MHz ($L = 20$ and $B_C = 5$ MHz). This figure can give a system designer the necessary tools to assess the requirements and viability of the sub-Nyquist multiband detector.

2.4.4 Remarks on NUS-Based Spectrum Sensing

Each NUS scheme exhibits distinct spectrum aliasing-suppression capabilities depending on the utilized spectral analysis tool; clearly, more advanced estimation methods can used [73]. However, the simplicity of the periodogram and the fact that its main building block is an FFT/DFT make it particularly appealing for spectrum sensing in CRs. A key limitation of RNUS is its implementation feasibility. Certain schemes are more amenable to be realized using off-the-shelf components than others. For example, the stratified sampling scheme can be implemented using two or more interleaved conventional ADCs each running at significantly low sub-Nyquist rate as shown in Figure 2.15. Instead of collecting a data sample every T_{US} seconds, the sampler collects measurements at nonuniformly distributed time instants dictated by a pseudorandom generator that drives the ADC. For SSEP with two ADCs in Figure 2.15, ADC 1 and 2 capture the sampling instants with odd, that is, $\{t_1,t_3,...\}$, and even, that is, $\{t_2,t_4,...\}$, indices, respectively. Synchronization among the interleaved ADCs can have a marginal effect on the detection quality because randomness is an integrated part of the proposed sampler. Novel ADC architectures that support random sampling are emerging, for example [71,76]. Figure 2.17 exhibits a sampler architecture that generates irregularly spaced signal measurements by appropriately driving a conventional ADC.

Similar to the classical energy detector, the proposed DASP-based sensing is not immune to interference and noise-level uncertainties. However, the threshold values that limit its $P_{FA,l}$ is

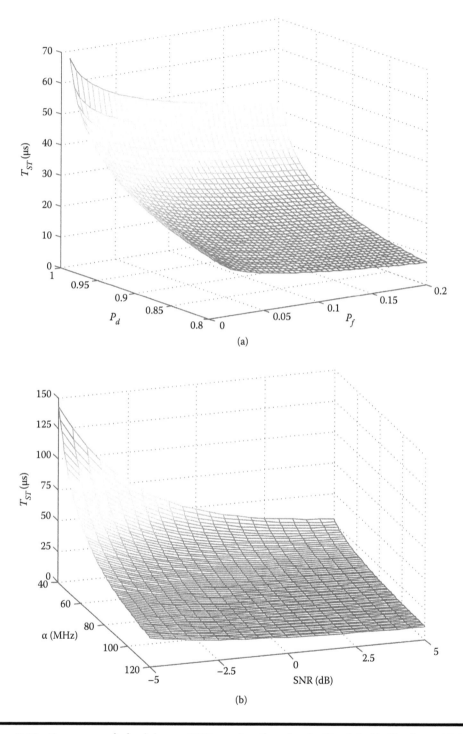

Figure 2.16 Recommended minimum SSEP sensing time for $B_A/B = 0.1$. (a) Fixed $\alpha = 56$ MHz and varying detection requirements; (b) $P_{FA,I} \leq 0.08$ and $P_{D,I} \geq 0.965$.

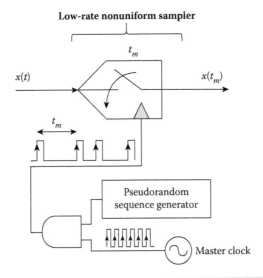

Figure 2.17   A method of realizing a nonuniform sampler using conventional ADCs driven by a clock with irregularly spaced rising edges.

a function of the combined overall signal and noise powers, that is, $\gamma_l = \mathfrak{F}\left(P_S + \sigma_w^2\right)$ in lieu of $\gamma_l = \mathfrak{F}\left(\sigma_w^2\right)$ in the uniform-sampling-based MBED [77]. This can make the DASP-based detectors more resilient to noise estimation errors because the combined signal plus noise powers can be continuously measured at the receiver using a cheap analog integrator. Note that NUS-based feature detectors that can be robust against noise and interference effects remain an explored area outside the CS framework.

To preserve the reconstructability of the detected transmissions, the sampling rates α should exceed at least twice the total bandwidth of the concurrently active subbands; that is, $\alpha \geq f_{Landau} = 2B_A$. Despite the fact that the DASP methods can operate at sub-Landau rates, in practice, α should be proportional to the joint bandwidth of the simultaneously active channels regardless of the total width of the monitored frequency range regardless of the total width of the monitored frequency range $\mathfrak{B}$. Prior to the emergence of the compressive sensing methodology, a range of signal reconstruction techniques were available to recover signals from their nonuniformly distributed measurements (e.g., see [78]). They are customarily based on minimizing the ℓ_2 Euclidean norm of the recovery error. Because the signal spectrum is typically sparse, CS reconstruction algorithms can be applied, and they are expected to outperform ℓ_2-based algorithms.

At this juncture, it can be argued that RNUS is a possible CS data acquisition approach and that DASP belongs to the general compressive sensing framework. This can be the case for certain sampling schemes, such as MCS and RSG. However, the impeccable CS performance guarantees can impose overconservative sampling requirements. They are often applicable to abstractly constructed sensing matrices. Above all, the fundamental difference between DASP and CS methodologies lies in their desired objectives and the utilized processing techniques to extract the required signal information. Whereas the former takes advantage of the incoherent spectrum aliasing of a nonuniformly sampled signal and uses a relatively simple spectrum estimator, compressive sensing focuses on the exact signal reconstruction and uses rather complex recovery techniques. This serves as an impetus to further investigate a unified sub-Nyquist framework for the multiband detection problem.

2.5 Comparison of Sub-Nyquist Spectrum Sensing Algorithms

We now succinctly compare the considered sub-Nyquist MSS approaches and evaluate the detection performance of a number of selected techniques.

2.5.1 Compressed versus Alias-Free Sampling for Multiband Detection

2.5.1.1 Performance Guarantees and Minimum Average Sampling Rates

Although CS provides performance guarantees in terms of the quality of the reconstructed spectrum (e.g., Fourier transform or PSD), the achieved multiband detection quality (e.g., in terms of probabilities of detection and false alarm) is not typically addressed. The time-consuming Monte Carlo simulations are commonly used to examine the performance of the CS-based wideband spectrum sensing algorithm. On the contrary, DASP offers clear guidelines on the attainable detection quality and equips the user with specific recommendations on how to ensure meeting certain sensing specifications, that is, a reliable MSS routine. Both CS and DASP sampling rates are affected by the level of spectrum occupancy, that is, the sparsity level. However, the DASP minimum admissible sampling rates can be arbitrarily low at a predetermined cost of longer sensing time, and vice versa. In CS, α has a lower bound, and the impact of the sensing time on the multiband detection operation is unpredictable. Hence, DASP-based detectors are more suitable candidates when substantial reductions in data acquisition rates are required.

2.5.1.2 Computational Complexity

The main attributes of DASP-based wideband sensing are simplicity and low computational complexity; it only involves DFT or optimized FFT-type operations. In contrast, CS-based techniques entail solving underdetermined sets of linear equations. This is usually computationally expensive even for state-of-the-art sparse recovery methods. Note that MCS and MWC techniques adopt a more efficient approach to spectrum sensing compared with other CS methods such as CS-1 and MASS. They utilize the CTF algorithm, where the processed matrices are approximately of size $m_b \times 2L$ instead of $M \times N$; $N \gg L \geqslant m_b$ is the number of Nyquist samples in the signal time analysis window, and it can be very large.

2.5.1.3 Postprocessing and Related CR Functionalities

Alias-free sampling facilitates spectrum sensing at low sub-Nyquist rates, and in its current formulation does not offer a means to estimate the signal PSD or power level. Nevertheless, it can be argued that the present smeared aliasing can be easily determined and subsequently removed to establish the underlying transmissions PSDs. On the other hand, CS-based multiband detectors can exactly recover the signal spectrum. Estimating the signal power level in a particular active system subband can be important in CR networks to characterize the PUs, to characterize SU transmission power control, and to prevent detrimental interference. Additionally, the received multiband signal at the CR incorporates PU and other CR opportunistic transmissions. The ability to reconstruct the signal from the collected sub-Nyquist samples enables the SU to both sense the spectrum and intercept/receive communications as in standard receivers. Advances in compressive sensing reconstruction algorithms can be leveraged, possibly even for the randomized-NUS based detection.

2.5.1.4 Implementation Complexity

CS and DASP face similar implementation challenges where pseudorandom sampling sequences are commonly used as a compression strategy. Certain CS approaches that do not utilize the aforementioned pseudorandom sampling were implemented, and prototype systems were produced, for example, MWC and RD. Nevertheless, they require complex specialized analog preconditioning modules prior to the low-rate sampling (see Figures 2.7 and 2.9). Their subsequent processing can be also sensitive to sampler mismodeling. Thus, such solutions are inflexible and are high size, weight, power, and cost (SWPAC). Designing flexible low-SWPAC sub-Nyquist samplers remains an open research question. Note that some NUS schemes, for example, RSG, are used in compressive sensing. The CS analog to the information converter in this case is a random time domain sampler, and the sensing matrix Y is a random partial Fourier matrix [42].

2.5.2 Numerical Examples

Consider a scenario that involves an SU monitoring $L = 160$ subbands, each of width $B_C = 7.5$ MHz and known central frequency, in search of a spectrum opportunity. The overseen frequency range is $\mathfrak{B} = [0, 1.2]$ GHz and the processed total single-sided bandwidth is 1.2 GHz. According to the Nyquist criterion, the sampling rate should be at least $f_{Nyq} = 2.4$ GHz. Let the maximum expected number of concurrently active subbands at any point in time or geographic location be $L_A = 8$, that is, the maximum occupancy is 5%. Here, an extensive set of Monte Carlo simulations are conducted to evaluate a selected number of the previously addressed sub-Nyquist wideband spectrum sensing algorithms. This is carried out in terms of the delivered probabilities of detection and false alarm. The objective is to gain an insight into their behavior for the available design resources and operation parameters. We are predominantly interested in the impact of the data acquisition rate, SNR, and sensing time T_{ST} on the multiband detection outcome. Whenever applicable, let $T_W = |T_j| = 0.2$ μs be the width of the individual signal time analysis window; J nonoverlapping and equal time windows are used, that is, $T_{ST} = JT_W$. The maximum spectrum occupancy is considered in all the experiments to represent the extreme system conditions; L_A QPSK or 16QAM transmissions with randomly selected carrier frequencies are present in $\mathfrak{B}$. For simplicity, all active channels are assumed to have equal power levels. To maximize the opportunistic use of a given subband $\mathcal{B}_l$ and minimize the interference to the PU, an adequate metric to assess the sensing quality is given by

$$\{\hat{P}_{D,l}, \hat{P}_{FA,l}\} = \underset{P_{D,l}(i) \in \mathbf{P}_{D,l}, P_{FA,l}(i) \in \mathbf{P}_{FA,l}}{\arg\max} P_{D,l}(i) + \left[1 - P_{FA,l}(i)\right] \tag{2.59}$$

where $\mathbf{P}_{D,l} = [P_{D,l}(1), P_{D,l}(2), \ldots, P_{D,l}(p)]^T$ and $\mathbf{P}_{FA,l} = [P_{FA,l}(1), P_{FA,l}(2), \ldots, P_{FA,l}(p)]^T$ are the attained probabilities for an extensive range of threshold values used to produce a complete ROC plot. Next, the basic CS-1 and CS-2 methods are examined along with the state-of-the-art MWC. The CS matrix Y in CS-1/CS-2 is a random partial Fourier matrix, and the greedy subspaces pursuit in [79] is employed to recover the sparse vector. RSG is the chosen RNUS scheme for the DASP approach.

Motivated by the goal of furnishing substantial savings on the data acquisition rates, Figure 2.18a depicts $\hat{P}_{D,l}$ and $\hat{P}_{FA,l}$ in Equation 2.59 for sub-Nyquist rates that achieve over 85% reductions on f_{Nyq} while SNR = 0 dB and the fixed total sensing time $T_{ST} = 15$ μs. It is noted that the MWC condition in Equation 2.36 is satisfied when $\alpha/f_{Nyq} \geq 0.1$; the more recent lower

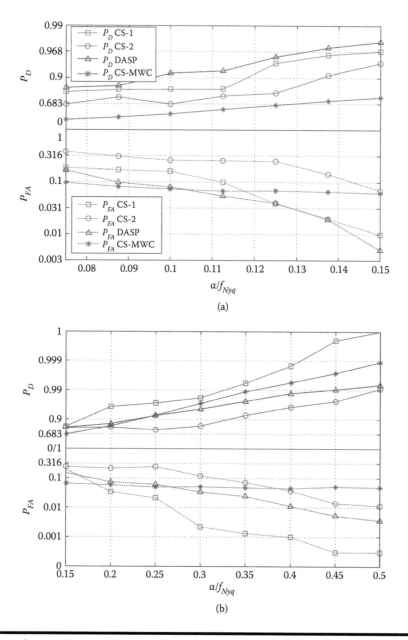

Figure 2.18 $\hat{P}_{D,l}$ and $\hat{P}_{FA,l}$ for selected sub-Nyquist wideband spectrum sensing algorithms with varying compression ratios; SNR = 0 dB. (a) Total sensing time is T_{ST} = 15 μs; (b) Total sensing time is T_{ST} = 3 μs.

rate limit in Equation 2.45 is exceeded when $\alpha/f_{Nyq} \geq 0.05$. It can be seen from the figure that DASP-based wideband spectrum sensing outperforms the compressive techniques, more noticeably MWC for low sampling rates. In Figure 2.18b, $\alpha/f_{Nyq} \in [0.15, 0.5]$ values are tested to assess the sub-Nyquist sensing methods response to higher sampling rates. Owing to the excessive high memory and computational requirements of the enormous sensing matrices associated with CS-1 for high α and T_{ST} and noting the large number of averaged Monte Carlo experiments, the sensing time is reduced to $T_{ST} = 3$ μs. In Figure 2.18b, it is evident that the CS-based algorithms deliver better detection quality compared with the DASP counterpart for higher sampling rates, for example, for $\alpha/f_{Nyq} \geq 0.2$ in Figure 2.18b. Simulations also illustrate that MWC exhibits a high sensing quality only when the operation sampling rate significantly exceeds the lower theoretical bound in Equation 2.36. Thus, Figure 2.18 shows that DASP-based algorithms can achieve competitive, if not superior, spectrum sensing performance compared with several compressive sensing approaches when tangible savings in the data acquisition rates are required. This advantage degrades as α increases owing to the substantial increase in the incurred computational cost for CS-based techniques.

In order to assess the effect of sensing time on the produced sensing results, Figure 2.19 shows the obtained probabilities of detection and false alarm for a changing T_{ST} when $\alpha/f_{Nyq} = 0.15$ and SNR = 0 dB. CS-1 was omitted because of its prohibitive computational complexity. It is apparent from the figure that DASP-based algorithms and CS-2 exploit the available sensing time to enhance their sensing capabilities, whereas MWC sensing probabilities remain nearly constant as the sensing time increases. This figure demonstrates that alias-free sampling and CS-2 approaches can effectively utilize or trade-off the sensing time to improve the spectrum sensing quality, and vice versa.

Finally, in Figure 2.20 the sub-Nyquist sensing algorithms are simulated for varying SNR values while the sensing time and the inverse of the compression ratio are kept fixed, that is, $T_{ST} = 15$ μs and $\alpha/f_{Nyq} = 0.15$. The figure illustrates that the DASP-based approach outperforms MWC and CS-2 as the SNR increases. MWC distinctively continues to deliver poor results compared with the other methods despite the increasing signal power. Thus, the sampling rate is a dominant limiting factor in MWC. In Figure 2.20, the sampling rate α is 1.5 times the MWC theoretical minimum admissible rate. The ability of MWC to achieve low $\hat{P}_{FA,l}$ for low α/f_{Nyq}, SNR, and T_{ST} in Figures 2.18 to 2.20 can be the result of the considerably low attained $\hat{P}_{D,l}$ in such ranges.

Therefore, exploiting the aliasing-suppression capabilities of random time domain sampling can lead to a low complexity and rather simple wideband spectrum sensing algorithms with competitive detection performance. They circumvent the need to undertake computationally intensive operations, for example, solving complex optimizations. Nevertheless, with the ever-expanding capability of DSP modules/cores and the emergence of new CS implementations, CS can still facilitate effective and yet efficient sub-Nyquist MSS solutions.

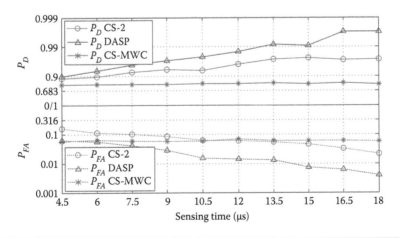

Figure 2.19 **Performance of sub-Nyquist wideband spectrum sensing algorithms for varying sensing times such that** α/f_{Nyq} = 0.15 and SNR = 0 dB.

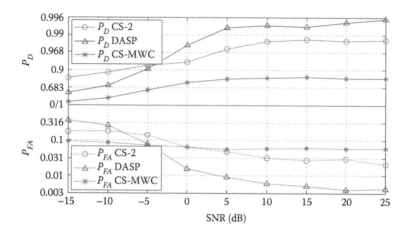

Figure 2.20 Performance of sub-Nyquist wideband spectrum sensing algorithms for varying signal-to-noise ratios (SNRs) such that $\alpha/f_{Nyq} = 0.15$ and $T_{ST} = 15$ μs.

2.6 Conclusions and Open Research Challenges

In this chapter, we first introduced various aspects of the spectrum sensing functionality in a CR. The ability of wideband spectrum sensing to promote multimedia communications over wireless links in CR networks was also highlighted. It enables the SUs to effectively exploit multiple spectral bands concurrently and thereby significantly improve the network opportunistic throughput to meet stringent QoS provisions. The design and implementation challenges of multiband detection were outlined, and special attention was paid to the data acquisition limitation in the wideband regimes of interest. Several wideband spectrum sensing algorithms were then discussed. Conventional parallel sensing methods that conform to the Nyquist sampling criterion commonly employ complex analog front-ends and a sweeping mechanism that can result in severe, intolerable delays. As an alternative, the sub-Nyquist techniques were addressed and categorized as being either CS based or NUS based. They offer new opportunities and mitigate the data acquisition bottleneck of digitally accomplishing the parallel multiband detection task. Both CS-based and NUS-based approaches have their own merits. Generally, DASP's main advantage is simplicity and low computational complexity compared to CS. However, CS offers a more concrete framework that can be used not only for spectrum sensing but also for subsequent CR functionalities such as PU characterization and transmission interception/decoding. Simulations demonstrate that for substantially low sub-Nyquist sampling rates, DASP-based sensing can produce higher-quality detections.

In most sub-Nyquist wideband sensing systems, the required sampling rate is proportional to the spectrum utilization (i.e., the sparsity level). Assuming maximum spectrum occupancy can lead to pessimistically overconservative measures to ensure the sensing reliability and cater for the worst-sense scenario. This approach can waste resources of value to portable devices such as power, space, and memory. In practice, the sparsity level of the wideband signal is time varying owing to the dynamic nature of PU transmissions. Future CR networks should be capable of performing efficient wideband spectrum sensing for unknown or time-varying spectrum occupancies. This calls for adaptive wideband detection techniques that can swiftly and efficiently select the appropriate resources, for example, sensing time, sub-Nyquist sampling rate and even a data acquisition scheme, without prior knowledge of the signal sparsity level. This can be a very challenging task, especially with the time-varying fading channels between the PUs and the CR.

Although the majority of CS-based multiband detectors assume knowledge of the sparsifying basis/frame (e.g., DFT/IDFT), a future research direction can focus on robust CS with an unknown basis/frame. This has become more pressing because the emerging CR networks are expected to alleviate spectrum underutilization by facilitating dynamic opportunistic spectrum access. Hence, the radio spectrum will no longer be sparse in the frequency domain, and adopting alternative sparsifying basis/frames will be mandatory for the use of the CS methodology. With regard to DASP, there is a similar challenge, where estimators other than those targeting the frequency representation will be required.

Owing to the hidden terminal problem and channel fading effects, practical, dependable wideband spectrum sensing will necessitate collaborative detection routines. Hence, sub-Nyquist multiband detection algorithms that promote collaborative sensing are highly desirable and are expected to be the focus in the future, for example, how to appropriately combine information from several CRs in real-time. Finally, realizing dynamic low-SWPAC sub-Nyquist samplers with their subsequent processing tasks is an open research question since portable devices supporting multimedia communications are expected to have limited power, space, and memory resources.

Glossary

List of Abbreviations

ADC	Analog-to-digital converter
AIC	Analog to information converter
ARS	Additive random sampling
AWGN	Additive white Gaussian noise
CR	Cognitive radio
CS	Compressed sampling or compressive sensing
CSD	Cyclic spectral density
CTF	Continuous to finite
DASP	Digital alias-free signal processing
DFT	Discrete Fourier transform
DSP	Digital signal processing
DTFT	Discrete-time Fourier transform
FFT	Fast Fourier transform
IFFT	Inverse fast Fourier transform
IID	Independent identically distributed
JRS	Jittered random sampling
LPF	Low-pass filter
MAC	Medium access control
MASS	Multirate asynchronous sub-Nyquist sampling
MBED	Multiband energy detector
MCS	Multicoset sampling

MJD	Multiple joint detection
MSS	Multiband spectrum sensing
MT-PSDE	Multitaper power spectral density estimator
MWC	Modulated wideband converter
NBSS	Narrowband spectrum sensing
NUS	Nonuniform sampling
PDF	Probability density function
PMNSS	Parallel multiband Nyquist spectrum sensing
PSD	Power spectral density
PU	Primary user
QoS	Quality of service
RD	Random demodulator
RF	Radio frequency
RNUS	Randomized nonuniform sampling
RSG	Random sampling on grid
ROC	Receiver operating characteristics
SMNSS	Sequential multiband Nyquist spectrum sensing
SNR	Signal-to-noise ratio
SS	Stratified sampling
SSEP	Stratified sampling with equal partitions
SU	Secondary user
TRS	Total random sampling
WSCS	Wide sense cyclostationary
WSS	Wide sense stationary

Special Notations, Operators, and Functions

Scalar variables are denoted by lowercase letters, vectors are denoted by bold lowercase letters, and matrices are denoted by bold uppercase letters:

x	Scalar variable
$\mathbf{x}$	Vector
$\mathbf{X}$	Matrix

Operators

$\mathbb{E}[x]$	Statistical expectation of x
$\mathbb{E}[x \mid y]$	Condition statistical expectation of x given y
$\lvert x \rvert$	Absolute value or magnitude of x
$\lfloor x \rfloor$	The largest integer less than or equal to x
$\lceil x \rceil$	The smallest integer greater than or equal to x
$Pr\{x\}$	Probability of a random variable x
$Pr\{x \mid y\}$	Probability of a random variable x conditioned on the value of y
$\lvert x \rvert$	Absolute value or magnitude of x
z^{*}	Conjugate of a complex variable z

*	Convolution
$\overset{m}{\circledast}$	The m-fold convolution operation
$\mathbf{x}^T$	Transpose of vector or matrix $\mathbf{x}$
$\mathbf{x}^H$	Conjugate transpose of vector or matrix $\mathbf{x}$
$x \ll y$	x is significantly smaller than y
$x \gg y$	x is significantly larger than y
$\mathbf{x} \preceq \mathbf{y}$	Each element in $\mathbf{x}$ is less than or equal to the corresponding element in $\mathbf{y}$
$\mathbf{x} \succeq \mathbf{y}$	Each element in $\mathbf{x}$ is greater than or equal to the corresponding element in $\mathbf{y}$

Special Functions

$\mathfrak{T}(\mathbf{y})$	Detector's test statistics to determine the subband status
Q-function	$Q(x) = \dfrac{1}{\sqrt{2\pi}} \displaystyle\int_{x}^{+\infty} e^{-\tau^2/2} d\tau$
$\mathfrak{F}(x)$	Generic function of a scalar, vector, or matrix
$\|\mathbf{x}\|_0$	Number of nonzero elements in vector $\mathbf{x}$

Principal Symbols

$\mathfrak{B}$	Overseen frequency range for $f \geq 0$ and $\mathfrak{B} = \cup_{l=1}^{L} \mathcal{B}_l$
$\mathcal{B}_l$	The frequency range of the lth system spectral subband for $f \geq 0$
B	Single-sided width of the overseen frequency range in Hertz
B_A	Joint width of the L_A concurrently active subbands in Hertz for $f \geq 0$
B_C	Width of a system spectral channel for $f \geq 0$ in Hertz
d_i	Decision of the ith collaborating CR and $d_i \in \{0,1\}$
f	Frequency point in Hertz
$f_{C,l}$	Carrier frequency of the lth transmission in Hertz
f_{Landau}	Landau sampling rate in Hertz
f_{min}	Initial frequency point of the monitored bandwidth in Hertz
f_{max}	Highest frequency of the monitored frequency range in Hertz
f_{Nyq}	Nyquist data acquisition rate in Hertz
f_{US}	Uniform sampling rate in Hertz
$f_{US,i}$	Uniform sampling rate in Hertz for the ith system branch
I	Number of collaborating SUs in cooperative sensing regime
J	Number of signal time analysis windows or averaged spectral estimates
K	Number of simultaneously active subbands $K \leq L_A$ for $f \geq 0$
K_S	Sparsity level
L	Number of monitored spectral subbands for $f \geq 0$
L_A	Maximum number of concurrently active system subbands for $f \geq 0$
m_b	Number of sampling branches in the MCS and MWC systems
M	Total number of captured data measurements
M_b	Number of grid points in a multicoset sampling block
M_i	Number of captured data measurements in the MASS i^{th} sampling branch
M_{PS}	Number of time slots within the period T_P of the modulating signal
N	Number of collected samples at rates exceeding Nyquist's
N_{MT}	Number of employed tapering functions in the multitaper PSD estimator

N_W	Number of components in the wavelet multiscale sum and product
$P_{D,l}$	Probability of detection in the lth subband
$\bar{P}_D$	Weighted sum of the probabilities of detection for the L system channels.
$P_{FA,l}$	Probability of false alarm in the lth subband
$\bar{P}_{FA}$	Weighted sum of the probabilities of false alarm for the L system channels.
$P_{M,l}$	Probability of missed detection in the lth subband
P_S	Power of the received signal
$P_{S,l}$	Power of the signal occupying the lth system subband
q	Number of sampling branches in the MASS system
R_O	Opportunistic throughput in bits/second
R_T	The total leveraged opportunistic throughput in bits/second
t	Time instant in seconds
t_m	Position of the mth sample point in seconds
T_{US}	The uniform sampling period in seconds
T_{OT}	The opportunistic transmission time slot in seconds
T_P	Fundamental period of a periodic function in seconds
T_{ST}	Total sensing time $T_{ST} = \left| \cup_{j=1}^{J} T_j \right|$ in seconds
T_{Total}	The total CR access time $T_{Total} = T_{ST} + T_{OT}$ in seconds
T_W	Width of the T_j signal analysis window in seconds
T_j	Signal time analysis window $T_j = \left[\tau_j, \tau_j + T_W \right]$ starting at the time instant τ_j
$x(t)$	The signal encompassing the K transmissions
$y(t)$	The received signal at the secondary user
$y_l(t)$	The signal transmitted over the lth system spectral channel
$w(t)$	The chosen tapering function template
$X(f)$	Spectrum of $x(t)$
$\hat{X}(f)$	Estimated spectrum of $x(t)$
α	Average sub-Nyquist sampling rate in hertz
β	Scaling factor of the periodogram-type estimator used for irregular sampling
$\delta(t)$	Dirac delta
$\delta[n]$	Kronecker delta
Δ_i	The delay equal to $\varrho_i T_{Nyq}$ in the ith instant of an MCS sequence in seconds
$\mathcal{D}$	The multicoset sampling pattern
$\mathcal{H}_{0,l}$	Hypothesis that signifies the idle state of the lth spectral subband
$\mathcal{H}_{1,l}$	Hypothesis that indicates the presence of a transmission in the lth channel
γ_l	Threshold value for the lth subband
η_l	Minimum desired probability of detection in the lth system subband
ρ_l	Maximum tolerated probability of false alarm in the lth spectral channel
$\bar{\rho}_i$	A prime number that sets the sub-Nyquist sampling rate in a MASS branch
λ_k	The kth eigenvalue
$v_k[m]$	The kth tapering function in the multitaper PSD estimator
M	The set of the numbers of samples in all of the MASS system branches
$\mathcal{P}_X(f)$	Power spectral density of the continuous-time signal $x(t)$
$\mathcal{P}_X(f)^d$	Power spectral density of the discrete-time signal $x(t_m)$
$\mathcal{W}_{\vartheta}^{X}(f)$	Continuous wavelet transform of signal $x(t)$ with ϑ as the dilation factor
ϑ	Wavelet transform dilation factor
$\varphi(f)$	Frequency response of the wavelet-smoothing function

$\mathbf{\Phi}$	Compressed sampling measurement matrix
$\mathbf{\Psi}$	Sparsifying basis/frame
Υ	CS sensing matrix $\Upsilon = \mathbf{\Phi}\mathbf{\Psi}^{-1}$
σ_w	Standard deviation of the present AWGN

References

1. FCC, *Spectrum Policy Task Force Report*, Federal Communications Commission, Tech. Report 02-134 [Online], 2002. Available from: http://hraunfoss.fcc.gov

2. M. McHenry, *Nfs Spectrum Occupancy Measurements Project Summary*, Shared Spectrum Co., Tech. Report 02-134 [Online], 2005. Available: http://sharedspectrum.com

3. J. Mitola, III, Software radios: Survey, critical evaluation and future directions, *IEEE Aerospace and Electronic Systems Magazine*, vol. 8, no. 4, pp. 25–36, 1993.

4. S. Haykin, Cognitive radio: Brain-empowered wireless communications, *IEEE Journal on Selected Areas in Communications*, vol. 23, no. 2, pp. 201–220, 2005.

5. Z. Quan, S. Cui, H. V. Poor, and A. H. Sayed, Collaborative wideband sensing for cognitive radios, *IEEE Signal Processing Magazine*, vol. 25, no. 6, pp. 60–73, 2008.

6. T. Yucek and H. Arslan, A survey of spectrum sensing algorithms for cognitive radio applications, *IEEE Communications Surveys and Tutorials*, vol. 11, no. 1, pp. 116–130, 2009.

7. S. Haykin, D. J. Thomson, and J. H. Reed, Spectrum sensing for cognitive radio, *Proceedings of the IEEE*, vol. 97, no. 5, pp. 849–877, 2009.

8. E. Axell, G. Leus, E. G. Larsson, and H. V. Poor, Spectrum sensing for cognitive radio: State-of-the-art and recent advances, *IEEE Signal Processing Magazine*, vol. 29, no. 3, pp. 101–116, 2012.

9. A. Goldsmith, S. A. Jafar, I. Maric, and S. Srinivasa, Breaking spectrum gridlock with cognitive radios: An information theoretic perspective, *Proceedings of the IEEE*, vol. 97, no. 5, pp. 894–914, 2009.

10. S. M. Kay, *Fundamentals of Statistical Signal Processing: Detection Theory*, NJ: Prentice-Hall, 1998.

11. H. V. Poor, *An Introduction to Signal Detection and Estimation*, New York: Springer, 1994.

12. W. A. Gardner, Exploitation of spectral redundancy in cyclostationary signals, *IEEE Signal Processing Magazine*, vol. 8, no. 2, pp. 14–36, 1991.

13. W. A. Gardner, A. Napolitano, and L. Paura, Cyclostationarity: Half a century of research, *Signal Processing*, vol. 86, no. 4, pp. 639–697, 2006.

14. Y. Zeng and Y.-C. Liang, Spectrum-sensing algorithms for cognitive radio based on statistical covariances, *IEEE Transactions on Vehicular Technology*, vol. 58, no. 4, pp. 1804–1815, 2009.

15. E. Axell and E. G. Larsson, A unified framework for GLRT-based spectrum sensing of signals with covariance matrices with known eigenvalue multiplicities, in *IEEE International Conference on Acoustics, Speech and Signal Processing (ICASSP '11)*, pp. 2956–2959, 2011.

16. R. Tandra and A. Sahai, SNR walls for signal detection, *IEEE Journal of Selected Topics in Signal Processing*, vol. 2, no. 1, pp. 4–17, 2008.

17. K. Letaief and W. Zhang, Cooperative communications for cognitive radio networks, *Proceedings of the IEEE*, vol. 97, no. 5, pp. 878–893, 2009.

18. M. Ibnkahla, *Cooperative Cognitive Radio Networks: The Complete Spectrum Cycle*, New York: CRC Press, 2014.

19. S. M. Mishra, A. Sahai, and R. W. Brodersen, Cooperative sensing among cognitive radios, in *IEEE International Conference on Communications (ICC'06)*, vol. 4, pp. 1658–1663, 2006.

20. J. Ma, G. Zhao, and Y. Li, Soft combination and detection for cooperative spectrum sensing in cognitive radio networks, *IEEE Transactions on Wireless Communications*, vol. 7, no. 11, pp. 4502–4507, 2008.

21. R. G. Vaughan, N. L. Scott, and D. R. White, The theory of bandpass sampling, *IEEE Transactions on Signal Processing*, vol. 39, no. 9, pp. 1973–1984, 1991.

22. Z.-C. Hao and J.-S. Hong, Highly selective ultra wideband bandpass filters with quasi-elliptic function response, *IET Microwaves, Antennas and Propagation*, vol. 5, no. 9, pp. 1103–1108, 2011.

23. J. M. de la Rosa, Sigma-delta modulators: Tutorial overview, design guide, and state-of-the-art survey, *IEEE Transactions on Circuits and Systems I: Regular Papers*, vol. 58, no. 1, pp. 1–21, 2011.

24. S. Balasubramanian, V. J. Patel, and W. Khalil, Current and emerging trends in the design of digital-to-analog converters, in P. Carbone, S. Kiaei, F. Xu, eds, *Design, Modeling and Testing of Data Converters*, pp. 83–118, Springer, Berlin, 2014.

25. Y. Xin, G. Yue, and L. Lai, Efficient channel search algorithms for cognitive radio in a multichannel system, in *IEEE Global Telecommunications Conference (GLOBECOM '10)*, pp. 1–5, 2010.

26. Z. Sun, *Performance Metrics, Sampling Schemes, and Detection Algorithms for Wideband Spectrum Sensing*, PhD thesis, University of Notre Dame, 2013.

27. S.-J. Kim, G. Li, and G. B. Giannakis, Multi-band cognitive radio spectrum sensing for quality-of-service traffic, *IEEE Transactions on Wireless Communications*, vol. 10, no. 10, pp. 3506–3515, 2011.

28. R. Fan and H. Jiang, Optimal multi-channel cooperative sensing in cognitive radio networks, *IEEE Transactions on Wireless Communications*, vol. 9, no. 3, pp. 1128–1138, 2010.

29. S. Stotas and A. Nallanathan, Overcoming the sensing-throughput tradeoff in cognitive radio networks, in *IEEE International Conference on Communications (ICC '10)*, pp. 1–5, 2010.

30. A. Ghassemi, S. Bavarian, and L. Lampe, Cognitive radio for smart grid communications, in *IEEE International Conference on Smart Grid Communications*, pp. 297–302, 2010.

31. P. Paysarvi-Hoseini and N. C. Beaulieu, Optimal wideband spectrum sensing framework for cognitive radio systems, *IEEE Transactions on Signal Processing*, vol. 59, no. 3, pp. 1170–1182, 2011.

32. L. Luo, N. M. Neihart, S. Roy, and D. J. Allstot, A two-stage sensing technique for dynamic spectrum access, *IEEE Transactions on Wireless Communications*, vol. 8, no. 6, pp. 3028–3037, 2009.

33. M. H. Hayes, *Statistical Digital Signal Processing and Modeling*, New York: John Wiley, 1996.

34. F. J. Harris, On the use of windows for harmonic analysis with the discrete Fourier transform, *Proceedings of the IEEE*, vol. 66, no. 1, pp. 51–83, 1978.

35. K. Hossain and B. Champagne, Wideband spectrum sensing for cognitive radios with correlated sub-band occupancy, *IEEE Signal Processing Letters*, vol. 18, no. 1, pp. 35–38, 2011.

36. C. Da Silva, B. Choi, and K. Kim, Distributed spectrum sensing for cognitive radio systems, in *Information Theory and Applications Workshop*, pp. 120–123, 2007.

37. Q. T. Zhang, Theoretical performance and thresholds of the multitaper method for spectrum sensing, *IEEE Transactions on Vehicular Technology*, vol. 60, no. 5, pp. 2128–2138, 2011.

38. D. J. Thomson, Spectrum estimation and harmonic analysis, *Proceedings of the IEEE*, vol. 70, no. 9, pp. 1055–1096, 1982.

39. B. Farhang-Boroujeny, Filter bank spectrum sensing for cognitive radios, *IEEE Transactions on Signal Processing*, vol. 56, no. 5, pp. 1801–1811, 2008.

40. Z. Tian and G. B. Giannakis, A wavelet approach to wideband spectrum sensing for cognitive radios, in *1st International Conference on Cognitive Radio Oriented Wireless Networks and Communications (CROWNCOM 06)*, pp. 1–5, 2006.

41. Y. Zeng, Y.-C. Liang, and M. W. Chia, Edge based wideband sensing for cognitive radio: Algorithm and performance evaluation, in *IEEE Symposium on New Frontiers in Dynamic Spectrum Access Networks (DySPAN 11)*, pp. 538–544, 2011.

42. M. F. Duarte and Y. C. Eldar, Structured compressed sensing: From theory to applications, *IEEE Transactions on Signal Processing*, vol. 59, no. 9, pp. 4053–4085, 2011.

43. S. Foucart and H. Rauhut, *A Mathematical Introduction to Compressive Sensing*. Applied and Numerical Harmonic Analysis, Basel, Switzerland: Birkhäuser, 2013.

44. J. A. Tropp and S. J. Wright, Computational methods for sparse solution of linear inverse problems, *Proceedings of the IEEE*, vol. 98, no. 6, pp. 948–958, 2010.

45. Y. C. Eldar and G. Kutyniok, *Compressed Sensing: Theory and Applications*, Cambridge: Cambridge University Press, 2012.

46. Z. Tian and G. B. Giannakis, Compressed sensing for wideband cognitive radios, in *IEEE International Conference on Acoustics, Speech and Signal Processing (ICASSP '07)*, vol. 4, pp. IV–1357, 2007.

47. Y. L. Polo, Y. Wang, A. Pandharipande, and G. Leus, Compressive wide-band spectrum sensing, in *IEEE International Conference on Acoustics, Speech and Signal Processing (ICASSP '09)*, pp. 2337–2340, 2009.

48. J. A. Tropp, J. N. Laska, M. F. Duarte, J. K. Romberg, and R. G. Baraniuk, Beyond Nyquist: Efficient sampling of sparse bandlimited signals, *IEEE Transactions on Information Theory*, vol. 56, no. 1, pp. 520–544, 2010.

49. M. F. Duarte and R. G. Baraniuk, Spectral compressive sensing, *Applied and Computational Harmonic Analysis*, vol. 35, no. 1, pp. 111–129, 2013.

50. P. J. Pankiewicz, T. Arildsen, and T. Larsen, Model-based calibration of filter imperfections in the random demodulator for compressive sensing, 2013.

51. Y. Wang, Z. Tian, and C. Feng, Sparsity order estimation and its application in compressive spectrum sensing for cognitive radios, *IEEE Transactions on Wireless Communications*, vol. 11, no. 6, pp. 2116–2125, 2012.

52. F. Zeng, C. Li, and Z. Tian, Distributed compressive spectrum sensing in cooperative multihop cognitive networks, *IEEE Journal of Selected Topics in Signal Processing*, vol. 5, no. 1, pp. 37–48, 2011.

53. M. Mishali and Y. C. Eldar, Blind multiband signal reconstruction: Compressed sensing for analog signals, *IEEE Transactions on Signal Processing*, vol. 57, no. 3, pp. 993–1009, 2009.

54. M. Mishali and Y. C. Eldar, From theory to practice: Sub-Nyquist sampling of sparse wideband analog signals, *IEEE Journal of Selected Topics in Signal Processing*, vol. 4, no. 2, pp. 375–391, 2010.

55. P. Feng, *Universal Minimum-Rate Sampling and Spectrum-Blind Reconstruction for Multiband Signals*, PhD thesis, University of Illinois at Urbana-Champaign, Urbana-Champaign, IL, 1997.

56. H. Landau, Necessary density conditions for sampling and interpolation of certain entire functions, *Acta Mathematica*, vol. 117, no. 1, pp. 37–52, 1967.

57. R. Venkataramani and Y. Bresler, Perfect reconstruction formulas and bounds on aliasing error in sub-Nyquist nonuniform sampling of multiband signals, *IEEE Transactions on Information Theory*, vol. 46, no. 6, pp. 2173–2183, 2000.

58. Y. Bresler, Spectrum-blind sampling and compressive sensing for continuous-index signals, in *Information Theory and Applications Workshop*, pp. 547–554, 2008.

59. M. A. Lexa, M. E. Davies, and J. S. Thompson, Reconciling compressive sampling systems for spectrally sparse continuous-time signals, *IEEE Transactions on Signal Processing*, vol. 60, no. 1, pp. 155–171, 2012.

60. H. Sun, W.-Y. Chiu, J. Jiang, A. Nallanathan, and H. V. Poor, Wideband spectrum sensing with sub-Nyquist sampling in cognitive radios, *IEEE Transactions on Signal Processing*, vol. 60, no. 1, pp. 6068–6073, 2009.

61. M. A. Lexa, M. E. Davies, J. S. Thompson, and J. Nikolic, Compressive power spectral density estimation, in *IEEE International Conference on Acoustics, Speech and Signal Processing (ICASSP '11)*, pp. 3884–3887, 2011.

62. D. D. Ariananda and G. Leus, Compressive wideband power spectrum estimation, *IEEE Transactions on Signal Processing*, vol. 60, no. 9, pp. 4775–4789, 2012.

63. D. Cohen and Y. C. Eldar, Sub-Nyquist sampling for power spectrum sensing in cognitive radios: A unified approach, 2013.

64. Z. Tian, Y. Tafesse, and B. M. Sadler, Cyclic feature detection with sub-Nyquist sampling for wideband spectrum sensing, *IEEE Journal of Selected Topics in Signal Processing*, vol. 6, no. 1, pp. 58–69, 2012.

65. B. I. Ahmad and A. Tarczynski, Reliable wideband multichannel spectrum sensing using randomized sampling schemes, *Signal Processing*, vol. 90, no. 7, pp. 2232–2242, 2010.

66. B. I. Ahmad and A. Tarczynski, Wideband spectrum sensing technique based on random sampling on grid: Achieving lower sampling rates, *Digital Signal Processing*, vol. 21, no. 3, pp. 466–476, 2011.

67. B. I. Ahmad and A. Tarczynski, A SARS method for reliable spectrum sensing in multiband communication systems, *IEEE Transactions on Signal Processing*, vol. 59, no. 12, pp. 6008–6020, 2011.

68. B. I. Ahmad and A. Tarczynski, Spectral analysis of stratified sampling: A means to perform efficient multiband spectrum sensing, *IEEE Transactions on Wireless Communications*, vol. 11, no. 1, pp. 178–187, 2012.

69. F. Marvasti, *Nonuniform Sampling: Theory and Practice*, vol. 1, Springer, 2001.

70. J. J. Wojtiuk, *Randomised Sampling for Radio Design*, PhD thesis, University of South Australia, Australia, 2000.

71. F. Papenfuss, *Digital Signal Processing of Nonuniform Sampled Signals*, Dissertation, Universitt Rostock, Shaker Verlag, 2007.

72. I. Bilinskis, *Digital Alias-Free Signal Processing*, West Sussex: John Wiley, 2007.

73. P. Babu and P. Stoica, Spectral analysis of nonuniformly sampled data—A review, *Digital Signal Processing*, vol. 20, no. 2, pp. 359–378, 2010.

74. H. S. Shapiro and R. A. Silverman, Alias-free sampling of random noise, *Journal of the Society for Industrial and Applied Mathematics*, vol. 8, no. 2, pp. 225–248, 1960.
75. A. Tarczynski and D. Qu, Optimal periodic sampling sequences for nearly-alias-free digital signal processing, in *IEEE International Symposium on Circuits and Systems (ISCAS '05)*, pp. 1425–1428, 2005.
76. M. Wakin, S. Becker, E. Nakamura, M. Grant, E. Sovero, D. Ching, J. Yoo, J. Romberg, A. Emami-Neyestanak, and E. Candes, A nonuniform sampler for wideband spectrally-sparse environments, *IEEE Journal on Emerging and Selected Topics in Circuits and Systems*, vol. 2, no. 3, pp. 516–529, 2012.
77. B. I. Ahmad and A. Tarczynski, A spectrum sensing method based on stratified sampling, in *IEEE International Symposium on Circuits and Systems (ISCAS '11)*, pp. 402–405, 2011.
78. A. Aldroubi and K. Gröchenig, Nonuniform sampling and reconstruction in shift-invariant spaces, *SIAM Review*, vol. 43, no. 4, pp. 585–620, 2001.
79. W. Dai and O. Milenkovic, Subspace pursuit for compressive sensing signal reconstruction, *IEEE Transactions on Information Theory*, vol. 55, no. 5, pp. 2230–2249, 2009.

Chapter 3

Bargaining-Based Spectrum Sharing for Broadband Multimedia Services in Cognitive Radio Network

Yang Yan,[1] Xiang Chen,[2] Xiaofeng Zhong,[1,3]
Ming Zhao,[1,3] and Jing Wang[1,3]

[1]*Research Institute of Information Technology (RIIT) of Tsinghua University and Tsinghua National Laboratory for Information Science and Technology (TNList), Beijing, China*

[2]*Aerospace Center, Tsinghua University, Beijing, China*

[3]*Department of Electronic Engineering, Tsinghua University, Beijing, China*

Contents

3.1 Introduction

Cognitive radio (CR) technology [1] can greatly improve spectrum efficiency by allowing secondary unlicensed users (SUs) to opportunistically obtain spectrum resources from primary licensed users (PUs), and thus can effectively alleviate the ever-increasing network pressure due to the rapid growth of wireless multimedia services. As a key component of CR technology, efficient *spectrum sharing* requires that CR network access be coordinated to prevent multiple SUs from colliding in overlapping portions of the spectrum, without causing interference with primary networks [2]. In this chapter, we consider a bilateral bargaining procedure between two SUs to achieve this goal, based on the spectrum resource availability from PUs.

As one of different functionalities required for spectrum management in CR networks, spectrum-sharing techniques generally focus on two types of solutions [2]:

- Intranetwork spectrum sharing, that is, spectrum sharing within a CR network, which is spectrum allocation between the entities of a CR network, without causing interference to the PUs.
- Internetwork spectrum sharing, that is, spectrum sharing among multiple coexisting CR networks, which ensures that the CR architecture enables multiple different systems to be deployed in overlapping locations and spectra.

This chapter mainly considers the *intranetwork* spectrum sharing, where the SUs of a CR network try to access the available spectrum resource from PUs.[1]

The concept of intranetwork spectrum sharing has only been proposed in the recent literature (e.g., see Refs. [3–5] and the references therein), some of which focus on *cooperative* intranetwork spectrum sharing. Cao and Zheng [3] considered cooperative local bargaining to provide both spectrum utilization and fairness. Local bargaining is performed by constructing local groups according to a poverty line that ensures a minimum spectrum allocation to each user. Jiang et al. [4] proposed a reinforcement-learning-based spectrum-sharing scheme. CR users can learn from the interaction between themselves and the environment to assess the success level of a particular action. Zheng and Cao [5], unlike the aforementioned references, considered *noncooperative* intranetwork spectrum sharing, in which an opportunistic spectrum management scheme was proposed. Users allocate channels based on their observations of interference patterns and neighbors.

In this chapter, we employ *noncooperative* bargaining theory[2] to realize intranetwork spectrum sharing between SUs. We describe in detail a proposed *two-sided alternating offers* spectrum bargaining mechanism where bargainers (the SUs) can be the offer proposer and offer responder; that is, they have equal bargaining power to decide how to divide the available spectrum resource.

---

[1] In fact, with considerations of different system attributes and operator policies, our proposed scheme can be adjusted and thus applied to internetwork spectrum sharing. We will discuss this in Section 3.2.

[2] Instead of the axiomatic approach to bargaining (such as the Nash Bargaining Solution [6]), which derives the results from a set of axioms, noncooperative bargaining theory involves writing down some particular sequences of moves (offers and replies) to be made over time in the course of negotiations, and then looking for a noncooperative equilibrium in the game [7].

This two-sided bargaining model better captures the characteristic of intranetwork spectrum sharing where SUs have the same *right* to access spectrum resource. In contrast, existing Refs. [8–11] considered cooperative spectrum sharing between PUs and SUs realized by a *one-sided* noncooperative bargaining scheme based on the fact that PUs own spectrum resources and thus have priority over spectrum usage, which indicates that PUs have more power during the bargaining process.

The main contributions of this chapter are summarized as follows:

■ *New modeling and solution technique:* To the best of our knowledge, we are the first to study intranetwork spectrum sharing using *noncooperative bargaining theory*. Moreover, we propose a *two-sided alternating offers* bargaining model that better captures the fact that SUs have equal bargaining powers in intranetwork spectrum sharing and has never been discussed in the previous literature on this topic.

■ *Finite-horizon and infinite-horizon dynamics and subgame perfect equilibrium:* We investigate the spectrum bargaining process under two different system scenarios: (1) a finite-horizon game where two SUs bargain over the partition of the spectrum during a fixed-length period of time and (2) an infinite-horizon game where two SUs do not know precisely when the bargaining will end. We employ a subgame perfect equilibrium (SPE) for the study of the interaction between SUs and fully characterize the corresponding SPEs. We also obtain some interesting insights into the outcomes through theoretical analysis and numerical results, and show that the proposed spectrum sharing scheme is fair, efficient, and incentive compatible.

The rest of this chapter is organized as follows. We introduce the system model and methodology in Section 3.2. In Section 3.3, we analyze the finite-horizon bargaining game. In Section 3.4, we extend the analysis to the infinite-horizon bargaining game. We discuss the bargaining outcomes through theoretical analysis and numerical results in Section 3.5. Finally, we conclude the chapter in Section 3.6.

3.2 System Model and Methodology

3.2.1 Spectrum-Sharing Bargaining Model

We consider an Orthogonal frequency-division multiplexing (OFDM)-based CR system with the network model shown in Figure 3.1. For ease of illustration, we simplify this figure by not elaborating the transmitters and receivers of the PU and SUs. The dashed ellipse denotes the secondary system where two SUs, that is, SU_1 and SU_2, coexist on the basis of an OFDM mode [12].[1] One PU exists in the network, and the solid lines with arrows between the PU and SUs indicate that two SUs can obtain feasible spectrum resource for broadband multimedia services from the PU through an appropriate spectrum sensing and decision mechanism.[2] The two SUs have to reach an agreement on how to divide this piece of spectrum, taking into consideration specific individual situations.

---

[1] The two-user secondary system assumption can greatly simplify the analysis of spectrum-sharing bargaining; however, applying Antoni's work [13,14], our framework can be easily extended to the general case with multiple SUs. We will provide more discussions about this extension next.

[2] This chapter focuses on analyzing the equilibrium strategies and engineering insights for the bargaining-based intranetwork spectrum sharing. For simplicity, we do not jointly consider spectrum sensing and decision issues, and will leave them for future investigation. All subsequent discussion in this chapter is based on the assumption that spectrum resources have been available for SUs.

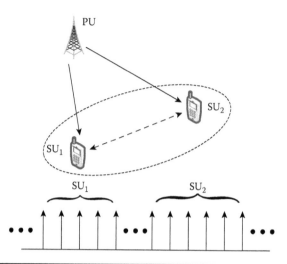

Figure 3.1 Spectrum-sharing bargaining in OFDM-based cognitive radio network.

Their bargaining interaction over spectrum division is indicated by the dashed line between them, which will be analyzed in detail in Sections 3.3 and 3.4. The channel gain at each subcarrier can be modeled as an AWGN channel, and we further assume that the channel remains fixed during the bargaining process. Furthermore, we allow SUs to bargain with each other via a common control channel [15] so as to reach an agreement. This (out-of-band) control channel can be used for the exchange of transmission parameters among different nodes and for delivering the decisions (for example, subcarrier allocation) within the secondary network. Once a possible agreement is reached, the two SUs work in OFDM mode for their own multimedia data transmissions.

The bargaining process is depicted as follows. In the first period (stage), SU_1 proposes a division of spectrum (that is, the number of subcarriers); after *observing* SU_1's offer, SU_2 decides whether to accept or reject this offer; if SU_2 accepts the offer, then the proposed division is implemented, and the game ends; otherwise, the bargaining moves into the second stage. In the second stage, SU_2 makes a counteroffer; after *observing* SU_2's offer, SU_1 decides whether to accept or reject this offer; if SU_1 accepts, then the proposed division is implemented, and the game ends; otherwise, the game moves into the third stage, and so on. Both SUs can perfectly observe all the offers that were proposed in the previous stages.[1]

The bargaining process can last for a fixed-length period of time (that is, a finite horizon where the number of stages is fixed), or an unlimited period of time (that is, an infinite horizon where SUs do know when the game will end). In any stage $t = 1,2,\ldots,T$(or ∞), if no division is accepted prior to t, and if t is odd (even), then

- SU_1 (SU_2) proposes a division.
- After observing SU_1's (SU_2's) offer, SU_2 (SU_1) decides whether to accept or reject this offer.
- If SU_2 (SU_1) accepts the offer, the spectrum resource is divided according to the proposal and the game ends; otherwise, the game moves into stage $t + 1$.

---

[1] To simplify the analysis and gain the first step in understanding the strategic bargaining mechanism for spectrum sharing, we consider the complete information scenario. The study of incomplete information related to SUs can be studied on the basis of a similar methodology and could be considered as a project for the future.

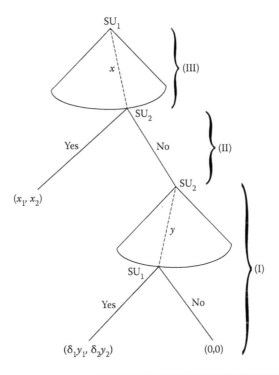

Figure 3.2 Two-period bargaining game with alternating offers.

The corresponding game trees of the two bargaining scenarios are depicted in Figures 3.2 and 3.3, respectively, which will be introduced shortly.

3.2.2 Utility Function

We assume that both SUs are rational players who make optimal strategic bargaining proposals in order to maximize their own benefits. For each SU, its achievable data rate (bits/second/Hz) can be viewed as a benefit that is given by

$$R_i(L^i) = \sum_{l=1}^{L^i} W \log_2(1 + \mathrm{SNR}_l^i) \quad \forall i \in N, \tag{3.1}$$

where i is the index of SU_i and $N = \{1,2\}$. W is the bandwidth of each subcarrier and is thus a constant. SNR_l^i is the signal-to-noise ratio at the l-th subcarrier of SU_i for the given time and location in the network. L^i is the number of the available subcarriers for SU_i. Given the spectrum resource, that is, all the available subcarriers L, we have $L^1 + L^2 = \mathcal{L}$.

For each subcarrier, $W \log_2(1 + \mathrm{SNR}_l^i)$ is positive, and thus it is obvious that the achievable data rate increases along with the number of subcarriers. This indicates that each SU aims at obtaining more subcarriers during the spectrum sharing in order to obtain better multimedia services. Thus, both SUs will bargain over how to divide the L subcarriers in order to maximize their own benefits. In order to simplify the analysis, we assume that any proposal from SU_i can be any scalar,

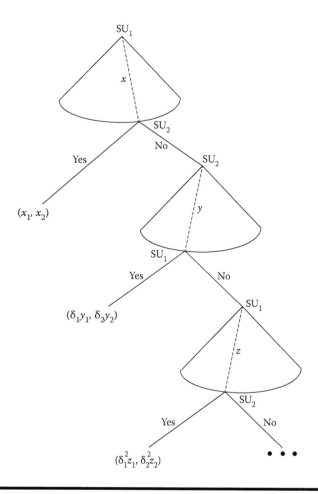

Figure 3.3   Infinite-horizon bargaining game with alternating offers.

not necessarily an integer.[1] We further normalize the number of subcarriers as $L = 1$. Moreover, let $\mathbf{x} = (x_1, x_2)$ with $x_1 + x_2 = 1$ denote the spectrum allocation in one stage if an agreement is reached. Therefore, the *SU's utility function* can be defined as

$$U_i(x_i) = x_i \quad \forall i \in N. \tag{3.2}$$

This utility function indicates that the more subcarriers an SU obtains, the higher the multimedia data rate it can achieve. Because we consider an intranetwork spectrum sharing where two SUs from the same system are *homogeneous*, the utility function in this chapter is simplified as in

---

[1] This assumption can greatly reduce the complexity of bargaining game analysis. In reality, the Long-Term Evolution (LTE) physical layer uses orthogonal frequency division multiple access (OFDMA) on the downlink and requires that bandwidth be scalable from 1.25 to 20 MHz. The difference of the center frequencies of the adjacent subcarriers is 15 kHz. The number of subcarriers varies from 128 to 2048. Therefore, this approximation is valid, especially when the bandwidth is larger.

Equation 3.2 without considering other factors such as the different bargaining powers attribut-able to the SUs' locations in the network, the channel gains, and SUs' transmission powers.[1]

In Section 3.2.1, we elaborate on the bargaining process where SU_1 and SU_2 are indifferent about the *timing* of an agreement. However, in reality, any bargaining process consumes time. In this chapter, we assume that SU is an *energy-constrained* device (for example, a wireless sensor or mobile device with a limited battery life), and thus SUs' bargaining preference should reflect the factor of time t. Therefore, the utility function should include this fact, that is, $U_i(x_i,t)$. As the bargaining proceeds, the battery life decreases. Thus, SUs value time and have preferences for *earlier* agreements by *discounting* the future utility at a rate. We can express the utility function taking the timing preference into consideration as follows [16,17]:

$$U_i(x_i,t) = \delta_i^t x_i \qquad \forall i \in N, \tag{3.3}$$

where $N = \{1,2\}$ and $\delta_i \in (0,1)$ is SU_i's *discount factor*, which is related to SU_i's battery status and can also measure the degree of SU_i's *patience* in bargaining. Intuitively, SUs with a larger δ tend to be more patient during the bargaining process.

In this subsection, we have revised utility functions from a complicated form to a simpler form, without influencing the bargaining results. If we take different system attributes as well as different operators' policies into account when specifying the utility function, our intranetwork bargaining approach can be similarly applied in the internetwork spectrum sharing case, with possible modifications if any.

3.2.3 Subgame Perfect Equilibrium

The bargaining process elaborated in Section 3.2.1 falls into the class of a dynamic game, more specifically, a dynamic multistage game with observed actions, which involves SUs' dynamic deci-sion making in multiple periods (a finite or an infinite horizon). Note that the classic Nash equilib-rium concept investigates players' *simultaneous* actions in a *static* game, and thus it is not applicable to the model we consider here [18]. Therefore, we employ a refined Nash equilibrium concept, that is, an SPE, for the bargaining analysis, which we define briefly as follows.[2]

Definition 1 A strategy profile s^* is an SPE in game Γ if for any subgame Γ' of Γ, $s^*_{|\Gamma'}$ is a Nash equilibrium of Γ', where $s^*_{|\Gamma'}$ is the restriction of strategy profile s^* to subgame Γ'.

Backward induction (BI) is a general technique for characterizing an SPE of a *finite-horizon* dynamic game with observed actions that we will employ in Section 3.3. For the infinite-horizon game, because the BI method cannot be applied, we will introduce the *one-stage deviation principle* to characterize the SPE in Section 3.4.

---

[1] In fact, these aforementioned factors can be modeled as different coefficients of x_i, which denote the SUs' dif-ferent bargaining positions for subcarrier sharing. However, they will not change the monotonicity of each SU's utility function on subcarrier number. Hence, the more general case with these factors can be similarly analyzed.

[2] The detailed definitions and discussions about Nash equilibrium, subgames, and other related concepts are beyond the scope of this chapter. See Ref. [18] for more details.

3.2.4 Bargaining Theory

Bargaining theory is an effective tool to study the situation where individual players can conclude a mutually beneficial agreement over scarce resource allocation [17]. In Section 3.2.4, we introduce some further background material on bargaining theory and explain the motivations behind our choice of the one-one bargaining model.

In general, bargaining theory can be classified into two categories. One is the axiomatic approach developed by Nash [6] using the framework of cooperative games. The other is the strategic approach [19,20], that is, noncooperative dynamic bargaining using noncooperative game theory. In this chapter, we assume that both the SUs are rational players who will maximize their own utilities through a process that can be analyzed by noncooperative bargaining theory.

In bargaining theory, bargaining with many players can proceed in a variety of ways. Many different models of *n*-player bargaining exist in the literature [13,14,21–25]. Multilateral noncooperative bargaining can be applied to the problem of *sharing a pie of a fixed size*, where *n* players bargain multilaterally to divide a piece of a resource [21–25]. In this chapter, we analyze bargaining-based spectrum sharing between two SUs. However, thanks to Antoni's work [13,14], general spectrum sharing with multiple SUs can be decomposed into multiple pairs of bilateral bargaining games between two SUs. Briefly, if no agreement is ever reached between two SUs, then each of the SUs can select other SUs as potential bargainers, in a random way or under an optimal partner selection scheme by considering network information such as the SUs' locations, channel conditions, and information on potential bargaining opponents (for example, the degree of bargaining patience). Therefore, a study of bilateral bargaining-based spectrum sharing serves as a basis for the general case involving multiple CR entities, and therefore is of great importance to us.

3.3 Finite-Horizon Bargaining Game

In this section, we consider the finite-horizon bargaining game where two SUs have to reach an agreement before or during Period $T < \infty$. If no agreement is reached within the finite T periods, the game ends with both players getting zero utility. As a first step, we study a two-period bargaining game. Then, the general finite T-period bargaining game will be similarly analyzed.

3.3.1 Two-Period Bargaining Game

Figure 3.2 illustrates a two-period bargaining game with alternating offers, where the bargaining procedure is the same as depicted in Section 3.2.1. We assume that SU_1 acts first. SU_i's utility in the second period is discounted using the discount factor $\delta_i \in (0,1)$. $\mathbf{x} = (x_1, x_2)$ with $x_1 + x_2 = 1$ denotes the subcarrier allocation proposed by SU_1 in the first period, which is also SU_1's and SU_2's utilities (see Equation 3.2). $\mathbf{y} = (y_1, y_2)$ with $y_1 + y_2 = 1$ denotes the allocation proposed by SU_2 in the second period. If no agreement is reached after two periods, the two SUs cannot divide the spectrum, and thus the utility profile is (0,0); that is, no one will obtain spectrum for their own transmissions.

We will find the SPE of this game by using BI. Since the subgame in Phase (I) in Figure 3.2 is also a dynamic game, we firstly consider this subgame by applying the BI method similarly. There is a different subgame for each value of $\mathbf{y}$, so we should find the optimal strategy of SU_1: if $y_1 > 0$, then SU_1 should accept this offer; if $y_1 = 0$, then SU_1 will be indifferent between accepting and rejecting this offer. We further divide the case where $y_1 = 0$ into two subcases:

1. Accept offer for all $y_1 \geq 0$
2. Accept offer if $y_1 > 0$, and reject it if $y_1 = 0$

For the first possible strategy of SU_1, it is obvious that SU_2's optimal offer will maximize $y_2 = 1 - y_1$ when $y_1 \geq 0$. Therefore, the optimal offer should be $\mathbf{y} = (y_1, y_2) = (0,1)$. Then, SU_1 accepts all offers.

For the second possible strategy of SU_1, SU_2 should offer $y_1 > 0$ so as to get a positive utility $1 - y_1$. Thus, its optimal offer will be

$$y_1^* = \arg\max_{y_1 > 0}(1 - y_1). \tag{3.4}$$

However, there is no optimal solution for Equation 3.4; hence, there is no SPE in this case. Therefore, the *unique* SPE of subgame of Phase (I) in Figure 3.2 is

■ SU_2 offers 0 to SU_1.
■ SU_1 accepts all $y_1 \geq 0$.

The *outcome* of the bargaining subgame (1) in Figure 3.2 is (0,1); that is, SU_2 obtains all the subcarriers.

Next, we track back to Phases (II) and (III) in Figure 3.2. Considering the discount factor, the outcome of subgame (1) is $(0,\delta_2)$. That is, SU_2 will get δ_2 if it rejects SU_1's offer in Phase (II). Similarly, there are two subcases for SU_2's strategies:

1. Accept the offer if $x_2 \geq \delta_2$, and reject it if $x_2 < \delta_2$.
2. Accept the offer if $x_2 > \delta_2$, and reject it if $x_2 \leq \delta_2$.

Combining the above steps with Phase (III), we can obtain SU_1's optimal offer in Phase (III) as $(1-\delta_2, \delta_2)$. Therefore, the *unique* SPE of the two-period bargaining game is as follows:

■ SU_1's initial proposal is $(1-\delta_2, \delta_2)$.
■ SU_2 accepts all offers when $x_2 \geq \delta_2$ and rejects all offers when $x_2 < \delta_2$.
■ SU_2 proposes (0,1) whenever it rejects SU_1's offer in the first period.
■ SU_1 accepts all proposals of SU_2 (after SU_2 rejects SU_1's opening proposal).

The *outcome* of this game is as follows:

■ SU_1 proposes $(1-\delta_2, \delta_2)$.
■ SU_2 accepts this offer.

From the SPE of this game, we can see that the two-period bargaining ends in the first period, with the utility profile $(1-\delta_2, \delta_2)$. Next, we will consider the general case where $T < \infty$. Because $T < \infty$, BI can still be applied for investigating the SPE of the game.

3.3.2 T-Period Bargaining Game

In the T-period bargaining game, SU_i's future utility is still discounted using the constant discount factor $\delta_i \in (0,1)$ in each period. We assume that SU_1 proposes first, and SU_2 makes the proposal in period T; that is, T is even. The case where T is odd can be analyzed similarly.

In period T, SU_2 will still make the proposal $(0,1)$. Therefore, in the penultimate period $T-1$, SU_1 will propose $(1-\delta_2, \delta_2)$. Going backward, in period $T-2$, SU_2 will propose $(\delta_1(1-\delta_2), 1-\delta_1(1-\delta_2))$.

The logic can be repeated until the first period, where SU_1 will propose (x_1^*, x_2^*) with $x_1^* + x_2^* = 1$, where

$$x_1^* = 1 - \delta_2\left(1 - \delta_1(1 - \delta_2(\cdots))\right).$$

According to the aforementioned equilibrium strategies of SU_1 and SU_2, the outcome of the T-period bargaining game is that SU_1 gets x_1^*, and SU_2 gets $x_2^* = 1 - x_1^*$. Next, let us calculate x_1^* as follows:

$$
\begin{aligned}
x_1^* &= 1 - \delta_2\left(1 - \delta_1(1 - \delta_2(\cdots))\right) \\
&= 1 - \delta_2 + \delta_1\delta_2 - \delta_1\delta_2^2 + \ldots + \delta_1^{T/2-1}\delta_2^{T/2-1} - \delta_1^{T/2-1}\delta_2^{T/2} \\
&= \sum_{t=0}^{T/2-1}(1-\delta_2)(\delta_1\delta_2)^t \qquad\qquad\qquad (3.5) \\
&= \frac{(1-\delta_2)[1-(\delta_1\delta_2)^{T/2}]}{1-\delta_1\delta_2}.
\end{aligned}
$$

SU_2 will get

$$x_2^* = 1 - x_1^* = \frac{\delta_2(1-\delta_1) - (1-\delta_2)(\delta_1\delta_2)^{T/2}}{1-\delta_1\delta_2}. \qquad (3.6)$$

In this section, we have investigated how SUs bargain with each other in a finite number of periods. Next, we extend our analysis to a more complicated scenario, that is, infinite-horizon bargaining, where two SUs do not know exactly when the bargaining will end.

3.4 Infinite-Horizon Bargaining Game

In this section, we extend our analysis to the infinite-horizon bargaining game, where both bargainers are not clear about the exact ending time of bargaining. Figure 3.3 illustrates the structure of this bargaining game. We still assume that SUs get zero utility if no agreement is ever reached. For the infinite-horizon game where T is infinity, the SPE analysis is much more complicated because the BI method cannot be applied as in Section 3.3. Instead, we will *conjecture* a strategy profile and then verify that this strategy profile can constitute an SPE by using the *one-stage deviation principle*.

3.4.1 One-Stage Deviation Principle

The one-stage deviation principle is a shortcut method to verify whether a strategy profile of a finite- or infinite-horizon game is an SPE. In this section, we use this principle to analyze the infinite-horizon bargaining game. First, we introduce the concept of the *history* of a dynamic multistage game, which will help the reader to better understand the one-stage deviation principle.

Definition 2 For period $t = 1,2,\ldots,\infty$, a t-period history h^t is a record of chosen actions from the beginning of the game up to period (stage) t, and is defined as

$$h^t = \left(a^1,\ldots,a^t\right) = \left((a_1^1,\ldots,a_n^1),\ldots,(a_1^t,\ldots,a_n^t)\right),$$

where a_i^t is player i's action in period t. $a^t = (a_1^t,\ldots,a_n^t)$ is the profile of all players' actions in period t.

From this definition, we can see that a t-period history is a list of t profiles of actions the players have already chosen, one for each period. Because each profile is a list of n actions, the total length of a t-period history is nt. It is obvious that this list of actions must be consistent with the game tree in the sense that each a_i^t must be an available action for i in period t after history $h^{t-1} = (a^1,\ldots,a^{t-1})$.

Next, we introduce another useful definition, that is, *continuity at infinity*, which is the sufficient condition for the one-stage deviation principle.

Definition 3 Consider an infinite-horizon multistage game with observed actions, denoted by Γ^∞. Let h^∞ denote an ∞-horizon history, that is, $h^\infty = (a^1,a^2,\ldots)$, an infinite sequence of actions. Let $h^t = (a^1,\ldots,a^t)$ be the restriction to first t periods. The game Γ^∞ is continuous at infinity if, for all players i, the utility function u_i satisfies

$$\lim_{t\to\infty}\ \sup_{h^\infty,\ \tilde{h}^\infty \in H^\infty(h^t) \text{ for some } h^t} \left|u_i(h^\infty) - u_i(\tilde{h}^\infty)\right| = 0,$$

where $H^\infty(h^t)$ is the set of infinite histories that follow h^t.

If a game is continuous at infinity, it indicates that if the interactions in the first t periods in both histories are the same for sufficiently large t, then player i's utility from the two histories are almost equal. It means that each player places a *negligible weight* on what will take place in periods far enough out in the future.

With this definition of continuity at infinity, we have the following lemma.

Lemma 1 The continuity at infinity condition is satisfied when the overall utilities are a discounted sum of utility in each stage, that is,

$$u_i = \sum_{t=1}^{\infty} \delta_i^{t-1} g_i^t(a^t),$$

where $g_i^t(a^t)$ is player i's single-stage utility in period t, δ_i is player i's discount factor, and the stage utilities are uniformly bounded; that is, there exists some B such that

$$\max_{t,a^t} |\, g_i^t(a^t)\,| < B.$$

Proof 1 (Sketch of Proof): *Let Γ^∞ denote an infinite-horizon multistage game with observed actions. Take any player $i \in N$, and let*

$$\max\left\{| g_i(a) - g_i(a') \| a, a' \in A \right\},$$

where A is the set of available actions for player i in a single stage, denote the difference between i's maximum utility and minimum utility in the stage game. From the condition in the lemma, we have

$$| g_i(a) - g_i(a') | \leq | g_i(a) | + | g_i(a') | < 2B.$$

Take any $t \in N$ and any history $h^t \in H^t$. Take any pair of infinite histories $h^\infty, \tilde{h}^\infty \in H^\infty(h^t)$ such that the initial segments of length nt of h^∞ and $\tilde{h}^\infty$ are both h^t. That is, if $h^t = (a^1, ..., a^t)$, we can write

$$h^\infty = (a^1, ..., a^t, a^{t+1}, a^{t+2}, ...)$$

and

$$\tilde{h}^\infty = (a^1, ..., a^t, \tilde{a}^{t+1}, \tilde{a}^{t+2}, ...).$$

Without loss of generality, we assume that $u_i(h^\infty) \geq u_i(\tilde{h}^\infty)$. Then,

$$\left| u_i(h^\infty) - u_i(\tilde{h}^\infty) \right| = u_i(h^\infty) - u_i(\tilde{h}^\infty)$$

$$= \left[\sum_{\tau=1}^{t} \delta_i^{\tau-1} g_i^\tau(a^\tau) + \sum_{\tau=t+1}^{\infty} \delta_i^{\tau-1} g_i^\tau(a^\tau) \right] - \left[\sum_{\tau=1}^{t} \delta_i^{\tau-1} g_i^\tau(a^\tau) + \sum_{\tau=t+1}^{\infty} \delta_i^{\tau-1} g_i^\tau(\tilde{a}^\tau) \right]$$

$$= \left(\sum_{\tau=t+1}^{\infty} \delta_i^{\tau-1} g_i^\tau(a^\tau) - \sum_{\tau=t+1}^{\infty} \delta_i^{\tau-1} g_i^\tau(\tilde{a}^\tau) \right)$$

$$\leq \sum_{\tau=t+1}^{\infty} \delta_i^{\tau-1} | g_i^\tau(a^\tau) - g_i^\tau(\tilde{a}^\tau) |$$

$$\leq 2B \sum_{\tau=t+1}^{\infty} \delta_i^{\tau-1}.$$

Because this holds for any h^∞ and $\tilde{h}^\infty$, we have

$$\sup | u_i(h^\infty) - u_i(\tilde{h}^\infty) | \leq 2B \sum_{\tau=t+1}^{\infty} \delta_i^{\tau-1}$$

Because $\delta_i \in (0,1)$, $2B \sum_{\tau=t+1}^{\infty} \delta_i^{\tau-1} \to 0$ as $t \to \infty$, we have

$$\lim_{t \to \infty} \sup_{h^{\infty}, \tilde{h}^{\infty} \in H^{\infty}(h^t) \text{ for some } h^t} \left| u_i(h^{\infty}) - u_i(\tilde{h}^{\infty}) \right| = 0.$$

This completes the proof of this lemma.

For the infinite-horizon bargaining game with a discount factor, we have assumed that SUs will get zero utility; that is, no media data are transmitted if no agreement is ever reached. Furthermore, if no agreement is reached in one period (and then bargaining moves into the next period), the SUs will get zero utility. Thus, for period t and SU_i, $g_i^t(a^t)$ can be zero if no proposal is accepted by both SUs, or some positive $x_i \in [0,1]$. It indicates that the period/stage utilities for both SUs are uniformly bounded. Moreover, we consider the time factor and incorporate a discount factor into the utility function. Therefore, Lemma 1 shows that the continuity at infinity condition applies in the infinite-horizon bargaining game.

With Lemma 1, we have the following theorem about the one-stage deviation principle for an infinite-horizon game.

Theorem 1 Consider an infinite-horizon game with observed actions Γ^{∞}, which are continuous at infinity. Then, the one-stage deviation principle holds; that is, the strategy profile s^* is an SPE if and only if for all i, h^t, and t, we have

$$u_i\left(s_i^*, s_{-i}^* \mid h^t\right) \geq u_i\left(s_i, s_{-i}^* \mid h^t\right),$$

for all s_i that satisfies $s_i(h^t) \neq s_i^*(h^t)$ and $s_{i|h^t}(h^{t+k}) = s_{i|h^t}^*(h^{t+k})$ for all $h^{t+k} \in \Gamma^{\infty}(h^t)$ and for all $k > 0$.

The proofs of Theorem 1 can be found in Ref. [18]. If one game satisfies the one-stage deviation principle, it is relatively easy to verify whether a strategy profile is an SPE or not by checking if there is any history h where some player i can increase its utility by deviating from $s_i^*(h)$ *only once* at h and conforming to s_i^* thereafter, given that all other players choose actions specified in s^* at history h and thereafter. Theorem 1 shows that we can use the one-stage deviation principle for investigating the SPE of the bargaining game with an infinite horizon.

3.4.2 SPE Analysis

The strategy of each SU in this infinite-horizon game includes

- Offer in period n
- Optimal response to its opponent's counteroffer in period $n+1$
- Counteroffer to its opponent in period $n+2$

Note that for each SU_i, all subgames that begin with SU_i's making an offer and SU_i's response to its opponent's offer are *structurally equivalent*.[1] Therefore, we consider a *stationary* strategy

---

[1] Informally, any pair of subgames are structurally equivalent if the sets of the available action sequences are the same, and the preferences of each player over the full plays of the game are the same in the two subgames. A detailed discussion on this topic is beyond the scope of this chapter. See Ref. [18] for more information.

profile where each SU always makes the same proposal, and the corresponding response to the other SU's offer depends only on the current proposal.

Define the following parameters:

$$x_1^* = \frac{1-\delta_2}{1-\delta_1\delta_2}, \quad x_2^* = 1 - x_1^* = \frac{\delta_2(1-\delta_1)}{1-\delta_1\delta_2}$$

and

$$y_1^* = \frac{\delta_1(1-\delta_2)}{1-\delta_1\delta_2}, \quad y_2^* = 1 - y_1^* = \frac{1-\delta_1}{1-\delta_1\delta_2}.$$

The following theorem summarizes the SPE of the infinite-horizon bargaining game.

Theorem 2 For the infinite-horizon bargaining game, the following strategy profile $s^* = (s_1^*, s_2^*)$

■ SU_1 proposes $x^* = (x_1^*, x_2^*)$ and accepts y_1 if and only if $y_1 \geq y_1^*$.

■ SU_2 proposes $y^* = (y_1^*, y_2^*)$ and accepts x_2 if and only if $x_2 \geq x_2^*$ constitutes an SPE.

Proof 2 (Proof): *We use the one-stage deviation principle to verify this theorem. Note that there are two types of subgame structures in this infinite-horizon game. The first type is the one in which the first action is an offer provided by SU_i. The second type is the one in which the first action is a response from SU_i to the offer provided by its opponent. We will verify the strategy profile in Theorem 2 in each type of subgame.*

First, let us consider the scenario where $i = 1$. For the first type of subgame, suppose that this offer is made by SU_1. Fix SU_2's strategy as s_2^*. If SU_1 takes actions based on s_1^*, then SU_2 will accept offer, and thus SU_1 will get $x_1^* = \dfrac{1-\delta_2}{1-\delta_1\delta_2}$. Next, we will consider the case where SU_1 deviates from s_1^*: (1) if SU_1 offers $x_2 > x_2^*$, then SU_2 will accept it, and thus SU_1 gets a lower utility than x_1^*; (2) if SU_1 offers $x_2 < x_2^*$, then SU_2 will reject it and propose y_1^* to SU_1. SU_1 will accept it (due to the one-stage deviation requirement) and get a discounted utility $\delta_1 y_1^*$. Because $\delta_1 y_1^* < x_1^*$, SU_1 will be worse off by deviating from s_1^*. For the second type of subgame, suppose that SU_1 responds to SU_2's offer. Fix SU_2's strategy as s_2^*; that is, SU_2 offers y_1^*. SU_1 will accept it and get y_1^* if conforms to s_1^*. If not, SU_1 will reject the offer even though $y_1 \geq y_1^*$ and propose a counteroffer $x^* = (x_1^*, x_2^*)$ (owing to the one-stage deviation requirement), and SU_2 will accept it. Thus, SU_1 will get a discounted utility $\delta_1 x_1^* = y_1^*$, which means that SU_1 cannot gain more by deviating from s_1^*.

Next, let us consider the case where $i = 2$. Similarly, for the first type of subgame, we assume that this offer is provided by SU_2. Fix SU_1's strategy as s_1^*. If SU_2 takes actions based on s_2^*, then SU_1 will accept the offer, and thus SU_2 will get $y_2^* = \dfrac{1-\delta_1}{1-\delta_1\delta_2}$. Next, we will consider the case where SU_2 deviates from s_2^*: (1) if SU_2 offers $y_2 < y_2^*$, then SU_1 will accept it, and thus SU_2 gets lower utility than y_2^*; (2) if SU_2 offers $y_2 > y_2^*$, then SU_1 will reject it and propose x_2^* to SU_2. SU_2 will *accept it (owing to the one-stage deviation requirement) and get a discounted utility $\delta_2 x_2^*$. Because $\delta_2 x_2^* < y_2^*$, SU_2 will be worse off by deviating from s_2^*. For the second type of subgame, suppose that SU_2 responds to SU_1's offer. Fix SU_1's strategy as s_1^*; that is, SU_1 offers x_2^*. SU_2 will accept and get x_2^* if conforming to s_2^*. If not, SU_2 will reject the offer and propose a counteroffer $y' = (y_1^*, y_2^*)$ (owing to the one-stage deviation requirement) in the next period, and SU_1 will accept y_1^*. Thus, SU_2 will get a discounted utility $\delta_2 y_2^* = x_2^*$, which means that SU_2 cannot benefit more by deviating from s_2^*.*

This completes the proof of Theorem 2.

From the SPE described in Theorem 2, we get the *outcome* of this infinite-horizon bargaining game as

$$
SU_1 \text{ proposes} \left(\frac{1-\delta_2}{1-\delta_1\delta_2}, \frac{\delta_2(1-\delta_1)}{1-\delta_1\delta_2} \right).
$$

■ SU_2 accepts this offer.

The final utilities that the SUs obtain are

$$
x_1^* = \frac{1-\delta_2}{1-\delta_1\delta_2}, \quad x_2^* = \frac{\delta_2(1-\delta_1)}{1-\delta_1\delta_2}. \tag{3.7}
$$

In the next section, we will provide several interesting observations and insights on the equilibria obtained in Sections 3.3 and 3.4.

3.5 Bargaining Equilibrium Summary and Numerical Results

In this section, we evaluate the bargaining equilibrium outcomes (SPEs) derived in Sections 3.3 and 3.4 from theoretical analysis and some numerical examples. Several interesting observations are as follows.

Observation 1 *We consider intranetwork spectrum sharing in cognitive radio networks, where multiple SUs equally share spectrum resource obtained from the PUs in order to transmit multimedia data. The SUs have the same spectrum sharing access rights because they are homogeneous in the same secondary network as that shown in Figure 3.1. Although there is a conflict of interest between SUs, no agreement would be imposed on any SU without its approval. Any bargaining participant has the rights to provide a proposal as well as to respond to any offer proposed by its opponent. Any agreement is made based on each SU's utility maximization decision making. Therefore, this proposed explicit bargaining process can provide fair spectrum sharing opportunities for SUs.*

Besides, we assert that this sharing scheme is incentive compatible. Generally, an allocation is called incentive compatible or self-enforcing if there is no incentive for an individual system to deviate from it. We use the framework of Nash equilibrium (more specifically, an SPE) to analyze the spectrum sharing results. From the definition of an SPE, we can see that if any player (SU) unilaterally deviates from the equilibrium strategy set, its benefit will decrease. From this, we can see that the sharing mechanism is compatible with the SUs' respective incentives.

Observation 2 *For both bargaining scenarios analyzed in Sections 3.3 and 3.4, we fully characterize the corresponding SPE strategies. Interestingly, each bargaining scenario ends in the first period when the SUs adopt the equilibrium strategies, although the game structure we design involves multiple period bargaining opportunities for the SUs (finite or infinite). The fact that the agreement can be reached at the very beginning of spectrum sharing indicates that there is no delay and thus no inefficiency due to delay in the SUs' bargaining process.*

In the situation where all previous SUs' proposals can be perfectly observed and all information about SUs' preference (e.g., discount factor related to its battery energy information) is known to both SUs, each SU can calculate its opponent's judgment about the offer, and thus find an offer that is acceptable for its opponent but optimal for itself. Furthermore, bargaining impatience plays an important role: the desirability of an earlier agreement yields a time-efficient bargaining result, which saves bargaining time (and thus energy) in the spectrum sharing process so as to enable the SUs to utilize spectrum.

Observation 3 *The bargaining patience (and thus the SUs' energy status) plays an important role in its benefit gains from the bargaining. Figure 3.4 illustrates how each SU's benefit (data rate) changes with the SUs' different discount factors (battery status or energy cost) in the infinite-horizon game.*

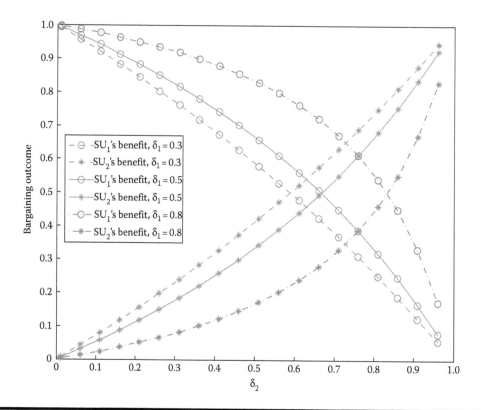

Figure 3.4 Benefits of infinite-horizon bargaining game with different discount factors.

The horizontal ordinate denotes different δ_2 values from 0 to 1. We choose three different values of δ_1: 0.3, 0.5, and 0.8. From Figure 3.4, we can see that

1. *For the same game parameter setting, SU_1's and SU_2's benefits are symmetrical about the straight line $x = \dfrac{1}{2}$. Because SU1 and SU2 divide a normalized piece of spectrum resource, the symmetry means that their interests conflict and their benefits change in opposite directions.*

2. *SU_1's (SU_2's) benefit increases as δ_1 (δ_2) increases and δ_2 (δ_1) decreases. This observation intuitively indicates that the more battery life (i.e., more energy) SU_i has, the higher the data rate it will achieve during the spectrum sharing. That is, the SU with the longer battery life has more bargaining power.*

3. *For different pairs of (δ_1, δ_2), the spectrum splitting result is not always $\left(\dfrac{1}{2}, \dfrac{1}{2}\right)$. The allocation $\left(\dfrac{1}{2}, \dfrac{1}{2}\right)$ seems attractive, however, if we consider each SU's specific situations, and the optimal allocation result depends on each SU's bargaining power, which is dependent on its energy status.*

Observation 4 *Compare the equilibrium outcomes in Equations 3.5 and 3.6 with the one in Equation 3.7. In the general finite-horizon bargaining game where $T < \infty$, when T approaches ∞, SU_1 will get*

$$\lim_{T \to \infty} \frac{(1-\delta_2)[1-(\delta_1\delta_2)^{T/2}]}{1-\delta_1\delta_2} = \frac{1-\delta_2}{1-\delta_1\delta_2}. \tag{3.8}$$

SU_2 will get

$$x_2^* = 1 - x_1^* = \frac{\delta_2(1-\delta_1)}{1-\delta_1\delta_2}. \tag{3.9}$$

Therefore, when T is approaching ∞, the bargaining outcome is the same as the one of the infinite-horizon bargaining game, although the analysis for the two types of game is quite different.

Figure 3.5 confirms the foregoing observation. In Figure 3.5, we fix δ_1 at 0.5 and choose three different values of δ_2: 0.3, 0.5, and 0.8. As the period number T increases, the benefit gap between the finite- and infinite-horizon game reduces. Besides, we can see that in the finite game model, the conclusion regarding bargaining patience in Observation 3 still applies; that is, the less patient a player's opponent is, the better bargaining result this player will obtain.

Observation 5 *From the final outcome of the infinite-horizon bargaining game as in Equation 3.7, we can obtain the bargaining outcome if SU_2 makes the first offer. From parameter symmetry, the utilities for SU_1 and SU_2 are*

$$x_1^* = \frac{\delta_1(1-\delta_2)}{1-\delta_1\delta_2}, \quad x_2^* = \frac{1-\delta_1}{1-\delta_1\delta_2}, \tag{3.10}$$

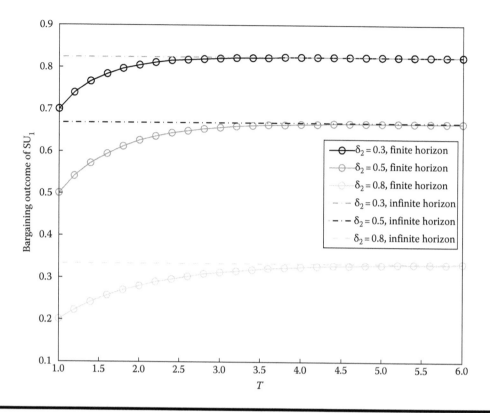

Figure 3.5   Comparison between infinite- and finite-horizon bargaining game.

which indicates that if SU_2 moves first, it will get

$$\frac{1-\delta_1}{1-\delta_1\delta_2} > \frac{\delta_2(1-\delta_1)}{1-\delta_1\delta_2}.$$

This means that at any point of the spectrum-sharing game, the proposer is in a better bargaining position than the responder because the proposer takes the advantage of the responder's bargaining impatience. Hence, being the first proposer must increase its benefit. We call this the first mover advantage (FMA). Correspondingly, in Figure 3.1, the SU who first makes proposals will dominate the spectrum-sharing bargaining.

The benefit difference Δ for SU_2 due to FMA is

$$\Delta = \frac{1-\delta_1}{1-\delta_1\delta_2} - \frac{\delta_2(1-\delta_1)}{1-\delta_1\delta_2} = \frac{(1-\delta_1)(1-\delta_2)}{1-\delta_1\delta_2}. \tag{3.11}$$

Figure 3.6 describes how the benefit difference evolves with different discount factors. We set different δ_1 values of 0.3, 0.5, and 0.8. For each δ_1, the curve is downward when δ_2 increases. When $\delta_2 = 1$, the difference is zero; that is, there is no FMA. This means that if SU_2 is not the first proposer of the

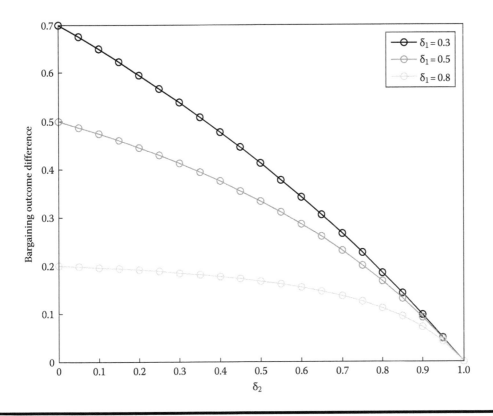

Figure 3.6 First mover advantage.

bargaining, its lower bargaining position will be improved when it becomes more patient. Meanwhile, for a given δ_2, *the benefit difference of* SU_2 *reduces when* SU_1 *is more patient. Therefore, the FMA will be gradually eliminated when both SUs have more bargaining patience (that is, more battery energy for the bargaining).*

3.6 Conclusion

This chapter investigates an intranetwork spectrum-sharing mechanism achieved by a bilateral bargaining between two SUs. The more general spectrum sharing with multiple SUs can be decomposed into several pairwise one-to-one bilateral bargaining games studied in this chapter. We discuss the proposed bargaining process under two different system scenarios, that is, finite- and infinite-horizon scenarios. By modeling such an explicit dynamic bargaining process as a multistage game with observed actions, we are able to characterize the corresponding subgame perfect equilibria under different system scenarios. Furthermore, we evaluate the properties of the equilibrium outcomes and verify that such a bargaining mechanism can fairly, effectively, and incentive compatibly achieve spectrum sharing between SUs by taking their individual network characteristics into account, which enables SUs to transmit data for their broadband multimedia services.

Acknowledgments

This work was supported in part by the National Basic Research Program of China (2012CB316002), the National Science and Technology Major Project (2011ZX03004 – 004), the National Natural Science Foundation of China (61132002), Tsinghua Research Funding *No.*2010*THZ*02–3, and the International Cooperation Program (2012DFG12010).

References

1. S. Haykin, Cognitive radio: Brain-empowered wireless communications, *IEEE Journal on Selected Areas in Communications*, vol. 23, no. 2, pp. 201–220, 2005.
2. I. F. Akyildiz, W.-Y. Lee, M. C. Vuran, and S. Mohanty, A survey on spectrum management in cognitive radio networks, *IEEE Communications Magazine*, vol. 46, no. 4, pp. 40–48, 2008.
3. L. Cao and H. Zheng, Distributed spectrum allocation via local bargaining, *IEEE SECON*, pp. 475–486, September 2005.
4. T. Jiang, D. Grace, and Y. Liu, Performance of cognitive radio reinforcement spectrum sharing using different weighting factors, *IEEE DySPAN*, pp. 1195–1199, August 2008.
5. H. Zheng and L. Cao, Device-centric spectrum management, *IEEE DySPAN*, pp. 56C65, November 2005.
6. J. Nash, The bargaining problem, *Econometrica*, vol. 18, pp. 155–162, 1950.
7. J. Sutton, Non-cooperative bargaining theory: An introduction, *The Review of Economic Studies*, vol. 53, no. 5, pp. 709–724, 1986.
8. Y. Yan, J. Huang, X. Zhong, and J. Wang, Dynamic spectrum negotiation with asymmetric information, *GameNets*, April 2011.
9. Y. Yan, J. Huang, X. Zhong, and J. Wang, Dynamic Bayesian spectrum bargaining with non-myopic users, *International ICST Conference on Wireless Internet*, Xi'An, China, October 2011.
10. Y. Yan, J. Huang, X. Zhong, M. Zhao, and J. Wang, Sequential bargaining in cooperative spectrum sharing: Incomplete information with reputation effect, *IEEE Globecom*, December 2011.
11. Y. Yan, J. Huang, and J. Wang, Dynamic bargaining for relay-based cooperative spectrum sharing, *IEEE Journal on Selected Areas in Communications*, vol. 31, no. 8, pp. 1480–1493, 2013.
12. T. Weiss and F. Jondral, Spectrum pooling: An innovative strategy for the enhancement of spectrum efficiency, *IEEE Communications Magazine*, vol. 42, no. 3, pp. S8C14, 2004.
13. C. Antoni, Bargaining power in communication networks, *Mathematical Social Sciences*, vol. 41, pp. 69–87, 2001.
14. C. Antoni, On bargaining partner selection when communication is restricted, *International Journal of Game Theory*, vol. 30, no. 4, pp. 503–515, 2001.
15. IEEE 802.22 Working Group on Wireless Regional Area Networks, IEEE P802.22/D0.1 Draft Standard for Wireless Regional Area Networks Part 22: Cognitive Wireless RAN Medium Access Control (MAC) and Physical Layer (PHY) specifications: Policies and procedures for operation in the TV Bands, *IEEE doc*: 22-06-0067-00-0000-*P*80222-*D*0.1, May 2006.
16. A. B. MacKenzie and L. A. DaSilva, *Game Theory for Wireless Engineers*, Morgan & Claypool, San Rafael, California, 2006.
17. A. Rubinstein and M. Osborne, *Bargaining and Markets*, Emerald Group Publishing, UK, 1990.
18. D. Fudenberg and J. Tirole, *Game Theory*, The MIT Press, Cambridge, MA, 1991.
19. I. Stahl, Bargaining Theory, *The Swedish Journal of Economics,* vol. 75, pp. 410–413, 1973.
20. A. Rubinstein, Perfect equilibrium in a bargaining model, *Econometrica*, vol. 50, pp. 97–109, 1982.
21. V. Krishna and R. Serrano, Multilateral bargaining, *Review of Economic Studies*, vol. 63, pp. 61–80, 1996.
22. H. Haller, Non-cooperative bargaining of $n \geq 3$ players, *Economics Letters*, vol. 22, pp. 11–13, 1986.

23. P. Torstensson, An n-person Rubinstein bargaining game, *International Game Theory Review*, vol. 11, no. 1, pp. 111–115, 2009.
24. K. G. Binmore, Bargaining and coalitions, *Game-Theoretic Models of Bargaining*, Alvin E. Roth (Ed.), Cambridge University Press, Cambridge, UK, pp. 269–304, 1986.
25. S. Hart, Axiomatic approaches to coalitional bargaining, *Game-Theoretic Models of Bargaining*, Alvin E. Roth (Ed.), Cambridge University Press, Cambridge, UK, pp. 305–320, 1986.

Chapter 4

Physical Layer Mobility Challenges for Multimedia Streaming QoS over Cognitive Radio Networks

Chihkai Chen and Kung Yao

Electrical Engineering Department, UCLA, Los Angeles, CA, USA

Contents

4.1 Introduction

The popularity of mobile wireless devices and smart devices in people's daily lives has resulted in increased demand for wireless spectrum. Cognitive radio network (CRN) technology provides a dynamic spectrum utilization paradigm to agilely exploit precious wireless spectrum resources [1]. Multimedia entertainment requires much spectrum to satisfy the quality of service (QoS) for streaming applications. CRN technology can facilitate access to unoccupied frequency bands, called white space or spectrum holes, and thereby exploit them for spectrum demands.

The fundamental task of a CR in a CRN is to detect the presence or absence of licensed users, also known as primary users (PUs). This is accomplished by sensing the radio frequency environment, a process called spectrum sensing [2]. The main objectives of spectrum sensing are not to cause harmful interference to PUs and also to efficiently identify and exploit the spectrum holes for the required throughput and QoS. Owing to random fading in wireless links, spectrum sensing has overcome the hidden PU issue. Distributed cooperative spectrum sensing can effectively resolve this issue. However, the sensing performance is affected not only by fading channels but also by many other factors, including noise power uncertainty, interference, collaboration trustworthiness issue, PU traffic, mobility, and quantization [3–6]. In addition, stochastic ordering theory has shown that randomization of the number of users degrades the system throughput. Moreover, the behavior of the CR users, and hence the dynamics of CRs, also influences the cooperative spectrum sensing performance.

In this chapter, we study the mobility of CRs by employing a semi-Markov stochastic process with random arrivals and random departures. We consider CRN systems in densely populated urban areas, where wireless devices experience shadowed fading wireless links. Because an energy-based spectrum sensing scheme is most widely used, we focus on analyzing the performance of spatially distributed, energy-based spectrum sensing over shadowed fading channels.

4.2 CR User Mobility

People use their devices at will and move around according to their needs and schedules. Hence, the CRN formed by the CR functionality in these devices is dynamically changing because of the CR user mobility.

4.2.1 CR Categorization

Some CR users tend to stay longer within the CRN coverage area. We call these Type A CR users, and the corresponding CRs are called Type A CRs. We can model the random amount of time Type A CRs stay in the system as a long-tailed Weibull distribution with two parameters, $\mu > 0$ and $k > 0$. The associated probability density function is given by

$$f(x) = \begin{cases} kx^{k-1}\lambda^{-k}\exp\left\{\left(-x/\lambda\right)^k\right\}, & x \geq 0 \\ 0, & x < 0 \end{cases} \tag{4.1}$$

However, Type B CRs are those carried by users who enter the CRN coverage area for a certain nonzero x_m amount of time and then leave the area quickly. The random amount of time Type B CRs stay in the system can be modeled by the Pareto distribution with two parameters $x_m > 0$ and $\alpha > 0$, which are given by

$$f(x) = \begin{cases} \alpha x_m^\alpha x^{-(\alpha+1)}, & x \geq x_m \\ 0, & x < x_m \end{cases} \tag{4.2}$$

4.2.2 Hidden Markov Model

Based on rush hours and normal hours, CR users join the system with different arrival rates. The arrival dynamics of Type A and Type B CRs can be modeled by a four-state hidden Markov model (HMM). An HMM is a directed graph or digraph, $G = \{V, E\}$. V is the set of all five nodes, S_1, S_2, S_3, S_4, and O. The first four nodes are the four hidden states in HMM, and the last node represents the observation node (see Figure 4.1). E contains all directed edges (S_i, S_j) in G. The directed edges are ordered pairs of vertices. A directed edge (S_i, S_j) indicates the edge that is directed from S_i to S_j. The associated transition probability $P(S_i|S_j)$ is denoted by p_{ij}. Therefore, the transition probability matrix is given by

$$P = \begin{bmatrix} p_{11} & p_{12} & p_{13} & p_{14} \\ p_{21} & p_{22} & p_{23} & p_{24} \\ p_{31} & p_{32} & p_{33} & p_{34} \\ p_{41} & p_{42} & p_{43} & p_{44} \end{bmatrix}. \tag{4.3}$$

Note that $p_{i1} + p_{i2} + p_{i3} + p_{i4} = 1$ because of the total probability property. State S_1 denotes that both Type A and Type B CRs are in the low-arrival-rate period. State S_2 denotes that Type A CRs arrive in the low-rate periods and Type B CRs join the system at a high rate. State S_3 denotes that Type A CRs' arrivals are in the high-rate period and Type B CRs' arrivals are in the low-rate period. State S_4 occurs when both Type A and Type B CRs are in the high-rate periods. Assume that Type A CRs join the network with the low-rate $\lambda_{A,1}$ and the high-rate $\lambda_{A,2}$; for Type B CRs, they arrive at a low-rate $\lambda_{B,1}$ and a high-rate $\lambda_{B,2}$.

4.2.3 Dynamic Probability Mass Function

Because of the mobility of CR users, the number of CRs in the system dynamically changes. This dynamic number $N(t)$ is indeed a semi-Markov stochastic process over an infinitely countable state space $S = \{0, 1, 2, ...\}$ as $\{N(t) = n \mid t \in R, n \in \{0, 1, 2, ...\}\}$ [7]. The arrival random

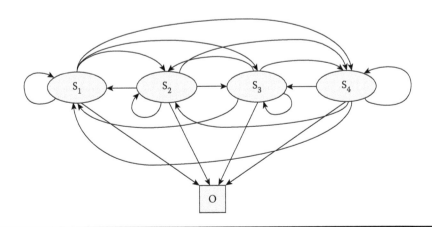

Figure 4.1   Four-state hidden Markov model.

process $\{A_i \mid i = 1, 2, \ldots\}$ is modeled by a composite Poisson arrival process. Under State S_1, the arrival process is the superposition of a Type A CR Poisson arrival process with rate $\lambda_{A,1}$ and a Type B CR Poisson arrival process with rate $\lambda_{B,1}$. Under State S_2, the arrival process is the superposition of a Type A CR Poisson arrival process with rate $\lambda_{A,1}$ and Type B CR Poisson arrival process with rate $\lambda_{B,2}$. Under State S_3, the arrival process is the superposition of a Type A CR Poisson arrival process with rate $\lambda_{A,2}$ and a Type B CR Poisson arrival process with rate $\lambda_{B,1}$. Under State S_4, the arrival process is the superposition of a Type A CR Poisson arrival process with rate $\lambda_{A,2}$ and a Type B CR Poisson arrival process with rate $\lambda_{B,2}$. The amount of time the ith CR stays in the system is T_i. T_i follows the Weibull distribution if the ith CR is a Type A CR; T_i follows the Pareto distribution if the ith CR belongs to Type B. Note that the $(i+1)$-th CR might leave the system before the ith CR. The associated departure random process $\{D_i \mid i = 1, 2, \ldots\} = \{A_i + T_i \mid i = 1, 2, \ldots\}$. Therefore, we can derive the time-varying probability mass function of the dynamic number of CRs in the system to capture the dynamics of this nonstationary semi-Markov stochastic process $\{N(t)\}$. Figure 4.2 shows one possible realization of $\{N(t)\}$.

To simplify the four-state HMM, we can form a supernode H_1 by merging S_1 and S_2. Similarly, we can merge the other two nodes, S_3 and S_4, to form a supernode H_2. The resulting graph can be viewed as an embedded graph G_1 in the original four-state HMM graph G as shown in Figure 4.3.

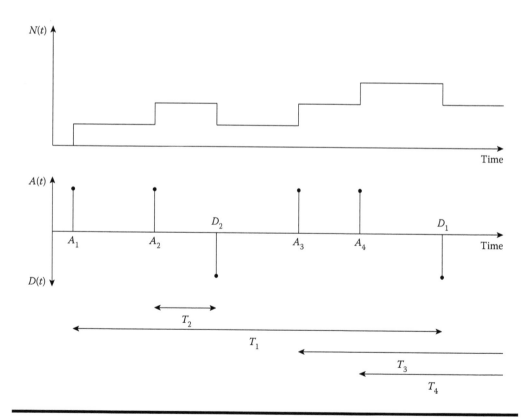

Figure 4.2 **Semi-Markov process $\{N(t)\}$, arrival/departure processes $\{A(t)\}/\{D(t)\}$, and cooperation time $\{T_i\}$.**

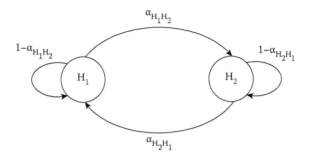

Figure 4.3 Embedded hidden Markov model for supernodes.

This is equivalent to clustering the original entries in the 4-by-4 transition probability matrix into four 2-by-2 submatrices as follows:

$$P = \begin{array}{c} \\ H_1 \\ \\ H_2 \\ \end{array} \overset{\begin{array}{cc} H_1 & \qquad H_2 \end{array}}{\left[\begin{array}{cccc} p_{11} & p_{12} & \vdots & p_{13} & p_{14} \\ p_{21} & p_{22} & \vdots & p_{23} & p_{24} \\ \cdots & \cdots & \cdot & \cdots & \cdots \\ p_{31} & p_{32} & \vdots & p_{33} & p_{34} \\ p_{41} & p_{42} & \vdots & p_{43} & p_{44} \end{array} \right]} = \left[\begin{array}{cc} \alpha_{H_1H_1} & \alpha_{H_1H_2} \\ \alpha_{H_2H_1} & \alpha_{H_2H_2} \end{array} \right]. \tag{4.4}$$

The top-left corner of a 2-by-2 submatrix represents the internal transition of supernode H_1 in the embedded graph G_1, and the top-right corner of a 2-by-2 submatrix represents the transition from supernode H_1 to the other supernode H_2. Similarly, the bottom left 2-by-2 submatrix represents the transition from supernode H_2 to supernode H_1, and the bottom-right 2-by-2 submatrix represents the internal transition from supernode H_2 back to itself. Therefore, we can obtain the transition probability matrix for the embedded graph G_1 as follows:

$$\alpha_{H_1H_2} = \frac{\left(p_{13} + p_{14} + p_{23} + p_{24} \right)}{\left(p_{11} + p_{12} + p_{21} + p_{22} \right) + \left(p_{13} + p_{14} + p_{23} + p_{24} \right)}$$

$$= \frac{1}{2} \left(p_{13} + p_{14} + p_{23} + p_{24} \right), \tag{4.5}$$

$$\alpha_{H_1H_1} = 1 - \alpha_{H_1H_2} = \frac{1}{2} \left(p_{11} + p_{12} + p_{21} + p_{22} \right). \tag{4.6}$$

Similarly,

$$\alpha_{H_2H_1} = \frac{1}{2}\left(p_{31} + p_{32} + p_{41} + p_{42}\right), \tag{4.7}$$

$$\alpha_{H_2H_2} = 1 - \alpha_{H_2H_1} = \frac{1}{2}\left(p_{33} + p_{34} + p_{43} + p_{44}\right). \tag{4.8}$$

Supernode H_1 represents the state of Type A CRs in the low-arrival period, and supernode H_2 is the state where Type A CRs stay in the high-arrival period. Therefore, we can obtain the dynamic probability mass function of the number of Type A CRs, $N_A(t)$ for $n = 0, 1, 2, \ldots$, as follows:

$$
\begin{aligned}
P_{N_A(t)}(t,n) = &\left(p_{31} + p_{32} + p_{41} + p_{42} + p_{13} + p_{14} + p_{23} + p_{24}\right)^{-1} \\
&\times (n!)^{-1}\left(t\sum_{i=0}^{\infty}\frac{(-t/\mu)^{ki}}{i!(ki+1)}\right)^n\left\{\left(p_{31} + p_{32} + p_{41} + p_{42}\right)\right. \\
&\times \lambda_{A,1}^n\exp\left[-\lambda_{A,1}t\sum_{i=0}^{\infty}\frac{(-t/\mu)^{ki}}{i!(ki+1)}\right] + \left(p_{13} + p_{14} + p_{23} + p_{24}\right) \\
&\left.\times \lambda_{A,2}^n\exp\left[-\lambda_{A,2}t\sum_{i=0}^{\infty}\frac{(-t/\mu)^{ki}}{i!(ki+1)}\right]\right\}.
\end{aligned} \tag{4.9}
$$

Following the same procedures as for Type A CRs, the dynamic probability mass function of the number of Type B CRs $N_B(t)$ can be obtained by, for $0 \le t < x_m$ and $n = 0, 1, 2, \ldots$,

$$
\begin{aligned}
P_{N_B(t)}(t,n) = &\left(p_{12} + p_{14} + p_{32} + p_{34} + p_{21} + p_{23} + p_{41} + p_{43}\right)^{-1}(n!)^{-1} \\
&\times t^n\left\{\left(p_{21} + p_{23} + p_{41} + p_{43}\right)\lambda_{B,1}^n\exp\left[-\lambda_{B,1}t\right]\right. \\
&\left. + \left(p_{12} + p_{14} + p_{32} + p_{34}\right)\lambda_{B,2}^n\exp\left[-\lambda_{B,2}t\right]\right\}.
\end{aligned} \tag{4.10}
$$

For $t \ge x_m$,

$$
\begin{aligned}
P_{N_B(t)}(t,n) = &\left(p_{12} + p_{14} + p_{32} + p_{34} + p_{21} + p_{23} + p_{41} + p_{43}\right)^{-1}(n!)^{-1} \\
&\times t^n\left(x_m^{\alpha_m-1}t^{-\alpha_m+1} - 1\right)^n\left(\frac{x_m}{1-\alpha_m}\right)^n\left\{\left(p_{21} + p_{23} + p_{41} + p_{43}\right)\right. \\
&\times \lambda_{B,1}^n\exp\left[-\lambda_{B,1}x_m\left(x_m^{\alpha_m-1}t^{-\alpha_m+1} - 1\right)/(1-\alpha_m)\right] \\
&\left. + \left(p_{12} + p_{14} + p_{32} + p_{34}\right)\lambda_{B,2}^n\exp\left[-\lambda_{B,2}x_m\left(x_m^{\alpha_m-1}t^{-\alpha_m+1} - 1\right)/(1-\alpha_m)\right]\right\}.
\end{aligned} \tag{4.11}
$$

The total number of CRs in the system is the summation of the number of Type A CRs and the number of Type B CRs, hence $\{N(t)\} = \{N_A(t) + N_B(t)\}$. Therefore, we can obtain the dynamic probability mass function of the number of CRs in the network as follows, for $0 \le t < x_m$ and $n = 0, 1, 2, \ldots$.

$$P_{N(t)}(t,n) = \frac{1}{n!}\Big\{ a_1\big[\beta(\lambda_{A,1},t) + \lambda_{B,1}t\big]^n e^{-[\beta(\lambda_{A,1},t)+\lambda_{B,1}t]}$$

$$+ a_2\big[\beta(\lambda_{A,2},t) + \lambda_{B,1}t\big]^n e^{-[\beta(\lambda_{A,2},t)+\lambda_{B,1}t]} + a_3\big[\beta(\lambda_{A,1},t) + \lambda_{B,2}t\big]^n e^{-[\beta(\lambda_{A,1},t)+\lambda_{B,2}t]} \quad (4.12)$$

$$+ a_4\big[\beta(\lambda_{A,2},t) + \lambda_{B,2}t\big]^n e^{-[\beta(\lambda_{A,2},t)+\lambda_{B,2}t]} \Big\},$$

where $\beta(y,t) = yt \displaystyle\sum_{i=0}^{\infty} \frac{(-t/\mu)^{ki}}{i!(ki+1)}$;

$$a_1 = \frac{\big(p_{31} + p_{32} + p_{41} + p_{42}\big)\big(p_{21} + p_{23} + p_{41} + p_{43}\big)}{\big(p_{13} + p_{14} + p_{23} + p_{24} + p_{31} + p_{32} + p_{41} + p_{42}\big)}$$

$$\times \big(p_{12} + p_{14} + p_{32} + p_{34} + p_{21} + p_{23} + p_{41} + p_{43}\big)^{-1};$$

$$a_2 = a_1 \times \big(p_{13} + p_{14} + p_{23} + p_{24}\big)/\big(p_{31} + p_{32} + p_{41} + p_{42}\big);$$

$$a_3 = a_1 \times \big(p_{12} + p_{14} + p_{32} + p_{34}\big)/\big(p_{21} + p_{23} + p_{41} + p_{43}\big);$$

$$a_4 = a_3 \times \big(p_{13} + p_{14} + p_{23} + p_{24}\big)/\big(p_{31} + p_{32} + p_{41} + p_{42}\big).$$

For $t \ge x_m$ and $n = 0, 1, 2, \ldots$,

$$P_{N(t)}(t,n) = \frac{1}{n!}\Big\{ a_1\big[\beta(\lambda_{A,1},t) + \kappa(\lambda_{B,1},t)\big]^n e^{-[\beta(\lambda_{A,1},t)+\kappa(\lambda_{B,1},t)]}$$

$$+ a_2\big[\beta(\lambda_{A,2},t) + \kappa(\lambda_{B,1},t)\big]^n e^{-[\beta(\lambda_{A,2},t)+\kappa(\lambda_{B,1},t)]}$$

$$+ a_3\big[\beta(\lambda_{A,1},t) + \kappa(\lambda_{B,2},t)\big]^n e^{-[\beta(\lambda_{A,1},t)+\kappa(\lambda_{B,2},t)]} \quad (4.13)$$

$$+ a_4\big[\beta(\lambda_{A,2},t) + \kappa(\lambda_{B,2},t)\big]^n e^{-[\beta(\lambda_{A,2},t)+\kappa(\lambda_{B,2},t)]} \Big\},$$

where $\kappa(y,t) = \dfrac{yx_m}{1-\alpha_m}\big(x_m^{\alpha_m-1}t^{-\alpha_m+1} - 1\big)$.

Based on this dynamic probability mass function, we analyze the dynamic performance for the cooperative spectrum sensing in the following section.

4.3 Spectrum Sensing Performance Analysis

Spectrum sensing can be formulated as binary hypothesis testing: H_0 denotes that the licensed frequency band is vacant, and H_1 denotes that the licensed PU is transmitting. Spectrum sensing either declares that the PU is idle as R_0 or claims that the frequency band is occupied as R_1. The probability of detection is the probability that the PU's occupancy is correctly detected as $P_d = P(R_1 \mid H_1)$. A high P_d means that the CRN can reliably detect the PU and cause less harmful interference to the licensed PUs with high confidence. The probability of a false alarm $P_{FA} = P(R_1 \mid H_0)$ represents the probability that spectrum sensing generates a false alarm while the licensed band is actually idle. A higher P_{FA} implies that the CRN is more likely to miss the opportunities to exploit the vacant frequency band, which degrades the CRN's own throughput performance.

When a CR user enters the CRN coverage area, it immediately joins the system formed by the other CRs in the area. A CR collaborates with all other CRs in the system to perform distributed cooperative spectrum sensing. Every CR collects the local energy statistic from the received PU signal by taking M samples over the observation time period T. Then, this local energy statistic is sent via a dedicated error-free channel to the fusion center. The fusion center calculates the fused energy statistic Z by summing all collected local energy statistics. It also makes spectrum sensing decisions by comparing the fused energy statistic Z against a threshold λ_T. If $Z \geq \lambda_T$, R_1 is declared; otherwise, the decision R_0 is made. The threshold λ_T can be decided by system design considerations.

A shadowed Nakagami-m fading wireless environment consists of multipath fading superimposed on shadowing. In this environment, the wireless receiver does not average out the envelope fading due to multipath but rather reacts to the instantaneous composite multipath or shadowed signal [8]. This is often the scenario in congested downtown areas with slow-moving pedestrians and vehicles. The population is dense in big cities, and the demand for extra bandwidth in these big cities is increasing rapidly these days. As a solution to the extra frequency demand in these big cities, the CR users will experience shadowed Nakagami-m fading channels, and the signal-to-noise ratio (SNR) for this shadowed Nakagami-m fading channel has the following gamma-lognormal probability density function:

$$f(\gamma) = \int_0^\infty \left(\frac{m}{w}\right)^m \frac{\gamma^{m-1}}{\Gamma(m)} e^{-m\gamma/w} \frac{10}{\sqrt{2\pi}\ln(10)\sigma_\Omega w} \exp\left\{-\left[10\log_{10}(w)-\mu_\Omega\right]^2 \Big/ 2\sigma_\Omega^2\right\} dw, \quad (4.14)$$

where m is the Nakagami-m parameter, and μ_Ω and σ_Ω are the parameters for the lognormal shadowing.

The collected energy statistic at a CR is affected by the shadowed Nakagami-m fading channel. The received primary signal power is not constant. Therefore, the probability of detection of the distributed cooperative spectrum sensing can be obtained by

$$P_d = \int_0^\infty Q_\mu\left(\sqrt{2\gamma}, \sqrt{\lambda_T}\right) f(\gamma)\, d\gamma, \quad (4.15)$$

where $Q_\mu(a,b)$ is the generalized Marcum Q-function, γ is the received SNR affected by the shadowed Nakagami-m fading channel, and $f(\gamma)$ is the gamma-lognormal probability density function of the received SNR. The probability of detection of the distributed cooperative spectrum sensing can also be represented as

$$P_d = \sum_{k=0}^\infty \frac{1}{k!} \frac{\Gamma(u+k, \lambda_T/2)}{\Gamma(u+k)} \int_0^\infty \gamma^k e^{-\gamma} f(\gamma)\, d\gamma, \quad (4.16)$$

where $\Gamma(s,x)$ is the upper incomplete gamma function, and $\Gamma(s)$ is the gamma function. After some routine mathematical manipulations, the probability of detection for cooperative spectrum sensing with N CRs over independent and identically distributed (i.i.d.) shadowed Nakagami-m fading channels can be obtained by

$$P_d = \sum_{k=0}^{\infty} \frac{(-1)^k \Gamma(m+k)\Gamma(Nu+k,\lambda_Z/2)}{k!\sqrt{\pi}m^{k+1}\Gamma(m)\Gamma(Nu+k)} \sum_{j=1}^{N_p} H_{x_j} 10^{\left(\sqrt{2}\sigma_\Omega x_j + \mu_\Omega\right)/10} \left(m - 10^{\left(\sqrt{2}\sigma_\Omega x_j + \mu_\Omega\right)/10}\right)^{-(m+k)}, \quad (4.17)$$

where x_j's are the zeros of the N_p-order Hermite polynomial, and H_{x_j}'s are the weight factors of the N_p-order Hermite polynomial. However, the probability of a false alarm for cooperative spectrum sensing with N CRs over i.i.d. shadowed Nakagami-m fading channels can be obtained by

$$P_{FA} = \frac{\Gamma(Nu,\lambda_T/2)}{\Gamma(Nu)}. \quad (4.18)$$

The CRs in the CRN are not fixed devices; they move depending on the users' needs. Considering mobility, for $0 \leq t < x_m$, the probability of detection is given by

$$P_d(t) = \sum_{n=1}^{\infty}\sum_{k=0}^{\infty} \frac{(-1)^k \Gamma(m+k)\Gamma(nu+k,\lambda_Z/2)}{k!\sqrt{\pi}m^{k+1}\Gamma(m)\Gamma(nu+k)}$$

$$\times \sum_{j=1}^{N_p} H_{x_j} 10^{\left(\sqrt{2}\sigma_\Omega x_j + \mu_\Omega\right)/10} \left(m - 10^{\left(\sqrt{2}\sigma_\Omega x_j + \mu_\Omega\right)/10}\right)^{-(m+k)}$$

$$\times \frac{1}{n!}\left\{a_1\left[\beta(\lambda_{A,1},t)+\lambda_{B,1}t\right]^n e^{-\left[\beta(\lambda_{A,1},t)+\lambda_{B,1}t\right]} + a_2\left[\beta(\lambda_{A,2},t)+\lambda_{B,1}t\right]^n e^{-\left[\beta(\lambda_{A,2},t)+\lambda_{B,1}t\right]}\right.$$

$$\left. + a_3\left[\beta(\lambda_{A,1},t)+\lambda_{B,2}t\right]^n e^{-\left[\beta(\lambda_{A,1},t)+\lambda_{B,2}t\right]} + a_4\left[\beta(\lambda_{A,2},t)+\lambda_{B,2}t\right]^n e^{-\left[\beta(\lambda_{A,2},t)+\lambda_{B,2}t\right]}\right\}. \quad (4.19)$$

Also, the probability of a false alarm is given by

$$P_{FA}(t) = \sum_{n=1}^{\infty} \frac{\Gamma(nu,\lambda_T/2)}{\Gamma(nu)} \frac{1}{n!}\left\{a_1\left[\beta(\lambda_{A,1},t)+\lambda_{B,1}t\right]^n e^{-\left[\beta(\lambda_{A,1},t)+\lambda_{B,1}t\right]}\right.$$

$$\left. + a_2\left[\beta(\lambda_{A,2},t)+\lambda_{B,1}t\right]^n e^{-\left[\beta(\lambda_{A,2},t)+\lambda_{B,1}t\right]} + a_3\left[\beta(\lambda_{A,1},t)+\lambda_{B,2}t\right]^n e^{-\left[\beta(\lambda_{A,1},t)+\lambda_{B,2}t\right]}\right.$$

$$\left. + a_4\left[\beta(\lambda_{A,2},t)+\lambda_{B,2}t\right]^n e^{-\left[\beta(\lambda_{A,2},t)+\lambda_{B,2}t\right]}\right\}. \quad (4.20)$$

For $t \geq x_m$, the probability of detection can be obtained by

$$
P_d(t) = \sum_{n=1}^{\infty} \sum_{k=0}^{\infty} \frac{(-1)^k \, \Gamma(m+k) \Gamma(nu+k, \lambda_Z/2)}{n! k! \sqrt{\pi} m^{k+1} \Gamma(m) \Gamma(nu+k)}
$$

$$
\times \sum_{j=1}^{N_p} H_{x_j} 10^{\left(\sqrt{2}\sigma_\Omega x_j + \mu_\Omega\right)/10} \left(m - 10^{\left(\sqrt{2}\sigma_\Omega x_j + \mu_\Omega\right)/10}\right)^{-(m+k)}
$$

$$
\times \Big\{ a_1 e^{-\left[\beta(\lambda_{A,1},t)+\kappa(\lambda_{B,1},t)\right]} \left[\beta(\lambda_{A,1},t)+\kappa(\lambda_{B,1},t)\right]^n + a_2 \left[\beta(\lambda_{A,2},t)+\kappa(\lambda_{B,1},t)\right]^n \quad (4.21)
$$

$$
\times e^{-\left[\beta(\lambda_{A,2},t)+\kappa(\lambda_{B,1},t)\right]} + a_3 \left[\beta(\lambda_{A,1},t)+\kappa(\lambda_{B,2},t)\right]^n e^{-\left[\beta(\lambda_{A,1},t)+\kappa(\lambda_{B,2},t)\right]}
$$

$$
+ a_4 \left[\beta(\lambda_{A,2},t)+\kappa(\lambda_{B,2},t)\right]^n e^{-\left[\beta(\lambda_{A,2},t)+\kappa(\lambda_{B,2},t)\right]} \Big\}.
$$

However, the probability of a false alarm can be obtained by

$$
P_{FA}(t) = \sum_{n=1}^{\infty} \frac{\Gamma(nu, \lambda_T/2)}{\Gamma(nu)} \frac{1}{n!} \Big\{ a_1 \left[\beta(\lambda_{A,1},t)+\kappa(\lambda_{B,1},t)\right]^n e^{-\left[\beta(\lambda_{A,1},t)+\kappa(\lambda_{B,1},t)\right]}
$$

$$
+ a_2 \left[\beta(\lambda_{A,2},t)+\kappa(\lambda_{B,1},t)\right]^n e^{-\left[\beta(\lambda_{A,2},t)+\kappa(\lambda_{B,1},t)\right]} + a_3 \left[\beta(\lambda_{A,1},t)+\kappa(\lambda_{B,2},t)\right]^n \quad (4.22)
$$

$$
\times e^{-\left[\beta(\lambda_{A,1},t)+\kappa(\lambda_{B,2},t)\right]} + a_4 \left[\beta(\lambda_{A,2},t)+\kappa(\lambda_{B,2},t)\right]^n e^{-\left[\beta(\lambda_{A,2},t)+\kappa(\lambda_{B,2},t)\right]} \Big\}.
$$

Taking into consideration QoS constraints, collision avoidance, interference control, throughput controls, and so on, the network has the system capacity to serve only up to L CRs in the system. Considering this limitation, for $0 \leq t < x_m$, the probability of detection is given by

$$
P_d(t) = \sum_{n=1}^{L} \sum_{k=0}^{\infty} \frac{(-1)^k \, L! \Gamma(m+k) \Gamma(nu+k, \lambda_Z/2)}{n! k! \sqrt{\pi} m^{k+1} \Gamma(m) \Gamma(nu+k)}
$$

$$
\times \sum_{j=1}^{N_p} H_{x_j} 10^{\left(\sqrt{2}\sigma_\Omega x_j + \mu_\Omega\right)/10} \left(m - 10^{\left(\sqrt{2}\sigma_\Omega x_j + \mu_\Omega\right)/10}\right)^{-(m+k)}
$$

$$
\times \Big\{ a_1 \left[\beta(\lambda_{A,1},t)+\lambda_{B,1}t\right]^n \left[\Gamma(L+1, \beta(\lambda_{A,1},t)+\lambda_{B,1}t)\right]^{-1} e^{-\left[\beta(\lambda_{A,1},t)+\lambda_{B,1}t\right]} + e^{-\left[\beta(\lambda_{A,2},t)+\lambda_{B,1}t\right]}
$$

$$
\times \frac{a_2}{\Gamma(L+1, \beta(\lambda_{A,2},t)+\lambda_{B,1}t)} \left[\beta(\lambda_{A,2},t)+\lambda_{B,1}t\right]^n + e^{-\left[\beta(\lambda_{A,1},t)+\lambda_{B,2}t\right]} \frac{a_3}{\Gamma(L+1, \beta(\lambda_{A,1},t)+\lambda_{B,2}t)}
$$

$$
\times \left[\beta(\lambda_{A,1},t)+\lambda_{B,2}t\right]^n + e^{-\left[\beta(\lambda_{A,2},t)+\lambda_{B,2}t\right]} \frac{a_4}{\Gamma(L+1, \beta(\lambda_{A,2},t)+\lambda_{B,2}t)} \left[\beta(\lambda_{A,2},t)+\lambda_{B,2}t\right]^n \Big\}.
$$

$$
(4.23)
$$

On the contrary, the probability of a false alarm is given by

$$
\begin{aligned}
P_{FA}(t) = \sum_{n=1}^{L} \frac{\Gamma(nu, \lambda_T/2)}{\Gamma(nu)} \frac{L!}{n!} \Bigg\{ & \frac{a_1}{\Gamma(L+1, \beta(\lambda_{A,1}, t) + \lambda_{B,1} t)} \\
\times \left[\beta(\lambda_{A,1}, t) + \lambda_{B,1} t \right]^n e^{-\left[\beta(\lambda_{A,1}, t) + \lambda_{B,1} t \right]} + & \frac{a_2}{\Gamma(L+1, \beta(\lambda_{A,2}, t) + \lambda_{B,1} t)} \\
\times \left[\beta(\lambda_{A,2}, t) + \lambda_{B,1} t \right]^n e^{-\left[\beta(\lambda_{A,2}, t) + \lambda_{B,1} t \right]} + & \frac{a_3}{\Gamma(L+1, \beta(\lambda_{A,1}, t) + \lambda_{B,2} t)} \\
\times \left[\beta(\lambda_{A,1}, t) + \lambda_{B,2} t \right]^n e^{-\left[\beta(\lambda_{A,1}, t) + \lambda_{B,2} t \right]} + & \frac{a_4}{\Gamma(L+1, \beta(\lambda_{A,2}, t) + \lambda_{B,2} t)} \\
\times \left[\beta(\lambda_{A,2}, t) + \lambda_{B,2} t \right]^n e^{-\left[\beta(\lambda_{A,2}, t) + \lambda_{B,2} t \right]} & \Bigg\}.
\end{aligned}
\tag{4.24}
$$

For $t \geq x_m$, the probability of detection is given by

$$
\begin{aligned}
P_d(t) = \sum_{n=1}^{L} \sum_{k=0}^{\infty} & \frac{(-1)^k L! \Gamma(m+k) \Gamma(nu+k, \lambda_Z/2)}{k! \sqrt{\pi} m^{k+1} \Gamma(m) \Gamma(nu+k)} \\
\times \sum_{j=1}^{N_p} & H_{x_j} 10^{\left(\sqrt{2} \sigma_\Omega x_j + \mu_\Omega \right)/10} \left(m - 10^{\left(\sqrt{2} \sigma_\Omega x_j + \mu_\Omega \right)/10} \right)^{-(m+k)} \frac{1}{n!} \Bigg\{ \left[\beta(\lambda_{A,1}, t) + \kappa(\lambda_{B,1}, t) \right]^n \\
\times & \frac{a_1}{\Gamma(L+1, \beta(\lambda_{A,1}, t) + \kappa(\lambda_{B,1}, t))} e^{-\left[\beta(\lambda_{A,1}, t) + \kappa(\lambda_{B,1}, t) \right]} + \left[\beta(\lambda_{A,2}, t) + \kappa(\lambda_{B,1}, t) \right]^n \\
\times & \frac{a_2}{\Gamma(L+1, \beta(\lambda_{A,2}, t) + \kappa(\lambda_{B,1}, t))} e^{-\left[\beta(\lambda_{A,2}, t) + \kappa(\lambda_{B,1}, t) \right]} + \left[\beta(\lambda_{A,1}, t) + \kappa(\lambda_{B,2}, t) \right]^n \\
\times & \frac{a_3}{\Gamma(L+1, \beta(\lambda_{A,1}, t) + \kappa(\lambda_{B,2}, t))} e^{-\left[\beta(\lambda_{A,1}, t) + \kappa(\lambda_{B,2}, t) \right]} + \left[\beta(\lambda_{A,2}, t) + \kappa(\lambda_{B,2}, t) \right]^n \\
\times & \frac{a_4}{\Gamma(L+1, \beta(\lambda_{A,2}, t) + \kappa(\lambda_{B,2}, t))} e^{-\left[\beta(\lambda_{A,2}, t) + \kappa(\lambda_{B,2}, t) \right]} \Bigg\}.
\end{aligned}
\tag{4.25}
$$

Furthermore, the probability of a false alarm is given by

$$
\begin{aligned}
P_{FA}(t) = \sum_{n=1}^{L} \frac{\Gamma(nu, \lambda_T/2)}{\Gamma(nu)} \frac{1}{n!} \Bigg\{ & \frac{a_1}{\Gamma(L+1, \beta(\lambda_{A,1}, t) + \kappa(\lambda_{B,1}, t))} \left[\beta(\lambda_{A,1}, t) \right. \\
\left. + \kappa(\lambda_{B,1}, t) \right]^n e^{-\left[\beta(\lambda_{A,1}, t) + \kappa(\lambda_{B,1}, t) \right]} + & \frac{a_2}{\Gamma(L+1, \beta(\lambda_{A,2}, t) + \kappa(\lambda_{B,1}, t))} \\
\times \left[\beta(\lambda_{A,2}, t) + \kappa(\lambda_{B,1}, t) \right]^n e^{-\left[\beta(\lambda_{A,2}, t) + \kappa(\lambda_{B,1}, t) \right]} + & \frac{a_3}{\Gamma(L+1, \beta(\lambda_{A,1}, t) + \kappa(\lambda_{B,2}, t))} \\
\times \left[\beta(\lambda_{A,1}, t) + \kappa(\lambda_{B,2}, t) \right]^n e^{-\left[\beta(\lambda_{A,1}, t) + \kappa(\lambda_{B,2}, t) \right]} + & \frac{a_4}{\Gamma(L+1, \beta(\lambda_{A,2}, t) + \kappa(\lambda_{B,2}, t))} \\
\times \left[\beta(\lambda_{A,2}, t) + \kappa(\lambda_{B,2}, t) \right]^n e^{-\left[\beta(\lambda_{A,2}, t) + \kappa(\lambda_{B,2}, t) \right]} & \Bigg\}.
\end{aligned}
\tag{4.26}
$$

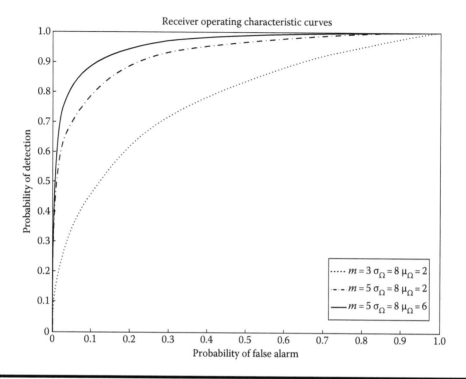

Figure 4.4 Receiver operating characteristic curves.

The performance of a detector is often described based on its receiver operating characteristic (ROC) curves (P_d versus P_{FA}). The ROC curves are shown in Figure 4.4 with different combinations of the Nakagami-m fading parameter, m, and lognormal shadowing parameters, μ_Ω and σ_Ω. A smaller m results in a more severe fading effect; for example, Nakagami-m fading degenerates to Rayleigh fading when $m = 1$. σ_Ω is typically observed between 5 and 12 dB in macrocellular applications and between 4 and 13 dB in microcellular applications.

4.4 Conclusions

CRNs are an emerging technology to efficiently use the spectrum. Multimedia streaming has high latency-sensitive QoS constraints. Mobility poses great challenges for the physical layer design that must be overcome for CRN technology to provide reliable transmission. The analytical closed-form expressions for cooperative spectrum sensing lay the foundation for engineers to research and develop CRNs to satisfy multimedia streaming QoS constraints.

References

1. S. Haykin, Cognitive radio: Brain-empowered wireless communications, *IEEE Journal on Selected Areas in Communications*, Vol. 23, No. 2, pp. 201–220, 2005.
2. B. Wang and K. J. R. Liu, Advances in cognitive radio networks: A survey, *IEEE Journal of Selected Topics in Signal Processing*, Vol. 5, No. 1, pp. 5–23, 2011.

3. R. Tandra and A. Sahai, SNR walls for signal detection, *IEEE Journal of Selected Topics in Signal Processing*, Vol. 2, No. 1, pp. 4–17, 2008.
4. G. Ganesan and Y. Li, Cooperative spectrum sensing in cognitive radio networks, in *Proceedings of IEEE International Symposium on New Frontiers in Dynamic Spectrum Access Networks (DySPAN)*, Baltimore, MD, USA, pp. 137–143, November 2005.
5. G. Baldini, T. Sturman, A. R. Biswas, R. Leschhorn, G. Gódor, and M. Street, Security aspects in software defined radio and cognitive radio networks: A survey and a way ahead, *IEEE Communication Surveys and Tutorials*, Vol. 14, No. 2, pp. 355–379, 2012.
6. A. W. Min and K. G. Shin, Impact of mobility on spectrum sensing in cognitive radio networks, in *Proceedings of the ACM Workshop on Cognitive Radio Networks*, New York, NY, 2009.
7. R. G. Gallager, *Discrete Stochastic Processes*. Kluwer Academic Publishers, Norwell, MA, 1995.
8. M. Ho and G. Stüber, Co-channel interference of microcellular systems on shadowed Nakagami fading channels, in *Proceedings of the IEEE Vehicular Technology Conference*, Secaucus, NJ, May 1993.

Chapter 5

Efficient Multimedia Services Provision over Cognitive Radio Networks Using a Traffic-Oriented Routing Scheme

---

Constandinos X. Mavromoustakis,[1] George Mastorakis,[2]
Athina Bourdena,[2] and Evangelos Pallis[2]

[1]*Department of Computer Science, University of Nicosia, Nicosia, Cyprus*

[2]*Department of Informatics Engineering, Technological Educational Institute of Crete, Heraklion, Greece*

Contents

5.1 Introduction

Emerging types of wireless applications and mobile services, rich in multimedia content with high demand for network resources and strict quality of experience (QoE) requirements, put more and more pressure for further research on efficient multimedia services provision over future-generation communication systems. Such communication systems will enable the provision of customized user-centric services, ranging from entertainment and lifestyle applications (e.g., mobile TV) to professional services such as video conferencing and real-time data exchange. Suitable platforms and gateways are under development to enable the provision of a wide array of new services to meet users' demands. The major challenge of such new services stems from the demands of wireless users' devices for energy consumption. Another vital issue is the guaranteed QoS of different multimedia services, in cases of unexpected delays or network failures. In addition, delays occur in networking architectures that exploit the "television white spaces" (TVWS) [1] for their operation. The opportunistic exploitation of TVWS can be performed by cognitive radio (CR) technology [2], which constitutes an emerging communication paradigm. CR networks are capable of interacting with the surrounding spectral environment by sensing the wide radio spectrum, dynamically identifying the local available frequencies, and efficiently exploiting them. However, the introduction of CR networks in TVWS hampers the current "command-and-control" paradigm of TV/UHF spectrum management; thus, the operation of CR technology in TVWS is highly associated with the regulation models that would be adopted [3–5]. For instance, Spectrum of Commons is a recent spectrum regulation model that permits the coexistence of primary and secondary users in the same spectrum band (i.e., UHF), under the condition that the latter do not cause interference to the former. It is especially applicable in distributed or ad hoc CR networks, where resource allocation is locally administered or handled by communication nodes, in contrast to adopt a centralized network approach that exploits a spectrum manager [6].

In addition, the opportunistic radio exploitation by ad hoc CR networks involves new challenges in the design of communication protocols, in several network layers. For instance, the design and adoption of novel routing schemes in the network layer are required for the efficient provision of emerging multimedia services, in order to avoid route failures and to establish and maintain fixed routing paths between secondary communication nodes. Beyond the route establishment, another challenge is to provide and guarantee the desired QoS level to mobile wireless end users that are also energy harvesting nodes.

Energy saving presents a significant element/aspect in high-performance deployment in ad hoc CR networks. More specifically, energy harvesting wireless nodes have to be tuned to the calculated values (i.e., capacity, traffic [7] of the nodes) of the energy saving/conservation scheme, while the latter (i.e., the scheme) has to take into account the bounded end-to-end delays of the nodes' transmissions. Considering the aforementioned issues, the conclusion is that network lifetime is firmly correlated/associated with the transmission characteristics [8] of a source node to a destination node, while the selected routing protocol [9] combines both the temporal traffic-aware behavior of the node [10] and the efficient routing scheme in an end-to-end path. This vital issue has already been studied in the study proposed in Ref. [8], whose authors designed a resource-intensive traffic-aware scheme for energy saving in wireless devices. More specifically, in the proposed scheme, a number of sleep-proxy nodes are exploited in order to evaluate the duration of the activity periods of each node. This evaluation is performed according to the capacity, as well as according to the estimated inter-cluster overall energy consumed within a time frame. Toward further investigating the scheme proposed in Ref. [8], the current work is an enhanced scheme that exploits the Fibonacci-based backward traffic difference (F-BTD), which aims to extend the lifetime of the secondary nodes of a CR network. The proposed F-BTD scheme recognizes the activity of the incoming traffic and accordingly permits the nodes to adaptively sleep. The proposed traffic-aware scheme examines the traffic-active moments of the nodes, as well as the volume of the traffic for a specified time window frame, in order to adjust the activity of each node based on the energy-efficient metrics of the CR secondary nodes. In order to minimize the energy consumption of each node, the backward traffic difference (BTD) measures the volume of the incoming traffic that is intended for each node within a time window frame. The BTD [10,11] considers the repetition of the traffic and estimates the backward difference of different moments, in accordance with the Fibonacci sequence for extracting the time duration for which the node is allowed to sleep.

In this context, this chapter elaborates on the design, development, and experimental evaluation of a resource-intensive traffic-aware scheme and an energy-efficient routing protocol for ad hoc CR network architectures, enabled for the efficient communication of secondary communication nodes that operate under the Spectrum of Commons regime, offering effective multimedia services provision. More specifically, a signaling mechanism combined with an energy-efficient scheme is proposed in order to support multimedia services in an end-to-end manner. The mechanism is based on the BTD estimation [7,8,10] in contrast to the end-to-end bounded delay of the transmission. Wireless ad hoc networks need to be supported by improved system reliability and availability through automatic traffic measurements and configuration of traffic-aware network parameters. Depending on the underlying routing scheme and the volume of traffic that each node receives/transmits, the proposed scheme aims at minimizing energy consumption by applying an asynchronous, nonperiodic sleep-time assignment slot to the secondary wireless nodes according to the F-BTD scheme. The F-BTD scheme is applied in order to enable delay-sensitive multimedia content in an end-to-end, reliable manner. Following this introductory section, Section 5.2 elaborates on related work and research motivation, while Section 5.3 presents the design and development of a novel green-aware routing protocol that offers energy-efficient data transition across secondary communication nodes with different TVWS availability. In order to achieve an energy-efficient methodology, the proposed framework uses a traffic-aware BTD scheme based on the Fibonacci sequence model for estimating the duration of the sleep time according to the nodal traversed traffic. With the BTD applied scheme, multimedia resources are treated in a prioritized manner so that requested traffic is forwarded to the destination node within a specified amount of time. The proposed scheme allows persistence to be ensured through a promiscuous caching mechanism. The persistence is enabled through the ability of data multimedia objects to survive through different resource-sharing environments and invocations. The proposed scheme guarantees the end-to-end availability of requested multimedia resources while significantly reducing energy consumption and maintains the requested scheduled transfers, during the resource-sharing process. For achieving energy conservation, the proposed traffic-aware policy can efficiently determine the ON and OFF duration/period of each node, which then effectively provides the reflective effects to the overall energy consumed by nodes. Section 5.4 elaborates on performance evaluation analysis of the proposed research approach, discussing experimental results, and Section 5.5 concludes this chapter by highlighting directions for future research.

5.2 Related Work and Research Motivation

Several research approaches have been proposed for efficient multimedia services provision over CR networks with the purpose of increasing radio spectrum availability, as well as avoiding delays and interference that are usually caused among secondary and primary nodes. CR technology aims to enhance spectrum utilization in the licensed frequencies and provide high-throughput communication by optimally selecting communication paths and channel assignments. Many approaches have been proposed to optimally share the radio spectrum, by exploiting CR technologies in different use-case scenarios. Research work in Ref. [12] analyzes multimedia transmissions over CR networks with appropriate admission control and channel selection. In Ref. [13], the authors investigate the impacts of dynamic resources on smooth video streaming in CR networks and propose a centralized channel allocation algorithm to achieve superior video delivery, by reducing the playback frozen probability. The authors in Ref. [14] manage the problem of multimedia service provisioning in the context of a multichannel CR network. In this regard, the proposed framework for multimedia services takes into consideration the channel heterogeneity among different secondary users, as well as the feature of multicast transmission. This effective framework is capable of incorporating cooperative transmission between users into direct transmission from the secondary base station. In addition, the authors in Ref. [15] study multihop communication by proposing an efficient routing scheme to improve the connectivity and spectrum efficiency of CR users. The correlation of multihop transmission with routing in CR networks is capable of reducing the transmission power, as well as improving the network connectivity and spectrum efficiency. In Ref. [16], a solution is proposed for the multimedia transmission problem over CR networks in lossy environments. The primary traffic interruptions are not the only reason for packets losses, which are also caused by collisions between secondary users due to concurrent access in spectral resources. The transmission performance of a multimedia stream over CR networks is affected by two critical factors: (1) interference caused by the primary user reclaims for resources, leading to more corrupted secondary packets, and (2) collisions caused by opportunistic transmissions of the secondary users. The authors in Ref. [17] elaborate on spectrum opportunities that vary over time and location in CR networks. The exploitation of multichannel routing schemes is capable of significantly improving the throughput of multihop wireless networks because interference can be avoided or reduced, and the network load can be balanced on different channels [18]. Nevertheless, the routing mechanism has to be strictly associated with energy efficiency when the CR networking architecture hosts wireless nodes requesting spectrum, through which the traffic will be transferred. Therefore, the routing mechanism in conjunction with an energy-efficient scheme should guarantee the end-to-end availability of requested resources, and it should also be able to significantly reduce energy consumption. In addition, the mechanism should be able to maintain the requested scheduled transfers and the entire end-to-end connectivity. In such networks, energy-efficient routing schemes achieve higher performance than the shortest routing-path schemes while at the same time reducing the power consumption of the relay users [19]. In addition, energy-efficient routing schemes alleviate the network-partitioning problem that is caused by the energy exhaustion of the relay nodes. In this regard, the researchers of Ref. [19] try to provide a high-throughput routing approach that involves not only route selection but also channel time slot allocation. Thus, the traffic is decomposed over different channels and time slots. Also, Bayhan and Alagoz [20] present promising CR networks with advanced functionalities that require sophisticated information-processing capabilities. CR networks need powerful energy sources to utilize all these functionalities. Current battery technologies cannot meet the enormous increase in power consumption stemming from the increasing traffic flow resulting from the improvement in fast semiconductor technologies [21]. In this way, energy efficiency may become a limiting factor in the development of advanced wireless communications technologies. Research on higher energy efficiency is primarily motivated by cost effectiveness, longer battery lifetime, and environmental concerns. Energy costs are constantly increasing, and the energy expenditure of a wireless network is a significant fraction (20%–30% [22]) of the total operator expenditures (site rental, licensing, etc.). Hence, energy should be effectively consumed for cost-effective systems. Reduction in energy consumption and energy-efficient operations are thus in the interests of the operators. From the user viewpoint, energy efficiency means longer battery lifetime. Short battery life annoys users and reduces the utility of wireless communications. Thus, energy efficiency is vital for both actors of wireless communications.

In addition, many recent measurement studies [11] have convincingly demonstrated the impact of traffic on end-to-end connectivity [23] and thus confirmed the impact on sleep-time duration and energy consumption. Measures extracted in real time using realistic traffic [11,23] have shown that the impact of the responsiveness of the routing scheme in regard to end-to-end transmission reliability is significant. Real-time communication networks and multimedia systems exhibit noticeable burstiness over a number of time scales, as explored in Refs. [24,25]. On the basis of stochastic traffic modeling, the variation of traffic over time in most cases exhibits fractal-like characteristics [26]. A scheme that, in conjunction with the routing mechanism used, utilizes the traffic pattern to determine the activity durations of the nodes with the purpose of conserving energy has not yet investigated. Such a scheme will be able to tune the wireless interfaces of the nodes into the sleep or active state according to the pattern of the incoming traffic and will be capable to reflect this pattern—through a model—to the next sleep-time duration of each secondary node. The existing scheduling strategies for wireless nodes could be classified into three categories: (1) coordinated sleeping [27], where nodes adjust their sleeping schedules; (2) random sleeping [26], where there is no certain adjustment mechanism between the nodes in the sleeping schedule, with all the attendant pros and cons [10]; and (3) on-demand adaptive mechanisms [28], where nodes enter into the sleep state depending on the environmental requirements, and outband signaling is used to notify a specific node that it should go to sleep in an on-demand manner.

Although many schemes have been proposed to handle energy conservation issues, the correlation of a traffic-aware scheduling scheme with the underlying routing protocol supported by the CR networking architecture has not yet been explored. This correlation scheme poses a promising opportunity for the deployment of new approaches with the association of different parameters of the communication mechanisms, in order to maximize energy saving. Such schemes are classified into active or passive mechanisms. Active techniques sustain energy by executing energy-aware operations, such as transmission scheduling and energy-aware routing. Mavromoustakis et al. [11] consider the correlation of energy conservation with a number of parameterized characteristics of the traffic (such as traffic prioritization) and enable a mechanism that tunes the interfaces' scheduler to expand in the sleep state based on the activity of the traffic of a certain node in the end-to-end path in real time. When nodes request multimedia resources, the underlying mechanisms should enable in a prioritized manner the guarantee of the end-to-end delay that will be fulfilled by the requested transfer. To this end, the mechanism should enable an adaptive forwarding capability in order to provide the destination and end nodes with the data requested within a specified time. The proposed scheme guarantees the end-to-end availability of requested multimedia resources while significantly reducing energy consumption and maintains the requested scheduled transfers while fulfilling the delay constraints and requirements of multimedia resources. The presented study deals with the minimization of the secondary nodes' energy consumption by deploying a temporal pattern of the incoming traffic. The proposed scheme tries to maximize the energy saving of each secondary node by exploiting the Fibonacci-based pattern of the traffic, as well as the delay limitation (bounded delay) of each transmission. The time-oriented sequence of the incoming traffic as well as the communication traffic volume (data and control packets) among peers is taken into consideration by the proposed scheme as a solution for the provision of energy conservation schedules of the communicating secondary nodes. To this end, a prototype traffic-aware methodology is exploited throughout the resource exchange process between nodes, by applying to the BTD scheme [29], a discrete Fibonacci backward estimation for deriving time-oriented differential traffic. Through the proposed approach, each node evaluates in the same way the differential dissimilar assignments of sleep–wake schedules for each of its traffic moments. Moreover, the sleep–wake schedules operate according to the traffic difference over time, the relative nodal capacity, and the associated traffic characteristics. The main challenges the proposed scheme tackles are the guaranteed end-to-end availability of resources and a significant reduction in energy consumption while maintaining the requested scheduled transfers during the resource distribution process. To further increase the availability of spectrum opportunities, the scheme exploits opportunistically the methodology of promiscuous caching [10], aiming to cache the packets destined for the node with turned-off interfaces (sleep state) in intermediate nodes. The proposed energy-saving framework, based on BTD estimation, with the proposed routing methodology enables the next sleep-time duration of the recipient node to be adjusted according to the activity duration and the volume of the traffic in conjunction with the consolidated routing mechanism.

5.3 Energy-Efficient Routing Scheme for Multimedia Services Provision over CR Networks

The main challenge in an ad hoc CR network environment is proper communication among secondary nodes by exploiting the available TVWS for the efficient provision of multimedia services. A routing protocol in this case has to consider the available spectrum opportunities in different geographical regions at local level. Challenges such as spectrum awareness, route quality, and route maintenance are crucial for the design and implementation of an efficient routing scheme. These attributes may enable optimum data delivery across locations and areas with dispersed radio spectrum opportunities, even if network failures in the path exist due to primary nodes' transmissions.

Figure 5.1 illustrates a simulation scenario in a heterogeneous spectrum environment. In this scenario, the incumbent systems (primary systems in Figure 5.1) operate over specific channels in

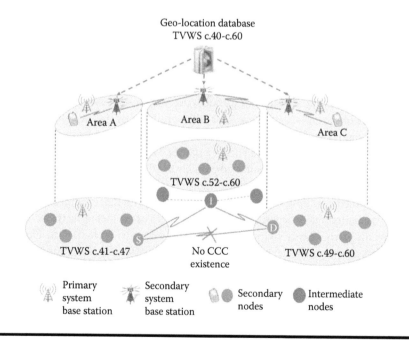

Figure 5.1   Secondary communication nodes operating over heterogeneous TVWS.

three geographical regions (i.e., Areas A, B, and C in Figure 5.1). On the other hand, secondary nodes are able to operate in an opportunistic manner, by exploiting the remaining vacant channels (i.e., TVWS) in each geographical region. The main challenge for proper communication by the secondary nodes is the absence of a common control channel (CCC). Secondary nodes of Area A cannot operate over a common channel with secondary nodes of areas B and C; thus, an inconsistency exists for the creation and maintenance of a routing path delivering multimedia streams. For this purpose, secondary nodes that are located in regions with higher TVWS availability (e.g., locations outside Areas A, B, and C) act as intermediate relay nodes, switching to alternative channels in order to enable the establishment of a routing path between secondary nodes of Areas A, B, and C.

Considering the aforementioned scenario in an ad hoc CR networking environment, spectrum awareness has to be examined in order to provide high-throughput data delivery and enable effective multimedia services provision by optimally choosing the most efficient routing paths between secondary nodes. In this framework, a prototype routing protocol has to be adopted in order to enable the discovery of routing paths taking into account the protection of incumbent systems, as well as TVWS heterogeneity in different regions. In addition, route quality issues are crucial in the design of a routing mechanism because the actual topology of such multihop CR networks is greatly affected by primary nodes' behavior, while the classical methods of evaluating the quality of end-to-end routes, such as bandwidth and throughput, are no longer adequate. In a general context, communication issues for efficient multimedia services provision in an ad hoc CR network over TVWS are an important and yet unexplored problem, especially when a multihop network architecture is examined. Hence, it is crucial for a novel routing protocol to be designed and implemented, in order to solve the problems described concerning the establishment and maintenance of efficient routing paths, among communication nodes with different radio spectrum availability.

5.3.1 *Proposed Underlying Routing Mechanism*

The relay/intermediate secondary nodes of the proposed scenario are permitted to operate over all the available TVWS (i.e., channel 40-channel 60) and act as bridge nodes, while they can gain information from the geolocation database regarding TVWS availability. These relay nodes are also enhanced with routing capabilities, enabling determination of optimum routing paths among the secondary nodes for efficient multimedia services provision. To facilitate the delivery of delay-sensitive multimedia services between secondary communication nodes, a novel routing protocol was designed, implemented, and evaluated by conducting experimental simulations. The proposed routing protocol exploits the exchange of ad hoc on-demand distance vector (AODV) messages [30] between secondary nodes, considering both route discovery and route reply processes. The exploitation of these hop-by-hop messages enables the prediction of vacant TVWS, and so these messages are broadcasted only when necessary. Throughout the route discovery process, an RREQ (route request) message that encompasses the available TVWS of the neighbor nodes is broadcasted by the source node in order to obtain a route path up to the destination node. Once the destination node receives the RREQ message, it has full knowledge of the available TVWS along the route path.

Finally, the destination node selects the optimum routing path, as a matter of minimum number of hops and delay (e.g., backoff delay, switching delay, queuing delay) and assigns a channel to each secondary node along the route. More specifically, each intermediate node conducts the evaluation of the delay metrics though an RREQ message that it forwards to the next node, while the metric evaluation is defined as En_i (see Table 5.1), where E is the end-to-end delay in milliseconds, and n_i represents the ith intermediate node that obtains the path. Further, En represents the delay that occurred during the RREQ message. In the next step of the proposed process, the destination node sends back an RREP (route reply) message to the source node that encompasses information regarding channel assignment.

Figure 5.2 presents the sequence diagram of the proposed routing mechanism for handling the communication among secondary nodes by swapping RREQ and RREP messages. More specifically, a secondary source node initializes a new flow by transmitting an RREQ message to the closest-neighbor secondary intermediate node that transmits in a common channel. As soon as each intermediate node receives an RREQ, it evaluates the delay of the flow (i.e., the queuing delay), and it is also informed about the status of the neighbor secondary nodes by the geolocation database. Then each intermediate node decides whether it will accommodate the flow, on the basis of the delay evaluation. In case the delay is decreased (comparing with the previous delay values in the path), the node accommodates the flow and forwards the RREQ message to the next hop (i.e., the next intermediate node). As soon as, the destination node receives the RREQ, it has full

Table 5.1   Pseudocode of the Basic Steps of the Proposed Message Exchange Process

1: Initiate new flow "f" by sending an RREQ
2: Update intermediate node "I" with neighbor status
//Calculate the delay of the flow "f" from the source node to the 1st intermediate node "I"
3: $D_n = D_{queuing} + D_{switching} + D_{backoff}$
4: k = number of intermediate nodes
//Decision of node "I"
5: **for (i = 1; i++; i = k){**
//Calculate the delay of the flow "f" from the 1st intermediate node "I" to next intermediate node "NI"
6: $D_{n_I} = D_{queuing,i} + D_{switching,i} + D_{backoff,i}$
7: **if** $Dn_i < D_n$
8: **then** flow accommodation
9: **until** (receive route acceptance)
10: generate and send RREP to source node
11: start data transmission
//Flow redirection, the ith intermediate node generates and broadcasts redirection information message
12: **else if** $Dn_i > D_n$
13: redirection process
14: go to steps 6–14
15: }

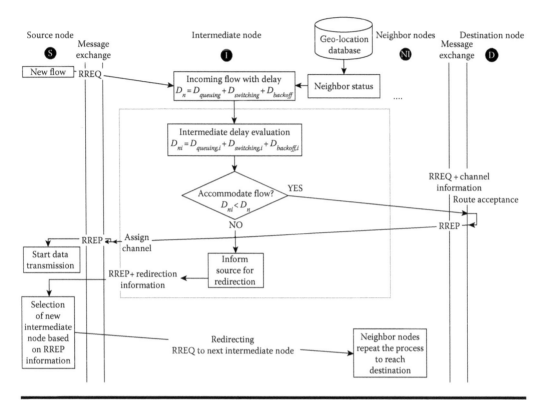

Figure 5.2 **Message exchange process of the proposed routing protocol.**

knowledge of the TVWS availability along the routing path. Following this, the destination node responds to the source by broadcasting an RREP message that includes relevant information concerning the channel allocation for the path. This process establishes the routing path, and useful data transmission is initiated.

Alternatively, if the intermediate node is not capable of serving the incoming flow, a redirection process (redirection in Figure 5.2) is in charge of informing the source node about the status of neighboring nodes, which could possibly act as an alternative intermediate node. The proposed routing protocol obtains a route, in case a source node wants to initiate data flow to a destination node. The obtained routes are maintained as long as they are necessary, while the use of sequence numbers in the exchange messages guarantees a loop-free routing process. The proposed routing protocol operates on a demand basis as it creates and maintains the routes when necessary; thus, it is characterized as a "reactive" protocol. In such a case, routing tables are exploited in order to maintain the routes, where each entry of the table contains information concerning the destination node, the next hop, the number of hops, the destination sequence number, the active neighboring nodes for this route, and the expiration time of the flow. Note that the number of RREQ messages sent by a source node per second is limited, while each RREQ message contains a time to live (TTL) value that specifies the number of times this message could be rebroadcasted. This value is set to a predefined value at the first transmission and is increased during retransmissions that may occur if no replies are received.

Toward further optimizing the proposed routing protocol, an assigning mechanism was implemented in order to relieve the intermediate nodes if the service load is high. The assigning mechanism is incorporated in each intermediate node, which is further able to determine if a neighbor node performs better during the process of routing path establishment. The process in the sequence diagram of Figure 5.2 is more or less the same; however, the main difference is that when an intermediate node receives the RREQ, it evaluates the obtained path and includes the delay evaluation in the RREQ message in order to forward it to all neighboring nodes. Once neighboring nodes receive the RREQ, they evaluate the path and send back to the intermediate node the measured delay evaluation through a redirecting reply message. Once the intermediate node receives the redirecting reply from several of its neighboring nodes, it then determines the optimum node in order to obtain the next hop in the path, considering the mitigation of the traffic in the nodes' queues. The pseudocode of the proposed assigning mechanism is summarized in Table 5.1. For enabling energy efficiency and achieving minimum delays for delay-sensitive multimedia services in the proposed framework, a BTD estimation [11] methodology is used. The main additional contribution is that in the proposed framework, the BTD estimation is bounded by the delay limitations of the transmission, whereas it takes into consideration the hop-by-hop link delay as well as the total end-to-end delay of the transmission. The bounded delay evaluation scheme has been incorporated in order to enable end-to-end transmission in an efficient manner, and it will also enable reliable and correct transmission of multimedia services. In addition, the Fibonacci-based association of the incoming traffic is measured using the BTD in an adaptive way, by measuring the discrete Fibonacci sequence of the incoming traffic of each node. According to the proposed algorithm, each node evaluates in real time the F-BTD and associates the optimal sleep-time duration of each node's life cycle. The latter should satisfy the delay requirements of the transmission. Considering the delay limitations in a multimedia service, it is difficult to maintain the QoS for certain multimedia applications, for example, voice. The designed model guarantees the end-to-end availability of requested resources while reducing energy consumption significantly and maintains the requested scheduled transfers in a mobility-enabled communication. In order to achieve a reasonable transmission guarantee and degree of reliability in the service provision by wireless devices, an adaptive scheme must be hosted in the end-to-end resource-exchanging process. The innovation adopted in this scheme is that each secondary mobile node uses different assignments of sleep–wake schedules based on the incoming traffic difference that each node receives through time in order to prolong each node's lifetime and allow multimedia services to be supported in a reliable manner. The scheme assigns the sleep-time duration according to the BTD scheme in a dissimilar manner in order to enhance nodes' lifetime, whereas it avoids mutation, which will result in network partitioning and resource-sharing losses. This configuration allows adequate provision of multimedia services to the end recipients.

Assume that a mobile secondary node has already used the depicted routing scheme of the previous section and established an end-to-end connection in order to transmit requested content/packets. Routing occurs on an end-to-end basis, and each node separately runs the traffic-aware mechanism using the BTD as described in the following section. The mechanism measures the traffic that traverses each one of the nodes where the F-BTD estimation through the assigned time window frame will affect the sleep-time duration and enable energy conservation in nodes as the conducted simulation experiments show.

5.3.2 Traffic-Aware Scheme for Energy-Efficient Transmission

5.3.2.1 Traffic-Driven Middleware and Supported Mechanisms

As wireless devices have many constraints in terms of processing, battery life, and achievable data rate, the proposed scheme aims at guaranteeing the reliability of the requested resources within a specified amount of time and within the allowed time frame in which the multimedia resources need to be transferred. In essence, the efficient mobile-exchanging process becomes complex because of the devices' characteristics and the change of components, over time as well as over space, in terms of connectivity, portability, accessibility/availability, and mobility. Toward reducing the impact of these changes, the resource-sharing application must have a context-aware adaptive behavior. Context-awareness through traffic-aware adaptation is a fundamental concept for pervasive and ubiquitous environments. In conjunction with the proposed routing methodology used, this chapter elaborates on traffic volume exploitation and manipulation. A traffic-aware policy requires an active scheme to be applied through which the traffic pattern will reflect a certain impact on the nodes. Wireless devices should take the incoming traffic into consideration in order to adapt and reflect feedback based on the traffic volume to the energy conservation mechanism. A middleware that hosts traffic changes and has a direct impact through the estimated scheme presented in the next section using a collaborative traffic-aware scheme is shown in Figure 5.3. This figure depicts a cross-layer interaction through a mechanism for traffic awareness in an end-to-end manner. In particular, real-time media traffic, such as voice and video, typically has high data rate requirements and stringent delay constraints, whereas wireless nodes generally have limited or momentary connectivity. The proposed middleware enables data packets to be traversed and manipulated through the utilized wireless data link, network, and transport layers by taking into consideration the traffic awareness mechanism and the model for volume estimation to be reflected on these layers. The proposed traffic-aware scheme and the associated mechanisms evaluate (after the bootstrap

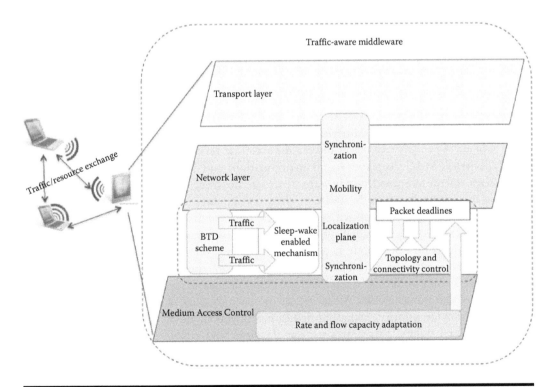

Figure 5.3 The traffic-aware interactive middleware mechanism with the associated influenced layers in the communication stack.

process of the system) the estimated (quantified as volume/capacity) traffic that is destined for each node. In Figure 5.3, V_k^i denotes the volume of traffic destined for node k and stored in node i using the promiscuous caching policy [10]. In this way, it enables—through the proposed mechanism—estimation of the next sleep duration of the node as presented in the following section. These mechanisms are performed to tune the wireless interface of each device to sleep or wake depending on the activity of each individual device in the resource-exchanging path.

A packet classification methodology was utilized as in Ref. [10] in order to mark the packets that are exchanged as delay sensitive or not. In turn, if packets are considered as delay sensitive, strict deadlines are applied by the sender, according to the specifications set in the network. If packet deadlines cannot be satisfied, then cached packets of nearby nodes enable recovery using promiscuous caching. This buffering mechanism for multimedia resources enables the resources' replication and increases resource-sharing reliability. The mechanisms shown in Figure 5.3 are described in the following sections with a quantitative analysis.

The proposed scheme introduces high-availability capabilities for resource sharing, allowing for continuous operation and smoother handling of system outages. The promiscuous caching mechanism estimates the volume of traffic that is cached on an intermediate (active state) node in the path, in order to measure the volume of traffic affected by the outage. The traffic-aware middleware that hosts the resource-intensive scheme allows a more flexible system infrastructure that can adapt to dynamic changes in resource-sharing application requirements and connectivity conditions. As the reflective middleware model is a principled and efficient way of dealing with highly dynamic environments, the proposed scheme also supports a reflective and flexible adaptation of the traffic volume V_k^i. Multimedia traffic is considered in terms of the repetition pattern by estimating the backward difference for extracting the time duration for which the node is allowed to reduce the energy consumption by entering the sleep state during the next time slot T. The middleware in conjunction with the proposed routing scheme enables secondary nodes to exchange efficiently the requested resources by evaluating within a time frame window the incoming traffic volume as well as the incoming traffic that is destined for these nodes. In order to enable recoverability of the incoming traffic, if a node is in the sleep state (outside the connectivity range), then the multimedia traffic is cached using the promiscuous caching concept in intermediate nodes in the path. The traffic-aware resource-sharing scheme expands a cross-layer interaction (see Figure 5.3) for Level 2 Medium Access Level (L2/wireless MAC sleep/active time manipulation) and L3 using the proposed routing methodology. In the proposed middleware, there are no strict associations among the tasks and the layers. The proposed traffic-aware mechanism evaluates (after the bootstrap process of the system) the estimated (quantified as volume/capacity) multimedia traffic that is destined for each node. In this way, it enables—through the proposed mechanism—estimation for the next slot sleep duration of the node as presented in the next section. In essence, the proposed mechanism enables efficient multimedia services by minimizing the end-to-end delay while allowing promiscuous buffering in intermediate nodes by measuring the limitations over time for each one of the transmissions. Moreover, as the primary goal of the QoS for multimedia services is to provide priority, including dedicated transmissions of certain peers that serve as buffering nodal points, parameters such as the jitter and latency (real-time and interactive traffic) should be significantly improved and kept under control. However, by providing priority for selected data flows, the proposed mechanism does not cause the failure of other flows provided neighboring nodes can host data multimedia packets for specified destination nodes for a specified amount of time. The framework in-focus takes into consideration the foregoing estimations, and through the volume of the traversed nodal traffic, it uses a model to estimate the next sleep-time duration of the destination node.

The power control is provided by determining the transmit periods and the associated power level such that the energy consumed is steadily reduced. To this end, by using the backward traffic-aware mechanism presented in the next section, the scheme aims to guarantee resource-sharing stability and at the same time offer energy conservation. Because nodes in wireless networks typically rely on their battery energy, the proposed framework encompassed in a traffic-aware middleware utilizes a reflective mechanism that hosts a traffic-aware scheme for conserving energy in CR wireless environments. The scheme evaluates the scheduled activity periods of each node in order to measure and estimate a "safe" forecast time duration for the scheduled time that each node can safely sleep in order to conserve energy.

The input nodal traffic is being considered in this work and estimated according to the BTD scheme. Wireless nodes, whether they are acting in the network as intermediate forwarding nodal points or as destinations, have to be self-aware in terms of power and processing as well as in terms of accurate participation in the transmission activity. There are many techniques such as the dynamic caching-oriented methods [11]. The present work utilizes a hybridized version of the proposed adaptive dynamic caching [31], which is considered to behave satisfactorily and enables simplicity in real-time implementation [11]. The performance of message delivery in devices communicating in a mobile peer-to-peer (MP2P) manner depends on two essential aspects: (1) the traffic volume and the impact on the devices' lifespan and (2) the mobility. Therefore, it is important to consider the traffic factor and enable, through the applied F-BTD scheme, a reflective action on the sleep-time duration of the nodes. The following section presents the related estimations, where according to the incoming traffic, each node evaluates the next sleep-time duration by using the F-BTD.

5.3.2.2 Selective F-BTD Estimation for Energy Conservation for Delay-Sensitive Multimedia Transmissions

Taking into account the fact that opportunistically connected nodes change their operational characteristics dynamically, when a source needs to send requested packets or stream of packets (file) to a destination where the destination nodes may have moved or are set in the sleep state, then the requested information will be missed and lost. For this purpose, a promiscuous caching mechanism is proposed [32] where nodes can passively cache/buffer their missed data packets/chunks in intermediate nodes in the end-to-end path. However, the volume of this cached was never used to impact the sleep-time duration of the node during the next sleep-time cycle. Within this context, this study proposes the F-BTD, where the volume of the cached traffic is considered in order to define the different Fibonacci traffic moments. The activity periods of a node are primarily dependent on the nature and the spikes of the incoming multimedia traffic destined for this node [32]. If the transmissions are performed on a periodic basis, then the nodes' lifetime can be forecasted and according to a model can be predicted and estimated [33]. However, this was not investigated in this work.

Each node admits multimedia traffic while in the active state, whereas if the node is in the sleep state it can cache the multimedia traffic in the 1-hop neighbor node (*Node(i–1)*) shown in Figure 5.4. As an example, this study employs the specifications of IEEE 802.16e [34], which recommend that the duration of the forwarding mechanism that takes place in a non-power-saving mode lies in the interval $1\ ns < \tau < 1\ ps$. This means that every ~0.125 μs (8 times in a millisecond),

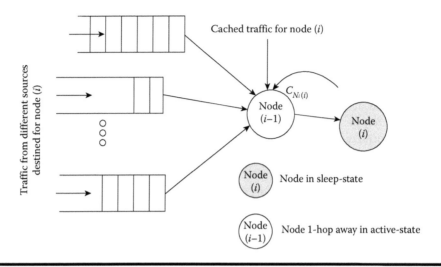

Figure 5.4 A schematic diagram of the promiscuous caching mechanism.

the communication triggering action between nodes may result in a problematic end-to-end accuracy. Adaptive dynamic caching [31,35] occurs and enables the packets to be "cached" in the 1-hop neighboring nodes. Correspondingly, if a node is no longer available because it entered the sleep state to conserve energy (in the interval slot $T = 0.125$ μs), then the packets are cached in an intermediate node with an adequate capacity equals to $C_{tf,k(s)}(t) > C_{tf,i}(t)$, where $C_{tf} > \alpha \cdot C_i$; α is the capacity adaptation degree based on the time duration of the capacity that is reserved on node N of C_k; and $C_{tf,k(s)}(t)$ is the required capacity, I is the destination node, and k is the buffering node (a hop before the destination via different paths).

This scheme is entirely based on the aggregated self-similarity nature of the incoming multimedia traffic, by associating the different traffic moments and the volume of the incoming traffic and increasing or decreasing the next sleep time of each node accordingly. This adaptive mechanism is achieved by using the F-BTD as shown in the next section.

5.3.2.3 Selective F-BTD Moments and Sleep-Time Duration Estimation

Let $C(t)$ be the capacity of the traffic that is destined for node i in the time slot (duration) t and $C_{N_i(t)}$ be the traffic volume capacity that is cached in node $(i-1)$ for time t. Then, as shown in Figure 5.5, according to the one-level BTD, the difference in the capacity measure can be estimated as the difference of the traffic while node (i) is set in the sleep state—and admits traffic—for a period, as follows:

$$F_{N_i(3)} \begin{bmatrix} T(\nabla C_{N_i(1)}) = T_2(\tau) - T_1(\tau - 1) \\ T(\nabla C_{N_i(2)}) = T_3(\tau - 1) - T_2(\tau - 2) \end{bmatrix}$$

$$\vdots \tag{5.1}$$

$$F_{N_i(i+n+3)} \left| T(\nabla C_{N_i(n+1)}) = T_n(\tau - (n-1)) - T_2(\tau - (n-2)) \right.$$

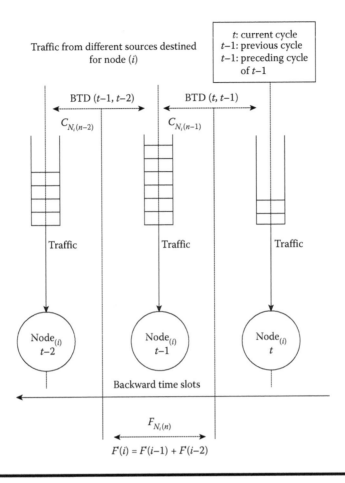

Figure 5.5 Two-level traffic moments for node (*i*) and the associated T_1, T_2, T_3 for obtaining $F_{N_i(n)}$.

where $\nabla C_{N_i(1)}$ denotes the first moment traffic/capacity difference that is destined for node (*i*) and it is cached in node (*i* – 1) for time τ, $T_2(t) - T_1(t-1)$ is the estimated traffic difference while packets are being cached/buffered in the (*i* – 1) hop for recoverability.

Equation 5.1 depicts the BTD estimation for one-level traffic comparisons, denoting that the moments are only being estimated for one-level ($T_2(t) - T_1(t-1)$).

The associated F-BTD moments can be evaluated by using the aggregated traffic of $\nabla C_{N_i(i-1)}$, $\nabla C_{N_i(i-2)}$ for $\nabla C_{N_i(i)}$ where it follows that

$$F_{N_i(n)} = T(\nabla C_{N_i(i-1)}) + T(\nabla C_{N_i(n-2)}), \quad \forall N_i(t) < t_s \tag{5.2}$$

where t_s is the maximum allowed end-to-end delay for the *s* stream to reach the destination, and $N_i(t)$ is the time that the specific packet chunks were cached in the intermediate node. $F_{N_i(n)}$ is the F-BTD moments (Figure 5.5) for the two previous BTD moments $\nabla C_{N_i(n-1)}$ and $\nabla C_{N_i(n-2)}$. Equation 5.2 is valid only if the F-BTD is greater than the next pair BTD evaluation as follows:

$$F_{N_i(n)} > F_{N_i(n+1)} \tag{5.3}$$

Let t be the duration for which node i should be in the active period, then the sleep-time duration can be measured using the following expression:

$$T_s(i)|^n = \frac{t(C_{N_i(\tau-1)}) + t(F_{N_i(\tau)})}{1 + \Delta T_s} \tag{5.4}$$

where $T_s(i)|^n$ is the estimated sleep time of the node during the next cycle, $t(C_{N_i(\tau-1)})$ is the time that the $C_{N_i(\tau-1)}$ capacity of the traffic needs to be processed, and $t(F_{N_i(\tau)})$ is the F-BTD if Equation 5.4 is satisfied. ΔT_s is measured according to

$$\Delta T_s = \frac{a_i(\tau)}{t_s} + T(\nabla C_{N_i(n)}) \tag{5.5}$$

where $\nabla C_{N_i(n)}$ is the normalized BTD, $a_i(t)$ is the incoming traffic for node i during the (current) time τ, and t_s is the maximum allowed end-to-end delay for the s stream to reach the destination. If Equation 5.4 is not satisfied, the sleep-time duration of the node is measured according to

$$T_s(i)|^n = \frac{t(C_{N_i(\tau-1)})}{1 + \Delta T_s} \tag{5.6}$$

The foregoing measures are in accordance with the algorithmic steps in Table 5.2.

The delay that the transmission experiences, δ_{ij}, should satisfy the constraint $\delta_{ij} < d_p$, where d_p is the maximum delay in the end-to-end path from a source to a destination and can be is evaluated as

$$d_p = \sum_{i=0}^{i-1} \delta_i + T_i \forall (\delta_i + T_i) < t_s \tag{5.7}$$

where δ_i is the duration that the requested data were hosted in node i, and T is the transmission delay.

Table 5.2 Basic Steps of the Proposed BTD Scheme

for Node(i) that there is $C(t) > 0$ in the best path in k intermediate nodes {
while ($C_{N_i(t)} > 0$) {//cached Traffic measurement
Evaluate ($T(\nabla C_{N_i(1)})$);
if (Activity_Period for node(i) > 0 && $C_{N_i(t)} \forall t(N_i) > 0$)
//Measure sleep-time duration
Evaluate $\nabla C_{N_i(1)}$ and $\nabla C_{N_i(2)}$;
if ($N_i(t) < t_s$)
estimate the $F_{N_i(n)}$ such that $F_{N_i(n)} > F_{N_i(n+1)}$
sleep for $T_s(i)|^n = \dfrac{t(C_{N_i(\tau-1)}) + t(F_{N_i(\tau)})}{1 + \Delta T_s}$; //during the next cycle
else
sleep for $T_s(i)|^n = \dfrac{t(C_{N_i(\tau-1)})}{1 + \Delta T_s}$; //during the next cycle
}//while
}//for

Taking into consideration the foregoing estimates, the energy efficiency EE_{t_f} can be defined as a measure of the capacity of *node(i)* divided by the *total power consumed by the node*, as

$$EE_{t_f}(T) = \frac{C_{t_f}(T)}{Total\ power} \forall \min(T_s(i)|^n)$$
(5.8)

Equation 5.8 can be defined as the primary metric for the lifespan extensibility of the wireless node in the system.

5.4 Performance Evaluation Analysis, Experimental Results, and Discussion

Several experimental tests were conducted to validate the efficiency of the proposed routing protocol and the resource-intensive traffic-aware scheme. Performance evaluation results were extracted by conducting exhaustive simulation runs and experimentation using NS-2 [36] and the generated real-traffic traces for implementing the proposed scenario. The energy consumption model used in the simulation for the calculation of energy consumption is based theoretically on the specifications of WiMax IEEE 802.16e (version 2005) [34]. The obtained results characterize the trade-off issues between the performance when deploying the discussed scenario and the energy consumption of each secondary CR node when using the proposed traffic-oriented scheme. The results also include comparisons with other existing schemes for the throughput, reliability, and accuracy offered by the proposed framework as well as EC efficiency conveying an estimated confidence interval (CI) of approximately $3\% < CI < 5\%$. All CIs were found to be less than 5% of the mean values of the examined parameters. The mobility model adopted in this work is based on the probabilistic mobility scenario derived by fractional random walk. The probabilistic random walk mobility model was derived from Brownian motion [37], where nodes move according to certain probabilities with respect to location and time.

According to such a simulation scenario, a number of data flows contend to pass through the same intermediate node. Thus, evaluation of delay metrics is crucial for an efficient performance of the proposed routing protocol. In this context, a number of delay metrics [38] are evaluated, such as the end-to-end delay, backoff delay, switching delay, and queuing delay. The end-to-end delay from the source node to the destination node is computed as the sum of the queuing delay and the node delay:

$$D_{End-to-End} = D_{queing} + D_{node}$$
(5.9)

The node delay at an intermediate node *i* is based on the switching delay and the backoff delay and is computed as follows:

$$D_{node} = \sum_{1}^{i} \left(D_{switching} + D_{backoff} \right)$$
(5.10)

A dynamic topology model was used, and the nodes used fractional random walk, where the BTD and F-BTD are applied. A common look-up application is being developed to enable users

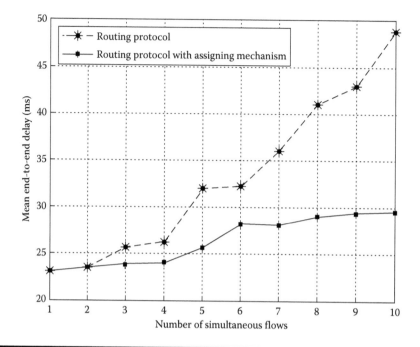

Figure 5.6 Mean end-to-end delay for different numbers of simultaneous flows.

to share resources on the move that are made available by peers for sharing. This application hosts files of different sizes that are requested by peers in an opportunistic manner. Each device has an asymmetrical storage capacity compared with the storages of the peer devices. The ranges of the capacities for which devices are supported are set in the interval 1–20 MB.[1]

Figure 5.6 shows simulation results pertaining to a performance comparison of the mean end-to-end delay, while the number of active flows increases for both versions of the proposed routing protocol. It is clear that when the routing protocol incorporates the assigning mechanism and the number of active flows in the network is small, there is no important advantage in terms of the mean end-to-end delay. However, when the number of active flows is greater than three, additional delays occur and flow redirection becomes necessary. Such results also show that the mean end-to-end delay is less in the case of the enhanced routing protocol, in comparison to the basic version of it that does not incorporate the assigning mechanism supporting the efficient provision of delay-sensitive multimedia streams over CR networks.

Figure 5.7 depicts the simulation results of the end-to-end delay for a single flow, when the probability of the primary user's presence increases, and Figure 5.8 presents the average end-to-end delay for ten simultaneous flows for a certain value of the probability of the primary user's presence. From both figures, it is clear that when the probability of the primary user presence is higher, the delay increases, while in the case of the basic routing protocol, the delay increase is more significant in comparison to the enhanced routing protocol incorporating the assigning mechanism. This result is reasonable because the probability of the presence of an incumbent system is detected as a route failure, introducing in this way an additional delay.

---

[1] The capacity of each device can be tuned according to the volume of the traffic in the configuration process.

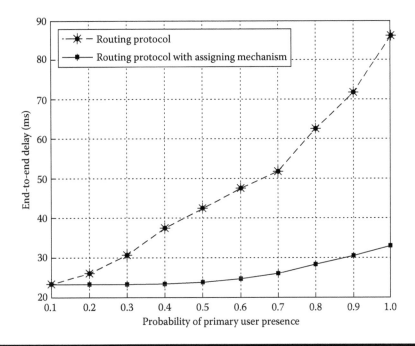

Figure 5.7   End-to-end delay for the first flow versus probability of PU presence.

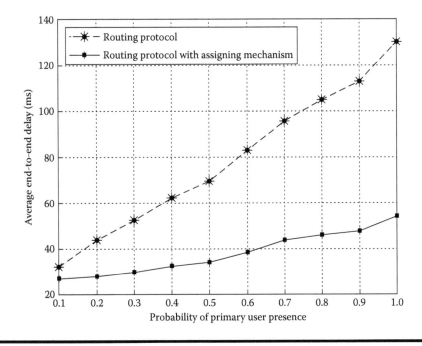

Figure 5.8   Average end-to-end delay for ten simultaneous flows versus probability of PU presence.

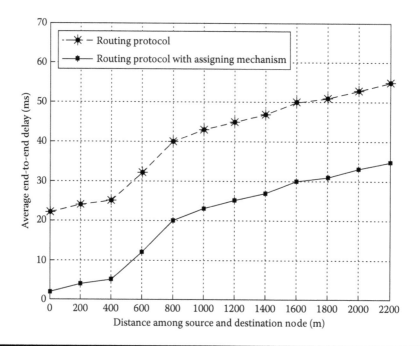

Figure 5.9 Average end-to-end delay versus node distance.

Figure 5.9 shows the average end-to-end delay that occurred among the source and destination nodes as the distance between them is increased. From this figure, it is clear that the distance affects the delay among nodes. This result is reasonable because the longer the routing path, the more numerous the primary nodes that affect the path, and the more significant the effects of the route range and route diversity. It is further observed that the initial version of the proposed routing protocol results in higher delays, as the distance is increasing in comparison to those occurred when assigning mechanism is introduced, resulting in the most optimal routing paths between the source and destination nodes. Consequently, the longer the path, the more significant the effects of the route diversity. In addition, Figure 5.10 depicts the comparison among both versions of the proposed routing protocol, under the number of hops that are required, in order to make feasible all routing paths between source and destination nodes, for each flow set according to the simulation scenario. This comparison shows that routing protocol incorporating the assigning mechanism performs better because it makes the decision for routing path establishment at every hop.

In this section, we demonstrate the effectiveness of the proposed BTD approach and validate the accuracy of the developed scheme by comparing the analytical results of the scheme to those obtained from extensive simulation experiments in work done in Refs. [8,11]. In Figure 5.11, the network lifetime in contrast to the number of mobile nodes for different schemes is presented. It is important to note that the network lifetime is significantly extended by using the proposed F-BTD for energy conservation. Figure 5.12 shows the average throughput plotted against the number of nodes during the resource exchange process for different schemes. It is undoubtedly true that the proposed scheme enables a higher average throughput response in the system by using the selective F-BTD scheme.

Figure 5.13 shows the average throughput with the total transfer delay in (microseconds) for different mobility degrees. The different throughput responses are shown in contrast to the

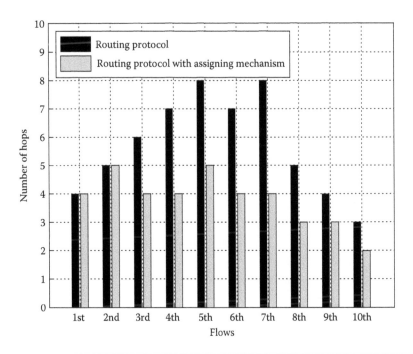

Figure 5.10   Number of hops per flow.

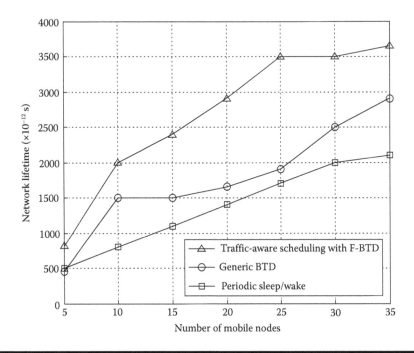

Figure 5.11   The variation in the average throughput with the number of nodes in the cluster zone for delay-bounded transmissions. Different comparable schemes are evaluated.

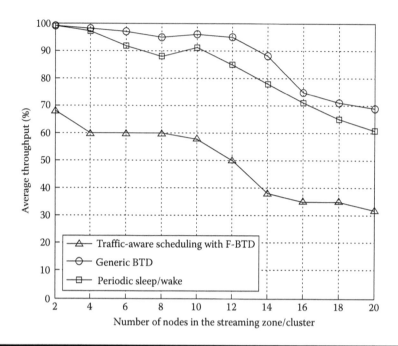

Figure 5.12 The variation in the average throughput with the number of nodes during the resource exchange process for different schemes.

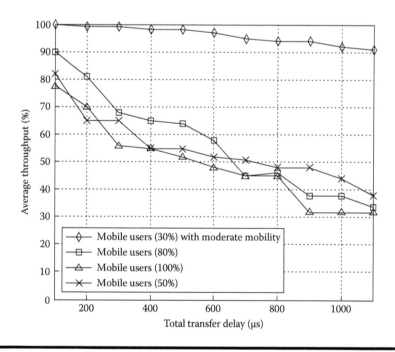

Figure 5.13 The variation in the average throughput with the total transfer delay (µs).

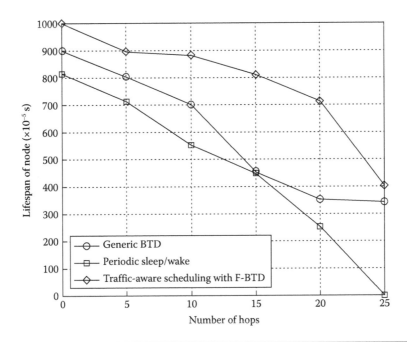

Figure 5.14   The variation in the lifespan of each node with the number of hops using different schemes under real-time evaluations.

mobility characteristics, for full node mobility, moderate mobility, and low (>30%) mobility. Figure 5.14 shows the variation in the lifespan of each node with the number of hops for different schemes evaluated in real time. Figure 5.15 shows the average energy consumed with the power ratio in decibels of the measured power referenced to 1 mW, and Figure 5.16 shows the fraction of the remaining energy through time for different schemes. The results obtained in Figure 5.16 show that the network lifetime can be significantly prolonged when the F-BTD is applied. By comparing the results obtained for the generic BTD scheme developed in Ref. [11] as well as with the periodic sleep/wake scheduling, obviously the proposed scheme offers greater energy efficiency while minimizing the delay per request.

Likewise, Figure 5.17 shows the respective complementary cumulative distribution function (CCDF) with the mean download time for requests over a certain capacity. The later evaluation was extracted in the presence of fading and no-fading communicating obstacles. The results obtained and presented in Figure 5.18 show the results for the variation in CCDF sharing reliability with the number of sharing peer users, which were extracted by conducting both simulation experiments and evaluations in the presence of fading and non-fading characteristics. Comparative results extracted by simulation and compared with existing schemes exhibit similarities in the measured values with the associated optimization variations (which are measured within the CI of 5%–7% for the conducted experiments).

The variation in the energy efficiency (in bytes/mW), which is defined as the service capacity/total energy consumed as in Equation 5.8, with the fraction of the remaining energy in each node is shown in Figure 5.19 for four different schemes. The proposed framework is shown to have the higher remaining energy for each node in the system, whereas, compared to the scheme in Ref. [8], it is shown to have an optimized energy efficiency as it allows greater energy efficiency in contrast to the remaining energy of each node.

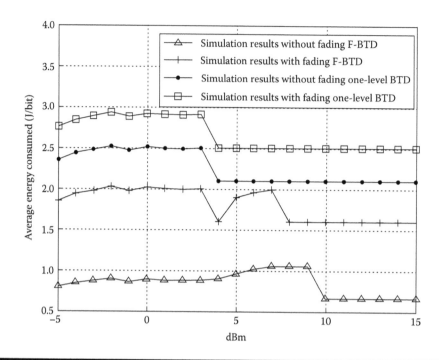

Figure 5.15 **The average energy consumed with the power ratio in milli-decibels (dBm) of the measured power referenced to 1 mW.**

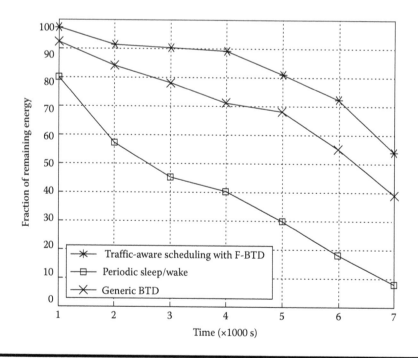

Figure 5.16 **The fraction of the remaining energy through time using real-time evaluation for different schemes.**

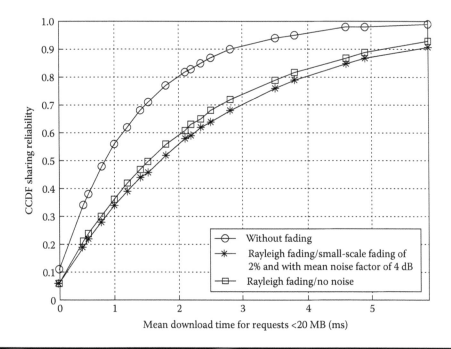

Figure 5.17 The CCDF for the sharing reliability with the mean download time for requests below a certain capacity.

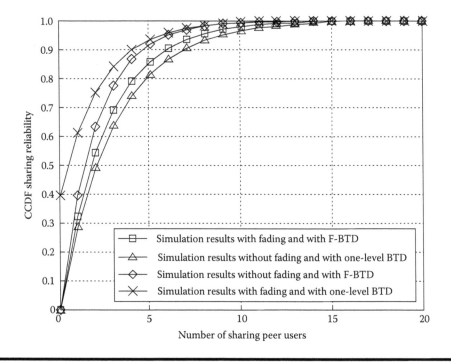

Figure 5.18 CCDF sharing reliability with the number of sharing peer users.

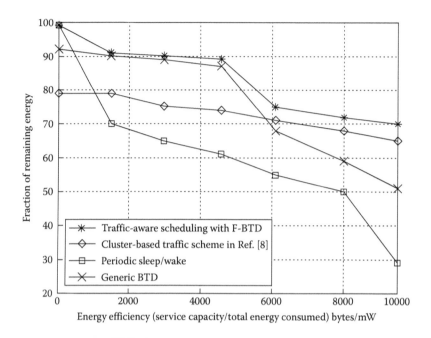

Figure 5.19 **Variation in energy efficiency (service capacity/total energy consumed; bytes/mW) with fraction of the remaining energy in each node.**

5.5 Conclusions and Further Research

In order to guarantee the end-to-end availability of requested multimedia resources, as well as significantly reduce energy consumption and maintain the requested scheduled transfers, a dynamic adaptive methodology is needed. This chapter presents an efficient routing mechanism that, in conjunction with the underlying F-BTD scheme, enables energy conservation and reliable data flow among secondary communication nodes with heterogeneous spectrum availability in CR systems, while it fulfills the delay constraints and requirements of multimedia resources. The proposed adaptive traffic-oriented mechanism takes into consideration the active moments and measures the incoming traffic by using the selective F-BTD scheme. The supported routing scheme establishes an end-to-end optimal path, and secondary nodes in CR systems can efficiently and, in a collaborative manner, share requested data/resources. The F-BTD scheme enables the self-tunability of the sleep schedule of each node, where it is applied through the traffic difference of the active moments of the traffic, measured within a certain transmission time frame. Within the proposed framework, the bounded end-to-end delay of the transmission is taken into consideration for each secondary node, aiming to impact the EC through the modeled traffic-aware mechanism. The performance evaluation through simulation shows that the proposed routing scheme, in conjunction with the F-BTD mechanism, manipulates the energy consumption of each secondary node/device effectively, and outperforms similar traffic-aware schemes. Moreover, the traffic-aware management scheme can significantly reduce the energy consumed and can keep the throughput response of the system at relatively high levels. The comparative measurements with other similar traffic-oriented and periodic schemes show that the proposed methodology can efficiently conserve energy, by offering at the same time significantly high Successful packet Delivery Ratio (SDRs), and can significantly extend the lifetime of each secondary node in the CR network. Further, the efficiency of the routing protocol operation, from the standpoint of the establishment of the maximum number of routing paths and minimum delays, was validated by adopting the proposed message exchange mechanism, which was developed based on the simulation scenario defined earlier. Toward evaluating the performance of the routing protocol in this respect, a large number set of experimental tests were conducted under controlled simulation conditions, in which various secondary systems were concurrently/simultaneously communicating over ad hoc connections, accessing the available TVWS. The obtained experimental results verified the validity of the proposed routing mechanism, in regard to enabling efficient communication among secondary nodes located in areas with different TVWS availabilities.

Further avenues in our ongoing research include the evaluation of our scheme using real-time measurements and real-time verification using the existing infrastructure. Issues to be considered are topology formation, using social collaboration as well as geographical profiles, in order to overcome potential partitioning problems. Moreover, we are investigating the usage of traffic engineering models to explicitly express the behavior of such dynamically changing scenarios. In addition, we are working toward combining our scheme with infinitesimal perturbation analysis and applying a stochastic algorithm to the performance gradient of the system. Moreover, within the current research context will be the expansion of this model into a multilevel Markov fractality model so that it emphasizes to the different moments of the active traffic to facilitate energy conservation. Finally, several optimization methods will be adopted toward minimizing delays occurring during the routing paths of data flows and maximizing the number of established paths. This comprises an open-ended research issue with many research concepts for future examination.

Acknowledgments

This project is implemented through the Operational Program "Education and Lifelong Learning", Action Archimedes III and is co-financed by the European Union (European Social Fund) and Greek national funds (National Strategic Reference Framework 2007–2013).

References

1. A. Bourdena, E. Pallis, G. Kormentzas, and G. Mastorakis, A prototype cognitive radio architecture for TVWS exploitation under the real time secondary spectrum market policy, *Physical Communication*, 2013.
2. I. F. Akyildiz, W. Y. Lee, M. C. Vuran, and S. Mohanty, NeXt generation/dynamic spectrum access/cognitive radio wireless networks: A survey, *Computer Networks*, vol. 50, no. 13, 2006, pp. 2127–2159.
3. A. Bourdena, E. Pallis, G. Kormentzas, C. Skianis, and G. Mastorakis, Real-time TVWS trading based on a centralised CR network architecture, in *Proceedings of the International Conference IEEE Globecom 2011*, IEEE International Workshop on Recent Advances in Cognitive Communications and Networking, Texas, Houston, USA, December 5–9, 2011, pp. 994–999.
4. A. Bourdena, E. Pallis, G. Kormentzas, and G. Mastorakis, A radio resource management framework for TVWS exploitation under an auction-based approach, in *Proceedings of the IEEE 8th International Conference on Network and Service Management (IEEE CNSM2012)*, Las Vegas, October 22–26, 2012, pp. 204–208.
5. A. Bourdena, E. Pallis, G. Kormentzas, H. Skianis, and G. Mastorakis, QoS provisioning and policy management in a broker-based CR network architecture, in *Proceedings of the International Conference IEEE Globecom 2012*, Anaheim, CA, December 3–7, 2012.

6. E. Hossain, D. Niyato, and Z. Han, *Dynamic Spectrum Access and Management in Cognitive Radio Networks*, 1st ed. Cambridge University Press, Cambridge, UK, 2009.

7. C. X. Mavromoustakis, Using backward traffic difference estimation for efficient energy saving schedules in wireless device, *IEEE CommSoft E-Letters*, vol. 1, no. 1, 2012, pp. 1–6.

8. M. C. Charalambous, C. X. Mavromoustakis, and M. B. Yassein, A resource intensive traffic-aware scheme for cluster-based energy conservation in wireless devices, in *Proceedings of the IEEE 14th International Conference on High Performance Computing and Communications (HPCC-2012) of the Third International Workshop on Wireless Networks and Multimedia (WNM-2012)*, to be held in conjunction, Liverpool, UK, June 25–27, 2012, pp. 879–884.

9. H. Shpungin, Energy efficient online routing in wireless ad hoc networks, in *Proceedings of the International Conference of the 8th Annual IEEE Communications Society Conference on Sensor, Mesh and Ad Hoc Communications and Networks (IEEE SECON'11)*, 2011, Salt Lake City, UT.

10. C. X. Mavromoustakis and K. G. Zerfiridis, On the diversity properties of wireless mobility with the user-centered temporal capacity awareness for EC in wireless devices, in *Proceedings of the International Conference of the Sixth IEEE International Conference on Wireless and Mobile Communications, ICWMC 2010*, Valencia, Spain, September 20–25, 2010, pp. 367–372.

11. C. X. Mavromoustakis, C. D. Dimitriou, and G. Mastorakis, Using real-time backward traffic difference estimation for energy conservation in wireless devices, in *Proceedings of the International Conference of the Fourth International Conference on Advances in P2P Systems (AP2PS 2012)*, Barcelona, Spain, September 23–28, 2012, pp. 18–23.

12. F. Wang, H. Jianwei, and Z. Yuping, Delay sensitive communications over cognitive radio networks, *IEEE Transactions on Wireless Communications*, vol. 11, no. 4, 2012, pp. 1402–1411.

13. S. Li, T. H. Luan, and X. Shen, Channel allocation for smooth video delivery over cognitive radio networks, in *Proceedings of the IEEE Global Telecommunications Conference (GLOBECOM 2010)*, 2010, pp. 1–5.

14. F. Hou, Z. Chen, J. Huang, Z. Li, and A. K. Katsaggelos, Multimedia multicast service provisioning in cognitive radio networks, in *Proceedings of the 9th International Wireless Communications and Mobile Computing Conference (IWCMC 2013)*, Sardinia, Italy, July 1–5, 2013, pp. 1175–1180.

15. F. Li, J. Zhang, and K. B. Letaief, Location-aware distributed routing in cognitive radio networks, in *Proceedings of the 1st IEEE Conference on Communications in China*, Beijing, China, August 15–17, 2012, pp. 733–738.

16. A. Chaoub and E. Ibn Elhaj, Modelling multimedia traffic over cognitive radio networks using Markov chain under collision errors, *International Journal of Wisdom Based Computing (IJWBC)*, vol. 1, no. 3, 2011, pp. 139–145.

17. S. M. Kamruzzaman, E. Kim, and D. G. Jeong, Spectrum and energy aware routing protocol for cognitive radio ad hoc networks, in *Proceedings of the IEEE International Conference on Communications (ICC)*, 2011, pp. 1–5.

18. A. Raniwala, K. Gopalan, and T. Chiueh, Centralized assignment and routing algorithms for multichannel wireless mesh networks, *ACM SIGMOBILE Mobile Computing and Communications Review*, vol. 8, no. 2, 2004, pp. 50–65.

19. M. Li, L. Zhang, V. Li, X. Shan, and Y. Ren, An energy-aware multipath routing protocol for mobile ad hoc networks, *ACM Sigcomm Asia*, vol. 5, 2005, pp. 10–12.

20. S. Bayhan and F. Alagoz, Scheduling in centralized cognitive radio networks for energy efficiency, *IEEE Transactions on Vehicular Technology*, vol. 62, no. 2, 2013, pp. 582–595.

21. G. Miao, N. Himayat, Y. G. Li, and A. Swami, Cross-layer optimization for energy-efficient wireless communications: A survey, *Wireless Communications and Mobile Computing*, vol. 9, no. 4, 2009, pp. 529–542.

22. I. Ashraf, F. Boccardi, and L. Ho, SLEEP mode techniques for small cell deployments, *IEEE Communications Magazine*, vol. 49, no. 8, 2011, pp. 72–79.

23. C. X. Mavromoustakis and H. D. Karatza, Real time performance evaluation of asynchronous time division traffic-aware and delay-tolerant scheme in ad-hoc sensor networks, *International Journal of Communication Systems (IJCS)*, vol. 23, no. 2, 2010, pp. 167–186.

24. J. Yu and A. P. Petropulu, Study of the effect of the wireless gateway on incoming self-similar traffic, *IEEE Transactions on Signal Processing*, vol. 54, no. 10, 2006, pp. 3741–3758.

25. X. Zhuang and S. Pande, A scalable priority queue architecture for high speed network processing, in *Proceedings of the 25th IEEE International Conference on Computer Communications (INFOCOM'06)*, 2006, pp. 1–12.
26. H. Cunqing and P. Y. Tak-Shink, Asynchronous random sleeping for sensor networks, *ACM Transactions on Sensor Networks (TOSN)*, vol. 3, 2007.
27. Q. Cao, T. Abdelzaher, T. He, and J. Stankonic, Towards optimal sleep scheduling in sensor networks for rare-event detection, in *Proceedings of the 4th International Symposium on Information Processing in Sensor Networks*, April 25–27, 2005.
28. I. Jawhar, J. Wu, and P. Agrawal, Resource scheduling in wireless networks using directional antennas, *IEEE Transactions on Parallel and Distributed Systems*, vol. 21, no. 9, 2010, pp. 1240–1253.
29. C. Dimitriou, C. X. Mavromoustakis, G. Mastorakis, and E. Pallis, On the performance response of delay-bounded energy-aware bandwidth allocation scheme in wireless networks, in *Proceedings of the International Workshop on Immersive and Interactive Multimedia Communications over the Future Internet*, organized in conjunction with IEEE International Communications Conference (ICC 2013), Budapest, Hungary, June 9–13, 2013.
30. Ad hoc, On-Demand Distance Vector (AODV) Routing, available at: RFC: http://www.ietf.org/rfc/rfc3561.txt, accessed: July 2013.
31. C. X. Mavromoustakis, On the impact of caching and a model for storage-capacity measurements for energy conservation in asymmetrical wireless devices, in *Proceedings of the 16th International Conference on Software, Telecommunications and Computer Networks (SoftCOM 2008)*, September 25–27, 2008, Dubrovnik, September 27, Split, pp. 243–247.
32. O. Bello, A. Bagula, and H. A. Chan, Multilayer traffic engineering in interworking multihop wireless networks, Special Issue on Smart Communication Protocols and Algorithms, *Journal Network Protocols and Algorithms*, vol. 4, no. 2, 2012, pp. 5–29.
33. F. Guo and J. Xu, Research on diffusion strategy about resource index of MP2P, *International Journal of Wireless and Microwave Technologies (IJWMT)*, vol. 2, no. 2, 2012, pp. 1–6.
34. WiMAX Forum Specification (2011-09-20), IEEE to WiMAX Forum, IEEE to WiMAX Forum.
35. C. X. Mavromoustakis and H. D. Karatza, A tiered-based asynchronous scheduling scheme for delay constrained energy efficient connectivity in asymmetrical wireless devices, *The Journal of Supercomputing*, vol. 59, no. 1, 2012, pp. 61–82.
36. NS-2 Simulator, http://www.isi.edu/nsnam/ns/, accessed June 2013.
37. T. Camp, J. Boleng, and V. Davies, A survey of mobility models for ad hoc network research, *Wireless Communication and Mobile Computing (WCMC): Special Issue on Mobile Ad Hoc Networking: Research Trends and Applications*, vol. 2, no. 5, 2002, pp. 483–502.
38. G. Cheng, W. Liu, Y. Li, and W. Cheng, Joint on-demand routing and spectrum assignment in cognitive radio networks, in *Proceedings of the IEEE International Conference on Communications*, ICC'07, 2007, pp. 6499–6503.

ADVANCED NETWORK PROTOCOLS FOR MULTIMEDIA OVER CRN

Chapter 6

Exploiting Cognitive Management for Supporting Multimedia Applications over Cognitive Radio Networks

Faouzi Bouali, Oriol Sallent, and Jordi Pérez-Romero

Signal Theory and Communications Department (TSC),
Universitat Politécnica de Catalunya (UPC), Barcelona, Spain

Contents

6.1 Context/Motivation

The cognitive radio (CR) paradigm has led to the emergence of intelligent radios that automatically adjust their behavior based on active monitoring of the environment [1,2]. In this context, the introduction of cognitive techniques for the management of wireless networks will lead to enhanced robustness by capitalizing on the learning capabilities intrinsic to cognitive systems. Strengthening these cognitive techniques is of great interest for optimizing spectrum management functions. Therefore, the technical requirements of new cognitive management systems have been considered in many studies [3–5]. In particular, many recent proposals have attempted to develop new models and efficient architectures for building new cognitive management systems in emerging environments, such as the Future Internet (FI) [6] or the home environment [7]. The proven usefulness of cognitive capabilities has stimulated the initiation of many research projects (e.g., Refs. [8,9]) and standardization activities (e.g., Refs. [10,11]) to further strengthen and promote the use of cognitive management systems.

The flexibility provided by CR is of paramount importance to enable the so-called dynamic spectrum access, a new communication paradigm that proposes the use and sharing of available spectrum in an opportunistic manner to increase its usage efficiency. Not surprisingly, this topic has received a lot of interest in the recent literature (e.g., Refs. [12–14]). The flexibility provided by spectrum agility has increased efficiency through proper spectrum selection criteria.

In this respect, we have proposed in Ref. [15] a framework based on the fittingness factor concept to assess the suitability of spectrum blocks to support the bit rate requirements of multimedia (MM) applications. The proposed architecture was extended in Ref. [16] with a reliability tester to detect relevant changes that may occur in radio and interference conditions and update a set of relevant statistics accordingly. Based on this extension, a spectrum management strategy has been developed to assess the resulting performance in a realistic digital home (DH) [16].

The proposed framework has been shown to efficiently support the bit rate requirements of delay-insensitive MM applications (e.g., file transfers) subject to the nonstationary conditions of cognitive radio networks (CRNs). Nevertheless, it may not be sufficient to support the tighter requirements of MM applications, such as the strict delay and jitter constraints associated with delay-sensitive MM applications (e.g., real-time video). This is because the variability of spectral resources may not significantly impact delay-insensitive applications, but has considerable consequences for delay-sensitive applications and often leads to unsatisfactory user experience.

Therefore, the first main contribution of this chapter is to extend the proposed knowledge management framework described in Ref. [16] to assess the suitability of spectrum resources to support the heterogeneous requirements of MM applications. In this respect, a generic formulation of the fittingness factor is developed, and all relevant functional blocks are implemented accordingly. The second contribution is the development of a proactive spectrum management strategy that exploits an estimation of the suitability of spectrum resources in relation to the various requirements of MM applications for assisting both spectrum selection (SS) and spectrum mobility (SM) functionalities. The third contribution is to benchmark the effectiveness of the proposed strategy to support a mixture of real-time and best-effort applications under the nonstationary conditions of CRNs.

The rest of this chapter is organized as follows. The considered cognitive management functional architecture for supporting MM applications over CRNs is presented in Section 6.2. In particular, a generic fittingness factor function formulation is developed in Section 6.3 to capture the various requirements of MM applications, and all relevant functional modules are implemented accordingly. A proactive fittingness factor spectrum management strategy is proposed in Section 6.4 to support both SS and SM functionalities. The results are presented in Section 6.5 to first assess performance under stationary conditions from a cost-benefit perspective and, second, to evaluate the robustness of the proposed methodology when conditions become nonstationary. The conclusions are presented in Section 6.6.

6.2 Considered Architecture

The problem considered here is the selection of the spectrum to be assigned to a set of L radio links that need to be established between pairs of terminals and/or network infrastructure nodes. The purpose of the lth radio link is to support the communication flow generated by a given MM application (e.g., voice, web browsing, or video) that is characterized by a set of K_l requirements to be achieved during session duration T_l. For each $k \in \{1,\dots,K_l\}$, $R_{req,l}^k$ denotes the kth requirement of the lth application. As shown in the example of Figure 6.1, the established radio links may connect a terminal node to a network infrastructure (links #1 and #2) or two terminal nodes (links #3 and #4). To reflect the various priorities that may be associated with MM applications, it is assumed that each application

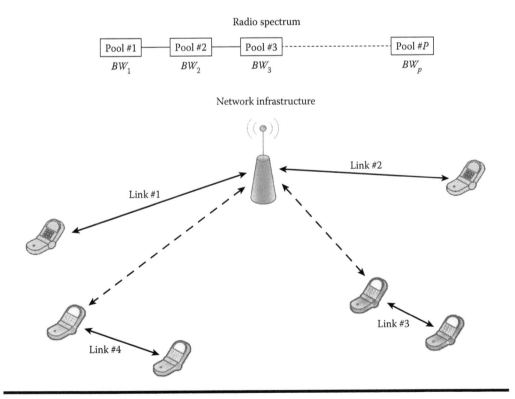

Figure 6.1   Example of established radio links.

may be either preemption-capable (*Preemp(l)* = *1*) or preemption-vulnerable (*Preemp(l)* = *0*) and that the former type may preempt the latter type if necessary. To meet these heterogeneous MM requirements, it is assumed that the network infrastructure is responsible for assigning suitable spectrum resources to each radio link. In this respect, the spectrum is modeled as a set of P blocks (termed *pools* in this chapter), each of bandwidth BW_p. The SS decisions made on the network infrastructure side are indicated to mobile terminals through suitable control channels (denoted in Figure 6.1 as dashed lines in case there is no data connection with the network infrastructure).

On the basis of the foregoing considerations, the generic functional architecture proposed in Ref. [15] and depicted in Figure 6.2 is instantiated in this chapter to support the various requirements of MM applications. Inspired by the European Telecommunications Standards Institute (ETSI)-Reconfigurable Radio System (RRS) architecture [17,18], two cognitive management entities are introduced on both the terminal and the network infrastructure sides to manage relevant knowledge about the environment. According to this architecture, most of the cognitive management functionalities are executed on the network infrastructure side with support from the terminal side. Specifically, as illustrated in Figure 6.2, the execution involves the following entities:

1. The knowledge management entity, which is responsible for storing and managing the relevant knowledge obtained from the radio environment that is used by the decision-making entity to make decisions. It is composed of a knowledge manager (KM) that monitors the suitability of existing spectral resources to support the set of MM applications.

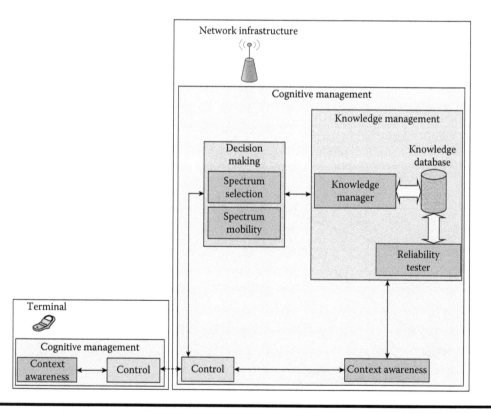

Figure 6.2　The considered cognitive management functional architecture for supporting MM over CRNs.

KM monitoring is based on information retrieved from a knowledge database (KD) that stores a set of relevant statistics about the environment and a list of available spectrum pools. To detect relevant changes in these statistics when operating in nonstationary environments, a reliability tester (RT) keeps analyzing a set of key performance indicators (KPIs) from the radio environment and may restart the process of generating KD statistics if necessary.

2. The decision-making entity, which is responsible for assigning the appropriate pools to different links. For that purpose, it interacts with the KM that will provide the relevant information for the decisions to be made. Decision making is split into two functional entities: (1) the SS functionality, which will pick up a suitable pool for each communication whenever a new request for establishing a radio link arrives, and (2) the SM functionality, which will perform the reconfiguration of assigned pools whenever changes occur in the environment and better pools can be found for some links.

3. The context awareness (CA) entity, which is responsible for taking measurements on both the terminal and network infrastructure sides. Actual measurements of the performance achieved in a given link when using a certain spectrum pool are exploited by the KM, which will retain these observations for supporting future SS decisions.

Finally, the control modules depicted in Figure 6.2 illustrate the need for the exchange of control messages between the terminal and network infrastructure to support the diverse cognitive management mechanisms and procedures (e.g., the establishment/release of radio links and measurement reporting). The interested reader is referred to Refs. [19,20] for details of possible implementation mechanisms for this exchange of control messages through the use of a cognitive control channel.

It is worth mentioning that the considered centralized setting mainly targets local environments (e.g., DH) where access to a central point is relatively easy. Nevertheless, the proposed framework also accepts decentralized or hybrid (e.g., centralized database for TV whitespace and decentralized for ISM band) architectures. In such scenarios, decision-making functionalities (i.e., SS and SM) would be executed at each terminal based on the support provided by a local knowledge management domain. More details on how to build such local knowledge domains are given at the end of Section 6.3.

6.3 Knowledge Management

In this section, the considered functional architecture is instantiated to monitor the time-varying suitability of spectrum resources to support the heterogeneous MM applications. To this end, a generic formulation of the fittingness factor is developed, and the relevant functional blocks are implemented accordingly for the considered centralized setting. Some initial guidelines about the applicability to decentralized scenarios are provided at the end of the section.

6.3.1 Fittingness Factor Definition

To assess the suitability of spectral resources to meet the requirements of different applications, the so-called fittingness factor ($F_{l,p}$) was considered in Ref. [21] as a metric capturing the suitability of each pth spectrum pool for the application supported by the lth radio link. $F_{l,p}$ was formulated to assess the suitability in terms of the bit rate that can be achieved while operating in the spectrum pool p versus the bit rate required by the corresponding application.

From the more general scope of MM applications, the fittingness factor $F_{l,p}$ can be formulated as a weighted sum of the marginal fittingness factors associated with each of the K_l requirements:

$$F_{l,p} = \sum_{k=1}^{K_l} \eta_k \times F_{l,p,k} \tag{6.1}$$

where $F_{l,p,k}$ denotes the marginal fittingness factor with respect to the kth requirement $R_{req,l}^k$, and $\{\eta_k\}_{1 \leq k \leq K_l}$ are the corresponding weights satisfying the following relationship:

$$\sum_{k=1}^{K_l} \eta_k = 1 \tag{6.2}$$

More specifically, the following marginal fittingness factor function is considered:

$$F_{l,p,k} = \frac{\left(\frac{R_k(l,p)}{R_{req,l}^k}\right)^\xi}{1 + \left(\frac{R_k(l,p)}{R_{req,l}^k}\right)^\xi} \tag{6.3}$$

where $R_k(l,p)$ denotes the achievable performance by link l using pool p with respect to the kth requirement ($R_{req,l}^k$), and ξ is a shaping parameter that allows the function to capture different degrees of elasticity of the application with respect to the kth requirement. Temporal variations of $F_{l,p,k}$ will be related to dynamics of the scenario (e.g., interference variability).

The proposed marginal fittingness factor formulation belongs to the family of sigmoid functions [22]. To better analyze its behavior, Figure 6.3 plots it as a function of the ratio $R_k(l,p) / R_{req,l}^k$ for different values of the shaping parameter ξ. It can be seen that $F_{l,p,k}$ is a monotonic increasing function of the achievable performance $R_k(l,p)$ that equals 0.5 at $R_k(l,p) = R_{req,l}^k$, and tends asymptotically to 1. The marginal increase in the marginal fittingness factor for large performances $R_k(l,p)$ well above the requirement becomes progressively smaller especially when intermediate values of ξ are used (e.g., $\xi = 5$). Therefore, $F_{l,p,k}$ provides a measure of the suitability of a given pool to support the kth link requirement ($R_{req,l}^k$), with values ranging from 0 (low suitability) to 1 (high suitability).

6.3.2 Knowledge Database

In order to characterize the suitability of a given pool p for a given link l based on the history of using this pool, it is proposed to retain in the KD observed $F_{l,p}$ values based on measurements extracted from the radio environment. In particular, the measurement of achieved performances $\{R_k(l,p)\}_{1 \leq k \leq K_l}$ of active link/pool pairs will be obtained, from which the associated marginal fittingness factor values are computed according to Equation 6.3. These marginal values are combined to derive the current value of $F_{l,p}$ from Equation 6.1 that is stored in the database together with a timer indicating the elapsed time since these measurements were obtained ($\Delta t_{l,p}$).

Considering that $F_{l,p}$ values can be associated with two states, LOW ($< \delta_{l,p}$) or HIGH ($\geq \delta_{l,p}$), and assuming a stationary behavior of the different pools during system operation, the following

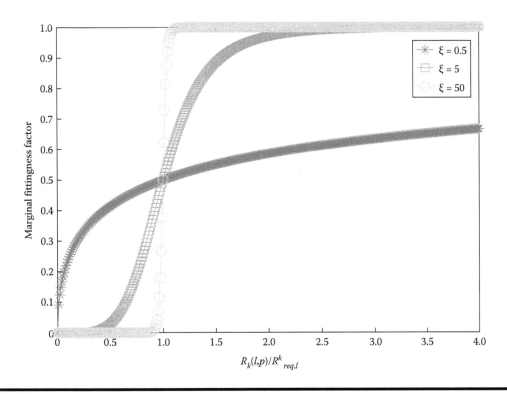

Figure 6.3   Behavior of the marginal fittingness factor ($F_{l,p,k}$).

statistics are also generated and stored in the database based on the previously obtained measurements of $F_{l,p}$:

■ The probability $P_L^{l,p}(\delta_{l,p})$ of observing a LOW fittingness factor:

$$P_L^{l,p}(\delta_{l,p}) = Prob\left[F_{l,p} < \delta_{l,p}\right] \tag{6.4}$$

■ The probability $P_H^{l,p}(\delta_{l,p})$ of observing a HIGH fittingness factor:

$$P_H^{l,p}(\delta_{l,p}) = 1 - P_L^{l,p}(\delta_{l,p}) \tag{6.5}$$

■ The average of observed LOW fittingness factor values:

$$\overline{F_L^{l,p}} = E\left(F_{l,p} \mid F_{l,p} < \delta_{l,p}\right) \tag{6.6}$$

where $E(X)$ denotes the expected value of random variable X.
■ The average of observed HIGH fittingness factor values:

$$\overline{F_H^{l,p}} = E\left(F_{l,p} \mid F_{l,p} \geq \delta_{l,p}\right) \tag{6.7}$$

Furthermore, to monitor fittingness factor variability, the following statistical metrics are considered to determine the extent to which the state of the fittingness factor is likely to remain constant for a certain time n:

■ Given $F_{l,p}$ is LOW at a given time instant t, the probability that $F_{l,p}$ will be LOW at each time instant up to time $t+n$; $n \in \mathbb{N}$ defined as follows:

$$P_{L,L}^{l,p}(n, \delta_{l,p}) = Prob\left[F_{l,p}(t+j) < \delta_{l,p}, \forall j \in \{1,\ldots,n\} \mid F_{l,p}(t) < \delta_{l,p} \right] \qquad (6.8)$$

where $F_{l,p}(t)$ denotes the observed $F_{l,p}$ value at time t.

■ Given $F_{l,p}$ is HIGH at a given time instant t, the probability that $F_{l,p}$ will be HIGH at each time instant up to time $t+n$; $n \in \mathbb{N}$ defined as follows:

$$P_{H,H}^{l,p}(n, \delta_{l,p}) = Prob\left[F_{l,p}(t+j) \geq \delta_{l,p}, \forall j \in \{1,\ldots,n\} \mid F_{l,p}(t) \geq \delta_{l,p} \right] \qquad (6.9)$$

All proposed statistics assume stationary behavior of the different pools so that they are generated only once at the beginning of system operation and are assumed to be valid at any later moment.

6.3.3 Knowledge Manager

The KM plays a key role between the knowledge management and decision-making domains of the proposed architecture. It manages the information retained in the KD to determine what information about the environment is most relevant to the decision-making entity.

The KM keeps an estimation of $F_{l,p}$ values based on the statistics available at the KD. These estimated values, denoted as $\hat{F}_{l,p}$, are obtained by following Algorithm 1 and are provided upon request to the decision-making module. The estimate $\hat{F}_{l,p}$ is determined based on whether the state of the $F_{l,p}$, stored in the KD $\Delta t_{l,p}$ time units before, is likely to be still valid at the KM execution time (this criterion is checked on lines 5 and 11 of Algorithm 1 with respect to the significance thresholds *Thr_LOW* and *Thr_HIGH*, respectively). In such a case, $\hat{F}_{l,p}$ is set to the last measured $F_{l,p}$ value (lines 6 and 12). Otherwise, $\hat{F}_{l,p}$ is randomly set to either $\overline{F_L^{l,p}}$ or $\overline{F_H^{l,p}}$, that is, the average $F_{l,p}$ values in the LOW and HIGH states, respectively, with probabilities $P_L^{l,p}(\delta_{l,p})$ and $1 - P_L^{l,p}(\delta_{l,p})$ (lines 8 and 14). Once all link/pool pairs have been explored, the list of all estimated fittingness factor values ($\{\hat{F}_{l,p}\}_{1 \leq l \leq L, 1 \leq p \leq P}$) is returned back to the decision-making entity (line 19).

The KM also captures relevant changes in these estimated values and informs the decision-making module for its consideration.

Algorithm 1, Knowledge Manager

1: Knowledge Manager
2: **for** l = 1 → L **do**
3: **for** p = 1 → P **do**
4:   **if** $F_{l,p}$ is LOW, **then**
5:    **if** $P_{L,L}^{l,p}(\Delta t_{l,p}, \delta_{l,p}) \geq Thr\_LOW$ **then**

6:     $\hat{F}_{l,p} \leftarrow F_{l,p}$;
7:   **else**
8:   Estimate $F_{l,p}$ as follows:

$$\hat{F}_{l,p} = \begin{cases} \overline{F_L^{l,p}} & \text{with probability } P_L^{l,p}(\delta_{l,p}), \\ \overline{F_H^{l,p}} & \text{with probability } 1-P_L^{l,p}(\delta_{l,p}). \end{cases}$$

9:   **end if**
10: **else**
11:    **if** $P_{H,H}^{l,p}(\Delta t_{l,p},\delta_{l,p}) \geq Thr\_HIGH$ **then**
12:      $\hat{F}_{l,p} \leftarrow F_{l,p}$;
13:    **else**
14:      Estimate $F_{l,p}$ as follows:

$$\hat{F}_{l,p} = \begin{cases} \overline{F_L^{l,p}} & \text{with probability } P_L^{l,p}(\delta_{l,p}), \\ \overline{F_H^{l,p}} & \text{with probability } 1-P_L^{l,p}(\delta_{l,p}). \end{cases}$$

15:    **end if**
16:   **end if**
17: **end for**
18: **end for**
19: return $(\{\hat{F}_{l,p}\}_{1\leq l\leq L,1\leq p\leq P})$

6.3.4 Reliability Tester

To detect relevant changes in KD statistics when operating in nonstationary environments, the RT is developed in this section to monitor the reliability of KD data. Whenever a relevant change is detected, KD statistics are regenerated under the new conditions.

In this respect, the RT detects relevant changes by monitoring the pools used by active link sessions. A change is judged as relevant if it has a significant impact on the perceived performance by the end user evaluated in terms of a set of M KPIs (e.g., the achieved bit rate and the number of pool reconfigurations).

Specifically, the RT procedure computes first, for all $m \in \{1,\dots,M\}$, an initial estimate of the mth KPI (denoted as KPI_m) based on its observed sample mean over established link sessions (let $\overline{KPI}_m$ denote this initial mean estimate). Then, the RT gradually increases the sample window size (denoted as S) as new link sessions are established, and updates the observed sample mean ($\overline{KPI}_m$), sample variance (σ_m), and γ confidence interval $\left[\overline{KPI}_{m,min}, \overline{KPI}_{m,max}\right]$ defined as the interval that satisfies the following relationship:

$$Prob\left[KPI_m \in \left[\overline{KPI}_{m,min}, \overline{KPI}_{m,max}\right]\right] = \gamma \tag{6.10}$$

Assuming large-sample conditions (typically on the order of $S > 30$), $\overline{KPI}_{m,min}$ and $\overline{KPI}_{m,max}$ are given by

$$\overline{KPI}_{m,min} = \overline{KPI}_m - z_{(1-\gamma)/2} \frac{\overline{\sigma}_m}{\sqrt{S}} \tag{6.11}$$

$$\overline{KPI}_{m,max} = \overline{KPI}_m + z_{(1-\gamma)/2} \frac{\overline{\sigma}_m}{\sqrt{S}} \tag{6.12}$$

where $z_{(1-\gamma)/2} = \phi^{-1}\left(1 - \dfrac{1-\gamma}{2}\right)$ and $\varphi(.)$ denotes the cumulative normal distribution function.

Note that as the sample window size S increases, $\overline{KPI}_m$ tends to converge, and the interval $\left[\overline{KPI}_{m,min}, \overline{KPI}_{m,max}\right]$ becomes narrower.

To achieve a good level of convergence in the initial estimate of KPI_m, the window size S is progressively increased until the width of the γ confidence interval becomes smaller than a fraction $0 < \rho < 1$ of the sample mean $(\overline{KPI}_m)$, that is,

$$\overline{KPI}_{m,max} - \overline{KPI}_{m,min} < \rho \cdot \overline{KPI}_m \tag{6.13}$$

Note that decreasing ρ would always improve convergence reliability, but would result in much more time before the condition in Equation 6.13 is met. Therefore, in practice, the smallest value of ρ that would achieve an acceptable level of convergence in each KPI_m can be used. For instance, for the scenario and set of KPIs considered in Sections 6.5.2 and 6.5.4, respectively, it has been checked that all considered KPIs achieve a good level of convergence at approximately $\rho = 0.1$.

After meeting the stopping rule of Equation 6.13, $\overline{KPI}_m$ becomes the initial estimate, and the current value S of the window size is maintained.

Then, the RT starts a procedure for monitoring possible changes in the average value of KPI_m based on the statistical technique known in the literature as binary hypothesis testing [23]. Specifically, a null hypothesis (H_0) is introduced to claim that there is no difference between the initial average value $(\overline{KPI}_m)$ and a new average value continuously updated based on a moving window of the same size S. Let $\widehat{KPI}_m$ denote this new average value, $\widehat{\sigma}_m$ be its sample variance, and $\left[\widehat{KPI}_{m,min}, \widehat{KPI}_{m,max}\right]$ be the corresponding confidence interval. As long as H_0 holds, KD statistics are assumed to be valid. On the contrary, an alternative hypothesis (H_1) claims that there is a difference between the initial and new average values, and thus KD statistics are no longer valid.

Nevertheless, the difference that may be observed between the two average values $\overline{KPI}_m$ and $\widehat{KPI}_m$ does not always imply invalidity of KD statistics but may be just the result of pure chance in the experiment. Therefore, the hypothesis-testing procedure should ensure with a certain probability that only those differences caused by an actual change in the scenario (e.g., the appearance of an external interferer or change in the position of the transmitter and/or receiver of a given link) are detected. This means, on the one hand, that the probability of selecting H_1 when H_0 actually holds (the so-called Type I error in hypothesis testing terminology [23]) should be kept below a maximum level α. On the other hand, the probability of selecting H_0 when H_1 actually holds (the so-called type II error) should also be kept below a maximum level β.

In general, different detection strategies may be followed to strike a balance between Type I and II errors. More specifically, it has been shown in Ref. [24] that the detection strategy should be designed based on the standardized difference (D_m) between $\overline{KPI}_m$ and $\widehat{KPI}_m$ and the ratio of the largest-to-smallest sample variance (k_m), defined as

$$D_m = \frac{|\overline{KPI}_m - \widehat{KPI}_m|}{\sqrt{\overline{KPI}_m^2 + \widehat{KPI}_m^2}} \tag{6.14}$$

$$k_m = \frac{max(\overline{\sigma}_m, \widehat{\sigma}_m)}{min(\overline{\sigma}_m, \widehat{\sigma}_m)} \tag{6.15}$$

where the $min(.,.)$ and $max(.,.)$ functions return the minimum and maximum of two real values, respectively.

For the scenario considered in Section 6.5.2, it has been checked that, on the one hand, the considered change results in high values of the standardized difference D_m. On the other hand, the ratio of sample variances k_m is typically close to 1 because both $\overline{KPI}_m$ and KPI_m are determined based on the same window size S. Under these conditions (high D_m and low k_m), the overlap decision method, which selects H_0 when the confidence intervals of the two average values overlap and H_1 otherwise, has been shown in Ref. [24] to exhibit a very low Type I error and an acceptable Type II error. Consequently, it has been retained in this chapter. The reader is referred to Ref. [24] for a detailed analysis of Type I and II errors.

Algorithm 2 describes the general RT procedure for detecting changes. After testing for changes in each of the converged KPIs in lines 5–11, the obtained hypothesis testing results are combined to determine the reliability of the entire KD data according to the procedure described in lines 13–17. Specifically, if H_1 is selected for at least one of the considered KPIs, the combined test selects H_1 (line 14).

Algorithm 2, The RT Procedure for Detecting Changes

1: **for** $m = 1 \rightarrow M$ **do**
    // Checking convergence of KPIs
2: **if** condition in Equation 6.13 holds for KPI_m **then**
3:      *Converge*(KPI_m) = *TRUE*
4: **end if**
    // Individual Tests for converged KPIs
5: **if** *Converge*(KPI_m) = *TRUE* **then**
6:    **if** $\left[\overline{KPI}_{m,min}, \overline{KPI}_{m,max}\right] \cap \left[\widehat{KPI}_{m,min}, \widehat{KPI}_{m,max}\right] = \varnothing$ **then**
7:      Select H_1;
8:    **else**
9:      Select H_0;
10:    **end if**
11: **end if**
12: **end for**
    // Multiple Tests for detecting changes
13: **if** H_1 is selected for at least one KPI_m **then**
14: Select H_1;

15: **else**
16:    Select H_0;
17: **end if**
18: **for** $l = 1 \rightarrow L$ **do**
19:    **for** $p = 1 \rightarrow P$ **do**
       // Detecting changes in inactive pools
20:       **if** (H_0 is selected) AND ($\Delta t_{l,p} > T\_Inact$) **then**
21:          count = 0;
22:          **for** $n = 1 \rightarrow N$ **do**
23:             Force a measurement of $F_{l,p}$;
24:             **if** $(F_{l,p} - \delta_{l,p}) \cdot (\widehat{F_{l,p}} - \delta_{l,p}) < 0$ **then**
25:                count++;
26:             **end if**
27:          **end for**
28:          **if** count = N **then**
29:          Select H_1;
30:          break;
31:          break;
32:          **else**
33:          Select H_0;
34:          **end if**
35:       **end if**
36:    **end for**
37: **end for**
    // Perform required updates if a change is detected
38:       **if** H_1 is selected **then**
39:          Regenerate KD statisics;
40:          Recalculate $\{KPI_m\}_{1 \leq m \leq M}$;
41:       **end if**

Note that the procedure described in lines 13–17 is adequate to detect changes that occur in pools that are used with some regularity by active links. However, it is not valid to detect changes occurring in pools that are never assigned. To overcome this problem, when H_0 is selected, the procedure described in lines 18–37 is carried out for each pool that remains unused by any of the considered links for more than a certain period of time named T_Inact. Specifically, a total of N forced measurements are taken to obtain the actual $F_{l,p}$ that is compared to the estimate $\hat{F}_{l,p}$ determined by the KM to test for possible changes (line 28) and select H_1 in case of a change (line 29).

The algorithm is concluded by regenerating KD data (line 39) if H_1 is selected following a change in the converged KPIs (line 14) or inactive pools (line 29). Then, the RT sets S to zero and loops back to the sequential analysis procedure described in the beginning of the section to recalculate all initial estimates $\{\overline{KPI_m}\}_{1 \leq m \leq M}$ under the new conditions (line 40). In this case, the KM will continue using the old KD statistics until the new statistics become available.

Note that to achieve the best performance in detecting changes, all RT-controllable parameters (e.g., ρ, γ, N, and T_Inact) should be tuned according to the specificities of the scenario at hand. In this respect, these parameters have been set in Section 6.5.2 according to the general guidelines provided in Ref. [16].

The knowledge management domain may also be applied to decentralized architectures. For such scenarios, a local KD can be constructed at each of the terminals. The local KD would retain $F_{l,p}$ values and the list of statistics defined in Section 6.3.2 obtained based on local observations made by each terminal. For small neighborhoods with homogeneous interference conditions, each terminal would cooperate with its neighbors via CA modules to get the most recent $F_{l,p}$ value observed in the neighborhood together with the timer $\Delta t_{l,p}$. Then, the KM would be executed at each of these terminals to determine a local $\hat{F}_{l,p}$ estimate based on the local KD statistics. Finally, the RT would be executed locally based on the KPIs observed by each terminal. In case a relevant change is detected, the terminal would regenerate its local KD and trigger via CA modules the procedure of updating all local KDs situated within the neighborhood.

6.4 Proposed Algorithmic Solution

The proposed fittingness factor function is applied to the SS decision-making process. The aim of this process is to allocate, for a given link l, the best spectrum pool $p^*(l)$ among the list of available pools (Av_Pools), that is, those that are not currently assigned to any other link, to support the corresponding MM application.

Enlightened by the comparative study of several SS criteria conducted in Ref. [21], the following proactive criterion is considered in this chapter:

$$p^*(l) = \arg \max_{p \in Av\_Pools} \left(g(\hat{F}_{l,p}) \right) \tag{6.16}$$

The function $g(\hat{F}_{l,p})$ assesses the likelihood of observing a HIGH $F_{l,p}$ value up to the end of session duration T_l as follows:

$$g(\hat{F}_{l,p}) = \begin{cases} \dfrac{1}{T_l} \sum_{t=1}^{T_l} P_{H,H}^{l,p}(\Delta t_{l,p} + t, \delta_{l,p}) & \text{if } \hat{F}_{l,p} \text{ is HIGH} \\ 0 & \text{if } \hat{F}_{l,p} \text{ is LOW} \end{cases} \tag{6.17}$$

recall that $\Delta t_{l,p}$ denotes the number of time units since the $F_{l,p}$ stored in the KD was obtained.

In the very specific case of multiple pools fulfilling the maximization, the pool with the highest $\hat{F}_{l,p}$ value is selected.

In what follows, a fittingness-factor-based implementation of the SS, SM, and CA functionalities is developed to support the considered MM applications.

6.4.1 Spectrum Selection

The proposed proactive fittingness-factor-based SS algorithm is described in Algorithm 3. It is executed each time the start of a new MM application requires the establishment of a radio link to support communication. Upon receiving a request for establishing a link l session, the algorithm first checks the list of available pools (Av_Pools). If there is at least one available pool, an estimation of $F_{l,p}$ values is obtained from the KM (line 3). On the basis of the provided data, the available spectrum pool $p^*(l)$ that maximizes the likelihood of observing a HIGH $F_{l,p}$ value up to the end of the link session duration is selected (lines 4 and 5). In turn, when there is no available pool, the SS decision depends on the "preemptability" of the MM application at hand. Specifically, for preemption-capable applications (i.e., *Preemp*(l) = 1), the algorithm first constructs the list of active preemption-vulnerable links (*Preemp_Links*) (line 8). Then, if there is at least one active preemption-vulnerable link, the best link l'^* in *Preemp_Links*, that is, the one whose pool $p(l'^*)$ better fits link l in terms of $g(\hat{F}_{l,p(l'^*)})$, is preempted and assigned to link l (line 11), thus causing the drop of the link l'^* session (line 12). Otherwise, the link request is rejected (line 14). As for preemption-vulnerable applications (i.e., *Preemp*(l) = 0), the link request is rejected without any preemption attempt (line 17).

Algorithm 3, Fittingness-Factor-Based SS

1: **if** link l establishment request **then**
   // At least one available pool
2:   **if** *Av_Pools*≠∅ **then**
3:     Get $\left\{\hat{F}_{l,p}\right\}_{p\in Av\_Pools}$ from the KM;

4:     Compute $\left\{g(\hat{F}_{l,p})\right\}_{p\in Av\_Pools}$;

5:     $p^*(l) \leftarrow \arg\max\limits_{p\in Av\_Pools}\left(g\left(\hat{F}_{l,p}\right)\right)$;

   // No available pool
6:   **else**
   // Preemption-capable application
7:     **if** *Preemp*(l) = 1 **then**
8:       *Preemp_Links* ← {active l'≠l, *Preemp*(l') = 0};
9:       **if** *Preemp_Links*≠∅ **then**
10:         $l'^* \leftarrow \arg\max\limits_{l'\in Preemp\_Links}\left(g\left(\hat{F}_{l,p(l')}\right)\right)$;
11:         $p^*(l) \leftarrow p(l'^*)$;
12:         Drop link l'^* session;
13:       **else**
14:         Reject link / request;
15:       **end if**
   // Preemption-vulnerable application
16:     **else**
17:       Reject link / request;
18:     **end if**
19:   **end if**
20: **end if**

6.4.2 Spectrum Mobility

The SM functionality attempts to ensure highly efficient allocation of available spectrum pools to each of the established radio links. Therefore, whenever an event that might have an influence on the SS decision-making process occurs, the SM will be executed to trigger possible changes in spectrum assignment, that is, spectrum handovers (SpHOs). Such events include (1) a spectrum pool being released owing to finalization of the corresponding MM application, or (2) a change in suitability of the available spectrum pools being detected.

The proposed proactive fittingness-factor-based SM algorithm is described in Algorithm 4. The algorithm is triggered whenever a pool previously selected by SS at link establishment is no longer the best in terms of $g(\hat{F}_{l,p})$ for the corresponding active link. This may happen whenever some active pools are released or experience some change in their $F_{l,p}$ values. Following both triggers, an estimation of $F_{l,p}$ values is obtained from the KM (line 2). The algorithm then explores the list of currently active links (*Active_Links*) in the increasing order of link indices assuming that prioritized MM applications have lower indices. The decision of whether to reconfigure each active link is based on a comparison between the in-use pool ($p^*(l)$) and the currently best pool in terms of $g(\hat{F}_{l,p})$ ($new\_p^*(l)$) (line 7). Specifically, if $\hat{F}_{l,p^*(l)}$ is LOW and $\hat{F}_{l,new\_p^*(l)}$ is HIGH, an SpHO from $p^*(l)$ to $new\_p^*(l)$ is performed because $new\_p^*(l)$ better fits link l. Finally, as reflected in the condition of line 8, the same SpHO should be performed if $p^*(l)$ is no longer available to link l after being reassigned to another active link in the previous iterations of the loop of line 4. Once all active links have been explored, the list of assigned pools is updated according to performed SpHOs (line 15).

Algorithm 4, Fittingness-Factor-Based SM

1: **if** (link l^* release) OR (change in any active $F_{l^*,p(l^*)}$) **then**

2:   Get $\left\{\hat{F}_{l,p}\right\}_{1\leq L\leq l,1\leq p\leq P}$ from the KM;

3:   *New_Assigned* ← ∅

4:   **for** $l = 1 \to |Active\_Links|$ **do**

5:     Compute $\left\{g(\hat{F}_{l,p})\right\}$;

6:     $new\_p^*(l) \leftarrow \arg\max\left(g\left(\hat{F}_{l,p}\right)\right)$;

7:     **if** (($\hat{F}_{l,p^*(l)}$ is LOW) AND ($\hat{F}_{l,new\_p^*(l)}$ is HIGH)) OR

8:       ($p^*(l) \in new\_Assigned$) **then**

9:       $p^*(l) \leftarrow new\_p^*(l)$;

10:       *New_Assigned* ← *New_Assigned* ∪ {*new* $p^*(l)$}

11:     **else**

12:       *New_Assigned* ← *New_Assigned* ∪ {$p^*(l)$};

13:     **end if**

14:   **end for**

15:   *Assigned* ← *New_Assigned*;

16: **end if**

6.4.3 Context Awareness

In general, two types of acquisition strategies can be considered, namely, a periodic strategy in which CA modules of Figure 6.2 periodically report measurements of the achieved performance to the KD, or an event-triggered strategy in which measurements reports are generated only when some relevant conditions are met. If the radio environment changes frequently, a simple periodic acquisition strategy can be used. Conversely, in less varying environments, an event-based acquisition strategy is preferred to avoid unnecessary signaling loads between CA modules. This chapter proposes the use of the event-triggered acquisition strategy described by Algorithm 5. In this case, measurement reports are generated only if the currently measured $F_{l,p}$ value (*current_$F_{l,p}$*) is LOW and the last reported $F_{l,p}$ value (*rep_$F_{l,p}$*) was HIGH or vice versa.

Algorithm 5, Event-Triggered Acquisition Strategy

1: **for** $p = 1 \rightarrow P$ **do**
2:    Compute *current_$F_{l,p}$* according to (6.1);
3:    **if** ((*current_$F_{l,p}$* is LOW) AND (*rep_$F_{l,p}$* is HIGH)) OR
4:       ((*current_$F_{l,p}$* is HIGH) AND (*rep_$F_{l,p}$* is LOW)) **then**
5:       *rep_$F_{l,p}$* ← *current_$F_{l,p}$*;
6:       Generate *Meas_Report(rep_$F_{l,p}$)*;
7:    **end if**
8: **end for**

6.5 Evaluation Study

This section aims at gaining an insight into the effectiveness of the proposed framework in assisting in the MM spectrum management decision-making process. In this respect, a mixture of real-time and best-effort applications is presented together with the corresponding fittingness factor formulations. A cost-benefit analysis of the proposed strategy is conducted to evaluate its rationality with respect to other strategies in a stationary environment. Finally, an assessment of robustness is performed when the conditions become nonstationary.

6.5.1 MM Traffic Mixture

To assess the relevance of the proposed framework to cope with the heterogeneous requirements of MM applications, the following MM traffic mixture is considered.

6.5.1.1 Real-Time Applications

■ *Set of requirements*: Assuming a constant bit rate (CBR) video traffic, real-time applications are characterized by a sustainable bit rate R_S that should be achieved over the session duration (T_l). Furthermore, real-time applications are delay- and jitter-sensitive and need to drop any packet not arriving within the delay or jitter boundaries. Even though the channel can support the sustainable bit rate R_S on average, packets can be delayed and finally discarded in the receiver owing to the variation in channel capacity, as shown in Figure 6.4. To focus on the delay specifically experienced in CRNs, a similar setting to that of Ref. [13] is considered, where an optimized buffering mechanism is assumed to absorb any delay experienced in conventional wireless networks, such as application layer, link layer, and transmission delays. Then, the additional delay factors uniquely introduced by CRNs (e.g., the time needed to sense spectrum or to switch it upon primary user appearance) would directly lead to data losses. Therefore, a target packet loss rate (L_T) will be used to reflect the delay or jitter requirements of real-time applications. In summary, real-time applications are characterized in terms of a sustainable bit rate R_S and packet loss rate L_T.
■ *Fittingness factor function*: The model described in Section 6.3.1 is applied for $K_l = 2$ with $R_{req,l}^1 = R_S$ and $R_{req,l}^2 = 1 / L_T$. Therefore, the fittingness factor function associated with real-time applications can be derived as

$$F_{l,p} = \eta_1 \times F_{l,p,1} + \eta_2 \times F_{l,p,2} \qquad (6.18)$$

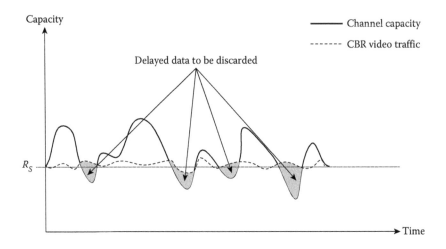

Figure 6.4 Data loss in real-time video applications.

where $F_{l,p,1}$ and $F_{l,p,2}$ are the marginal fittingness factors from the sustainable bit rate and packet loss rate perspectives, respectively, given by

$$F_{l,p,1} = \frac{\left(\frac{R(l,p)}{R_S}\right)^{\xi}}{1+\left(\frac{R(l,p)}{R_S}\right)^{\xi}} \tag{6.19}$$

$$F_{l,p,2} = \frac{\left(\frac{1/L(l,p)}{1/L_T}\right)^{\xi}}{1+\left(\frac{1/L(l,p)}{1/L_T}\right)^{\xi}} = \frac{\left(\frac{L_T}{L(l,p)}\right)^{\xi}}{1+\left(\frac{L_T}{L(l,p)}\right)^{\xi}} \tag{6.20}$$

where $R(l,p)$ and $L(l,p)$ denote the average bit rate and packet loss rate achieved by link l using pool p over the last N_{Avg} packets, respectively.

6.5.1.2 Best-Effort Applications

- *Set of requirements*: Best-effort applications aim at maximizing the achieved bit rate over link session (T_l) without any requirement in terms of delay or jitter. Therefore, only a loose requirement in terms of a minimum bit rate R_{min} is considered for such applications.
- *Fittingness factor function*: The model described in Section 6.3.1 is applied for $K_l = 1$ with $R_{req,l}^1 = R_{min}$. Therefore, the fittingness factor function associated with best-effort applications can be derived as

$$F_{l,p} = F_{l,p,1} = \frac{\left(\frac{R(l,p)}{R_{min}}\right)^{\xi}}{1+\left(\frac{R(l,p)}{R_{min}}\right)^{\xi}} \tag{6.21}$$

6.5.2 Simulation Model

$L = 2$ radio links are considered with independent traffic loads for each link, λ_l being the arrival rate over the lth link that is varied during simulations.

Performance is evaluated using a system-level simulator operating in steps of 0.01 s under the following assumptions:

- The radio environment is modeled as a set of $P = 4$ spectrum pools. The available bandwidth at each pool is $BW_1 = BW_2 = 0.4$ MHz and $BW_3 = BW_4 = 1.2$ MHz.
- Link #1 is associated with real-time applications characterized in terms of a sustainable bit rate $R_S = 1$ Mbps and a target packet loss rate $L_T = 0.05$ to be met during session duration $T_1 = 20$ minute. Assuming a CBR video traffic and a constant packet size of P_{size}, packets are generated periodically each P_{size}/R_Ss. The generated packets are transmitted if the capacity of the in-use pool ($R(l,p)$) is enough to support R_S ($R(l,p) \geq R_S$), and dropped otherwise.
- Link #2 is associated with best-effort applications that are characterized in terms of a minimum bit rate $R_{min} = 1$ kbps to be achieved during the session duration $T_2 = 2$ minute. To model the elastic behavior of best-effort applications, the generation process is assumed to be adaptive to take full advantage of the available capacity $R(l,p)$.
- A simple interference model that captures the most relevant features affecting SS has been retained. A heterogeneous interference situation is considered in which the sum of the noise and interference power spectral density $I(p)$ experienced in each pool $p \in \{1,\ldots,P\}$ follows a two-state discrete-time Markov chain jumping between a state of low-interference $I_0(p)$ and a state of high-interference $I_1(p)$ with transition probabilities $P_{01}(p)$ (i.e., the probability of moving from state I_0 to I_1 in one simulation step) and $P_{10}(p)$ (i.e., the probability of moving from state I_1 to I_0), as described by Figure 6.5a. An example of the temporal evolution of $I(p)$ in a given pool p is shown in Figure 6.5b. To support packet-based transmissions, $I(p)$ is assumed to remain constant during each packet duration.

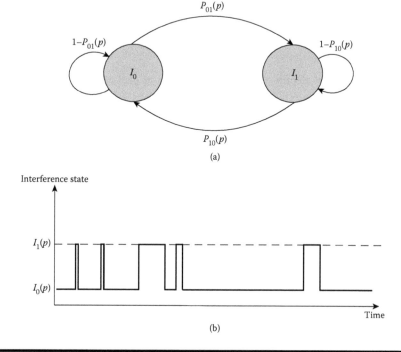

Figure 6.5 Evolution of the interference power spectral density: (a) Markov chain and (b) example of temporal interference evolution.

Table 6.1 Interference Conditions of the Different Pools before and after the Change

Pool	Before the Change				After the Change			
	I_0		I_1		I_0		I_1	
	Duration	R(l,p) (kbps)	Duration	R(l,p) (kbps)	Duration	R(l,p) (kbps)	Duration	R(l,p) (kbps)
1	∞	512	—	—	7.5 h	1536	0.5 h	96
2	∞	512	—	—	∞	512	—	—
3	7.5 h	1536	0.5 h	96	∞	512	—	—
4	0.5 h	1536	2 min	96	0.5 h	1536	2 min	96

■ On the basis of these probabilities, the average durations of the $I_0(p)$ and $I_1(p)$ states are, respectively, given by

$$\overline{I_0(p)} = \frac{1}{P_{01}(p)} \tag{6.22}$$

$$\overline{I_1(p)} = \frac{1}{P_{10}(p)} \tag{6.23}$$

■ To assess robustness to changes in interference conditions, the simulation is assumed to start with a given statistical pattern of $I(p)$ in each of the considered spectrum pools. Then, at a certain point of time, some of these interference patterns are altered. Table 6.1 shows the durations of the two possible states of $I(p)$ with the corresponding bit rates before and after the change. Note that, after the change, the statistical patterns of pools #1 and #3 are swapped.

The system is observed until 14,000 sessions are established for each link. The change in interference conditions occurs after establishing 5000 sessions for each link. The other simulation parameters are $\xi = 5$, $K = 1$, $\delta_{l,p} = 0.5$, *Thr_LOW* = 0.9 and *Thr_HIGH* = 0.1, $\eta_l = 0.5$, $\rho = 0.1$, $\gamma = 0.95$, $N = 200$, $T\_Inact = 10$ h, $N_{Avg} = 100$, and $P_{size} = 1250$ bytes.

6.5.3 Benchmarking

To assess the influence of the proposed RT, the following variants are compared:

■ *SS*: This approach uses the proactive fittingness-factor-based SS supported by only the KD (i.e., the decision-making entity of Figure 6.2 is assumed to bypass the KM and retrieve only the last measured $F_{l,p}$ value from the KD).
■ *SS+KM*: This approach considers the proactive fittingness-factor-based SS supported by the KM module. No SM support is considered. In comparison to SS, the use of KM will allow for a better capability to track changes in the interference conditions of each pool owing to the consideration of the temporal properties of $F_{l,p}$ statistics in addition to the most recent reported values.

- *SS+KM+SM*: This approach is the proposed proactive strategy in Section 6.4 that implements the fittingness-factor-based SS supported by both KM and SM. This strategy incorporates the track-change benefits of KM together with the reallocation flexibility associated with SM.
- *SS+KM+SM+RT*: This is the complete approach that includes the SS, KM, SM, and RT functionalities. In comparison to *SS+KM+SM*, the use of an RT enables detection of relevant changes that may occur in the scenario and updation of KD statistics accordingly.

Apart from the considered variants, the following reference schemes are introduced for benchmarking purposes:

- *Rand*: This scheme implements only the SS module of Figure 6.2 and performs a random selection among available pools. Neither SM nor KM modules are used.
- *Optim*: This scheme is an upper-bound theoretical reference. As described by Algorithm 6, the scheme starts by averaging, for each link l, performance metrics $R_k(l,p)$ over the last N_{avg} packets of the currently generated packets ($P_{count}(l)$) (i.e., when the condition on line 2 is satisfied, where A MOD B returns the remainder of the division of A by B). Then, the list of active link sessions is explored in increasing index order, assuming that lower indices correspond to higher-priority MM applications (line 7). Specifically, the procedure first assigns the pool that better satisfies all requirements of the highest-priority MM application (line 8), marks it as occupied (line 9), and loops back to perform assignments of lower-priority MM applications. In case several pools are fulfilling the maximization, the pool with the highest sum of achieved performances $\sum_{k=1}^{K_l} \left(\theta_{l,p}^k \cdot R_k(l,p) \right)$ is selected. Note that this theoretical scheme assumes that the actual performances $R_k(l,p)$ of all link/pool pairs are continuously sent to the decision-making entity through measurement reports.

Algorithm 6, Upper-Bound Theoretical Reference

1: **for** $l = 1 \rightarrow L$ **do**
2:   **if** $P_{count}(l)\ MOD\ N_{Avg} = 0$ **then**
3:     Update $\{R_k(l,p)\}_{1 \leq k \leq K_l}$;
4:   **end if**
5: **end for**
6: $Av\_Pools \leftarrow \{1,..,P\}$;
7: **for** $l = 1 \rightarrow L$ **do**
8:   Selects the pool that maximizes the total achievement of the K_l requirements as follows:

$$p^*(l) = \arg \max_{p \in Av\_Pools} \sum_{k=1}^{K_l} \left(\theta_{l,p}^k \right); \theta_{l,p}^k = \begin{array}{ll} 1 & if\ R_k(l,p) \geq R_{req,l}^k \\ 0 & otherwise \end{array} \quad (6.24)$$

9:   $Av\_Pools \backslash \{p^*(l)\}$;
10: **end for**

6.5.4 Performance Indicators

To assess the performance of the proposed strategy, the following indicators are considered:

- Real-time applications (i.e., link #1): Average bit rate in bits per second ($R(1)$), packet loss rate ($Loss(1)$), and blocking probability ($Bloc(1)$).

■ Best-effort applications (i.e., link #2): Average bit rate in bits per second ($R(2)$) and percentage of preempted sessions by real-time sessions (*Preemp(2)*).
■ The SpHO signaling introduced by the SM functionality evaluated in terms of the total number of SpHOs performed per link session (*Nb. SpHOs/session*).
■ The total signaling overhead in the radio interface experienced on average per link session given by

$$Overhead = \frac{1}{Nb\_Succ\_Estab}(Nb\_Estab\_Req \cdot \overline{Estab}$$

$$+ Nb\_Succ\_Estab \cdot \overline{Release} + Nb\_SpHO \cdot \overline{SpHO} + Nb\_Rep \cdot \overline{Rep}) \tag{6.25}$$

where *Nb_Estab_Req*, *Nb_Succ_Estab*, *Nb_SpHO*, and *Nb_Rep* denote the number of link establishment requests, successful link establishments, executed SpHOs, and measurement reports generated by CA modules during the simulation duration, respectively. The corresponding costs for establishing a link session, releasing it, performing an SpHO, and generating a measurement report are set to $\overline{Estab} = 266$ *Bytes*, $\overline{Release} = 64$ *Bytes*, $\overline{SpHO} = 167$ *Bytes*, and $\overline{Rep} = 43$ *Bytes*, respectively, following the message formats given in Ref. [25].

Finally, the procedure of detecting changes described in Section 6.3.4 is performed by the RT based on monitoring the following two KPIs for each link l (which yields a total of $M = 2L = 4$ KPIs):

■ The dissatisfaction probability, defined as the probability of not satisfying at least one of the K_l requirements, *Dissf* (l) defined as

$$Dissf(l) = Prob\left[\bigcup_{k=1}^{K_l}\left\{R_k(l, p) < R_{req,l}^k\right\}\right] \tag{6.26}$$

■ The average number of SpHOs performed per session, *SpHO(l)*.

6.5.5 Performance Evaluation

This section presents the performance evaluation of the different schemes introduced in Section 6.5.3. The goal of this analysis is twofold: (1) to identify which of the functional elements of the considered architecture have the most significant impact on performance depending on the system operation conditions and (2) to benchmark the performance of the proposed spectrum management strategy with respect to the reference *Rand* and *Optim* schemes. For all schemes, the event-triggered acquisition strategy described in Algorithm 5 is considered.

Figure 6.6a plots the average bit rate achieved by real-time applications ($R(1)$) with the requirement R_S shown in the dotted line as a function of the total offered traffic load in bits per second defined as

$$\text{Offered traffic (bps)} = L_1 \cdot R_S + L_2 \cdot R_{min} \tag{6.27}$$

where $L_l = \lambda_l \cdot T_l$ denotes the offered traffic load to link l in erlangs.

Figure 6.6b plots the corresponding packet loss rate (*Loss(1)*) defined in Section 6.5.4 with respect to the requirement L_T. Only traffic loads for which *Loss(1)* is below L_T are considered, which corresponds to *offered traffic* (bps) = *1.3* Mbps. For the sake of simplicity, equal traffic loads are considered for both links (i.e., $L_1 = L_2$).

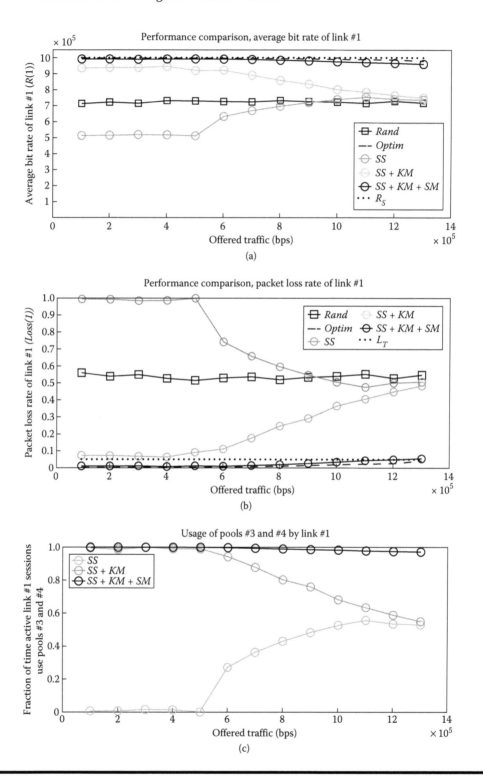

Figure 6.6 **SS performance of real-time applications (i.e., link #1) in terms of (a) average bit rate (*R*(1)), (b) packet loss rate (*Loss*(1)), and (c) usage fraction of pools #3 and #4.**

For a better understanding of the obtained results, Figure 6.6c plots the fraction of time during which link #1 uses pools #3 and #4 under the consideration that link #1 will be satisfied only when using these pools during the low-interference state, while it will not be satisfied whenever it is allocated pools #1 or #2 or pools #3 or #4 during the high-interference state.

As seen in Figure 6.6a, when comparing *SS* against *SS+KM*, for low traffic loads below 0.6 Mbps, the introduction of KM leads to a very substantial increase in the achieved bit rate. The reason for this improvement is that whenever interference increases in pools #3 and #4 (i.e., they move to state I_1), the corresponding measured value of $F_{l,p}$ will be LOW. As a result, the *SS* strategy that just keeps the last reported value of $F_{l,p}$ will decide in the future to allocate only pools #1 and #2, which provide a much lower bit rate than the requirement R_S (Figure 6.6a). As a result, many packets get discarded owing to excessive buffering delays, which strongly penalizes the packet loss rate (Figure 6.6b). Therefore, the network will not be able to exploit pools #3 and 4 when they move again to the low-interference state I_0 and become able to support the requirements of real-time applications (e.g., link #1). As an aside, Figure 6.6c indicates that the fraction of time during which these pools are allocated to link #1 is approximately 0. In contrast, the use of the KM component considers the temporal properties of $F_{l,p}$ statistics to disregard the last reported value and uses an estimated value instead whenever a certain amount of time has elapsed since this value was reported (see the conditions of lines 5 and 11 in Algorithm 1). Correspondingly, sometime after the interference increase, the network will reallocate pools #3 and #4 to link #1, and thus will be able to identify that they have re-entered the low-interference state. Note that in Figure 6.6c, the fraction of time that link #1 uses pools #3 or #4 is close to 1 when KM is supported, thus resulting in much higher bit rates (Figure 6.6a) and much lower packet loss rates (Figure 6.6b). In summary, for low loads, the KM allows for a better exploration of the different pools to identify the changes in their interference conditions, which improves performance. In turn, when the load increases above 0.6 Mbps, pools #1 or #2 will tend to be occupied by link #2 sessions most of the time, which forces the system to assign pools #3 and #4 to link #1 sessions even with reduced $F_{l,p}$. In this case, interference reductions in pools #3 and #4 are detected without the use of KM. This situation is reflected in Figures 6.6a and 6.6b, where the performances of *SS* and *SS+KM* become equivalent for high loads.

In regard to the role of SM, its use leads to small improvements for low loads (see the comparison between *SS+KM* and *SS+KM+SM* in Figures 6.6a and 6.6b). The reason for this slight improvement is that for low loads, it is rare that a link is not assigned the pool with the highest fittingness factor because of contention with another link. Consequently, there is no need to perform SpHOs toward a better pool except when the interference amount increases in the allocated pool, which is responsible for the small improvement observed when comparing *SS+KM* and *SS+KM+SM*. In contrast, when the traffic load increases, it is more likely that the preferred pool becomes occupied by another link, and thus, a pool offering a lower bit rate is allocated. In this case, the execution of SM after the release of the link occupying the preferred pool will lead to improved performance. Note that in Figure 6.6c, with *SS+KM+SM*, link #1 is typically allocated pools #3 or #4, leading to the significant increase in the achieved bit rate and reduction in packet loss rate achieved by *SS+KM+SM* with respect to *SS+KM* in Figures 6.6a and 6.6b for high loads, respectively.

The gain observed by *SS+KM+SM* with respect to the *Rand* scheme ranges from 25% to 30% in terms of bit rate improvement (see Figure 6.6a) and from 90% to 100% in terms of packet loss rate reduction (see Figure 6.6b). The proposed strategy meets the requirements of real-time applications in terms of sustainable bit rate (R_S) and packer loss rate (L_T) and performs very similarly to the upper-bound optimal scheme for most of the considered traffic loads, mainly owing to the support of the KM and SM components for relatively low and high loads, respectively. Nevertheless, the slight improvement introduced by the upper-bound scheme comes at a much higher cost in terms of signaling overhead, as will be shown in the next section.

Next, the performance of best-effort applications (i.e., link #2) is evaluated. Performance is assessed in terms of the achieved bit rate and the percentage of preempted sessions.

Figure 6.7a plots the average bit rate achieved by best-effort applications ($R(2)$) with the requirement R_{min} shown in the dotted line as a function of the total offered traffic load. To evaluate the impact of the SS preemption capability described in lines 7–16 of Algorithm 3,

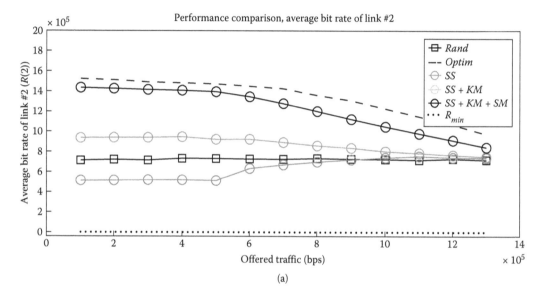

(a)

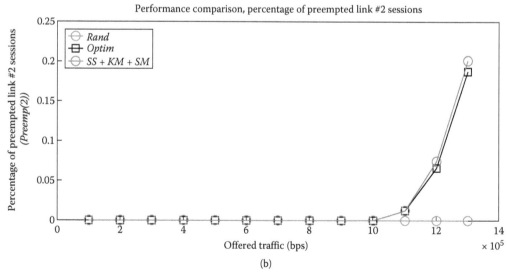

(b)

Figure 6.7 SS performance of best-effort applications (i.e., link #2) in terms of (a) average bit rate ($R(2)$) and (b) percentage of preempted sessions (*Preemp(2)*).

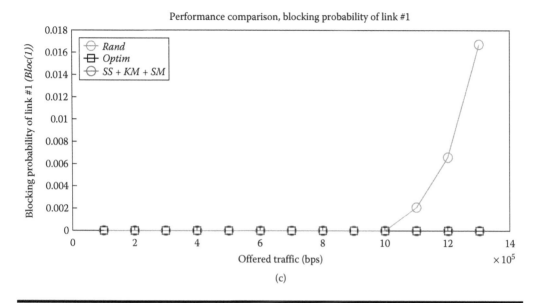

Figure 6.7 (Continued) SS performance of best-effort applications (i.e., link #2) in terms of (c) impact on blocking probability of real-time sessions (*Bloc*(1)).

Figures 6.7b and 6.7c plot the percentage of preempted best-effort sessions and blocking probability of real-time sessions, respectively. Note that the *SS* and *SS+KM* variants are not shown in the last two Figures (6.7b and 6.7c) because their performance is the same as that of *SS+KM+SM*.

The results show that all schemes experience a bit rate well above the requirement R_{min} (Figure 6.7a). The relative performance of each scheme with respect to the others shows a similar behavior to the case of real-time applications previously observed in Figure 6.6a. The proposed strategy (i.e., *SS+KM+SM*) significantly outperforms the *Rand* scheme thanks to the support of the KM and SM components for low and high loads, respectively. The gain observed by *SS+KM+SM* relative to the *Rand* scheme ranges from 10% to 100% (see Figure 6.7a). Note that a larger deviation is observed with respect to the *Optim* scheme compared to the case of real-time applications because all pools are providing a HIGH fittingness factor value to best-effort applications regardless of interference conditions. While *Optim* may change assigned pools to maximize the bit rate, the SM functionality performs reassignments only to push the fittingness factor from LOW to HIGH, which justifies the higher bit rate achieved by *Optim*. Nevertheless, this comes at the cost of a much higher signaling overhead, as will be shown in the next session.

With respect to the retainability of established link #2 sessions, the results show that as the traffic load increases, both *SS+KM+SM* and *Optim* schemes start to preempt link #2 sessions, while the *Rand* does not (Figure 6.7b). This is because real-time sessions (i.e., link #1) have priority over best-effort sessions (i.e., link #2). At high traffic loads, an available pool at the time of establishing real-time sessions is less likely to be found, and thus a best-effort session is more likely to be preeempted in order not to reject real-time sessions. This is better illustrated in Figure 6.7c, where it can be observed that the blocking probability of real-time sessions is maintained equal to zero for the *SS+KM+SM* and *Optim* schemes, while it gradually increases for *Rand*.

6.5.6 Associated Signaling Cost at the Radio Interface

To assess the signaling cost introduced by the proposed strategy, this section makes an analysis of the impact of SM reconfigurations on the radio interface. The impact is evaluated in terms of the signaling overhead (*Overhead*) and amount of SpHO signaling (*Nb. SpHOs/session*) introduced in Section 6.5.4.

Figure 6.8a plots the signaling overhead introduced by *SS+KM*, *SS+KM+SM*, and *Optim* as a function of the total offered traffic load in bits per second defined in Equation 6.27 with a vertical axis in the logarithmic scale for improved visualization. Figure 6.8b plots the corresponding amount of SpHO signaling *SS+KM+SM* introduces for each of the two possible triggers of SM (i.e., a link release or change in $F_{l,p}$).

The results show that the overhead introduced by both *SS+KM* and *SS+KM+SM* is much smaller than that of the reference scheme *Optim* (Figure 6.8a). The reason behind this significant

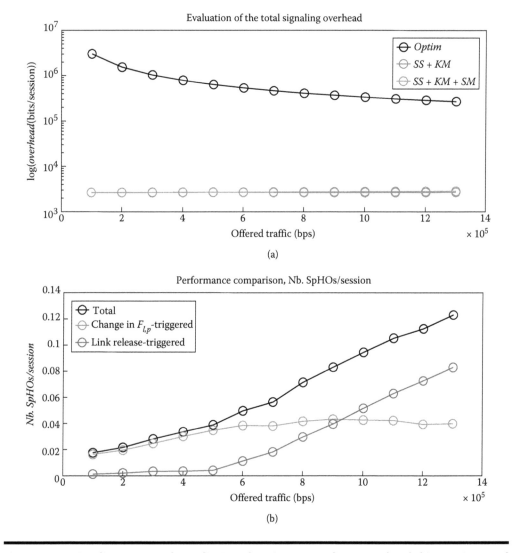

(a)

(b)

Figure 6.8 Signaling cost at the radio interface in terms of (a) *Overhead* (bits/session) and (b) number of SpHOs/session with **SS+KM+SM**.

reduction is that, on the one hand, *Optim* requires to continuously report the actual $R_k(l,p)$ performances of all link/pool pairs to the decision-making entity, which strongly increases the amount of report signaling per session. On the other hand, *Optim* reconfigures pools assigned to best-effort sessions (i.e., link #2) whenever an increase in the achievable bit rate $R(2,p)$ may be obtained even when the current fittingness factor is already HIGH. In contrast, the additional overhead introduced by the proposed strategy (i.e., *SS+KM+SM*) compared to *SS+KM* is below 5% for all the considered traffic load owing to the low number of SpHOs incurred by *SS+KM+SM* (note in Figure 6.8b that the average number of executed SpHOs is below 0.12 for all considered traffic loads). The analysis of the amount of SpHO signaling associated with each trigger in Figure 6.8b shows that, for very low traffic loads, most of performed SpHOs are triggered by changes in $F_{l,p}$ values because at such loads, both links #1 and #2 sessions are typically assigned their preferred pools according to the fittingness factor function of (6.1) (i.e., pools #3 or #4)). As the traffic load increases, the trigger of SM by link releases becomes more relevant with more SpHO executions. The increase in the number of executed SpHO is justified by the fact that at this high traffic load, a link is more likely to find its preferred pool occupied by another link at link establishment.

In summary, the proposed strategy approximates very well the upper-bound scheme with much less reconfiguration and reporting events and thus much lower signaling overhead, which strongly supports its practicality.

6.5.7 RT Capability to Detect Changes

This section evaluates the capability of the RT to detect relevant changes in the radio and interference conditions of the different spectrum pools.

Figure 6.9a plots the temporal evolution of the initial RT estimate of the number of performed SpHOs per link #1 session ($\overline{SpHO(1)}$) with the corresponding confidence interval shown in dashed lines for an intermediate traffic load of $L_l = \lambda_l \times T_l = 0.8Er, l \in \{1,2\}$. The time evolution on the *x*-axis is shown in terms of the number of established link #1 sessions according to the generation process described in Section 6.5.2. Note that only $SpHO(1)$ is considered because this is the KPI for which H_1 is selected by the RT after the change occurs.

The results show that $\overline{SpHO(1)}$ tends to converge as more link #1 sessions are established. Specifically, the stopping rule of Equation 6.13 is met after establishing around 2,800 sessions. At this point in time, the initial $\overline{SpHO(1)}$ estimate and its confidence interval are frozen, and the RT starts to watch the second moving average estimate $\left(\widehat{SpHO(1)}\right)$.

Figure 6.9b plots the temporal evolution of $\widehat{SpHO(1)}$ and its corresponding confidence interval around the considered change. The initial $\overline{SpHO(1)}$ estimate is also shown for comparison purposes. The results show that, before the change occurs, the moving average estimate $\widehat{SpHO(1)}$ oscillates close to $\overline{SpHO(1)}$ owing to the intrinsic randomness of radio conditions. Nevertheless, the RT disregards these oscillations and selects H_0 because the confidence intervals of the two estimates still overlap. After the change occurs, $\widehat{SpHO(1)}$ starts to deviate greatly from $\overline{SpHO(1)}$ until their confidence intervals no longer overlap. At this moment, H_1 is selected for $SpHO(1)$, and the considered change is detected by the RT. It is noteworthy that no wrong change (Type I error) is detected during the entire simulation, thus showing the efficiency of the proposed hypothesis testing strategy in Section 6.3.4 in minimizing useless regeneration of KD data.

In summary, the RT probabilistic detector, based on hypothesis testing, enables proper filtering of those changes related to the intrinsic randomness of the radio environment and efficient detection of relevant changes in the scenario.

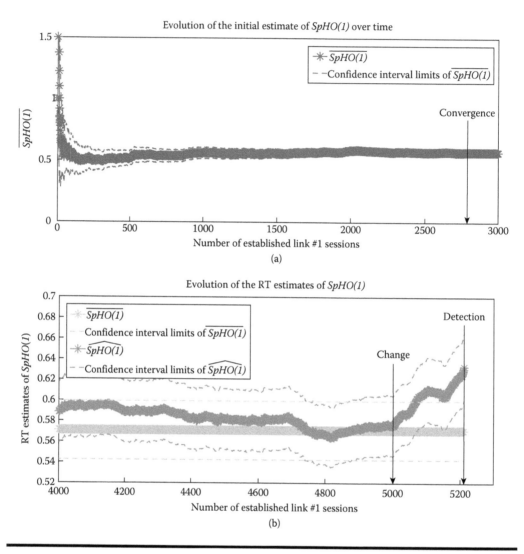

Figure 6.9 Evolution of the RT estimates of *SpHO(1)*, (a) initially and (b) around the considered change.

6.5.8 Robustness to Changes in the Scenario

This section evaluates the performance of the proposed strategy to changes in the interference conditions under which KD statistics were generated.

Figure 6.10a shows the online evolution of the packet loss rate of link #1 sessions (*Loss(1)*). An intermediate traffic load of $L_l = 0.8Er$, $l \in \{1,2\}$ is considered, with the change in interference conditions occurring after establishing 5,000 sessions. To analyze the impact of the RT, both variants *SS+KM+SM* and *SS+KM+SM+RT* are considered. Figure 6.10b plots the corresponding average number of performed SpHOs (*SpHO(1)*).

The results show that, after the change in interference conditions, the use of the RT functionality results in a significant reduction in both the packet loss rate (Figure 6.10a) and the number of performed SpHOs (Figure 6.10b). The reason for this improvement is that, before

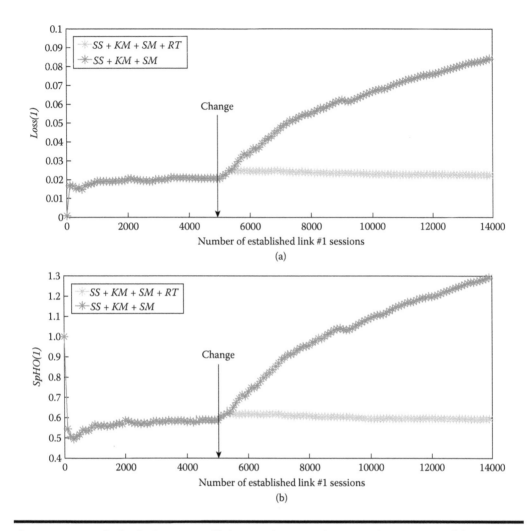

Figure 6.10   Analysis of robustness of link #1 performance in terms of (a) *Loss(1)* and (b) Sp*HO(1)*.

the change, both strategies mainly assign pools #3 and #4 to link #1. After the change, the strategy *SS+KM+SM* without any support from the RT still relies on the out-of-date KD statistics previously generated, so it continues to exclude pool #1 and assign pool #3 to link #1 sessions. This turns out to be a wrong decision because, according to the new conditions after the change, pool #1 should be assigned instead of pool #3, which from now on becomes unable to support the packet loss rate and bit rate requirements of link #1. Correspondingly, this degrades the packet loss rate (Figure 6.10a). In addition, much more frequent SpHOs are required to change pool #3 whenever it is assigned (Figure 6.10b). On the contrary, when the RT is used (i.e., *SS+KM+SM+RT*), new KD statistics are generated after detecting the change. As a consequence, the $\hat{F}_{l,p}$ estimates provided by the KM prevent the SS from assigning pool #3 to link #1 in the future. Instead, pool #1, which allows much better performance $\{R_k(l,p)\}_{k\in\{1,\dots,K_l\}}$ under the new interference conditions, is assigned to link #1. This results in a significant gain in terms of both the packet loss rate (Figure 6.10a) and the number of performed SpHOs (Figure 6.10b).

Note that the performance obtained with RT after the change remains approximately the same as before the change. This identical performance is because the change shown in Table 6.1 does not change the overall set of pools, but just swaps the interference patterns of pools #1 and #3. Consequently, if KD statistics are properly updated after the change, the strategy *SS+KM+SM+RT* should be able to find the proper combination of pools and active links that maintains the same performance that existed before the change.

6.6 Chapter Summary

This chapter has proposed a new cognitive management framework for supporting multimedia applications over CRNs. The proposed framework integrates a generic formulation of the fittingness factor to capture the suitability of spectrum resources to support various requirements of multimedia applications subject to unknown changes in interference conditions. A set of advanced statistics capturing the hidden dependence between the actual radio environment and fittingness factor behavior are retained in a KD. These statistics are used by the KM to extract the most relevant knowledge to assist in a novel proactive strategy for both the SS and SM decision-making processes. To support the nonstationarity of CRNs, an RT has been developed to detect, based on hypothesis testing, relevant changes in the scenario subject to the intrinsic randomness of radio conditions and perform, when necessary, the required updates. The proposed framework has been applied to support a mixture of real-time and best-effort applications. The results have shown that the proposed strategy, assisted by the KM and SM functionality, efficiently exploits the KM support at low loads and the SM functionality at high loads to support the strict requirements of real-time applications in terms of a sustainable bit rate and packet loss rate, while increasing as much as possible the bit rate achieved by best-effort applications. For both application types, significant performance gains have been observed with respect to a random selection with a very similar behavior to the upper-bound optimal scheme. An analysis of the practicality in terms of signaling requirements has demonstrated that the proposed strategy exploits the estimated suitability levels provided by the KM and the event-based triggering of SM to significantly reduce the amount of measurement reporting and pool reconfigurations, respectively, with respect to the optimal solution, which strongly supports the practicality of the proposed approach. An analysis of robustness under the nonstationary conditions of CRNs has demonstrated that the proposed RT is able to exploit the probabilistic detection rule of hypothesis testing to filter most of the changes due to the intrinsic radio randomness and detect actual changes in the scenario. Thanks to the RT support, the proposed spectrum management strategy exhibits substantial robustness to unexpected changes, thus showing the practicality of the proposed strategy in such realistic environments.

As part of future work, it is intended to further develop the applicability of the proposed framework to decentralized scenarios, including the adaptation of decision-making strategies.

References

1. J. Mitola, III and G. Q. Maguire, Jr., Cognitive radio: Making software radios more personal, *IEEE Personal Communications*, vol. 6, no. 4, pp. 13–18, 1999.
2. S. Haykin, Cognitive radio: Brain-empowered wireless communications, *IEEE Journal on Selected Areas in Communications*, vol. 23, no. 2, pp. 201–220, 2005.
3. R. W. Thomas, D. H. Friend, L. A. Dasilva, and A. B. Mackenzie, Cognitive networks: Adaptation and learning to achieve end-to-end performance objectives, *IEEE Communications Magazine*, vol. 44, no. 12, pp. 51–57, 2006.

4. J. Kephart and D. Chess, The vision of autonomic computing, *Computer*, vol. 36, no. 1, pp. 41–50, 2003.
5. P. Demestichas, G. Dimitrakopoulos, J. Strassner, and D. Bourse, Introducing reconfigurability and cognitive networks concepts in the wireless world, *IEEE Vehicular Technology Magazine*, vol. 1, no. 2, pp. 32–39, 2006.
6. P. Demestichas, K. Tsagkaris, and V. Stavroulaki, Cognitive management systems for supporting operators in the emerging Future Internet era, in *2010 IEEE 21st International Symposium on Personal, Indoor and Mobile Radio Communications Workshops (PIMRC Workshops)*, September 2010, pp. 21–25.
7. E. Meshkova, Z. Wang, J. Nasreddine, D. Denkovski, C. Zhao, K. Rerkrai, T. Farnham, et al., Using cognitive radio principles for wireless resource management in home networking, in *2011 IEEE Consumer Communications and Networking Conference (CCNC)*, January 2011, pp. 669–673.
8. The end-to-end-efficiency (E3) project [Online]. Available from: http://www.ict-e3.eu/
9. The ICT EU (Opportunistic Networks and Cognitive Management Systems for Efficient Application Provision in the Future InterneT) OneFIT Project [Online]. Available from: http://www.ict-onefit.eu/
10. IEEE Standard Coordinating Committee 41 (SCC41) [Online]. Available from: http://grouper.ieee.org/groups/scc41/index.html/
11. ETSI Technical Committee (TC) on RRS [Online]. Available from: http://www.etsi.org/WebSite/technologies/RRS.aspx
12. Y. Li, Y. Dong, H. Zhang, H. Zhao, H. Shi, and X. Zhao, Qos provisioning spectrum decision algorithm based on predictions in cognitive radio networks, in *Proceeding of the Wireless Communications Networking and Mobile Computing (WiCOM) 2010*, September 2010, pp. 1–4.
13. W.-Y. Lee and I. Akyldiz, A spectrum decision framework for cognitive radio networks, *IEEE Transactions on Mobile Computing*, vol. 10, no. 2, pp. 161–174, 2011.
14. V. Mishra, L. C. Tong, and C. Syin, Deterministic time pattern based channel selection in cognitive radio network, in *2013 International Conference on Information Networking (ICOIN)*, 2013, pp. 169–174.
15. F. Bouali, O. Sallent, J. Pérez-Romero, and R. Agustí, A cognitive management framework for spectrum selection, *Computer Networks*, vol. 57, no. 14, pp. 2752–2765, 2013.
16. F. Bouali, O. Sallent, J. Perez-Romero, and R. Agusti, Exploiting knowledge management for supporting multi-band spectrum selection in non-stationary environments, *IEEE Transactions on Wireless Communications*, vol. 12, no. 12, pp. 6228–6243, 2013.
17. *European Telecommunications Standards Institute (ETSI)*, Technical Report (TR) 102 682 V1.1.1 Reconfigurable Radio Systems (RRS); Functional Architecture (FA) for the Management and Control of Reconfigurable Radio Systems, 2009.
18. One FIT Deliverable D2.2, OneFIT functional and system architecture [Online], 2011. Available from: http://www.ict-onefit.eu/
19. *European Telecommunications Standards Institute (ETSI)*, Technical Report (TR) 102 684 V1.1.1 Reconfigurable Radio Systems (RRS); Feasibility Study on Control Channels for Cognitive Radio Systems, 2012.
20. V. Stavroulaki, K. Tsagkaris, P. Demestichas, J. Gebert, M. Mueck, A. Schmidt, R. Ferrus, et al., Cognitive control channels: From concept to identification of implementation options, *IEEE Communications Magazine*, vol. 50, no. 7, pp. 96–108, 2012.
21. F. Bouali, O. Sallent, J. Pérez-Romero, and R. Agustí, A fittingness factor-based spectrum management framework for cognitive radio networks, *Wireless Personal Communications* [Online], pp. 1–15, 2013. Available from: http://dx.doi.org/10.1007/s11277-013-1128-6
22. L. Badia, M. Lindström, J. Zander, and M. Zorzi, Demand and pricing effects on the radio resource allocation of multimedia communication systems, in *GLOBECOM '03. IEEE*, vol. 7, pp. 4116–4121, 2003.
23. E. L. Lehmann and J. P. Romano, *Testing Statistical Hypotheses*, 3rd ed., ser. Springer Texts in Statistics, Springer, New York, 2005.
24. N. Schenker and J. F. Gentleman, On judging the significance of differences by examining the overlap between confidence intervals, *The American Statistician*, vol. 55, no. 3, pp. 182–186, 2001.
25. One FIT Deliverable D3.3, Protocols, performance assessment and consolidation on interfaces for standardization [Online], 2012. Available from: http://www.ict-onefit.eu/

Chapter 7

Cross-Layer MIMO Cognitive Communications for QoS-/ QoE-Demanding Applications

Reema Imran,[1] Maha Odeh,[2] Nizar Zorba,[3] and Christos Verikoukis[4]

[1]*Electrical Engineering department, University of Jordan, Jordan*

[2]*Interdisciplinary Centre for Security, Reliability and Trust (SNT), University of Luxemburg, Luxemburg*

[3]*Electrical Engineering department, Qatar University, Qatar*

[4]*CTTC, Castelldefels, Barcelona, Spain*

Contents

7.1 Introduction

The radio spectrum is becoming more crowded as the demand for wireless applications is explosively increasing, which presents a significant challenge for the operators to obtain sufficient bandwidth to handle them. Several operators (e.g., Vodafone, ATT) have recently replaced the traditional flat-rate data download option by more restrictive ones, as they are unable to offer the rate requested by all the users. On the other hand, several surveys have revealed that this precious resource is widely underutilized [1,2].

The conflict between spectrum scarcity and service demands has motivated research on innovative schemes that can solve the problem of the inherited allocation techniques, by adding some kind of intelligence to their implementations. The notion of cognitive radio [3] is considered as an intelligent method of permitting medium sharing, by allowing opportunistic transmissions from lower-priority users without threatening the connection quality for the higher-priority ones.

Therefore, cognitive radio seems to be a very promising technique [3]. However, many obstacles have to be overcome in order to introduce the cognitive radio in the market, and to accelerate its adoption by commercial systems. Such a deployment requires flexible, allocation of the appropriate band and adaptation of the transmission parameters to suit the surrounding environment/channel, and prevention of any harmful interference on the higher-priority users' receivers, in order to fulfill the ultimate goal of employing cognitive radio [3].

From a different perspective, recent studies have focused on the space dimension utilization by including more transmission/reception antennas in the communication process [4,5]. They inject a new degree of freedom into cellular systems that can boost the operator's efficiency, increase the throughput, decrease the error probability, and enable interference cancellation for simultaneous multiuser transmissions. Another application of multiple-input multiple-output (MIMO) systems is to enable cognitive radio in the spatial domain, the so-called spatial sharing, where a user can be serviced through the first antenna/stream while another user is serviced via the second antenna/stream as long as it does not affect the quality of service (QoS)/quality of experience (QoE) satisfaction [6–8] of the first user. One of the most important and feasible MIMO transmission techniques is zero-forcing beamforming (ZFB), which has a proven track record of excellent performance in realistic systems [5]. The concept of cognition is widely known [3], but its application within the spatial domain is relatively unexplored. Its consideration is mainly motivated by the inclusion of MIMO antennas in the broadband wireless communication standards [9].

The core concept of spatial cognition is that users are separated by their channel impulse response, and they share the same time, frequency, and code. Therefore, well-defined priority levels should be established to satisfy each user's demands: primary users (PUs) always have the privilege to access the medium, and secondary users (SUs) will be allocated the system resource when it is possible (i.e., they are the cognitive users), as long as they do not interfere with the PU's transmission. This means that two conflicting goals should be optimized: maximizing the throughput of the cognitive users while reducing the interference to the PUs.

In multiuser channels, the inherited systems deal with delay and throughput separately, and as a consequence, access methods such as ALOHA [10] were suggested to prevent collisions without utilizing the multiuser diversity concept. However, this problem was solved in Ref. [11] by incorporating the channel conditions (channel state information) in their scheduling scheme, while providing delay constraints for packets. Moreover, Zorba et al. [12] and Weifei et al. [13] evaluate the maximum delay that faces the worst channel user in the system. Nevertheless, these studies do not consider the priority levels imposed by the cognition concept. Few studies have focused on QoS/QoE provisioning in CR systems with multiple antennas [14], and their preliminary results focus on SU satisfaction from the delay perspective. Interesting results about QoE are obtained in Ref. [15] when considering the cognition effect on the service acknowledgment and in Ref. [16] if routing is tackled in the cognition effect on QoE.

The heterogeneity of the users (i.e., PU and SU) and applications imposes different demands on the operator in terms of satisfaction of its users' QoS/QoE needs. While QoS [6] refers to the engineering design metrics of throughput, delay, and so on, QoE [7], on the other hand, refers to how the end user is affected by the throughput and delay increase/decrease. QoE can be evaluated using objective and subjective methods [17,18]. Subjective methods depend on human observations to evaluate quality levels, and they are usually set up in real-life (controlled) environments [19,20] and depend on real users' experience that could be biased [19,21]. On the other hand, objective methods give more consistent evaluations [22] from the point of view of design and engineering, as they are based on engineering metrics from the network itself [22], where some predefined metrics on the basis of users' satisfaction and the price that the users pay for the service are set up, and any new scheme/algorithm should meet predefined values. Obviously, both subjective and objective methods have to be developed in order to attain the final objective of QoE for users, with mathematical expressions that link the two approaches [23,24]. In this chapter, we focus on the objective method and how to achieve predefined QoE metrics through the implementation of a hybrid MIMO-cognitive communication strategy.

Our goal is to provide quantitative QoS results that are based on the network performance satisfaction metrics, in terms of the resultant QoE from the minimum guaranteed rate and also from the maximum allowed scheduling delay and jitter. Many studies have shown that the user's satisfaction is insignificantly raised by a performance higher than the requested demands, while on the other hand, the satisfaction drastically reduces when the user's requirements are not fulfilled [25]. To satisfy the delay requirements for the PUs and SUs, we investigate the worst-case delay. Obviously, the worst case highly depends on the scheduling policy. In this chapter, we employ the opportunistic scheduling approach [26], and the obtained results depend on different variables such as the number of the PUs and SUs, the requested demands, and the users' measured signal-to-noise ratio (SNR) values.

In this chapter, QoS and QoE in the context of spatial cognitive transmission are investigated. We focus on cognition in multiple-antenna scenarios where the selection scheme dramatically influences the users' performance. The statistical distributions for both random and the opportunistic selection are derived to judge the feasibility of cognition. The main contributions of the chapter are as follows:

- A spatial cognition scenario is analyzed where the ZFB strategy is employed. Opportunistic scheduling is considered in order to select the serviced users at each time.
- The statistical distributions (probability density function [PDF] and cumulative density function [CDF]) are obtained within the proposed multiuser scenario.
- QoS and QoE metrics [22] based on the maximum allowed scheduling delay, minimum guaranteed throughput, and mean jitter are obtained in closed-form expressions, to show the impact of each parameter on system performance, and to optimize it.

The rest of this chapter is organized as follows: in Section 7.2 the system model is presented, while in Section 7.3 a revision for ZFB is tackled, in which derivations of the PDF and CDF for both scheduling strategies are obtained. Section 7.4 discusses the QoS and QoE metrics in the cognitive scenarios, while the numerical and simulation results are displayed in Section 7.5, followed by the chapter conclusions in Section 7.6.

7.2 System Model

We consider a single-cell multiple-antenna downlink scenario, where a single base station (BS) equipped with n_t transmit antennas offers traffic to N user terminals, each one being equipped with a single receiving antenna. We assume a quasi-static block fading channel $\mathbf{h}_{[1 \times n_t]}$ between the BS antennas and each of the users, where the channel from each transmit antenna to a user is characterized by independent and identically distributed (i.i.d.) complex Gaussian entries ~CN(0,1). Let $\mathbf{x}(t)$ be the $n_t \times 1$ transmitted vector, and denote $y_n(t)$ as the nth user received signal given by

$$y_n(t) = \mathbf{h}_n(t)\mathbf{x}(t) + z_n(t) \tag{7.1}$$

where $z_n(t)$ is an additive i.i.d. complex noise component with zero mean and $E\{|z_n|^2\} = \sigma^2$.

The transmitter delivers service to M users simultaneously, where $M \leq n_t$, so that a more compact system formulation is obtained by stacking the received signals into an $M \times 1$ vector $\mathbf{y}(t)$ and the noise components $\mathbf{z}(t)_{[M \times 1]}$ for the set of M selected users as $\mathbf{y}(t) = \mathbf{H}(t)\mathbf{x}(t) + \mathbf{z}(t)$, with $\mathbf{H}(t) = [\mathbf{h}_1(t); \ldots; \mathbf{h}_M(t)]$ as the compound channel. Note that the transmitted signal $\mathbf{x}(t)$ encloses the uncorrelated data symbols $s_m(t)$ to each one of the selected users, where $E\{|s_m|^2\} = 1$ is employed. We assume that perfect channel state information is available at the transmitter side, and the total transmission power is $P_t = 1$. For simplicity of notation, the time index is dropped whenever possible.

7.3 Zero-Forcing Beamforming

Transmit beamforming has been utilized as a technique that can simultaneously deliver service to many users through several transmitting antennas, by aligning the transmitting beams to the direction that ensures the best signal reception to the targeted users, while the interference among them is mitigated. Several strategies have been proposed for transmit beamforming, and one of the simplest and most commercially implementable techniques is ZFB. ZFB has the ability to null the interference among the scheduled users, by decoupling the downlink channels and transforming them into orthogonal and independent subchannels. The simultaneous transmission to M users modifies the transmitted signal $\mathbf{x}$ in Equation 7.1 as follows:

$$\mathbf{x} = \mathbf{B}\mathbf{s} \tag{7.2}$$

where $\mathbf{s}$ is the sequence of transmitting symbols, and $\mathbf{B}$ is the beamforming matrix, which for the ZFB is obtained through the Moore–Penrose pseudoinverse [27]:

$$\mathbf{B} = \mathbf{H}^H (\mathbf{H}\mathbf{H}^H)^{-1} \tag{7.3}$$

A uniform power allocation strategy is considered, so that the sum power of the transmitting vectors has to be normalized to $P_t = 1$, as follows:

$$|\mathbf{B}|^2 = \alpha^2 \mathbf{I} \tag{7.4}$$

where α is the normalization parameter [5] defined as

$$\alpha_m^2 = \gamma_m = \frac{1}{n_t \left(\mathbf{H}\mathbf{H}^H \right)_{m,m}^{-1}} \tag{7.5}$$

so that the equivalent channel (**HB**) is now diagonal, which guarantees no interference among the users as follows:

$$\mathbf{HB} = \mathbf{D} = \mathrm{diag}\,(\alpha_1, \alpha_2, \ldots, \alpha_M) \tag{7.6}$$

Therefore, the SNR of each serviced user can be stated as [27]

$$SNR_m = \frac{\alpha_m^2}{\sigma_m^2} \tag{7.7}$$

The most attractive characteristic of ZFB is the whipped-out interference among the selected users, which represents the ideal technique to be used in spatial cognition, to prevent any interference from the secondary (cognitive) users toward the primary user. To increase the ZFB performance in multiuser scenarios, ZFB employment is more suitable with an algorithm that selects the most orthogonal users, as a pre-amended step before the spatial processing [5,28]. Any number of antennas n_t can be employed within ZFB, but a value of $n_t > 2$ does not show attractive performance owing to the effort expended on interference cancellation (which will push down the whole system behavior), making $n_t = 2$ the optimal value that shows the largest benefit of ZFB in comparison to single-antenna systems [27]. Therefore, we assume this from now on, and we will only obtain the mathematical expressions for $n_t = 2$.

In the next section, we will show how to employ ZFB within the spatial cognition philosophy, and we will obtain its performance characterization.

7.3.1 Spatial Cognition in ZFB

Many studies have proposed cognitive radio on decentralized networks [29], but an interesting scenario for commercial operators is to apply cognition on the same BS to match the current cellular deployment, so that one of the ZFB beams targets the PUs while the other beam targets the SUs. Such a mechanism will enable operators to increase the total number of served customers and as a result their profits will increase. A major issue that has to be considered in cognitive systems is that QoS/QoE satisfaction for the PU has the highest priority, and an SU will not be scheduled if the PU does not fulfill such a QoS requirement (defined on the basis of the SNR network metric [22], as later discussed) when the SU is provided service.

For the single-user case (i.e., no cognition), if the user fails to satisfy his or her own demands (the SNR threshold value) owing to bad instantaneous channel characteristics, the user will be declared in outage [30]. This is also the case in cognitive scenarios, but allowing an SU to be serviced should not increase the outage for the PU, so that a study of the impact of cognition on the PU's performance and the incurred outage is required. To proceed with the mathematical formulations, the distribution of the SNR expression in Equation 7.5 is required, which we reformulated as [31]

$$\gamma_m = \alpha_m^2 = \frac{det\left(\mathbf{HH}^H\right)}{n_t A_{mm}} \tag{7.8}$$

where $det(.)$ stands for the matrix determinant, and A_{mm} stands for the determinant resulting from striking out the mth row and the mth column of $det(.\mathbf{HH}^H)$. From [31], we have that

$$det\left(\mathbf{HH}^H\right) = \|\mathbf{h}_1\|^2 \|\mathbf{h}_2\|^2 - \left|\mathbf{h}_1^h\mathbf{h}_2\right|^2 = \|\mathbf{h}_1\|^2 \|\mathbf{h}_2\|^2 \left(1 - \delta^2\right) \tag{7.9}$$

with $\delta^2 = \dfrac{\left| \mathbf{h}_1^{\flat} \mathbf{h}_2 \right|^2}{\left\| \mathbf{h}_1 \right\|^2 \left\| \mathbf{h}_2 \right\|^2}$ as the normalized scalar product between $\mathbf{h}_1$ and $\mathbf{h}_2$, $0 \leq \delta^2 \leq 1$, where 0 occurs when the users' channels are wholly orthogonal and 1 when they are collinear. Therefore, the resultant SNR value [31] for the mth selected user will be

$$\gamma_m = \frac{\left\| \mathbf{h}_m \right\|^2 \left(1 - \delta^2 \right)}{2} \tag{7.10}$$

To thoroughly study the impact of cognition, Equation 7.10 should be further investigated, where the statistical formulation for the PU–SU coexistence is explained by the interaction of each term of $\|\mathbf{h}_m\|^2$ and $1 - \delta^2$, where $\|\mathbf{h}_m\|^2$ is the addition of two i.i.d. complex Gaussian random variables that have a $\Gamma(2, \gamma_\circ)$ distribution, while $\gamma_\circ$ is the average system SNR.

Therefore, the PDF of the $\mathbf{h}_m^2$ term can be expressed as follows [32]:

$$q(\gamma) = \frac{\gamma}{\gamma_\circ^2} e^{-\frac{\gamma}{\gamma_\circ}} \tag{7.11}$$

As a result, the corresponding CDF is interpreted as

$$Q(\gamma) = 1 - e^{-\frac{\gamma}{\gamma_\circ}} \left(1 + \frac{\gamma}{\gamma_\circ} \right) \tag{7.12}$$

Likewise, δ^2 has a beta distribution $\beta(m, n_t - m)$ [5,28], with the corresponding PDF [32]:

$$f_\delta\left(\delta; m, n_t - m \right) = \frac{\delta^{m-1} \left(1 - \delta \right)^{n_t - m - 1}}{\displaystyle\int_0^1 \delta^{m-1} \left(1 - \delta \right)^{n_t - m - 1}} \tag{7.13}$$

where m is the number of each scheduled user.

Consequently, the δ^2 term has a CDF that we obtain as a regularized incomplete beta function $I_\delta(m, n_t - m)$ as

$$I_\delta\left(m, n_t - m \right) = \frac{\beta(\delta, m, M - m)}{\beta(m, M - m)} \tag{7.14}$$

Now we want to characterize the foregoing distributions and to know if they match any widely known distribution. One of the familiar characteristics of the beta distribution is that if X has $\beta(a, b)$, then $1 - X$ has $\beta(b, a)$. For the case of $n_t = 2$, both δ^2 and $1 - \delta^2$ have $\beta(1,1)$. The distribution of the multiplication of the two terms can be concluded from the relation between the beta and gamma distributions; if X and Y are independently distributed with $\Gamma(a, \gamma_\circ)$ and $\Gamma(b, \gamma_\circ)$, respectively, their sum, $X + Y$, has the distribution $\Gamma(a + b, \gamma_\circ)$, which is represented by $\|\mathbf{h}_1\|^2$. Another important transformation of the gamma distribution is that $\dfrac{X}{Y + X}$ has $\beta(a, b)$ [33]. With all these obtained results, because $\|\mathbf{h}_1\|^2$ has a $\Gamma(2, \gamma_\circ)$ distribution and $(1 - \delta^2)$ has $\beta(1,1)$, their product will have $\Gamma(1,\gamma_\circ)$, which presents an exponential distribution, so that the term $\dfrac{\|\mathbf{h}_m\|^2\left(1-\delta^2\right)}{2}$ will have the following PDF and CDF:

$$f_\gamma(\gamma) = \frac{2}{\gamma_\circ} e^{-2\frac{\gamma}{\gamma_\circ}} \tag{7.15}$$

$$F_\gamma(\gamma) = 1 - e^{-\frac{2\gamma}{\gamma_\circ}} \tag{7.16}$$

which will be employed in later sections.

7.3.2 Opportunistic Transmission in ZFB Cognitive Scenarios

The rate optimal transmission scheme in multiuser scenarios is opportunistic transmission [26], where the BS extracts the diversity gain by rewarding the service to the user with the best channel condition at any time slot. This concept cannot be directly generalized to ZFB because selecting the users with highest channel norm $\|\mathbf{h}\|^2$ does not mean that they will obtain the highest quality owing to the spatial correlation among the serviced users, which will drastically degrade the quality performance, as shown by Equation 7.5. To overcome this hurdle, a preliminary step should be defined to avoid the collinearity and limit the correlation among the selected users [5,28], so as to select the users with best channel characteristics and with the lowest correlation among them. The scheduling scheme can be implemented as follows:

■ Select the PU (if there is more than one PU) that has the highest channel norm $\|\mathbf{h}_{PU}\|^2$.

■ Evaluate $\delta_s^2 = \dfrac{\left|\mathbf{h}_{PU}^h \mathbf{h}_{SU}\right|^2}{\|\mathbf{h}_{PU}\|^2 \|\mathbf{h}_{SU}\|^2}$ for all the SUs.

■ Pick the users S who satisfy the condition $\delta_s \le \delta_\circ$, where $\delta_\circ$ is the correlation threshold that is predefined by the system administrator.

■ Select the user with highest $\|\mathbf{h}_s\|^2\left(1-\delta_s^2\right)$ among all the users who satisfy the previous condition.

Therefore, the users with the best channel conditions and the lowest spatial correlation among them are selected.

This preliminary step to increase the orthogonality definitely changes the PDF and CDF. The new distributions are derived from Equation 7.10 by imposing $\delta_s \in [0,\delta_\circ]$. The statistical characteristic of the SNR distribution should change to adapt to the modifications, as the term $1 - \delta^2$ in Equation 7.9 no longer has a $\beta(1,1)$ distribution owing to the restriction limit $\delta_\circ$. It now follows a uniform distribution $U\left(1-\delta_\circ^2,1\right)$ as presented in Ref. [32], so we are able to obtain the CDF of the obtained SNR for our scenario as

$$Y_\gamma(\gamma) = \int_{1-\delta_\circ^2}^1 \int_0^{\frac{\gamma}{\delta^2}} \frac{z}{\gamma_\circ^2} e^{-\frac{z}{\gamma_\circ}} dz d\delta^2 = 1 + \frac{1-\delta_\circ^2}{\delta_\circ^2} e^{-\frac{2\gamma}{\gamma_\circ(1-\delta_\circ^2)}} - \frac{1}{\delta_\circ^2} e^{-\frac{2\gamma}{\gamma_\circ}} \tag{7.17}$$

with its related PDF as

$$y_\gamma(\gamma) = \frac{2}{\delta_\circ^2 \gamma_\circ} \left(e^{-\frac{2\gamma}{\gamma_\circ}} - e^{-\frac{2\gamma}{\gamma_\circ(1-\delta_\circ^2)}} \right) \tag{7.18}$$

Therefore, using the large number law [32] in an iterative way, we formulate the number of users S that can survive the $\delta_\circ$ selection, for the gth iteration as

$$S_g = N_{SU} \left(1 - \left(1 - \delta_\circ^2 \right)^{(n_t-1)} \right)^g \tag{7.19}$$

where N_{SU} is the number of available SUs in the scenario. With $n_t = 2$ antennas (i.e., one single iteration is required), the number S states is

$$S = N_{SU} \delta_\circ^2 \tag{7.20}$$

From these equations, it can be concluded that the value of $\delta_\circ$ cannot be randomly selected, as it has a lower bound to guarantee at least one selection to achieve this orthogonality

$$\delta_\circ \geq \sqrt{\frac{1}{N_{SU}}} \tag{7.21}$$

With the foregoing presented opportunistic transmission, the serving SNR is the maximum among all other users' SNRs, so the CDF for the PU is derived from Equation 7.17 as

$$OY_\gamma^{PU}(\gamma) = \left(1 + \frac{1-\delta_\circ^2}{\delta_\circ^2} e^{-\frac{2\gamma}{\gamma_\circ(1-\delta_\circ^2)}} - \frac{1}{\delta_\circ^2} e^{-\frac{2\gamma}{\gamma_\circ}} \right)^{N_{PU}} \tag{7.22}$$

where N_{PU} denotes the number of available PU users in the scenario. The corresponding PDF is therefore obtained as

$$Oy_\gamma^{PU}(\gamma) = \frac{2N_{PU}}{\delta_\circ^2 \gamma_\circ} \left(e^{-\frac{2\gamma}{\gamma_\circ}} - e^{-\frac{2\gamma}{\gamma_\circ(1-\delta_\circ^2)}} \right) \left(1 + \frac{1-\delta_\circ^2}{\delta_\circ^2} e^{-\frac{2\gamma}{\gamma_\circ(1-\delta_\circ^2)}} - \frac{1}{\delta_\circ^2} e^{-\frac{2\gamma}{\gamma_\circ}} \right)^{N_{PU}-1} \tag{7.23}$$

On the other hand, the PDF for the opportunistic selection tackling the SU users is

$$Oy_\gamma^{SU}(\gamma) = \frac{2S}{\delta_\circ^2 \gamma_\circ} \left(e^{-\frac{2\gamma}{\gamma_\circ}} - e^{-\frac{2\gamma}{\gamma_\circ(1-\delta_\circ^2)}} \right) \left(1 + \frac{1-\delta_\circ^2}{\delta_\circ^2} e^{-\frac{2\gamma}{\gamma_\circ(1-\delta_\circ^2)}} - \frac{1}{\delta_\circ^2} e^{-\frac{2\gamma}{\gamma_\circ}} \right)^{S-1} \tag{7.24}$$

which is very similar to Equation 7.23 with N_{PU} replaced by S, which is the number of SUs satisfying the $\delta_\circ$ correlation restriction (i.e., $S \leq N_{SU}$).

Even though the opportunistic scheme can achieve the highest possible throughput, it has many disadvantages, especially if the system supports diverse QoS demands [25] in terms of the maximum delay network metric, as we will see later.

7.4 QoS/QoE Performance in Spatial Cognitive Scenarios

QoS can be evaluated from the engineering metrics available for operators such as throughput and delay, among others. On the other hand, ITU has proposed the mean opinion score (MOS) to evaluate the QoE [34], where five increasing levels for its quality characterization have been defined, and MOS = 5 indicates the highest QoE. Because our analysis is based on objective QoE characterization, several network metrics such as the minimum required rate for correct reception and the maximum tolerable delay for delay-sensitive applications can be considered, among others. This chapter tackles the aforementioned QoE metric in a cognitive scenario, where the PUs have strict QoS/QoE demands [28], while the SUs QoS/QoE demands are satisfied when possible.

It is practically impossible to provide QoS/QoE 100% of the time, owing to the incapability of the channel (i.e., fading) to provide the required SNR to correctly decode the received packets. As a consequence, the concept of outage [30], which is widely considered in cellular systems, is defined not only for the SUs but also for the PUs. This chapter deals with two kinds of outage [12] that impact the QoE: access delay outage and rate outage. The first outage refers to the channel opportunistic access instant and whether the user is selected within the maximum tolerable time delay K. The rate outage is related to the received data rate of the selected user and whether its rate requirement is achieved or not.

7.4.1 QoS/QoE Related to the Maximum Tolerable Access Delay

In the delay optimal scheduling scheme (i.e., random) with N users, the maximum scheduling delay is N time slots [35]. Nevertheless, the efficiency changes as the selection of the served user is done at each slot, based on its channel conditions and/or spatial correlation with other users. Therefore, the access and service acquisition are not guaranteed.

The access delay is a very important indicator of system performance and QoS, where its maximum access delay is the largest number of slots that a user's packet waits since it is ready to be transmitted once the user gets access to the channel [12]. If the user is not scheduled within this period, the user is in delay outage. Then the probability that a maximum of K time slots are needed to select a certain user from a group of users follows a geometric distribution [12,32], as

$$\text{Prob}\,[X \le K] = 1 - (1 - P_{ac})^K \tag{7.25}$$

Then the access outage ζ_{ac} is defined as follows:

$$\zeta_{ac} = (1 - P_{ac})^K \tag{7.26}$$

where P_{ac} is the probability of accessing the channel, which in a cognitive scenario depends on the kind of served user and the chosen threshold for both PU and SU. This study displays the delay faced by the primary and secondary subsystems separately, and a unified framework is established later to judge the overall performance of the whole system.

From Equation 7.26, the maximum scheduling delay over all the users (PUs and SUs) is

$$K = \frac{\log_2\left(\zeta_{ac}\right)}{\log_2\left(1 - P_{ac}\right)} \tag{7.27}$$

The access outage for the primary users, P_{ac}^{PU}, does not only rely on the number of the PUs, but it also depends on the selection scheme, whether it is just random or opportunistic selection.

For the former case, we formulate the access outage under the random scheduling philosophy from Equations 7.12 and 7.26 as

$$P_{ac}^{PU_{RAN}} = \frac{e^{-\frac{\gamma_{th}^{PU}}{\gamma^{\circ}}}\left(1+\frac{\gamma_{th}^{PU}}{\gamma^{\circ}}\right)}{N_{PU}} \tag{7.28}$$

while for opportunistic selection, the probability of access outage, $P_{ac}^{PU_{OPP}}$, is now obtained from Equations 7.12 and 7.26 as

$$P_{ac}^{PU_{OPP}} = \frac{1-\left(1-e^{-\frac{\gamma_{th}^{PU}}{\gamma^{\circ}}}\left(1+\frac{\gamma_{th}^{PU}}{\gamma^{\circ}}\right)\right)^{N_{PU}}}{N_{PU}} \tag{7.29}$$

Note that owing to the cognition policy, the SU's activities should not affect any PU's transmission. Therefore, if PUs cannot afford the required quality, none of the SUs, independent of their channel conditions, will be scheduled for transmission. Thus, the access outage for SUs, P_{access}^{SU}, depends on the quality for the PU, the threshold values (the value γ_{th} below which the user will not be awarded any kind of service), and the number of SUs that exists in the system. For the case of random selection, we obtain the resultant access outage, $P_{ac}^{SU_{RAN}}$, using Equations 7.16 and 7.26 as

$$P_{ac}^{SU_{RAN}} = \frac{e^{-2\frac{\gamma_{thPU}+\gamma_{thSU}}{\gamma^{\circ}}}}{N_{SU}} \tag{7.30}$$

Taking into account the fact that $\delta_{\circ}$ is the correlation threshold previously defined in Equation 7.21, the access outage for SUs in the opportunistic scheduling strategy, $P_{ac}^{SU_{OPP}}$, using Equations 7.22 and 7.26 is

$$P_{ac}^{SU_{OPP}} = \frac{\left(1-\left(1+\frac{1-\delta_{\circ}^2}{\delta_{\circ}^2}e^{-\frac{2\gamma_{thPU}}{\gamma^{\circ}(1-\delta_{\circ}^2)}}-\frac{1}{\delta_{\circ}^2}e^{-\frac{2\gamma_{thPU}}{\gamma^{\circ}}}\right)^{N_{PU}}\right)\left(1-\left(1+\frac{1-\delta_{\circ}^2}{\delta_{\circ}^2}e^{-\frac{2\gamma_{thSU}}{\gamma^{\circ}(1-\delta_{\circ}^2)}}-\frac{1}{\delta_{\circ}^2}e^{-\frac{2\gamma_{thSU}}{\gamma^{\circ}}}\right)^{S}\right)}{N_{SU}} \tag{7.31}$$

To identify the whole system delay into a single parameter, a unification procedure should take place by evaluating the average performance [32] as

$$\tilde{n} = \sum_i P_i N_i = P_{ac}^{PU} * N_{PU} + P_{ac}^{SU} * N_{SU} \tag{7.32}$$

enabling us to present the resultant total scheduling delay for the cognitive scenario, using Equations 7.27 and 7.32, as follows:

$$K_{res} = \frac{\log_2\left(\zeta_{ac}\right)}{\log_2\left(1-\frac{\tilde{n}}{N_{PU}+N_{SU}}\right)} \tag{7.33}$$

where the impact of the scheduling probability is also seen through the value of ζ_{ac} as in Equation 7.26. Once this access delay QoS network performance is obtained, its QoE metric is formulated by the IQX hypothesis [36], which defines an exponential relationship between the QoE representing the user perception and QoS network performance, so that we obtain the resultant QoE metric related to the access delay as

$$QoE_{ac} = \eta_{ac}\exp\left(-\mu_{ac}K_{res}\right) + \nu_{ac} \tag{7.34}$$

where η_{ac}, μ_{ac}, and ν_{ac} are constants that are selected on the basis on the scenario and application, as later shown in the simulations section.

7.4.2 QoS/QoE Related to the Minimum Guaranteed Rate

One of the basic requirements of digital transmission is the reliability of the delivered data. To guarantee such a requirement, a minimum rate (or the corresponding SNR) depending on the type of application should be guaranteed to ensure the error-free reception of the packets. Thus, the user who gets the permission to access the medium will not be allocated any service unless his or her rate is above the minimum required rate (even if he or she has the largest SNR over all the other users). Otherwise, the user is in rate outage.

The difference between the users' demands leads to different outage for a given minimum rate. An SU suffers from rate outage in two cases: first, if the supported rate is less than the minimum required rate (i.e., a predefined threshold), and second, when the primary user is in outage state. As the PUs channel norm $\|\mathbf{h}_m\|^2$ has a $\Gamma(n_t, \gamma_o)$ distribution, then if the norm is below the threshold value, the PU will be declared in an outage state, and consequently the whole system suffers from an outage. Instead of basing the study on the minimum rate demand, it would be easier—and the results would be exactly the same—if the study were performed over the minimum SNR value, so that the threshold would be predefined over an SNR value. As the SNR value depends greatly on the scheduling approach, the value of the rate outage within the random selection philosophy, ζ_{rate}^{PU}, is based on the SNR distribution, and we obtain it using Equation 7.12 as

$$\zeta_{rate}^{PU\,RAN} = \mathrm{Prob}\left[\gamma < \gamma_{th}^{PU}\right] = \left(1 - e^{-\frac{\gamma_{th}^{PU}}{\gamma_\circ}}\left(1 + \frac{\gamma_{th}^{PU}}{\gamma_\circ}\right)\right) \tag{7.35}$$

where by manipulating the previous expression, the $\gamma_{th}^{PU\,RAN}$ can be expressed as

$$\gamma_{th}^{PU\,RAN} = -\gamma_\circ\left(\left(-\left(1 - \zeta_{rate}^{PU}\right)e^{-1}\right)e^{-\left(1 - \zeta_{rate}^{PU}\right)e^{-1}} + 1\right) \tag{7.36}$$

while the rate cognition outage for the SU users, $\zeta_{rate}^{SU\,RAN}$, is given by using Equation 7.16 as

$$\zeta_{rate}^{SU\,RAN} = 1 - e^{-2\frac{\gamma_{th}^{PU} + \gamma_{th}^{SU}}{\gamma_\circ}} \tag{7.37}$$

where it can be seen that the only effective parameters are the threshold values for both kinds of users; the number of PUs or SUs has no effect.

On the other hand, using Equation 7.12 for the opportunistic selection approach, we express the resultant rate outage $\zeta_{rate}^{PU_{OPP}}$ as

$$\zeta_{rate}^{PU_{OPP}} = \left(1 - e^{-\frac{\gamma_{th}^{PU}}{\gamma_\circ}}\left(1 + \frac{\gamma_{th}^{PU}}{\gamma_\circ}\right)\right)^{N_{PU}} \tag{7.38}$$

and to formulate the service threshold, we obtain it by rewriting Equation 7.38 as

$$\gamma_{th}^{PU} = -\gamma_\circ\left(\left(-\left(1 - \sqrt[N_{PU}]{\zeta_{rate}^{PU_{OPP}}}\right)e^{-1}\right)e^{-\left(1 - \sqrt[N_{PU}]{\zeta_{rate}^{PU_{OPP}}}\right)e^{-1}} + 1\right) \tag{7.39}$$

With the network QoS performance defined, the QoE metric related to the cognition rate is

$$QoE_{rt} = \eta_{rt}\exp\left(\mu_{rt}\gamma_{th}^{PU}\right) + \nu_{rt} \tag{7.40}$$

where η_{rt}, μ_{rt}, and ν_{rt} are constant values that depend on the application and the scenario. In contrast to Equation 7.34, where an increase in the access delay decreases the QoE of the user, the foregoing QoE metric exponentially rises, because a higher rate leads to an increase in the QoE for the user, and thus the positive relation in Equation 7.40.

7.4.3 QoS/QoE Related to the Minimum Guaranteed Throughput

In multiuser scenarios, the wireless medium is shared among different users. This means that some users are waiting some slots (i.e., getting zero rate) until getting serviced. Therefore, to discuss such a performance, a descriptive metric of the user's activity over the time is known in the literature as the minimum guaranteed throughput [12], and we study it for our scenario. This term refers to the normalized minimum guaranteed rate to a certain user per time slot, which helps operators to provide applications and services that are tailor-made to avoid extra dimensioning in the network, ensuring the satisfaction of the QoS/QoE demands of the customers over wireless systems [37]. The minimum guaranteed throughput is denoted as *Th*, in bps, and is given [12] by

$$Th = \frac{B_w\log_2\left(1 + \gamma_{th}\right)}{K} \tag{7.41}$$

From Equations 7.36 and 7.41, the PU's minimum guaranteed throughput for the PU users within the random selection, Th_{RAN}^{PU}, can be obtained as

$$Th_{RAN}^{PU} = \frac{B_w\log_2\left(1 - \frac{e^{-\frac{\gamma_{th}^{PU}}{\gamma_\circ}}\left(1 + \frac{\gamma_{th}^{PU}}{\gamma_\circ}\right)}{N_{PU}}\right)\log_2\left(1 - \gamma_\circ\left(\left(-\left(1 - \zeta_{rate_{RAN}}^{PU}\right)e^{-1}\right)e^{-\left(1 - \zeta_{rate_{RAN}}^{PU}\right)e^{-1}} + 1\right)\right)}{\log_2\left(\zeta_{ac}\right)} \tag{7.42}$$

while for the opportunistic scheduling philosophy using Equations 7.39 and 7.41, the minimum guaranteed throughput, Th_{OPP}^{PU}, is derived as

$$Th_{OPP}^{PU} = \frac{B_w \log_2 \left[1 - \frac{1 - \left(1 - e^{-\frac{\gamma_{th}^{PU}}{\gamma_\circ}} \left(1 + \frac{\gamma_{th}^{PU}}{\gamma_\circ} \right) \right)^{N_{PU}}}{N_{PU}} \right] \log_2 \left(1 - \gamma_\circ \left(\left(-(1-\Omega)e^{-1} \right) e^{-(1-\Omega)e^{-1}} + 1 \right) \right)}{\log_2 \left(\zeta_{ac} \right)} \tag{7.43}$$

where $\Omega = \sqrt{[N_{PU}]\zeta_{rate_{OPP}}^{PU}}$.

For both network QoS-guaranteed-throughput performance metrics, and through the IQX relation [36], we present the QoE for PUs through an exponential increasing metric as

$$QoE_{th(w)} = \eta_{th(w)} \exp\left(\mu_{th(w)} Th_w^{PU} \right) + v_{th(w)} \tag{7.44}$$

where w relates to the users' scheduling mechanism, even random from Equation 7.42 or opportunistic in Equation 7.43.

For the random users' selection, a degradation in the throughput occurs as the number of users N_{PU} increases owing to the multiuser diversity absence as a higher delay is obtained without any rate increase. On the other hand, this is different for the opportunistic scheme, as the impact of N_{PU} is not clear, because both the multiuser gain and the scheduling delay are increased.

7.4.4 QoS/QoE Related to the Maximum Allowed Jitter

The delay jitter describes the variation in the delays that the transmitter packets suffer until they reach their destination [38]. In multiuser systems, employing real-time applications (e.g., Voice over IP [VoIP]) can be challenging because packets should be received with a fixed delay between them, to make the voice stream continuous to the user. The average jitter metric [32] for a user to be scheduled in any time $\in (1, K_{max})$ is

$$J = \sum_{k=1}^{K_{max}} k P_{ac} \left(1 - P_{ac} \right)^{k-1} \tag{7.45}$$

so that for each kind (i.e., PU or SU) and for each scheduling philosophy (i.e., random or opportunistic), we can substitute the specific P_{ac} value from section 7.4.1. With further mathematical manipulations, we reformulate previous equation as

$$J = \frac{1 - \left(1 - P_{ac} \right)^{K_{max}+1}}{P_{ac}} - \left(K_{max} + 1 \right) \left(1 - P_{ac} \right)^{K_{max}} \tag{7.46}$$

obtaining a closed-form expression for the resultant system QoS jitter indicator, which, through the IQX formula, yields a QoE metric for the jitter performance as

$$QoE_{jt} = \eta_{jt} \exp\left(-\mu_{jt} J \right) + v_{jt} \tag{7.47}$$

where η_{jt}, μ_{jt}, and v_{jt} are again constants dependent on the application and scenario.

7.5 Simulations and Results

The performance of the QoS/QoE in a spatial cognitive scenario is presented by means of computer simulations and analytical formulations. The aim of this analysis is to display the impact of cognition on the total system performance. As previously explained, PUs have strict demands to be satisfied, while SUs are serviced as long as PUs' QoS/QoE requirements are not degraded. The performance of the proposed scheme is evaluated by Monte Carlo simulations, where a wireless scenario is considered with $n_t = 2$ transmitting antenna in a cell with a variable number of active users, each one of them equipped with a single receiving antenna, where a total transmitted power $P_t = 1$ and noise power $\sigma^2 = 1$ are imposed. The simulation parameters are denoted in Table 7.1.

Figure 7.1 presents the QoE in terms of the access delay and plots the comparison between the cognitive scenario and the single-user scenario (i.e., no cognition) in the case of the opportunistic selection for $\gamma_{th} = 1$ and an access outage equal to 10%. It can be seen that as the number of users increases, it becomes harder for each user to be selected, and consequently its QoE decreases. The cognition induces a higher QoE over all the considered number of users. Note the perfect match between the simulations and the mathematically obtained results in Equation 7.34, as all the closed-form expressions in this chapter are based on exact mathematical formulations. The same match will be shown in all the results obtained in this chapter.

One of the useful metrics that can help an operator to support the applications without oversizing the network is the QoS indicator related to the minimum guaranteed throughput, which differs according to the user type (PU or SU). In order to assess the cognitive radio benefits versus the transmission to a single user, Figure 7.2 shows the performance of both approaches in terms of the guaranteed throughput. A total of eight available users are in

Table 7.1   Simulation Parameters

Parameter	Value
Iterations	10,000
Number of antennas	2
Transmit power	0 dB
Noise variance	1
η_{ac}	1
μ_{ac}	0.9
v_{ac}	0.2
η_{th} (both PU and SU)	0.1
μ_{th} (both PU and SU)	0.2
v_{th} (both PU and SU)	0.5
η_{jt}	1
μ_{jt}	0.9
v_{jt}	0.2

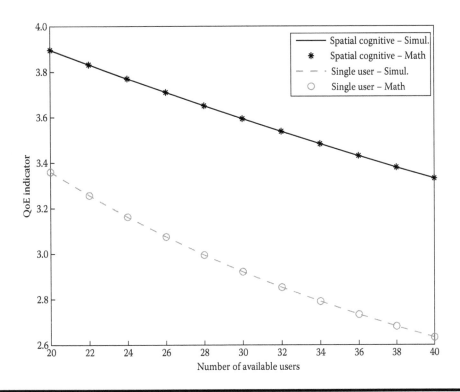

Figure 7.1 **QoE$_{ac}$ for a variable number of users in the system, both for the spatial cognitive and the single-user service.**

the system, all of whom have the same threshold value. Notice that as we increase the SNR threshold value, two contradictory effects happen: By increasing the threshold, the minimum allocated rate gets higher. At the same time, it makes it harder for the user to be above the threshold to enable its access to the system. At low threshold values, the dominant factor is the rate effect so that the guaranteed rate increases, and as we increase the SNR value, the delay effect increases and the guaranteed rate starts to decrease. We can see that the cognitive system outperforms the single-user service; this allows the operator to employ more diverse applications and to please the users.

The three factors affecting the QoE for the PUs are depicted in Figure 7.3. It is a 3D plot that shows the impact of the available number of users and the SNR threshold on the resultant QoE for the PUs, both for the case of random or opportunistic selection. When the number of users increases, the chances for a user to get access to the channel decrease, so its QoE is accordingly decreased. The figure helps to understand the choice of the optimum threshold that can maximize the QoE for a certain number of available PUs. On the other hand, Figure 7.4 shows a similar plot for the SUs, where a smaller value for the QoE is presented as they have a lower probability of accessing the system. Because the good match between simulations and theory has been shown in the previous figures, only the simulations are presented in both Figures 7.3 and 7.4.

In the alternative scenario, the total outage of the system is considered to be fixed, and all the users in each subsystem suffer from the same total outage. The objective is to investigate the maximum jitter performance of each subsystem, and it can be seen from Figure 7.5 that the maximum

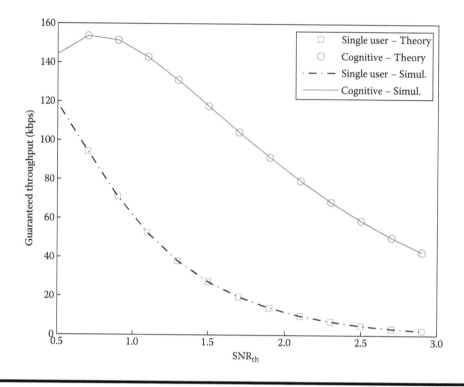

Figure 7.2 Minimum guaranteed throughput for a variable SNR threshold value, displayed for both PUs and SUs.

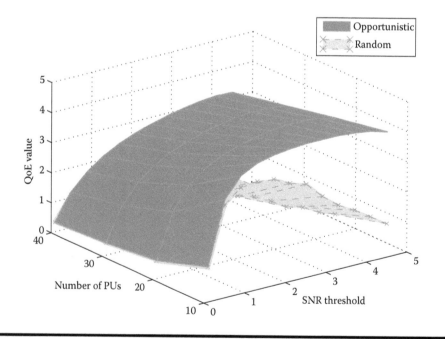

Figure 7.3 $QoE_{th(PU)}$ **with a variable SNR threshold and available number of PUs.**

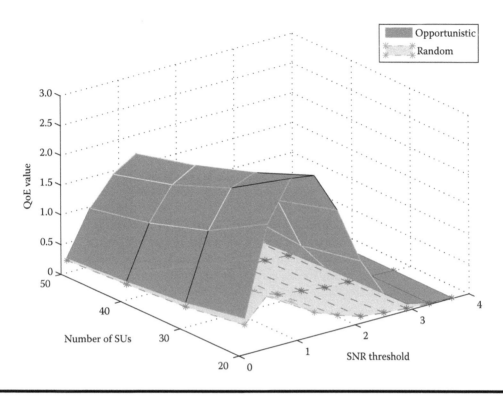

Figure 7.4 $QoE_{th(SU)}$ with a variable SNR threshold and available number of SUs.

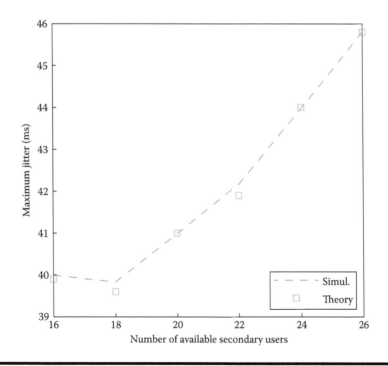

Figure 7.5 The maximum obtained jitter for each SU for a variable number of SUs.

jitter shows very small values for a certain number of users and for a given threshold value (i.e., the optimum number of SUs for a given threshold). This concept is very beneficial in network scaling, as the number of the SUs that can exist in the cell can be determined. Because there is a conflict between access outage and rate outage increasing the number of the users will reduce the rate outage value while overcoming the access outage.

7.6 Conclusion

Employing cognition within operators is a tempting choice because it enhances the total system QoS and QoE performance, increasing the throughput and decreasing system delay. The system QoS/QoE enhancement in MIMO cognition through ZFB is the core of the study, where statistical optimization is being employed within such scenarios. The users' (whether PUs or SUs) QoS performance in terms of the delay, minimum guaranteed throughput, and access jitter as well as their mapping to the QoE were presented in closed-form expressions to characterize each parameter effect in the system. Simulations have been done to display the benefits of the results under different conditions, where a perfect match between the theoretically obtained results and the simulations is obtained.

Acknowledgments

This work was partially supported by the Research Projects Jordan University DAR 1313 and Qatar University QUSG-CENG-DEE-13/14-3

References

1. FCC (Federal Communication Commission), *Spectrum Policy Task Force*, ET document no. 02-135, Federal Communication Commission, 2002.
2. M. Mchenry, *Spectrum White Space Measurements*, New America Foundation Broadband Forum, US, 2003.
3. S. Haykin, Cognitive radio: Brain-empowered wireless communication, *IEEE Journal on Selected Areas in Communications*, vol. 23, pp. 201–222, 2005.
4. L. Guofeng, H. Yahui, C. Song, and T. Hui, QoE-driven cross-layer optimized video transmission scheme over MIMO-OFDM systems, in *IEEE ICON*, Singapore, December 2012.
5. T. Yoo and A. Goldsmith, On the optimality of multi-antenna broadcast scheduling using zero-forcing beamforming, *IEEE Journal on Selected Areas in Communications*, vol. 24, pp. 528–541, 2006.
6. N. A. Ali, A. M. Taha, and H. S. Hassanein, Quality of service in 3GPP R12 LTE-advanced, *IEEE Communications Magazine*, vol. 51, no. 8, pp. 103–109, 2013.
7. L. Atzori, C. W. Chen, T. Dagiuklas, and H. R. Wu, QoE management in emerging multimedia services, *IEEE Communications Magazine*, vol. 50, pp. 18–20, 2013.
8. S. Misra, S. Das, M. Khatua, and M. Obaidat, QoS-guaranteed bandwidth shifting and redistribution in mobile cloud environment, *IEEE Transactions on Cloud Computing*, vol. 12, pp. 181–193, 2013.
9. 3GPP TS 36.104, E-UTRA Base Station (BS) Radio Transmission and Reception. v8.8.0, December 2009.
10. N. Abramson, The ALOHA systems—Another alternative for computer communications, in *Fall Joint Computer*, TX, USA, November 1970.
11. A. Gopalan, C. Caramanis, and S. Shakkottai, Low-delay wireless scheduling with partial channel-state information, in *IEEE Infocom*, Orlando, FL, March 2012.
12. N. Zorba, A. I. Perez-Neira, A. Foglar, and C. Verikoukis, Cross layer QoS guarantees in multiuser WLAN systems, *Springer Wireless Personal Communications*, vol. 51, pp. 549–563, 2009.

13. Z. Weifei, C. Ng, and M. Medard, Joint coding and scheduling optimization in wireless systems with varying delay sensitivities, in *IEEE SECON*, Seoul, June 2012.

14. J. Tigang, W. Honggang, and A. Vasilakos, QoE-driven channel allocation schemes for multimedia transmission of priority-based secondary users over cognitive radio networks, *IEEE Journal on Selected Areas in Communications*, vol. 30, pp. 1215–1224, 2012.

15. T. Hung, H. Zepernick, M. Fiedler, and P. Hoc, Outage probability, average transmission time, and quality of experience for cognitive radio networks over general fading channels, in *EURO-NGI Conference on Next Generation Internet (NGI)*, Karlskrona, Sweden, 2012.

16. A. Popescu and M. Fiedler, On routing in cognitive radio networks, in *International Conference on Communications (COMM)*, Bucharest, Romania, June 2012.

17. I. Martinez-Yelmo, I. Seoane, and C. Guerrero, Fair quality of experience (QoE) measurements related with networking technologies, in *WWIC 2010*, Lulea, Sweden, June 2010.

18. A. Vishwanath, P. Dutta, M. Chetlu, P. Gupta, S. Kalyanaraman, and A. Ghosh, Perspectives on quality of experience for video streaming over WiMAX, *ACM Sigmobile Mobile Computing and Communications Review*, vol. 13, pp.15–25, 2009.

19. N. Staelens, S. Moens, W. Van den Broeck, I. Marien, B. Vermeulen, P. Lambert, R. Van de Walle, and P. Demeester, Assessing quality of experience of IPTV and video on demand services in real life environments, *IEEE Transactions on Broadcasting*, vol. 56, pp. 458–466, 2010.

20. I. Ketyko, K. Moor, W. Joseph, L. Martens, and L. Marez, Performing QoE-measurements in actual 3G network, in *IEEE BMSB*, Shanghai, China, March 2010.

21. C. N. Ververidis, J. Riihijarvi and P. Mahonen, Evaluation of quality of experience for video streaming over dynamic spectrum access systems, in *IEEE WoWMoM*, Montreal, Canada, June 2010.

22. R. Serral-Graci, E. Cerqueira, M. Curado, M. Yannuzzi, and E. Monteiro, An overview of quality of experience measurement challenges for video applications in IP networks, in *WWIC 2010*, Lulea, Sweden, June 2010.

23. S. Jelassi, G. Rubino, H. Melvin, H. Youssef, and G. Pujolle, Quality of experience of VoIP service: A survey of assessment approaches and open issues, *IEEE Communications Surveys & Tutorials*, vol. 14, no. 2, pp. 491–513, 2012.

24. J. Jailton, T. Carvalho, W. Valente, C. Natalino, R. Frances, and K. Dias, A quality of experience handover architecture for heterogeneous mobile wireless multimedia networks, *IEEE Communications Magazine*, vol. 51, no. 6, pp. 152–159, 2013.

25. P. Svedman, S. K. Wilson, L. J. Cimini, and B. Ottersten, Opportunistic beamforming and scheduling for OFDMA systems, *IEEE Transactions on Communications*, vol. 55, pp. 941–952, 2007.

26. R. Knopp and P. Humblet, Information capacity and power control in single cell multiuser communications, in *IEEE ICC*, Seattle, WA, June 1995.

27. D. Bartolomé and A. I. Pérez-Neira, Multiuser spatial scheduling in the downlink of wireless systems, in *IEEE SAM*, Barcelona, Spain, July 2004.

28. K. Hamdi, W. Zhang, and K. Ben Letaief, Opportunistic spectrum sharing in cognitive MIMO wireless networks, *IEEE Transaction on Wireless Communication*, vol. 9, pp. 4098–4109, 2009.

29. R. Mochaourab and E. Jorswieck, Exchange economy in two-user multiple-input single-output interference channels, *IEEE Journal of Selected Topics in Signal Processing*, vol. 6, pp. 151–164, 2012.

30. B. K. Chalise and A. Czylwik, Robust downlink beamforming based upon outage probability criterion, in *IEEE VTC Fall*, Los Angeles, CA, September 2004.

31. M. Odeh, N. Zorba, and C. Verikoukis, Power consumption in spatial cognitive scenarios, in *IEEE-ICC*, Kyoto, Japan, June 2011.

32. A. Papoulis and S. U. Pillai, *Probability, Random Variables and Stochastic Processes*, 4th ed., McGraw Hill, New York, 2002.

33. I. S. Gradshteyn and I. M. Ryzhik, *Table of Integrals, Series and Products*, 7th ed., Academic Press, San Diego, CA, 2007.

34. L. Zhou, Z. Yang, Y. Wen, H. Wang, and M. Guizani, Resource allocation with incomplete information for QoE-driven multimedia communications, *IEEE Transactions on Wireless Communications*, vol. 12, no. 8, pp. 3733–3745, 2013.

35. E. L. Hahne and R. G. Gallager, Round robin scheduling for fair flow control in data communication networks, in *IEEE ICC*, Toronto, Canada, June 1986.

36. M. Fiedler, T. Hossfeld, and P. Tran-Gia, A generic quantitative relationship between quality of experience and quality of service, *IEEE Networks*, vol. 24, pp. 36–41, 2010.

37. M. G. Martini, C. W. Chen, Z. Chen, T. Dagiuklas, L. Sun, and X. Zhu, Guest editorial QoE-aware wireless multimedia systems, *IEEE Journal on Selected Areas in Communications*, vol. 30, pp. 1153–1156, 2012.

38. C. Demichelis and P. Chimento, *IP Packet Delay Variation Metric for IP Performance Metrics (IPPM)*, The Internet Engineering Task Force, IETF-RFC, Florida, USA, 2002.

Chapter 8

Cross-Layer Performance Analysis of Cognitive Radio Networks: A QoS Provisioning Perspective

Yakim Y. Mihov

Faculty of Telecommunications, Technical University of Sofia, Bulgaria

Contents

8.1 Introduction

The increasing demand for bandwidth to support existing and newly emerging wireless services and applications makes efficient spectrum use a matter of great importance. The traditional static command-and-control approach to spectrum regulation has resulted in low spectrum utilization because much of the licensed spectrum is relatively unused across time and frequency. Dynamic spectrum access (DSA) offers a new paradigm for spectrum regulation that is expected to solve the problem with inefficient spectrum use due to static spectrum regulation.

Cognitive radio (CR) is considered as the key enabling technology for DSA. A comprehensive definition of the term CR is given in Ref. [1]: (1) a type of radio in which communication systems are aware of their environment and internal state and can make decisions about their radio operating behavior based on that information and predefined objectives; (2) a radio [as defined in item (1)] that uses software-defined radio, adaptive radio, and other technologies to adjust automatically its behavior or operations to achieve desired objectives. The term *CR network* is defined as a type of radio network in which the behavior of each radio is controlled by a cognitive control mechanism to adapt to changes in topology, operating conditions, or user needs [1]. The cognitive control mechanism is a component of a CR that assesses inputs such as environmental, spectral, and communications channel conditions and predefined objectives to make decisions about radio operating behavior. It should be noted that a node in a CR network may have its own cognitive capabilities, or it may receive instructions from another node with such capabilities. The cognitive capabilities potentially include awareness of the network environment, network state and topology, and shared awareness obtained by exchanging information with other nodes (typically, neighboring nodes) or other network-accessible information sources. Cognitive decision making considers this collective information, and this decision making may be performed in coordination or collaboration with other nodes. Definitions and explanations of key concepts in the fields of spectrum management, CR, policy-defined radio, adaptive radio, software-defined radio, and related technologies are provided in Ref. [1]. A comprehensive survey of the state-of-the-art trends in CR technology and its application to DSA is presented in Ref. [2]. Recent developments and basic open research issues related to spectrum management in CR networks are discussed in Ref. [3].

DSA is expected to be one of the most prominent applications of CR technology. The term DSA can be defined as the real-time adjustment of spectrum utilization in response to changing circumstances and objectives [1]. It should be noted that the changing circumstances and objectives include (but are not limited to) energy conservation, changes in the state of the radio (operational mode, battery life, location, etc.), interference avoidance (either suffered or inflicted), changes in environmental/external constraints (spectrum, propagation, operational policies, etc.), spectrum usage efficiency targets, quality of service (QoS), graceful degradation guidelines, and maximization of radio lifetime. DSA strategies can be broadly categorized under three essential models: dynamic exclusive-use model with two approaches: spectrum property rights and dynamic spectrum allocation; open sharing model (a.k.a. spectrum commons model); hierarchical access model with two approaches: hierarchical spectrum underlay and hierarchical spectrum overlay [4]. The dynamic exclusive-use model maintains the basic structure of the current spectrum regulation policy; that is, spectrum bands are licensed to services for exclusive use. The main idea is to introduce flexibility to improve spectrum efficiency. Two approaches have been proposed under this model: spectrum property rights and dynamic spectrum allocation. The former approach allows licensees to sell and trade spectrum and to freely choose technology. The economy and the market will thus play a key role in driving toward the most profitable use of the limited spectrum resources. The latter approach, dynamic spectrum allocation, was studied by the European DRiVE project. It aims to improve spectrum efficiency through dynamic spectrum assignment by exploiting the spatial and temporal traffic statistics of different services. The exclusive-use model is incapable of eliminating white space in spectrum resulting from the bursty nature of wireless traffic. The spectrum commons model employs open horizontal[1] sharing among peer users as the

---

[1] Horizontal spectrum sharing is defined as spectrum sharing between users that have equal regulatory spectrum access rights [1].

basis for managing a spectral region. Advocates of this model draw support from the phenomenal success of wireless services operating in the unlicensed industrial, scientific, and medical (ISM) radio band (e.g., Wi-Fi). The hierarchical access model is generally based on the concept of hierarchical spectrum access. Hierarchical spectrum access is defined as a type of spectrum access in which a hierarchy of radio users or radio applications determines which radios have precedence [1]. It should be noted that the most common hierarchy proposed today (in the context of DSA) is one that distinguishes between primary users (PUs) and secondary users (SUs). In this hierarchy, SUs may only access spectrum when PUs are not occupying it. However, other hierarchies are possible, including the existence of tertiary users or hierarchies based on the type or the criticality of the communication. The hierarchy may be determined by a central authority, such as a regulator, or through active collaboration among affected systems. The hierarchy may be static or it may be established dynamically based on the current environment. The main idea behind the hierarchical access model (in the context of DSA) is to open licensed spectrum to unlicensed SUs in a manner that guarantees that the interference perceived by licensed PUs (i.e., licensees) is within allowable/ tolerable limits [4]. Two basic approaches to vertical[2] spectrum sharing between PUs and SUs have been proposed: hierarchical spectrum underlay and hierarchical spectrum overlay. The former approach imposes severe constraints on the transmission power of the SUs (e.g., radiated power limits, power spectral density limits, etc.) in order to protect the PUs from interference. Typically, at least one of the systems employs a spread spectrum technique with a large processing gain to ensure that the undesired signal power observed by an incumbent licensed user (i.e., a PU) is below a designated threshold. Based on the worst-case assumption that PUs transmit all the time, this approach does not rely on detection and exploitation of spectrum white space. The latter approach, hierarchical spectrum overlay (a.k.a. opportunistic spectrum access [OSA]), envisions the exploitation of spatial and temporal spectral opportunities in a noninterfering manner. Differing from the hierarchical spectrum underlay, this approach does not necessarily impose severe restrictions on the transmission power of the SUs, but rather on when and where they may transmit. It directly targets spatial and temporal spectrum white space by allowing SUs to identify and exploit local and instantaneous spectrum availability in a nonintrusive manner.

This chapter addresses issues related to QoS provisioning in secondary CR networks operating in compliance with the hierarchical spectrum overlay approach for DSA. In such a secondary CR network, the SUs are allowed to transmit only on momentarily unoccupied spectrum segments (spectrum white spaces) of the spectrum resources assigned to the primary network, provided that no harmful interference is inflicted on the PUs. Owing to the fluctuating nature of the spectrum segments available to the SUs, QoS provisioning in the secondary CR network is a challenging and demanding task.

8.2 Cross-Layer QoS-Related Analysis of CR Networks: An Overview

The CR concept has been recognized worldwide by standardization and regulation bodies because it proposes to push efficiency in spectrum access and resource allocation beyond its traditional limits, and a number of wireless technologies could benefit from it (e.g., see Refs. [5,6], and the references therein). Open research issues cover a wide range of system aspects, from the hardware

---

[2] Vertical spectrum sharing is defined as spectrum sharing between users that have different regulatory spectrum access rights [1].

components up to the network layer design. A large number of unresolved technical, regulatory, and economic challenges need to be addressed.

The performance evaluation in regard to QoS provisioning in CR networks operating in compliance with the hierarchical spectrum overlay approach for DSA is a complex and difficult task owing to the potentially high and unpredictable (or hardly predictable) fluctuations in the available spectrum (idle frequency bands) and owing to specific implementation details of the spectrum management functionality. Because the spectrum resources available to a CR network depend on the current (instantaneous) traffic load in the incumbent primary network, the traffic capacity of the CR network inherently fluctuates and may vary to a large extent, which further complicates QoS provisioning of the SUs. The spectrum management functionality of a CR network is a key prerequisite for dependable spectrum opportunity identification and efficient spectrum opportunity exploitation. Spectrum opportunity identification involves accurately identifying and intelligently tracking idle frequency bands that are dynamic in both time and space. Spectrum opportunity exploitation involves transmitting on identified idle frequency bands. It should be noted that spectrum opportunity exploitation must conform to the enforced regulatory policy, which defines the basic etiquette for SUs to ensure compatibility with incumbent PUs. The CR network should provide sufficient benefit to the SUs while protecting the spectrum licensees (i.e., the PUs) from harmful or intolerable interference. The tension between the desire for performance of the SUs and the need for protection of the PUs dictates the interaction across spectrum opportunity identification, spectrum opportunity exploitation, and regulatory policy.

8.2.1 Concise Overview of Essential Spectrum Management Functions in CR Networks

A CR network must be able to determine which segments of the spectrum are available, to select the most favorable idle channels for transmission, to coordinate spectrum access among the SUs, and to vacate channels on which incumbent PUs begin to transmit in order not to violate the interference constraints imposed by the primary network [3]. These capabilities can be realized via the following essential spectrum management functions: spectrum sensing, spectrum analysis, spectrum decision, spectrum sharing, and spectrum mobility (a.k.a. spectrum handover or spectrum hand-off).

Spectrum sensing in very broad terms involves the detection (by a given receiver) of the presence of a transmitted signal of interest [7, chapter 4]. The ability of the CR network to dynamically access spectrum holes that appear dynamically is predicated upon its ability to detect these spectrum holes in the first place. Moreover, while SUs are occupying available spectrum holes, they must be on the lookout for the return of PUs; that is, continuous monitoring of the spectrum may be necessary. The fundamental design features of any spectrum sensing system are its accuracy (i.e., probability of detection and probability of a false alarm), sensitivity, timeliness, and resilience to interference and noise. Spectrum sensing can be noncooperative or cooperative [3]. Cooperative spectrum sensing implies collaborative processing of spectrum sensing data, which affects the precision of the spectrum sensing results (e.g., increases the probability of detection or decreases the probability of a false alarm), and is generally more accurate (especially under fading propagation conditions) than noncooperative spectrum sensing. In general, spectrum sensing techniques can be classified into three major groups: primary transmitter detection, primary receiver detection, and interference temperature management, as described in the following. Three schemes are generally used for primary transmitter detection: energy detection, matched filter detection, and cyclostationary feature detection (see, e.g., Refs. [3], [8, chapter 7], [9, chapter 1], and [10, chapter 1]). The energy detection approach is optimal for detecting any unknown zero-mean constellation signal. However, the performance of the energy detector is susceptible to uncertainty in noise power. Furthermore, energy detectors are prone to generating false alarms triggered by unintended signals because they cannot differentiate signal types. The matched filter detection approach is optimal in the presence of stationary Gaussian noise, provided that sufficient a priori knowledge of the PU signals (at physical [PHY] and/or medium access control [MAC] layers) is available to the CR network. However, such a priori knowledge may not always be readily available. Moreover, the use of matched filters predicates the need for a dedicated receiver for every kind of primary system, which increases the implementation complexity, when the target system is one out of multiple possible systems. The cyclostationary feature detection approach exploits the built-in periodicity of the modulated PU signals, for example, sine wave carriers, pulse trains, hopping sequences, cyclic prefixes, etc. The main advantage of cyclostationary detection is that it can distinguish the noise energy from the signal energy. Thus, a cyclostationary detector is more robust to noise uncertainty than an energy detector. Also, it can work with lower signal-to-noise ratio (SNR) than an energy detector. However, the implementation of cyclostationary detection is more complicated than the implementation of energy detection, and cyclostationary detectors need a longer observation time than energy detectors, which means that some spectrum holes with a short time duration may not be efficiently exploited. Primary receiver detection may be feasible for TV receivers only, or further hardware, such as a supporting sensor network, in the area with the primary receivers is required (see Refs. [9, chapter 1, p. 14], [11,12]). The objective of this approach is to detect the PUs that are receiving data within the communication range of an SU. The primary receiver usually emits local oscillator leakage power from its radio frequency front-end when it receives signals from a primary transmitter. In order to determine the spectrum availability, the primary receiver detection method exploits this local oscillator leakage power, instead of the signal from the primary transmitter, and detects the presence of the primary receiver directly. The interference temperature management concept is based on a new model for measuring interference, referred to as interference temperature (see Refs. [9, chapter 1, pp. 14–16], [13–15]). The interference temperature limit is the amount of new interference that the primary receiver could tolerate. As long as SUs do not exceed this limit, they can use the spectrum band. Although this model is best fitted for the objective of spectrum sensing, the difficulty lies in accurately determining the interference temperature limit for each location-specific case. There is no practical way for an SU to measure or estimate the interference temperature because SUs have difficulty in distinguishing between actual signals from PUs and noise or interference. Moreover, with the increase in the interference temperature limit, the SNR at the primary receiver decreases, resulting in a decrease in the coverage or the capacity of the primary network. It should be noted that the identification of spectrum opportunities is not limited only to spectrum sensing performed by the SUs. Other possible approaches to identifying spectrum opportunities are based on database registry and beacon signals [16, chapter 7]. For example, the application of a database registry, that is, a radio environment map, is described in Ref. [17, chapter 11]. However, such approaches suffer from high infrastructure cost and may be infeasible for some application scenarios. For further information on issues related to spectrum sensing, the interested reader is referred to the following Refs. [18–23].

In spectrum analysis, information from spectrum sensing is analyzed to gain knowledge about the spectrum holes (e.g., interference estimation, duration of availability, probability of collision with a licensed user due to sensing error, etc.); that is, spectrum analysis is performed to estimate spectrum quality [16, chapter 2]. One of the issues here is how to quantify the quality of the spectrum opportunity. The quality can be characterized by the SNR of the target frequency band, the average duration of the spectrum holes, the correlation of the availability of spectrum holes, and so on. The information about the spectrum quality available to an SU can be imprecise and noisy. The application of learning algorithms from artificial intelligence is one of the candidate techniques that could be employed by the CR network for efficient spectrum analysis.

Based on the spectrum analysis, a spectrum decision to access the spectrum (e.g., frequency, bandwidth, modulation mode, transmit power, and time duration) is made by optimizing the system performance given the desired objective (e.g., to maximize the throughput of the SUs) and constraints (e.g., to maintain the interference caused to licensed PUs below the target threshold) [16, chapter 2]. The spectrum decision does not depend only on the spectrum availability, but is also determined based on internal (and possibly external) policies (see Refs. [3,24,25]). The complexity of the decision model depends on the parameters considered during spectrum analysis (e.g., the average duration of the spectrum holes, the SNR of the target frequency band, the utility of the SUs obtained through accessing the spectrum holes). The decision model becomes more complex when an SU has multiple objectives. Stochastic optimization methods (e.g., Markov decision processes) would be an attractive tool to model and solve the spectrum access decision problem in a secondary CR network environment.

After a decision on spectrum access has been made based on the spectrum analysis, the spectrum holes are accessed by SUs [16, chapter 2]. Spectrum access is coordinated and managed by a cognitive medium access control (CMAC) protocol, which intends to avoid collisions with licensed PUs and also with other unlicensed SUs. A CMAC protocol could be based on a fixed allocation MAC (e.g., FDMA, TDMA, CDMA) or on a random access MAC (e.g., ALOHA, CSMA/CA). Spectrum sharing should include much of the general functionality of any MAC protocol [3]. However, the unique characteristics of CR networks, such as the coexistence of SUs with PUs and the wide range of available spectrum, impose additional specific requirements. Spectrum sharing can be centralized or distributed, cooperative or noncooperative. For further information on spectrum sharing (in regard to CMAC protocols), the interested reader is referred to the following Refs. [9, chapter 7], [26–30].

Spectrum mobility is a function related to the change of the operating frequency band of SUs [16, chapter 2]. When a licensed PU starts accessing a radio channel that is currently being occupied by an SU, the SU must immediately vacate this channel and move to an idle spectrum band. This change in the operating frequency band is referred to as spectrum mobility (a.k.a. spectrum hand-off or spectrum handover). During spectrum handover, the protocol parameters at different layers of the protocol stack have to be adjusted accordingly in order to match the new operating frequency band. The purpose of spectrum handover is to try to ensure that the transmission of the SU can continue on a new available spectrum band with minimum performance degradation. Because the incurred transmission delay during spectrum handover could be high, some cross-layer modifications or adaptations of components in the protocol stack may be necessary for achieving efficient QoS provisioning in the secondary CR network. When an SU performs spectrum handover, two major issues have to be taken into account: the target idle channel must not be occupied by any other SU (i.e., the self-coexistence requirement), and the receiver of the corresponding cognitive link must be notified of the spectrum handover (i.e., the synchronization requirement). In order to satisfy these requirements, the CMAC protocol must be designed with provision for spectrum handover information exchange. For further information on spectrum mobility (spectrum handover), the interested reader is referred to the following Refs. [31–35].

8.2.2 Concise Overview of Existing Cross-Layer Approaches for QoS Provisioning in CR Networks

QoS provisioning in CR networks operating in compliance with the hierarchical spectrum overlay approach for DSA is a complex and challenging open research issue. Fluctuations in the availability of unoccupied spectrum segments of the spectrum resources assigned to the primary network pose serious difficulties for achieving and guaranteeing a predetermined desired level of QoS for the SUs. Because the actual spectrum resources available to the secondary CR network depend on the current spectrum resource utilization in the primary network, the capacity of the CR network generally fluctuates and may be largely variable, which further exacerbates the need for effective means of providing QoS for the SUs. Because of the high complexity of contriving and implementing a unified, feasible, and efficient solution to the issues of QoS provisioning in CR networks, many different approaches (e.g., methods, algorithms, protocols, strategies, etc.) for QoS provisioning of the SUs have been proposed and investigated in the literature.

The delivery of a Voice over IP (VoIP) service via secondary CR networks has been analyzed in Refs. [36,37]. In Ref. [36], two CMAC schemes have been proposed: a contention-based CMAC scheme and a contention-free CMAC scheme. Both independent and correlated channel busy/idle state models for primary activity are considered. An analytical model based on a discrete-time Markov chain (DTMC) is developed to obtain the voice-service capacity (i.e., the maximum number of VoIP SUs that can be supported with a QoS guarantee) of the two proposed schemes, taking into account the impact of the activity of the PUs. The analytical analysis is verified via simulation. The performance evaluation reveals how the activity of the PUs and the different CMAC schemes affect the cognitive voice-service capacity. The analytical model (along with the CMAC schemes) could be incorporated into radio resource management and call admission control (CAC) of VoIP CR networks. In Ref. [37], the VoIP capacity of a CR network has been analyzed via a queuing model based on a DTMC, a Markov-modulated Poisson process traffic model, and a Markov channel model. It is assumed that spectrum sensing is based on energy detection. VoIP performance (including packet-dropping probability) is investigated on the basis of various analytical and simulation results. For further information on providing VoIP services over CR networks, the interested reader is referred to the following Refs. [38–41].

A novel delay-oriented continuous spectrum sensing (DO-CSS) scheme, which is especially suitable for delay-sensitive services in CR networks, has been suggested in Ref. [42]. According to the proposed DO-CSS scheme, a licensed PU band is divided into two parts (subbands): one subband only for spectrum sensing and the other subband only for SU data transmission. In this way, instead of performing conventional periodic spectrum sensing over the entire PU band, which always interrupts SU data transmissions during the sensing intervals (as a result of which SUs incur transmission delay), continuous spectrum sensing is being performed incessantly over the subband designated for spectrum sensing, whereas SU transmissions are allowed to occupy only the subband designated for data transmission (and thus no spectrum sensing delay is incurred). The optimal bandwidth for sensing a PU signal that maximizes the SU throughput (under the constraint that the required interference protection to the PUs is guaranteed) is obtained. Moreover, a throughput-oriented CSS (TO-CSS) scheme has been suggested. In the DO-CSS scheme, the SU makes a decision on the PU activity within each frame, while the TO-CSS scheme needs to make only one decision on the PU activity over multiple consecutive frames. Provided that the required protection for the PUs is guaranteed, the optimal sensing bandwidth that maximizes the SU throughput is also derived for the TO-CSS scheme. Both theoretical analyses and simulation results show the advantages of the proposed schemes (DO-CSS and TO-CSS) in comparison with the conventional periodic spectrum sensing. However, to implement simultaneous spectrum sensing and data transmission, the SU transmitter needs to be equipped with two radios, one for spectrum sensing and the other for data transmission, which can be considered as a shortcoming of these two schemes.

A cross-layer design scheme for improvement of Transmission Control Protocol (TCP) performance in CR networks has been investigated in Ref. [43]. According to the proposed scheme, spectrum sensing, access decision, the PHY layer modulation and coding scheme (MCS), and the data link layer frame size are jointly optimized to maximize the TCP throughput. The system model of the CR network is represented and formulated as a partially observable Markov decision process.

Simulation results illustrate the effectiveness of the cross-layer design scheme. It is shown that lower layer design parameters have a significant impact on the TCP throughput in CR networks, and the TCP throughput can be substantially improved via the suggested cross-layer design scheme.

In Ref. [44], a cross-layer spectrum allocation framework has been proposed that jointly considers QoS provisioning for heterogeneous real-time and non-real-time SUs, spectrum sensing, spectrum access decision, channel allocation, and CAC in distributed cooperative CR networks. Based on the statistical information of the available licensed channels that can be learned over time, a number of the idle channels identified via spectrum sensing are allocated to the optimum admissible number of real-time SUs considering their blocking and dropping probability requirements. The remaining idle channels can be allocated adaptively to the optimum admissible number of non-real-time SUs considering the spectrum sensing and utilization indispensability. Extensive analytical and simulation results demonstrate the effectiveness of the proposed QoS-based spectrum resource allocation framework. For further information on providing delay-sensitive or delay-tolerant services over CR networks, the interested reader is referred to the following Refs. [45,46] and [47,48], respectively.

A novel spectrum hand-off scheme with spectrum admission control (SAC) has been suggested in Ref. [49]. In the proposed scheme, SUs make up secondary user groups (SUGs) to achieve the appointed detection probability for PU signals with cooperative spectrum sensing. The suggested SAC (for the SUGs) is based on the concept of channel reservation; that is, SUGs performing spectrum hand-off have priority over newly arriving SUGs to access the reserved channels. A Markov model is developed for performance analysis of the spectrum hand-off scheme in terms of blocking probability, forced termination probability, and system throughput. Numerical results demonstrate the effectiveness of the proposed spectrum hand-off scheme.

In Ref. [50], centralized and distributed DSA schemes with SU class prioritization (i.e., high-priority SUs and low-priority SUs) have been investigated. Priority-based spectrum hand-off is considered. The use of hand-off buffer is also taken into consideration. The impact of subchannel reservation for high-priority SUs is studied in detail. Each of the proposed schemes is analyzed via a continuous-time Markov chain. For performance evaluation, the blocking probability, the forced termination probability, the call completion rate, and the mean hand-off delay for the high-priority SUs and for the low-priority SUs are obtained and discussed.

Based on the concise overview presented here and based on other relevant publications in the literature (e.g., Refs. [51–54]), it can be concluded that, in general, a common feature of the existing cross-layer approaches for QoS provisioning in CR networks is their specificity in the context of a particular analyzed system model of a CR network. Therefore, these approaches have a limited scope of applicability within the context of the specific CR system model under consideration.

In this chapter, a novel cross-layer approach for QoS provisioning in CR networks is proposed [55]. Because a general system model of a CR network is analyzed, the suggested new cross-layer approach has a wide scope of applicability, which determines its theoretical significance.

8.3 Novel Cross-Layer Approach for QoS Provisioning in CR Networks

A cross-layer analytical model is developed for investigation of QoS provisioning in CR networks operating in compliance with the hierarchical spectrum overlay strategy for DSA. A new cross-layer approach for QoS provisioning of the SUs is proposed [55]. Essential QoS-related performance issues are addressed and analyzed in detail. Various cross-layer interdependencies relevant to QoS provisioning of SUs are investigated.

8.3.1 Cross-Layer Analytical Model

In the analyzed system model, the primary network is considered to have n channels of equal bandwidth. The PU call arrival and departure streams are modeled by Poisson random processes with rates λ_p and μ_p, respectively. The offered PU traffic is $A_p = \lambda_p/\mu_p$.

The probability B_p for PU call blocking is given by the Erlang loss formula:

$$B_p = E_n(A_p) = \frac{\dfrac{A_p^n}{n!}}{\displaystyle\sum_{m=0}^{n} \frac{A_p^m}{m!}}. \tag{8.1}$$

The selection of an idle PU channel for the service of a new PU call is performed on a random basis. Thus, the mean PU channel utilization η is

$$\eta = \frac{A_p(1-B_p)}{n}. \tag{8.2}$$

The bandwidth of a PU channel is denoted by BW. The maximum tolerable interference duration in the primary network is denoted by T_{int}.

The CR network considered here is infrastructure-based and operates in compliance with the hierarchical spectrum overlay approach for DSA. An SU call occupies one subchannel of bandwidth BW/k; that is, one PU channel comprises k subchannels. Hence, the total number of subchannels in the system is nk.

Spectrum sensing is performed periodically in accordance with predefined spectrum sensing periods by SU subscriber stations that are registered on the CR network (not necessarily serving ongoing SU calls). Each spectrum sensing period consists of a quiet period (QP) and a transmission period (TP). During a QP, SUs perform spectrum sensing and are not allowed to transmit or receive. During a TP, SUs are allowed to transmit or receive.

The characteristics of the PU signals are assumed to be a priori known by the CR network, and the spectrum sensing is based on cyclostationary feature detection. Each SU performing spectrum sensing is assumed to sense only one PU channel during a QP. The detection threshold is assumed to be optimal, which means that the probability of missed detection is equal to the probability of a false alarm:

$$1 - p_d = p_f = p_e, \tag{8.3}$$

where p_d is the probability of detection, p_f is the probability of a false alarm, and p_e is the probability of a detection error.

On the basis of these assumptions, the probability of a detection error can be expressed in terms of the Q-function:

$$p_e = Q\left(\sqrt{2T_{ss}BW} \, \frac{\sqrt{SNR}}{\sqrt{(\alpha-1)SNR + \dfrac{1}{2}} + \sqrt{\dfrac{1}{2}}} \right), \tag{8.4}$$

where T_{ss} is the duration of the spectrum sensing procedure, SNR is the SNR of the PU signal at the sensing SU, and α is an intrinsic PU signal parameter that relates to its randomness

$(1 \leq \alpha \leq 2; \alpha = 1$ for constant-amplitude signals, e.g., BPSK, QPSK; $\alpha = 2$ for complex Gaussian signals).

Furthermore, cooperative spectrum sensing is considered. The spectrum sensing data fusion is done at the CR base station in accordance with the OR-rule fusion scheme. The number of SU subscriber stations that sense one and the same PU channel is denoted by g $(2 \leq g \leq k)$. At least ng SU subscriber stations are assumed to be registered on the CR network and to perform spectrum sensing at the discretion of the CR base station. The individual detections among a group of cooperating sensing SUs are assumed to be independent. The cooperative detection probability P_d and the cooperative false alarm probability P_f are given by

$$P_d = 1 - \prod_{m=1}^{g} \left(1 - p_{d_m}\right), \tag{8.5}$$

and

$$P_f = 1 - \prod_{m=1}^{g} \left(1 - p_{f_m}\right), \tag{8.6}$$

where p_{d_m} is the individual detection probability, and p_{f_m} is the individual false alarm probability of the m-th cooperating sensing SU, which can be evaluated using Equations 8.3 and 8.4.

The CMAC protocol is assumed to provide perfect spectrum handover and perfect spectrum sharing.

The nominal transmission time t of an SU within a spectrum sensing period, that is, the duration of the TPs, is

$$t = T_p - T_{QP}, \tag{8.7}$$

where T_p is the duration of the spectrum sensing periods, and T_{QP} is the duration of the QPs. It should be noted that $T_{ss} \leq T_{QP}$. To satisfy the interference requirements imposed by the primary network, the following relation has to be satisfied:

$$t \leq T_{int}. \tag{8.8}$$

The QoS provisioning of the SUs is investigated in terms of SU call blocking, SU call dropping, and the maximum tolerable transmission delay in the CR network. The new cross-layer approach for SU QoS provisioning proposed in this chapter is the *concurrent* application of limitation, delay bound, and scheduling. Limitation sets the SU CAC threshold l; that is, no more than l concurrent SU calls are admitted to the CR network. The delay bound sets the maximum tolerable transmission delay D in the CR network; that is, an SU call is dropped if it incurs a transmission delay greater than D in the CR network. Scheduling is applied to handle the transmissions of the SUs when the number of SU calls exceeds the number of currently available subchannels; that is, if not all of the SUs can transmit because of lack of enough idle subchannels, the scheduling service determines the SUs to transmit in the current spectrum sensing period. The objective of the suggested novel approach is to achieve and guarantee a predefined desired level of QoS for the SUs in the CR network.

It should be noted that different scheduling algorithms could be employed based on different design criteria. The application of the so-called smallest delay first (SDF) scheduling algorithm is proposed and investigated here. It gives transmission priority to the SUs with the smallest incurred transmission delay; that is, SUs are selected for transmission depending on (and in ascending order of) their incurred transmission delay. Obviously, the SDF algorithm facilitates the delay performance of the secondary CR network.

The SU call arrival and departure streams are modeled by Poisson random processes with rates λ_s and μ_s, respectively. The offered SU traffic is $A_s = \lambda_s/\mu_s$. Considering the application of limitation, the probability P_j^s that there are j ($0 \le j \le l$) SU calls in the CR network is given by

$$P_j^s = \frac{\dfrac{A_s^j}{j!}}{\displaystyle\sum_{m=0}^{l} \frac{A_s^m}{m!}}, \tag{8.9}$$

and the probability B_s for SU call blocking is given by the Erlang loss formula:

$$B_s = E_l(A_s) = \frac{\dfrac{A_s^l}{l!}}{\displaystyle\sum_{m=0}^{l} \frac{A_s^m}{m!}}. \tag{8.10}$$

Owing to the delay bound D in the proposed approach for QoS provisioning, SU call dropping occurs if an SU does not have any successful transmissions during q consecutive spectrum sensing periods:

$$q = \text{floor}\left(\frac{D}{T_p}\right) + 1, \tag{8.11}$$

where *floor* is a function that rounds its argument to the nearest integer toward negative infinity.

SU transmissions are successful if and only if subchannels of unoccupied PU channels are selected and no new PU calls occupy them during the ongoing spectrum sensing period.

The probability β that a PU channel is unoccupied and correctly determined as idle is given by

$$\beta = (1 - \eta)(1 - P_f). \tag{8.12}$$

The probability $Y(z, n, \beta)$ that exactly z ($0 \le z \le n$) PU channels are unoccupied and determined as idle and remain unoccupied by PU calls during the current spectrum sensing period can be derived using the binomial and exponential distributions:

$$Y(z, n, \beta) = \frac{n!}{z!(n-z)!} \beta^z (1-\beta)^{n-z} e^{-\lambda_p T_p}. \tag{8.13}$$

If the number of SU calls j exceeds the number of currently available subchannels zk, then $j - zk$ SU calls cannot be served during the current spectrum sensing period and incur transmission delay. These $j - zk$ SU calls will be dropped if they are not served for $q - 1$ more consecutive spectrum sensing periods. Considering the SDF algorithm, these $j - zk$ SUs are not going to be served during the next $q - 1$ consecutive spectrum sensing periods if and only if no more than z PU channels are available to the CR network during these periods. Thus, the conditional probability that $j - zk$ SU calls are dropped, provided that there are j SU calls and z ($z < j/k$) available PU channels in the current spectrum sensing period, is $\left[\sum_{m=0}^{z} Y(m,n,\beta)\right]^{q-1}$. The probability B_d for SU call dropping can be evaluated as the ratio of the mean number of dropped SU calls to the mean number of SU calls in the CR network:

$$B_d = \frac{\sum_{j=0}^{l}\sum_{z=0}^{\text{floor}\left(\frac{j}{k}\right)} (j-zk)P_j^s Y(z,n,\beta)\left[\sum_{m=0}^{z} Y(m,n,\beta)\right]^{q-1}}{\sum_{j=0}^{l} jP_j^s}. \tag{8.14}$$

Taking into account that $\sum_{j=0}^{l} jP_j^s = A_s(1-B_s)$, Equation 8.14 can be written as follows:

$$B_d = \frac{\sum_{j=0}^{l} P_j^s \sum_{z=0}^{\text{floor}\left(\frac{j}{k}\right)} (j-zk)Y(z,n,\beta)\left[\sum_{m=0}^{z} Y(m,n,\beta)\right]^{q-1}}{A_s(1-B_s)}. \tag{8.15}$$

8.3.2 Analysis of Numerical Results

Analytical and simulation results are presented (see Figures 8.1 through 8.3)[3] in order to prove the feasibility and analyze the performance of the proposed new cross-layer approach for QoS provisioning in CR networks.

An elaborate and sophisticated simulation model of the analyzed CR system has been developed. It incorporates all relevant system characteristics and includes all parameters introduced in the previous subsection (Section 8.3.1). The novel cross-layer approach for QoS provisioning of the SUs and the SDF scheduling algorithm have also been implemented and incorporated in the simulation model. Numerous simulation experiments have been performed, and the analytical results have always coincided with the simulation results. Therefore, the analytical analysis is completely consistent with the simulation analysis.

---

[3] The values of the parameters enclosed in parentheses are derived.

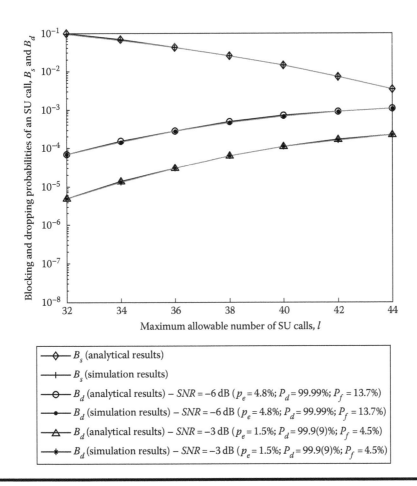

Figure 8.1 **SU call blocking and dropping probabilities versus the CAC threshold for SU calls.** Parameters: T_p = 20 ms; T_{ss} = 38 μs; T_{int} = 25 ms; BW = 360 kHz; α = 2; g = 3; n = 30; k = 3; D = 50 ms; A_s = 30 Erl; A_p = 15 Erl; $μ_p$ = 0.02 s⁻¹; (B_p = 2.10⁻⁴). The SDF scheduling algorithm is applied.

For the numerical results presented in this subsection (i.e., Figures 8.1 through 8.3), it was assumed for simplicity (but without loss of generality) that the SNRs of the PU signals perceived by the sensing SUs are equal.

First, the effect of limitation on SU QoS provisioning is analyzed. As the SU CAC threshold decreases, the SU call-blocking probability B_s increases and the SU call-dropping probability B_d decreases, and vice versa (see Figure 8.1). In general, this effect is advantageous because users are much more sensitive to interruption of an ongoing call than to rejection of a new call; that is, for achieving good QoS provisioning, B_d should be maintained reasonably much smaller than B_s. If this limitation is prudently applied, this QoS requirement can be satisfied. Moreover, Figure 8.1 illustrates the effect of the perceived SNR of the PU signals on the SU call-dropping probability B_d. As *SNR* increases, B_d decreases, and vice versa. Hence, PU channels with higher SNR should be preferred for spectrum sensing (if possible).

Next, the effect of the delay bound D on the SU call-dropping probability B_d is analyzed. As D increases, B_d decreases, and vice versa (see Figure 8.2). In general, D should be determined and set

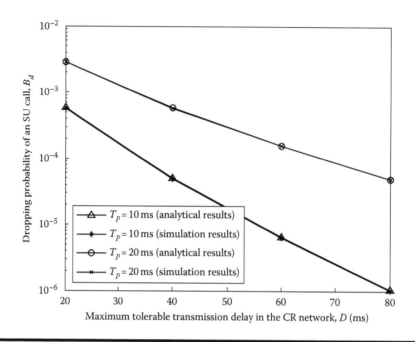

Figure 8.2 SU call-dropping probability versus the maximum allowable transmission delay in the CR network. Parameters: $T_{ss} = 38$ μs; $T_{int} = 25$ ms; $BW = 360$ kHz; $SNR = -6$ dB; $\alpha = 2$; $g = 3$; $n = 30$; $k = 3$; $(p_e = 4.8\%$; $P_d = 99.99\%$; $P_f = 13.7\%)$; $l = 39$; $A_s = 30$ Erl; $A_p = 15$ Erl; $\mu_p = 0.02$ s^{-1}; $(B_s = 2.10^{-2}$; $B_p = 2.10^{-4})$. The SDF scheduling algorithm is applied.

according to the type of service being delivered to the SUs. Therefore, CR networks are inherently efficient in delivering non-real-time (delay-tolerant) services. Furthermore, Figure 8.2 shows the effect of the duration of the spectrum sensing period T_p on the SU call-dropping probability B_d. As T_p decreases, B_d also decreases (and vice versa). However, it should be noted that as T_p decreases, the duration of the TPs decreases as well, which results in reduced CR throughput. Consequently, a possible way to decrease B_d is to decrease T_p, but this comes at the cost of reducing (to some extent) the throughput of the CR network.

Finally, the effect of the offered PU traffic A_p on the traffic capacity of the CR network A_s is analyzed when given QoS requirements have to be guaranteed. As A_p increases, A_s decreases, and vice versa (see Figure 8.3). It is obvious that the traffic capacity of the CR network strongly depends on the PU traffic load. The interplay of PU traffic and SU traffic should always be carefully considered. Thus, as a general rule, the use of CR for DSA is reasonable only if the spectrum resources of the primary network are sufficiently underutilized. Moreover, Figure 8.3 illustrates the effect of the delay bound D on the traffic capacity of the CR network. As D decreases, A_s also decreases (and vice versa). Therefore, the proposed approach for QoS provisioning can support real-time services (with stringent delay requirements) over CR at the price of reduced traffic capacity.

The numerical results presented in this subsection prove the feasibility and demonstrate the effectiveness of the suggested new cross-layer approach for QoS provisioning of the SUs. Insights into QoS provisioning in CR networks have been gained. It is evident that numerous cross-layer interdependencies should be carefully considered in order to achieve optimal or near-optimal performance of the CR network.

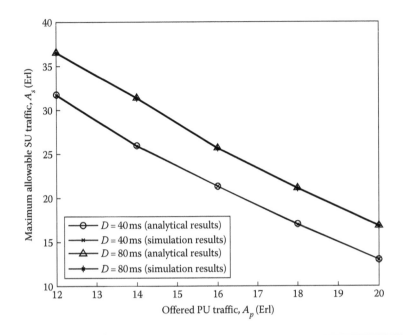

Figure 8.3   Cognitive traffic capacity versus the offered PU traffic load. Parameters: T_p = 20 ms; T_{ss} = 38 μs; T_{int} = 25 ms; *BW* = 360 kHz; *SNR* = −6 dB; α = 2; *g* = 3; *n* = 30; *k* = 3; (p_e = 4.8%; P_d = 99.99%; P_f = 13.7%); *l* = 46; $μ_p$ = 0.02 s⁻¹; B_s ≤ 2.10⁻²; B_d ≤ 1.10⁻⁴; (B_p ≤ 1.10⁻²). **The SDF scheduling algorithm is applied.**

8.4 Summary

In this chapter, QoS provisioning in CR networks operating in compliance with the hierarchical spectrum overlay approach for DSA has been investigated. Spectrum management functions essential to the operation of the CR networks have been considered. Existing state-of-the-art cross-layer approaches for QoS provisioning in CR networks have been reviewed and discussed. A common feature of all these approaches is that they have a limited scope of applicability within the context of the specific CR system model under consideration. Hence, a novel cross-layer approach for QoS provisioning in CR networks has been proposed, namely, the concurrent application of limitation, delay bound, and scheduling. An analytical cross-layer model of an infrastructure-based secondary CR network has been developed in order to investigate the feasibility of the suggested new approach for QoS provisioning of the SUs. Numerous cross-layer interdependencies relevant to QoS provisioning in CR networks have been investigated and analyzed in depth. The proposed novel cross-layer approach for QoS provisioning is effective, general (i.e., based on a nonspecific system model of a CR network), and relatively simple, which determines its wide scope of applicability and theoretical significance.

References

1. IEEE Std. 1900.1-2008, IEEE standard definitions and concepts for dynamic spectrum access: Terminology relating to emerging wireless networks, system functionality, and spectrum management, pp. c1–48, 2008.

2. Shin, K. G., H. Kim, A. W. Min, and A. Kumar, Cognitive radios for dynamic spectrum access: From concept to reality, *IEEE Wireless Communications*, vol. 17, no. 6, pp. 64–74, 2010.

3. Akyildiz, I. F., W.-Y. Lee, M. C. Vuran, and S. Mohanty, A survey on spectrum management in cognitive radio networks, *IEEE Communications Magazine*, vol. 46, no. 4, pp. 40–48, 2008.

4. Zhao, Q. and B. Sadler, A survey of dynamic spectrum access, *IEEE Signal Processing Magazine*, vol. 24, no. 3, pp. 79–89, 2007.

5. Harada, H., Y. Alemseged, S. Filin, M. Riegel, M. Gundlach, O. Holland, B. Bochow, M. Ariyoshi, and L. Grande, IEEE dynamic spectrum access networks standards committee, *IEEE Communications Magazine*, vol. 51, no. 3, pp. 104–111, 2013.

6. Sum, C.-S., G. P. Villardi, M. A. Rahman, T. Baykas, H. N. Tran, Z. Lan, C. Sun, et al., Cognitive communication in TV white spaces: An overview of regulations, standards, and technology, *IEEE Communications Magazine*, vol. 51, no. 7, pp. 138–145, 2013.

7. Doyle, L. E., *Essentials of Cognitive Radio*, Cambridge University Press, New York, 2009.

8. Chen, K.-C. and R. Prasad, *Cognitive Radio Networks*, John Wiley, Chippenham, UK, 2009.

9. Xiao, Y. and F. Hu (Eds), *Cognitive Radio Networks*, CRC Press, Boca Raton, FL, 2009.

10. Zhang, Y., J. Zheng, and H.-H. Chen (Eds), *Cognitive Radio Networks: Architectures, Protocols, and Standards*, CRC Press, Boca Raton, FL, 2010.

11. Rawat, D. B. and G. Yan, Spectrum sensing methods and dynamic spectrum sharing in cognitive radio networks: A survey, *International Journal of Research and Reviews in Wireless Sensor Networks*, vol. 1, no. 1, pp. 1–13, 2011.

12. Garhwal, A. and P. P. Bhattacharya, A survey on spectrum sensing techniques in cognitive radio, *International Journal of Computer Science & Communication Networks*, vol. 1, no. 2, pp. 196–205, 2011.

13. Clancy, T. C., Dynamic spectrum access using the interference temperature model, *Annals of Telecommunications*, vol. 64, no. 7–8, pp. 573–592, 2009.

14. Federal Communications Commission Spectrum Policy Task Force, *Report of the Interference Protection Working Group*, FCC, Washington, DC, USA, pp. 1–49, 2002.

15. Wang, H. and G. Bi, A novel algorithm for interference temperature estimation of cognition domain, *The 2nd International Conference on Consumer Electronics, Communications and Networks (CECNet)*, pp. 57–60, Yichang, Hubei, China, April 2012.

16. Hossain, E., D. Niyato, and Z. Han, *Dynamic Spectrum Access and Management in Cognitive Radio Networks*, Cambridge University Press, New York, 2009.

17. Fette, B. A. (Ed.), *Cognitive Radio Technology*, 2nd ed., Academic Press, Burlington, MA, USA, 2009.

18. Yücek, T. and H. Arslan, A survey of spectrum sensing algorithms for cognitive radio applications, *IEEE Communications Surveys and Tutorials*, vol. 11, no. 1, pp. 116–130, 2009.

19. Axell, E., G. Leus, E. G. Larsson, and H. V. Poor, Spectrum sensing for cognitive radio: State-of-the-art and recent advances, *IEEE Signal Processing Magazine*, vol. 29, no. 3, pp. 101–116, 2012.

20. Sun, H., A. Nallanathan, C.-X. Wang, and Y. Chen, Wideband spectrum sensing for cognitive radio networks: A survey, *IEEE Wireless Communications*, vol. 20, no. 2, pp. 74–81, 2013.

21. Laska, J. N., W. F. Bradley, T. W. Rondeau, K. E. Nolan, and B. Vigoda, Compressive sensing for dynamic spectrum access networks: Techniques and tradeoffs, *IEEE International Symposium on New Frontiers in Dynamic Spectrum Access Networks (IEEE DySPAN)*, pp. 156–163, Aachen, Germany, May 2011.

22. Sun, H., W.-Y. Chiu, and A. Nallanathan, Adaptive compressive spectrum sensing for wideband cognitive radios, *IEEE Communications Letters*, vol. 16, no. 11, pp. 1812–1815, 2012.

23. Tian, Z., Y. Tafesse, and B. M. Sadler, Cyclic feature detection with sub-Nyquist sampling for wideband spectrum sensing, *IEEE Journal of Selected Topics in Signal Processing*, vol. 6, no. 1, pp. 58–69, 2012.

24. Yao, Y., S. R. Ngoga, and A. Popescu, Cognitive radio spectrum decision based on channel usage prediction, *The 8th Euro-NF Conference on Next Generation Internet (NGI)*, pp. 41–48, Karlskrona, Sweden, June 2012.

25. Masonta, M. T., M. Mzyece, and N. Ntlatlapa, Spectrum decision in cognitive radio networks: A survey, *IEEE Communications Surveys and Tutorials*, vol. 15, no. 3, pp. 1088–1107, 2013.

26. Liang, Y.-C., K.-C. Chen, G. Y. Li, and P. Mähönen, Cognitive radio networking and communications: An overview, *IEEE Transactions on Vehicular Technology*, vol. 60, no. 7, pp. 3386–3407, 2011.

27. Zhang, Z., K. Long, and J. Wang, Self-organization paradigms and optimization approaches for cognitive radio technologies: A survey, *IEEE Wireless Communications*, vol. 20, no. 2, pp. 36–42, 2013.
28. Agarwal, S., R. K. Shakya, Y. N. Singh, and A. Roy, DSAT-MAC: Dynamic slot allocation based TDMA MAC protocol for cognitive radio networks, *The 9th International Conference on Wireless and Optical Communications Networks (WOCN)*, pp. 1–6, Indore, India, September 2012.
29. Kunert, K., M. Jonsson, and U. Bilstrup, Deterministic real-time medium access for cognitive industrial radio networks, *The 9th IEEE International Workshop on Factory Communication Systems (IEEE WFCS)*, pp. 91–94, Lemgo, Germany, May 2012.
30. Xu, D., E. Jung, and X. Liu, Efficient and fair bandwidth allocation in multichannel cognitive radio networks, *IEEE Transactions on Mobile Computing*, vol. 11, no. 8, pp. 1372–1385, 2012.
31. Christian, I., S. Moh, I. Chung, and J. Lee, Spectrum mobility in cognitive radio networks, *IEEE Communications Magazine*, vol. 50, no. 6, pp. 114–121, 2012.
32. Lee, W.-Y. and I. F. Akyildiz, Spectrum-aware mobility management in cognitive radio cellular networks, *IEEE Transactions on Mobile Computing*, vol. 11, no. 4, pp. 529–542, 2012.
33. Wang, L.-C., C.-W. Wang, and C.-J. Chang, Modeling and analysis for spectrum handoffs in cognitive radio networks, *IEEE Transactions on Mobile Computing*, vol. 11, no. 9, pp. 1499–1513, 2012.
34. Song, Y. and J. Xie, ProSpect: A proactive spectrum handoff framework for cognitive radio ad hoc networks without common control channel, *IEEE Transactions on Mobile Computing*, vol. 11, no. 7, pp. 1127–1139, 2012.
35. Nejatian, S., S. K. Syed-Yusof, N. M. Abdul Latiff, and V. Asadpour, Proactive integrated handoff management in CR-MANETs: A conceptual model, *IEEE Symposium on Wireless Technology and Applications (ISWTA)*, pp. 32–37, Bandung, Indonesia, September 2012.
36. Wang, P., D. Niyato, and H. Jiang, Voice-service capacity analysis for cognitive radio networks, *IEEE Transactions on Vehicular Technology*, vol. 59, no. 4, pp. 1779–1790, 2010.
37. Lee, H. and D.-H. Cho, Capacity improvement and analysis of VoIP service in a cognitive radio system, *IEEE Transactions on Vehicular Technology*, vol. 59, no. 4, pp. 1646–1651, 2010.
38. Wang, Z., T. Jiang, L. Jiang, and X. He, VoIP capacity analysis in cognitive radio system with single/ multiple channels, *The 6th International Conference on Wireless Communications, Networking and Mobile Computing (WiCOM)*, pp. 1–4, Chengdu, China, September 2010.
39. Hassanein, H. S., G. H. Badawy, and T. D. Todd, Secondary user VoIP capacity in opportunistic spectrum access networks with friendly scheduling, *IEEE Wireless Communications and Networking Conference (IEEE WCNC)*, pp. 1760–1765, Paris, France, April 2012.
40. Gunawardena, S. and W. Zhuang, Voice capacity of cognitive radio networks, *IEEE International Conference on Communications (IEEE ICC)*, pp. 1–5, Cape Town, South Africa, May 2010.
41. Gunawardena, S. and W. Zhuang, On-off voice capacity of single-hop cognitive radio networks with distributed channel access control, *IEEE International Conference on Communications (IEEE ICC)*, pp. 5216–5220, Ottawa, Canada, June 2012.
42. Yin, W., P. Ren, Q. Du, and Y. Wang, Delay and throughput oriented continuous spectrum sensing schemes in cognitive radio networks, *IEEE Transactions on Wireless Communications*, vol. 11, no. 6, pp. 2148–2159, 2012.
43. Luo, C., F. R. Yu, H. Ji, and V. C. M. Leung, Cross-layer design for TCP performance improvement in cognitive radio networks, *IEEE Transactions on Vehicular Technology*, vol. 59, no. 5, pp. 2485–2495, 2010.
44. Alshamrani, A., X. S. Shen, and L.-L. Xie, QoS provisioning for heterogeneous services in cooperative cognitive radio networks, *IEEE Journal on Selected Areas in Communications*, vol. 29, no. 4, pp. 819–830, 2011.
45. Wang, F., J. Huang, and Y. Zhao, Delay sensitive communications over cognitive radio networks, *IEEE Transactions on Wireless Communications*, vol. 11, no. 4, pp. 1402–1411, 2012.
46. Jha, S. C., U. Phuyal, M. M. Rashid, and V. K. Bhargava, Design of OMC-MAC: An opportunistic multi-channel MAC with QoS provisioning for distributed cognitive radio networks, *IEEE Transactions on Wireless Communications*, vol. 10, no. 10, pp. 3414–3425, 2011.

47. Martinez-Bauset, J., A. Popescu, V. Pla, and A. Popescu, Cognitive radio networks with elastic traffic, *The 8th Euro-NF Conference on Next Generation Internet (NGI)*, pp. 17–24, Karlskrona, Sweden, June 2012.

48. Gunawardena, S. and W. Zhuang, Service response time of elastic data traffic in cognitive radio networks, *IEEE Journal on Selected Areas in Communications*, vol. 31, no. 3, pp. 559–570, 2013.

49. Wu, C., C. He, L. Jiang, and Y. Chen, A novel spectrum handoff scheme with spectrum admission control in cognitive radio networks, *IEEE Global Telecommunications Conference (IEEE GLOBECOM)*, pp. 1–5, Houston, TX, December 2011.

50. Tumuluru, V. K., P. Wang, D. Niyato, and W. Song, Performance analysis of cognitive radio spectrum access with prioritized traffic, *IEEE Transactions on Vehicular Technology*, vol. 61, no. 4, pp. 1895–1906, 2012.

51. Lim, H.-J., D.-Y. Seol, and G.-H. Im, Joint sensing adaptation and resource allocation for cognitive radio with imperfect sensing, *IEEE Transactions on Communications*, vol. 60, no. 4, pp. 1091–1100, 2012.

52. Gözüpek, D., B. Eraslan, and F. Alagöz, Throughput satisfaction based scheduling for cognitive radio networks, *IEEE Transactions on Vehicular Technology*, vol. 61, no. 9, pp. 4079–4094, 2012.

53. Jiang, T., H. Wang, and A. V. Vasilakos, QoE-driven channel allocation schemes for multimedia transmission of priority-based secondary users over cognitive radio networks, *IEEE Journal on Selected Areas in Communications*, vol. 30, no. 7, pp. 1215–1224, 2012.

54. Hong, Y., X. Luo, and L. Sun, Cross-layer resource scheduling and allocation with QoS support for cognitive radio systems, *The 8th International Conference on Wireless Communications, Networking and Mobile Computing (WiCOM)*, pp. 1–5, Shanghai, China, September 2012.

55. Mihov, Y. Y., Cross-layer QoS provisioning in cognitive radio networks, *IEEE Communications Letters*, vol. 16, no. 5, pp. 678–681, 2012.

Chapter 9

Reliable Multicast Video Transmission over Cognitive Radio Networks: Equal or Unequal Loss Protection?

Abdelaali Chaoub[1] and Elhassane Ibn-Elhaj[2]

[1]*Laboratory of Electronic and Communication, Mohammadia School of Engineers, Mohammed V-Agdal University, Rabat, Morocco*

[2]*Department of Telecommunication, National Institute of Posts and Telecommunications, Rabat, Morocco*

Contents

9.1 Introduction

In this chapter, a recurrent use case of multimedia applications over CR networks, namely, video transmission, has been considered. More importantly, we focus on applications requiring a timely delivery of information with a tolerable amount of distortion. We focus on multicasting scenarios where one participant is providing access to a video application directly available to a given population of clients with heterogeneous reception bandwidths and QoS requirements. The stochastic profile of the licensed traffic to coexist with is the Poisson distribution. A special emphasis will be given to ULP and ELP assignment. In particular, a detailed description of the ULP scheme is given. The proposed ULP framework takes advantage of MDC properties, and makes use of the priority encoding transmission (PET) packetization framework of Albanese et al. This technique assigns different amounts FEC to different descriptions according to their importance, and Reed Solomon (RS) codes are employed to generate FEC symbols. The used MDC has the property that each packet acts as an independent block of the original message. For the ELP scheme, we recommend the use of LT codes for multicasting operations as they can generate on the fly as many encoding symbols as are needed to decode the transmitted data. In the case of erroneous transmission, the receiver needs just to wait for more packets to be received instead of acknowledging the transmission and requesting retransmission of the missed packets. Afterward, we develop an analytic expression for the success probability in both cases, which will be used to derive the PSNR formula of the finale stream as a function of the loss pattern. In addition, we recall two PSNR-based maximization algorithms. The outcome of the first algorithm is the length of the FEC coefficients associated with the ULP-based MDC scheme, and the amount of redundancy needed in the ELP case is investigated through the second algorithm, followed by a comparative study between the ULP and ELP modes to identify the scheme that best tackles the packet loss pattern. Later, a number of simulations have been performed to assess the effectiveness of the designed multicast architecture.

The rest of this chapter is structured as follows. Section 9.2 presents the background of this research and its motivations. A brief introduction to multimedia transmission over CR networks has been discussed. Section 9.3 raises some interesting issues to deal with the packet loss in the context of fast applications such as multimedia transmissions. In the beginning of Section 9.4, we justify the choice of a centralized architecture instead of an ad hoc one, followed by a brief description of the network topology that provides the infrastructure for the secondary multicasting use case. Besides, a novel time frame structure, adapted to the scenario being treated, is introduced. Next, an ULP multicast system is introduced and discussed along with the conventional equal loss protection multicast design, with the goal of performing a comparative evaluation of both methods in a realistic CR context. We also compute the new expression for the success probability taking into account both primary interruptions and SC impairments and recall some PSNR maximization algorithms and procedures that have been developed to adjust the transmission parameters with respect to the environmental conditions and in response to network and application constraints. Section 9.5 illustrates some numerical results for a real-video multicast session to corroborate the theoretical design and gives a concise summary of the results related to the comparative study conducted between the ULP and ELP schemes. Finally, Section 9.6 presents the conclusions of this chapter.

9.2 Cognitive Radio Meets Multimedia: Introduction

During the past few decades, multimedia applications over wireless networks have grown greatly along with the spread of modern smartphones and the proliferation of new applications. This recent trend has led to a surge in demand for spectrum and bandwidth, and hence the need for new paradigms to overcome the scarcity of spectrum resources. These reasons, among many others, have rendered necessary a reconsideration of the traditional spectrum management process in order to meet the expectations of the new era of services.

9.2.1 Cognitive Radio: Emerging Solution for Spectrum Crisis

Cognitive radio (CR) technology [1] has emerged as a new solution that aims to increase spectral efficiency by leveraging spectrum holes. The licensed spectrum is not complete either in the time or the spatial domain. Accordingly, CR uses opportunistic spectrum sharing (OSS) to exploit temporarily the white spaces. Every CR-based communication system will define two classes of spectrum users. The first is the primary network comprising primary users (PUs), which is the holder of the spectrum license, and the second is the secondary network comprising secondary users (SUs), who are allowed to use the available spectral gaps and must evacuate promptly as soon as the corresponding PU is detected to avoid mutual collisions.

Using access technologies based on selfish sensing raises some technical challenges such as sensing reliability and the hidden node problem (Figure 9.1). Geographic coordination through a central database to identify the vacant subchannels (SCs) is a good substitute for selfish spectral sensing, and a combination of both methods may also be envisaged. Each cognitive peer conducts local spectrum sensing and reports the results to the central unit, and the latter consolidates the local decisions into a global decision [2] (Figure 9.2). This information is exchanged using a dedicated control channel called the group control channel (GCC) [3]. The central peer executes this operation at the start of every time frame t.

The CR for Virtual Unlicensed Spectrum (CORVUS) model is a CR-based approach to create and use virtual unlicensed spectrum [4]. This approach allows the cognitive device to use different noncontiguous SCs scattered over a set of poorly used frequency bands. The selected SCs are exploited to build an SU link (SUL) to transmit SU data. The CORVUS approach relies on a centralized spectrum pooling architecture [4], which helps improve the reliability of spectrum access and reduces the number of jammed SCs if one or more PUs return during the lifetime of

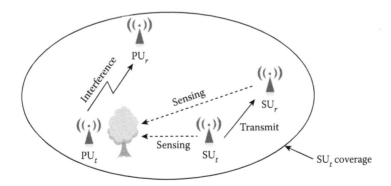

Figure 9.1   An example of hidden node configurations.

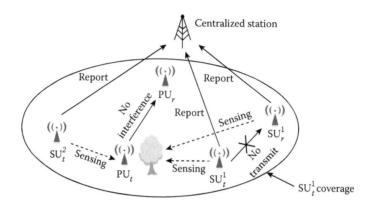

Figure 9.2 Detection of the hidden terminal using centralized architecture.

the constructed SUL. For low interference with the primary infrastructure, the orthogonal frequency division multiplexing (OFDM) scheme has found great popularity in license-exempt systems [2]. The secondary signal is sent over several subcarriers of the OFDM modulation process, and the remaining subcarriers of the inverse fast Fourier transform component may be fed by zeros to avoid radio power emission on the occupied SCs, thus keeping the interference level acceptable and bounded.

The CORVUS approach has generated much interest in recent studies. The authors in Ref. [5] exploited the CORVUS architecture to distribute a fountain-encoded multimedia stream over CR networks with a Poissonian primary traffic pattern. Luby transform (LT) codes are used as channel correcting codes and also to avoid the problem of coordination between SCs. A spectral efficiency performance analysis was conducted to identify the optimal coding overhead and the required number of SCs. The same approach was adopted in Ref. [6], have considering that multiple cognitive nodes can share the same CR infrastructure by separating the SUs in time, that is, TDMA. In addition to interference caused by primary traffic reclaims Ref. [6] Chaoub et al. [6] have also considered the risk of collisions between SUs due to concurrent access. Some authors such as Willkomm et al. [3] have modeled SC availability using the binomial process and have suggested the use of a general model for link maintenance to enable the provision of the secondary service even if the SUL get broken owing to primary interruptions. The proposed link maintenance model relies on the redundancy principle. Numerical simulations have been performed in terms of the average goodput to evaluate the robustness of the proposed model.

9.2.2 CR and Multimedia Communications

CR brings a paradigm shift as a well-regarded agile technology for managing, controlling, and optimizing frequency spectrum allocation. It allows users from overutilized bands to operate in nearby empty bands so that customer connections last long enough to satisfy the required services.

This emerging approach promises to facilitate vast technological advances and seems to meet future spectrum needs. With the tremendous proliferation of smartphones, it is widely expected that this technology will be involved in revolutionizing spectrum-consuming technologies, especially those that are "quality hungry," such as multimedia applications [7].

Multimedia services [5] are a new field that has gained high user acceptance during the last few decades. Most recent systems are concerned with the processing of multimedia data such as image, audio, video, and animation. The integration of multimedia in recent systems has allowed for more comprehensive and easier information exchange and interpretation, owing to the portability of multimedia files and the large amount of information contained in them. Regarding the spectacular use of multimedia that has led to an increasing demand for more quality of service (QoS) and bandwidth requirements, CR networks are expected to shape the future of multimedia services. CR has big potential to enhance multimedia applications with improved end-user perceived quality.

9.3 Multimedia over CR-Based Networks: Challenges and Problem Statement

Recent advances in research on CR are expected to provide the basis for innovative and reliable large-scale multimedia communications. The tricky part in this topic is tailoring the legacy mul-·timedia services to suit the specific context of CR networks to be able to tackle certain peculiar characteristics of cognitive communications, which renders the problem of studying multimedia traffic transmission over CR networks an innovative, exciting, and challenging idea [6].

9.3.1 Multicasting Multimedia Content over CR Networks

SUs should maintain a peaceful coexistence with the primary infrastructure. Once the PU captures its own SC, the secondary transmission may experience a short period of interference before switching to another vacant SC, causing some data to be lost, and delay resulting from the spectrum handoff may cause some codewords to be received after the deadline. In addition, the secondary signal may be attenuated as a result of fading due to propagation across frequency channels. That is, the operation of heterogeneous primary and secondary devices in the same frequency band, as well as fading and noise across SCs, simultaneously contribute to packet losses in CR networks.

Because the secondary network is operating with low transmit power to protect the payload data of the incumbent users, multicast transmissions have great potential to help efficient data delivery to multiple destinations over a single stream in order to preserve precious spectral resources, bandwidth, and transmit power (Figure 9.3). The basic idea of multicast

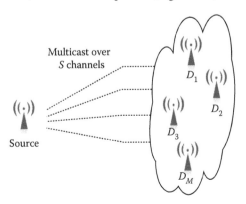

Figure 9.3   Multicasting-based CR network.

communications is to create receive diversity via multiuser diversity by gathering the same signals in a one-to-many fashion. This emerging approach has gained considerable maturity during the last years and seems to match increasing expectations of today's customers to meet the massive multimedia demands, especially applications with image and video content that are tricky to send over the network.

9.3.2 Loss Protection Schemes for Cognitive Multimedia Transmissions

Layered multimedia transmission over CR networks is among the most affected by the problem of packet loss owing to its stringent QoS constraints. Because cognitive transmissions impose tight constraints on delay due to the temporary availability of licensed spectrum, the coding approach has been chosen to handle the packet loss instead of other solutions such as retransmission and feedback channels. Lost packets may be modeled as erasures; thus, we can employ some error-correcting code to recover the erased information. Given the fact that layered multimedia transmissions face a dynamic and unpredictable loss pattern over CR networks, the manner in which the stream layers are treated and protected may have a substantial impact on transmission reliability. A family of loss protection schemes usually desired for time-sensitive and layered multimedia content over lossy networks is the unequal loss protection (ULP) frameworks [8,9]. The use of such schemes allows the generation of multiple levels of quality and alleviates packet losses.

It is commonly believed that ULP schemes are immune to lossy and rapidly varying conditions. Nevertheless, a pertinent question to ask is whether the ULP always performs better than the standard equal loss protection schemes in CR networks under delay-constrained circumstances. One of the most commonly adopted solutions for dealing with such questions is to purposely design and deploy a Medium Access Control (MAC) protocol capable of handling different CR system constraints and make decisions about the most suitable coding strategy. Choosing the best protection scheme is of significant interest, especially in dynamic and error-prone environments such as CR networks, and for layered multimedia transmissions, which usually have high QoS demands, under connectionless transport protocols.

The peak signal-to-noise ratio (PSNR) represents an efficient distortion measure that quantifies the perceived quality of the reconstructed media stream. Therefore, maximizing the PSNR while controlling packet losses improves bandwidth utilization and guarantees user satisfaction in terms of perceived quality. To this extent, the forward error correction (FEC) parameters for the ULP encoders may be determined by maximizing the PSNR of the received stream as a function of the loss pattern characterizing the considered transmission system. Afterward, the effectiveness of the PET-based ULP framework is demonstrated by a comparative study with the traditional ELP technique. Comparing both techniques in a secondary context is an appealing idea to discover the most appropriate mode according to the network state.

9.3.3 Prior Work on Multimedia Transmissions over CR Networks

Nowadays, the topic of CR has attracted great interest among academics. However, little research effort has been devoted to the problem of ULP-centered proposals for CR contexts.

In Ref. [10], Kushwaha et al. have studied the coding aspect over CR networks, and a brief summary of the spectrum pooling concept has been introduced. Next, different coding types have been presented, and their applications in the context of secondary use have been analyzed and discussed. Principally, the paper has given an overview of the multiple description coding (MDC) as source coding well suited for use in CR networks. For simulation results, the paper has adopted LT codes to combat transmission errors. This study was an attempt to give a general analysis of MDC applications on CR networks, and no numerical results have been presented for this specific coding scheme. In Ref. [11], Li has investigated the use of MDC in CR systems to overcome the losses due to primary traffic arrival on secondary applications that are delay sensitive and distortion tolerable. Using a Gaussian source, he has proposed an algorithm to transform the selection of rates and distortions problem into an optimization problem for the expected utility. The PUs' occupancy over each frequency channel was modeled as a Markov chain. Numerical results have been presented for real-time image coding. However, this contribution has not considered the noise and fading aspect of networks, and consequently there is an additional packets lost average due to the lossy environment, which degrades considerably the performance of the selected SUL. Moreover, this study has considered only the Gaussian sources and needs to be generalized to more source types. The use of ULP schemes in CR networks was discussed in greater depth in Refs. [8,9,12,13]. The problem of image transmission over lossy CR networks using progressive source codes associated with error-correcting codes where the stream delivery is reinforced by the use of unequal error protection has been tackled the first time in Ref. [8]. In Ref. [12], characterizing reliability in CR networks has been shown to be one of the major bottlenecks in the performance evaluation of multimedia traffic transmissions in unlicensed usage situations. The study [9] focused on a recurring problem in the cellular infrastructures, namely, the nonreciprocity between uplink and downlink in interfering with the primary network and also in mutual collisions. A general cross-layer model is proposed in Ref. [13] to ensure end-to-end multimedia content delivery in realistic CR contexts.

As shown in the previous studies on the topic, loss protection schemes for CR networks and their potential to enhance the design of multimedia transmissions have not been sufficiently investigated.

In this chapter, a recurrent use case of multimedia applications over CR networks, namely, video transmission, has been considered. More importantly, we focus on applications requiring a timely delivery of information with a tolerable amount of distortion. We focus on multicasting scenarios where one participant is providing access to a video application directly available to a given population of clients with heterogeneous reception bandwidths and QoS requirements. The stochastic profile of the licensed traffic to coexist with is the Poisson distribution. A special emphasis will be given to ULP and ELP assignment. In particular, a detailed description of the ULP scheme is given. The proposed ULP framework takes advantage of MDC properties, and makes use of the priority encoding transmission (PET) packetization framework of Albanese et al. This technique assigns different amounts FEC to different descriptions according to their importance, and Reed Solomon (RS) codes are employed to generate FEC symbols. The used MDC has the property that each packet acts as an independent block of the original message. For the ELP scheme, we recommend the use of LT codes for multicasting operations as they can generate on the fly as many encoding symbols as are needed to decode the transmitted data. In the case of erroneous transmission, the receiver needs just to wait for more packets to be received instead of acknowledging the transmitter and requesting retransmission of the missed packets. Afterward, we develop an analytic expression for the success probability in both cases, which will be used to derive the PSNR formula of the finale stream as a function of the loss pattern. In addition, we recall two PSNR-based maximization algorithms. The outcome of the first algorithm is the length of the FEC coefficients associated with the ULP-based MDC scheme, and the amount of redundancy needed in the ELP case is investigated through the second algorithm, followed by a comparative study between the ULP and ELP modes to identify the scheme that best tackles the packet loss pattern. Later, a number of simulations have been performed to assess the effectiveness of the designed multicast architecture.

The rest of this chapter is structured as follows. Section 9.2 presents the background of this research and its motivations. A brief introduction to multimedia transmission over CR networks has been discussed. Section 9.3 raises some interesting issues to deal with the packet loss in the context of fast applications such as multimedia transmissions. In the beginning of Section 9.4, we justify the choice of a centralized architecture instead of an ad hoc one, followed by a brief description of the network topology that provides the infrastructure for the secondary multicasting use case. Besides, a novel time frame structure, adapted to the scenario being treated, is introduced. Next, an ULP multicast system is introduced and discussed along with the conventional equal loss protection multicast design, with the goal of performing a comparative evaluation of both methods in a realistic CR context. We also compute the new expression for the success probability taking into account both primary interruptions and SC impairments and recall some PSNR maximization algorithms and procedures that have been developed to adjust the transmission parameters with respect to the environmental conditions and in response to network and application constraints. Section 9.5 illustrates some numerical results for a real video multicast session to corroborate the theoretical design and gives a concise summary of the results related to the comparative study conducted between the ULP and ELP schemes. Finally, Section 9.6 presents the conclusions of this chapter.

9.4 Proposed Network Model

9.4.1 General Analysis

Real-spectrum measurements taken on the band (3 GHz, 6 GHz) in downtown Berkeley [4] reveals the underutilization of this spectrum portion over time, which results in a total system bandwidth of $W = 3$ GHz directly available for unlicensed usage.

We suppose that our secondary scenario operates in this band, and we adopt the concept of OSS. The benefit of such a scheme is that it allows the SUs to maximize their transmission rate when a spectrum opportunity is detected.

Concretely, we consider a CR network hosting a common cognitive source S multicasting a single video in parallel to a set of M cognitive destinations $(D_i)_{i \in \{1, \cdots, M\}}$ forming a multicast group (MG) (Figure 9.4). There is a strict deadline for all codewords, denoted by $T_{\max}$. Only packets entirely received before the deadline are involved in the process of reconstruction of the output media.

Video applications feature rich multimedia content combining sound, image, and text, which make them large media files, leading to delicate transmission scenarios. Multicasting is a key cost-effective solution that reduces the complexity of such contexts and a good replacement for most traditional transmission techniques.

The SUs may not have all the necessary information about the cognitive network. As a result, we opt for a centralized control of spectral and temporal resources. A third trusted party continuously collects the observations reported by different secondary devices. This centralized infrastructure is more efficient than ad hoc architecture, which is inherently limited in its knowledge of the exclusive users of the resources. The ad hoc approach requires additional hardware resources to be implemented at the user level, which is very costly, in practice, in terms of deployment.

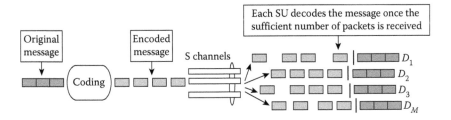

Figure 9.4   Multicast using LT codes.

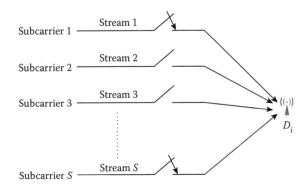

Figure 9.5   Unequal loss protection based communication.

In error-prone environments such as CR networks, the provision of service could be disrupted at any point as a result of packet drops. Robust protection schemes are required to reduce the stream erasures. The central idea is to be able to utilize a low-quality stream in case of unreliable cognitive transmissions with respect to the packet loss pattern (Figure 9.5). In such situations, ULP schemes are devised to provide better service provision and adaptability to fast-varying contexts.

For delay-tolerant transmissions in environments where retransmission is guaranteed, lost packets can be resent until the successful reception of all the media. In this case, the conventional ELP schemes are still applied to lessen the stream erasures.

For the purpose of designing efficient and reliable multimedia transmissions in CR networks, efficient ULP and ELP frameworks making use of an appropriate progressive compression in conjunction with a powerful channel coding are very suitable for this type of network. The channel coding is used to compensate for the corrupted packets, and source compression permits recovering of the secondary content up to a certain quality commensurate with the number of packets received.

Set partitioning in hierarchical trees (SPIHT) is a high-quality source coder based on wavelets. The algorithm codes the most important wavelet transform coefficients first, and transmits the bits progressively. It produces a fully scalable compression, which means that if the transmission is stopped at any point, a lower-bit-rate media file can still be decompressed and reconstructed. The method also compresses video very efficiently.

RS [15] and Fountain [16] codes have attracted significant interest as erasure codes to provide the FEC property in ULP and ELP schemes. Fountain codes [16] are a class of rateless codes with infinite efficiency. Recently, these codes have gained high user acceptance [17], especially

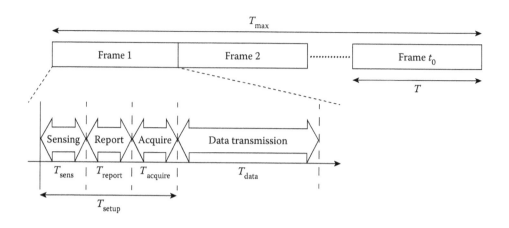

Figure 9.6   Extended time frame structure.

for time-sensitive applications such as multimedia streaming. Multimedia Broadcast Multicast Service (MBMS) [18] and Digital Video Broadcasting-IPTV (DVB-IPTV) [19] are concrete examples of standards opting for fountain codes to overcome the packet loss problem in delay-constrained multimedia applications. Within this family, we find the well-known LT codes [20] and Raptor codes [21]. In this study, the video is source-encoded using SPIHT, different layers of the ULP scheme are protected using RS codes [15], and LT codes [22] are used to generate FEC codes for the ELP case.

Both ULP and ELP techniques will be discussed in a secondary use scenario so that the MAC layer may be able to make a decision about the protection scheme it deems best for the current cognitive transmission. In what follows, a performance comparison is carried out between ULP and ELP techniques through assessing the PSNR of the received video as a function of the experienced loss rate. The PSNR represents an efficient distortion measure to quantify the perceived quality of the recovered media stream.

We start by extending the legacy structure of the time frame in order to support the newly designed MAC protocol. As in Ref. [3], we divide each frame T_t into four parts: a sensing block T_{sens}, a reporting block T_{report}, an acquire block $T_{acquire}$, and a data transmission block T_{data} (Figure 9.6). The first three blocks refer to the setup phase $T_{setup} = T_{sens} + T_{report} + T_{acquire}$. In the sensing block, all users conduct local spectrum sensing simultaneously. During the reporting block, the local sensing results are reported to the nearest centralized peer through a common channel, and the collected data will be treated and processed to extract the transmission parameters that are the most appropriate for the current context. Throughout the acquire block, results returned by the intelligent MAC algorithms must be disseminated to the remaining peers, after which new SCs need to be acquired to compensate for the lost ones or to serve new incoming SUs. Finally, the next stream is ready to transmission over the CR network in a maximum delay of T_{data}.

9.4.2 ULP Framework for CR Networks

Currently, many contributions have been made toward the problem of ULP techniques. The one introduced in Ref. [15] is based on RS codes and is among the most practical approaches. It is based on the PET method of Albanese et al. [14]. In this study, the technique used in Ref. [15] is applied (Figure 9.7).

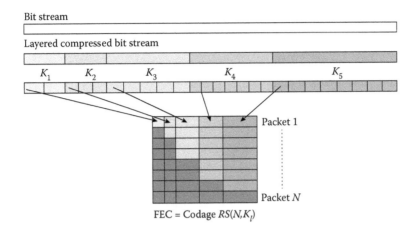

Figure 9.7 MDC framework based priority encoding transmission.

Using the initial message, a scalable bit stream is formed by applying SPIHT compression [23] on the stream. The bit stream source is then partitioned into L fragments $(R_l)_{1 \leq l \leq L}$ indexed in order of decreasing importance because the most important data are emitted first in a progressive source coder. Next, each layer R_l is blocked into K_l source blocks. The lth source block will be expanded into channel codewords of length N using the RS code, $RS(N,K_l)$, as an FEC mechanism. FEC_l is the FEC value assigned to the description l. FEC symbols are added to each message fragment for protection against packet losses resulting from both primary traffic interruptions and SC characteristics, such that the sub-stream R_l and its corresponding FEC_l form a description D_l. RS codes have the ability to decode the transmitted description D_l using any set of K_l received packets. Then, a total of L descriptions $(D_l)_{1 \leq l \leq L}$ is obtained, and each description D_l has an RS rate of K_l / N (Figure 9.7).

Let FEC be an L-tuple whose entries are the length of the FEC assigned to each stream; that is, $FEC = (FEC_1, FEC_2, ..., FEC_L)$, where FEC_l is the FEC amount assigned to the description l where $l \in \{1, \cdots, L\}$. We state that $N > FEC_1 \geq FEC_2 \geq ... \geq FEC_L$.

The expected PSNR of the received stream is the sum of the different PSNR amounts corresponding to the received descriptions, each weighted by the probability of receiving its corresponding description, and is given by

$$psnr\left((D_l)_{1 \leq l \leq L}\right) = \sum_{l=1}^{L} P(D_l) \, psnr(D_l) \tag{9.1}$$

where $P(D_l)$ is the probability of successfully decoding the description D_l at the receiver, and the quantity $psnr(D_l)$ is the additional PSNR amount gained when the receiver decodes the lth description given that $l - 1$ descriptions have already been successfully decoded.

There are mainly two events that markedly affect the traffic distribution on the selected SUL, and consequently two goals must be fulfilled simultaneously: (1) cope with packet losses caused by the unforeseen PU return, and (2) counteract packet erasures due to SC fading and noise. Thus, the description D_l can be decoded at the receiver if two conditions are achieved, namely: the number of PET-encoded packets needed to recover the original description D_l could be successfully transmitted over the S SCs and, on the other hand, at most FEC_l packets get lost due to SC fading and noise.

Each subcarrier s, where $s \in \{1, \cdots, S\}$, is assumed to be a frequency-flat block-fading Rayleigh channel. Let h_s be the sth SC gain. SC gains are supposed to be independent and exponentially distributed, each with a mean of $\overline{h_s}$. At each member of the MG, there is an SNR ξ_m below which a transmitted packet will be considered to be corrupted. Let P be the transmit power of the source node and ω be the noise power at the MG.

In multicasting transmissions, receiving members of the same MG may be seen and tackled in a holistic and unified manner because all the group members behave as a single destination from the source node viewpoint.

The error probability of the sth link, denoted as π_s, is defined as the probability that at least one node fails to reconstitute the conveyed message, and is simply expressed as

$$
\begin{aligned}
\pi_s &= Prob\left(\bigcup_{m=1}^{M} \left\{ \frac{h_s P}{\omega} < \xi_m \right\} \right) \\
&= 1 - Prob\left(\frac{h_s P}{\omega} \geq \max\left((\xi_m)_{1 \leq l \leq M} \right) \right) \\
&= 1 - \exp\left(-\frac{\max\left((\xi_m)_{1 \leq l \leq M} \right) \omega}{P \overline{h_s}} \right).
\end{aligned}
\tag{9.2}
$$

In addition to the aforementioned effects of fading and noise, SUs could be evicted from their SCs at any point in time whenever the corresponding PU returns. We assume that each subcarrier s has a channel capacity R_s and is licensed to a Poissonian primary traffic with arrival rate λ_s (Figure 9.8). Before packet transmission, all the cognitive nodes conduct local spectrum sensing simultaneously to determine a set of unoccupied SCs. Then according to the novel metric $Q_s = (1 - \pi_s)^{\left(\frac{R_s \times T_{\max}}{B} \right)} \times exp(-\lambda_s T_{\max})$, the SUL is established using a concatenation of the highest-quality metric SCs, where B is the packet size and $\frac{R_s \times T_{\max}}{B}$ is the maximum number of packets that can be transmitted over SC s during $T_{\max}$. The metric Q_s quantifies the ability of SC s to combat the effects of primary interruptions and SC impairments. Afterward, the multicast source

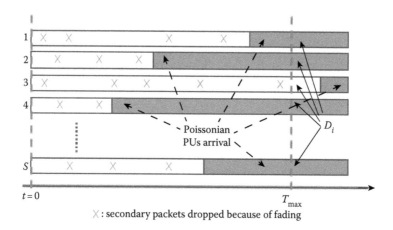

Figure 9.8 Subchannel access model.

node starts conveying its packets over the chosen SUL during the maximum tolerable delay T_{max}. Let N_{PU} be the maximum number of packets that can be sent over the set of SCs S before the Poissonian primary traffic arrival, and let N_{SC} be the number of secondary packets corrupted as a result of fading and noise over the same set of SCs S (Figure 9.8). It is worth recalling that the minimum number of PET-encoded packets needed to reconstruct the layer l at the MG is $N - FEC_l$. Then, the expression of $P(D_l)$ is

$$P(D_l) = Prob\ (\{N_{PU} - N_{SC} \geq N - FEC_l\}). \tag{9.3}$$

It is clearly seen that both random variables N_{PU} and N_{SC} are dependent, which makes the calculation of the probability $P(D_l)$ very cumbersome. For the sake of simplicity, we use the expectation of N_{SC} as an approximation for the average packet loss due to erasure aspects of fading SCs encountered, as long as the corresponding PUs have not yet claimed their channels. Thus, a simplified expression for the probability $P(D_l)$ is

$$P(D_l) = Prob\ (\{N_{PU} - E[N_{SC}] \geq N - FEC_l\}). \tag{9.4}$$

9.4.2.1 An Analytical Expression for N_{PU}

The variable N_{PU} could be expressed as a sum of S independent random variables

$$N_{PU} = \sum_{s=1}^{S} N_{PU}^s. \tag{9.5}$$

N_{PU}^s denotes the number of packets transmitted over SC s before the PU return, with $s \in \{1,\ldots,S\}$. The random variable N_{PU}^s is proportional to the available time on the sth SC (white color in Figure 9.8) denoted by the random variable T_{PU}^s. For a Poisson primary traffic model and given the delay constraints, the variable T_{PU}^s is derived as

$$T_{PU}^s = \begin{cases} \tau_s, & \text{if } \tau_s \leq T_{max} \\ T_{max}, & \text{if } \tau_s > T_{max} \end{cases} \tag{9.6}$$

where $\tau_s \sim exp\lambda_i$. Hence, the probability density function (PDF) of T_{PU}^s is written as

$$f_{T_{PU}^s}(t) = \lambda_s \exp(-\lambda_s t)(u(t) - u(t - T_{max}))$$
$$+ \exp(-\lambda_s T_{max})\delta(t - T_{max}) \tag{9.7}$$

where $u(n)$ and $\delta(n)$ are, respectively, the unit step function and the Dirac delta function. The proportionality between N_{PU}^s and T_{PU}^s is described as

$$N_{PU}^s = \frac{R_s \times T_{PU}^s}{B}. \tag{9.8}$$

For the sake of simplicity of the mathematical formulas, we set $M_s = R_s/B$, and we then obtain

$$f_{N_{PU}^s}(n) = \frac{\lambda_s}{M_s} \exp\left(\frac{-\lambda_s n}{M_s}\right)(u(n) - u(n - M_s T_{max}))$$

$$+ \exp(-\lambda_s T_{max})\delta(n - M_s T_{max}). \tag{9.9}$$

Using the definition of the PDF of a sum of independent random variables as the convolution of their individual PDFs, $f_{N_{PU}}(n)$ could be subsequently inferred as

$$f_{N_{PU}}(n) = \bigotimes_{s=1}^{S} f_{N_{PU}^s}(n). \tag{9.10}$$

9.4.2.2 An Analytical Expression for N_{sc}

The pdf of N_{SC} is the convolution of the PDFs of S binomial variables N_{SC}^s, each denoting the number of lost packets due to fading and noise on SC s with $s \in \{1, \cdots, S\}$.

The losses caused by SC characteristics affect the number of packets that can be transmitted over the SC before the PU arrival given by T_{PU}^i (the x-marks in Figure 9.8). Hence, assuming an independent loss process, the PDF of N_{SC}^s can be expressed as

$$f_{N_{SC}^s}(n) = Prob(N_{SC}^s = n) = \binom{N_{PU}^s}{n} \times \pi_s^n \times (1 - \pi_s)^{N_{PU}^s - n}, \tag{9.11}$$

where π_s and N_{PU}^s are defined, respectively, in Equations 9.2 and 9.9.

9.4.2.3 An Analytical Expression for $E[N_{SC}^s]$

From Equation 9.11, the dependency between the variables N_{PU}^s and N_{SC}^s is explicitly demonstrated. Thus, we proceed by computing the conditional expectation as

$$E[N_{SC}^s] = E[E[N_{SC}^s / N_{PU}^s]]. \tag{9.12}$$

Using the properties of the conditional expectation, there exists a function Ψ such that

$$E[N_{SC}^s / N_{PU}^s] = \Psi(N_{PU}^s). \tag{9.13}$$

To find Ψ, we have to evaluate the expression $\Psi(n) = E[N_{SC}^s / N_{PU}^s = n]$

$$\Psi(n) = \sum_{m=0}^{n} m Prob\left(N_{SC}^s = m / N_{PU}^s = n\right). \tag{9.14}$$

We substitute Equation 9.11 into Equation 9.14, and we get

$$\Psi(n) = \sum_{m=0}^{n} m \binom{n}{m} \times \pi_s^m \times (1 - \pi_s)^{n-m}. \tag{9.15}$$

Using some simple algebra: $m\binom{n}{m} = n\binom{n-1}{m-1}$, we can write

$$\Psi(n) = n \times \sum_{m=1}^{n} \binom{n-1}{m-1} \times \pi_s^m \times (1-\pi_s)^{n-m}. \tag{9.16}$$

Finally, $\Psi(n)$ is succinctly described as

$$\Psi(n) = n\pi_i. \tag{9.17}$$

The expectation of N_{SC}^s can be represented as

$$E[N_{SC}^s] = E[\Psi(N_{PU}^s)] = \pi_s E[N_{PU}^s]. \tag{9.18}$$

We need to compute $E[N_{PU}^s]$.
The mean of N_{PU}^s can be written as follows:

$$E[N_{PU}^s] = \int_0^{+\infty} n f_{N_{PU}^s}(n)\,dn. \tag{9.19}$$

Then, substituting Equation 9.9 into Equation 9.19

$$E[N_{PU}^s] = \int_0^{M_s T_{max}} \frac{\lambda_s n}{M_s} \exp\left(\frac{-\lambda_s n}{M_s}\right) dn$$

$$+ \exp(-\lambda_s T_{max}) \int_0^{+\infty} n\delta(n - M_s T_{max})\,dn. \tag{9.20}$$

Using integration by parts, the first integral could be evaluated as

$$\int_0^{M_s T_{max}} \frac{\lambda_s n}{M_s} \exp\left(\frac{-\lambda_s n}{M_s}\right) dn$$

$$= \frac{M_s}{\lambda_s}\left(1 - \exp(-\lambda_s T_{max})\right) - M_s T_{max} \exp(-\lambda_s T_{max}) \tag{9.21}$$

and the second integral is computed as

$$\int_0^{+\infty} n\delta(n - M_s T_{max}) = M_s T_{max}. \tag{9.22}$$

By substituting these expressions of integrals into Equation 9.20, and using Equation 9.18, we get

$$E[N_{SC}^s] = \frac{M_s \pi_s}{\lambda_s}\left(1 - \exp(-\lambda_s T_{max})\right). \tag{9.23}$$

At the end, $E[N_{SC}]$ is determined by the following expression:

$$E[N_{SC}] = \sum_{s=1}^{S} \frac{M_s \pi_s}{\lambda_s} \left(1 - \exp\left(-\lambda_s T_{max}\right)\right). \tag{9.24}$$

Given the PDF of N_{PU} in Equation 9.10 and the closed form of $E[N_{sc}]$ given above, the probability $P(D_j)$ could be readily inferred.

The value of FEC to assign to each layer is determined by the outcome of the PSNR maximization problem implemented in Ref. [8].

The algorithm introduced in Ref. [15] can be exploited to optimize this ULP scheme under some constraints [13].

9.4.3 Equal Loss Protection Scheme for Secondary Transmissions

The ELP method is among the most classical coding approaches applied to convey a stream on a lossy network. In the current study, the ELP is an LT code–based scheme that protects different layers using the same LT encoder and decoder. The ELP scheme provides equal packet loss prevention to all layers, regardless of the contribution of each layer in the displayed quality at the receiver side. Thus, the base and enhancement layers are equally protected, and all stream parts, $(R_l)_{1 \le l \le L}$, consume the same amount of redundancy (Figure 9.9).

The source stream is progressively encoded using a progressive compression scheme, namely, SPIHT, and divided into L fragments $(R_l)_{1 \le l \le L}$. Each fragment R_l is blocked into the same number of source blocks K. Then, the kth source block of each segment R_l, where $1 \le l \le L$, is placed into the kth packet, where $k = 1,\ldots,K$, to generate K packets. The LT decoder needs at least N' packets to recover the original K packets with probability $1 - DEP$, where N' is slightly greater than K. An overhead of 5% achieves low decoding error probability values using some specially designed degree distributions such as the robust soliton distribution. To compensate for the eventual discarded packets, some redundancy X is inserted into the coded stream. Redundant packets are constructed by arranging the media packets in the columns of a matrix and generating the

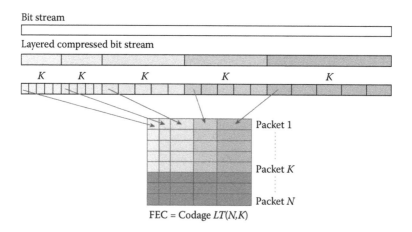

Figure 9.9 LT codes-based equal loss protection framework.

redundant bytes along the rows using an LT code. It follows that the total number of transmitted packets is $N = N' + X$, and the LT coder/decoder used is $LT(N,K)$ (Figure 9.9).

In this case, the expected PSNR of the received stream is the PSNR corresponding to K SPIHT-encoded packets weighted by the probability of receiving at least N' packets. It can be written as follows:

$$psnr(N \, packets) = P(N' \, packets) \times (K \, packets). \qquad (9.25)$$

$P(N' \, packets)$ is the probability of successfully receiving at least N', and the quantity $psnr(K \, packets)$ is the PSNR gained when the receiver successfully decodes the K transmitted packets.

To deliver the stream $(R_l)_{1 \le l \le L}$ intact to the receiver, a sufficient number of LT-encoded packets, subjected to primary interruptions as well as SC losses, need to be collected over the unused portions of the spectrum. In particular, the probability $P(N' \, packets)$ is formulated as

$$P(N' \, packets) = Prob \, (N_{PU} - E[N_{SC}] \ge N - X) \qquad (9.26)$$

where both random variables N_{PU} and N_{SC} have been previously developed in Equations 9.10 and 9.24).

The optimal X is then computed theoretically via the PSNR maximization framework developed in Ref. [13].

Concretely, the MAC layer may choose the most suitable coding technique for the current communication context. The coding scheme choice relies on the packet loss rate history estimated either as a prerequisite of the class of service encompassing the multimedia service to be fulfilled [24] or by using a packet sniffer to compare the measured number of packets at both the transmitter and the receiver sides for the previous frames, $(Ti')_{t' \le t-1}$, to quantify the amount of erroneous packets and predict the fraction of packets lost. Afterward, redundancy parameters are fixed by solving the aforementioned optimization algorithms, and then a fair performance comparison between ELP and ULP strategies can be carried out to extract the most efficient protection scheme with respect to the estimated loss factor.

9.5 Numerical Results and Discussions

In this section, numerical simulations have been undertaken to investigate the PSNR benefits of the proposed MAC protocol. We seek to outline the achieved gains when using the proposed ULP-based multicast design to ensure the secondary video traffic transmission.

9.5.1 General Simulations

Consider an uncompressed CIF video sequence (Container) with a frame rate of 30 frames/s and a resolution of 352 × 288 pixels. Each pixel consists of 24 bits, while each byte represents one of the red, green, and blue (RGB) components. Because of its efficiency in extremely lossy conditions as in the case of CR networks, information bits are carried on OFDM symbols using an SC bandwidth of $W = 312.5$ kHz. We assume one GOP transmission, SPIHT compressed [23] using a bit rate of $r = 0.2$ bit/pixel resulting in a number of packets $N = 324$ to be transmitted within a strict delay of $T_{max} = 20$ ms. The packet size is $B = 1000$ bits. For simplicity of analysis and without loss of generality, we assume that the maximum delay corresponds to the time for one frame.

In a wireless system using the IEEE Standard 802.11a [25], each SC represents an OFDM subcarrier modulated with 64-Quadrature Amplitude Modulation (QAM) associated with a convolutional coder having a rate of 2/3, giving rise to a greater common channel capacity of R_0 = 1 Mbps.

After the sensing phase and using the spectrum pooling concept, a pool of S = 10 vacant SCs is constituted out of the 48 available data subcarriers in the 802.11a-based system.

The Poisson distribution with parameter λ is well suited to analyzing events occurring on an average of $J = \lambda/T_{max}$ times per period T_{max}. So, for frequent primary returns, we consider the following set of arrival rates: λ = [30, 20, 10, 25, 36, 40, 60, 24, 32, 15].

Further, we take the following distribution of fading losses:

$$\bar{h} = [6.9e^{-9}, 5.2e^{-9}, 21e^{-9}, 10.4e^{-9}, 4e^{-9}, 8e^{-9}, 3.4e^{-9}, 21e^{-9}, 7e^{-9}, 13.9e^{-9}].$$

The transmit power for the source node is fixed at P = 46, and the maximum SNR threshold value at the MG is ξ = 0.35. The noise power is calculated using Boltzmann's formula for the considered bandwidth W = 3 GHz.

The parameters for the robust soliton distribution used in LT code are c_{LT} = 0.1 and δ_{LT} = 0.5, and the decoding error probability, *DEP*, is 1%.

For the visual quality assessment, we display the first frame of the transmitted GOP.

In Figure 9.10, FEC is given for each description to maximize the PSNR for each received video frame subject to primary interruptions, λ, and the SC loss process, π, in the case of the MDC scheme. Base layers are heavily protected, and enhancement layers are less protected, implementing the ULP for the video stream.

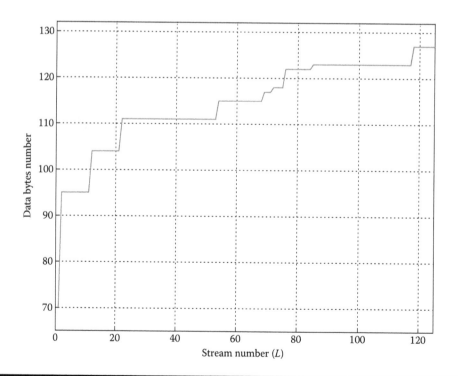

Figure 9.10   Data bytes number for each stream.

Figure 9.11 shows the influence of adding some redundant packets X to the transmitted video frame on *PSNR(N packets)* subject to primary interruptions λ and the SC loss process π. The resulting graph has an optimal point, where the achieved PSNR is maximum for $X = 219$. When the protection level is increased, it does not necessarily result in better quality on the client side. This behavior emphasizes the importance of finding a balance that meets stream protection and expected quality.

Figure 9.12 depicts the impact of the fraction of packets lost on the obtained PSNR under both ULP and ELP. As intuitively expected, ULP surpasses ELP in ranges [0, 0.62] and [0.67, 0.71]. However, ELP outperforms ULP outside the given ranges. ULP degrades gradually when the packet loss rate is increased, unlike the ELP approach, where the transition is sharp.

Figure 9.13 shows the reconstructed Container video frame in both the ULP and ELP cases. In Figures 9.13c through 9.13g, and depending on the number of successfully received descriptions, good overall video quality is obtained despite the presence of primary traffic interruptions and SC losses. Under extremely lossy conditions, where only three descriptions are received, the proposed MDC scheme still exhibits acceptable perceived video quality. However, the conveyed video frame is useless in the case where only one description is received (Figure 9.13b). When some enhancement layers, for example, descriptions 9 and 10, are successfully received, the ULP yields better results than the ELP (Figure 9.13l). When at most eight descriptions are received, ELP performs better than ULP.

Similar video processing has been applied on other video sequences such as "Bus", "News" and "Mother daughter", and the recovered video show good visual quality even for low data rates.

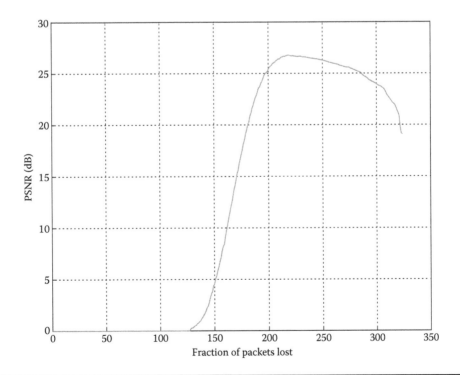

Figure 9.11   Computed PSNR plotted against the redundancy, X.

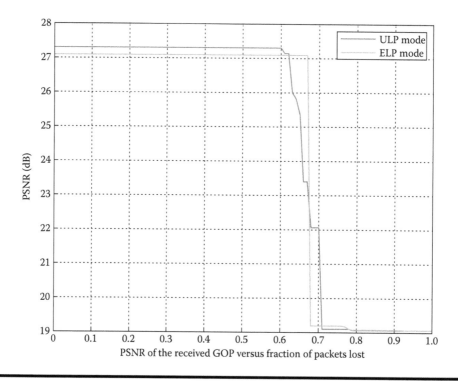

Figure 9.12 PSNR of the received video frame versus fraction of packets lost.

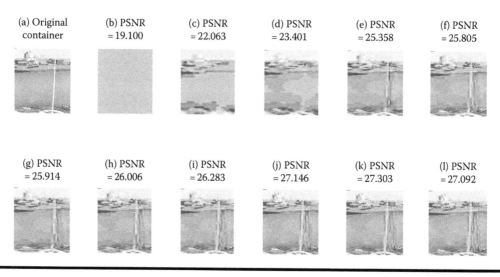

Figure 9.13   (a) The original 352 × 288 container frame. The reconstructed container frame: (b) one received description, (c) two received descriptions, (d) three received descriptions, (e) four received descriptions, (f) five received descriptions, (g) six received descriptions, (h) seven received descriptions, (i) eight received descriptions, (j) nine received descriptions, (k) all received descriptions, (l) ELP case.

9.5.2 Analysis and Interpretation of Results

In the case of very low packet loss rates, the proposed ULP increases the likelihood of successfully delivering all the descriptions, and the transmitted quality is displayed at the receiver side. In contrast, the conventional ELP scheme is limited in the number of packets it can recover, that is, the original K SPIHT-encoded packets. Thus, ULP will surpass ELP with respect to average packet losses.

For slightly higher packet loss rates, it is intuitively expected that the ELP will outperform the ULP because the latter distributes the FEC amounts in proportion to the importance of the layers; thus, the base layer receives better protection than the enhancement layers. As a result, some enhancement layers will soon become useless, whereas the ELP scheme will be more robust to this effect, because all stream segments are equally protected.

At higher packet loss rates, the ULP results in better efficacy compared to the ELP with respect to the achieved PSNR because ELP has a sharp transition at loss rates close to the added redundancy amount X, in contrast to the ULP method, which is characterized by a gradual degradation in quality.

In extreme packet loss conditions, it is expected that the ELP will yield better results than the ULP. For loss rates exceeding the FEC value assigned to the first description FEC_1, the base layer is irrecoverable in the ULP case. However, using the ELP approach, the bits emitted first in the SPIHT compression process can still be recovered.

9.6 Conclusions

Recently, multicast communications over OFDM-based CR networks have generated great interest as a key cost-effective solution to finding strategies for sustaining the enormous proliferation of new services with limited power and spectrum resources. Throughout the chapter, we have consolidated several contributions on this research area into a logical structure. The resulting multicast system design for CR networks captures different system parameters as well as summarizes many relevant results and builds a solid foundation for future improvement. Meanwhile, we have derived the formula of the PSNR at the MG and suggested the reuse of some recent algorithms to transform the choice of the best coding scheme and the selection of the redundancy parameters issue into a challenging optimization problem for a given expected utility, namely, the PSNR. The theoretical analysis of the proposed scheme has been supported by practical simulations for a real-video transmission in the context of CR. The test case demonstrates that the ULP approach provides a gradual degradation of video quality as the fraction of lost packets increases, and thus the ULP allocation outperforms the ELP scheme over a wide range of packet loss rate values. The gradual degradation property makes the ULP mode the most suitable in cases of unknown or unpredictable network conditions. The ELP approach yields some interesting results at very high loss rates. For CR networks with extremely lossy conditions, the ELP approach is advised. The topic of multimedia transmission over CR networks using multicast technology has converged into a multidimensional optimization issue on the MAC layer side.

References

1. Mitola, J., III and Maguire, G. Q., Cognitive radio: Making software radios more personal, *IEEE Personal Communications*, 6(4), 13–18, 1999.
2. Weiss, T. and Jondral, F., Spectrum pooling: An innovative strategy for the enhancement of spectrum efficiency, *IEEE Communications Magazine*, 42, 8–14, 2004.

3. Willkomm, D., Gross, J., and Wolisz, A., Reliable link maintenance in cognitive radio systems, in *IEEE International Symposium on New Frontiers in Dynamic Spectrum Access Networks*, November 2005, pp. 371–378.

4. Broderson, R. W., Wolisz, A., Cabric, D., Mishra, S. M., and Willkomm, D., *Corvus: A Cognitive Radio Approach for Usage of Virtual Unlicensed Spectrum*, White Paper, Technical Report, University of California, Berkeley, 2004.

5. Kushwaha, H., Xing, Y., Chandramouli, R., and Heffes, H., Reliable multimedia transmission over cognitive radio networks using fountain codes, *Proceedings of the IEEE*, 96(1), 155–165, 2008.

6. Chaoub, A., IbnElhaj, E., and El Abbadi, J., Multimedia traffic transmission over TDMA shared cognitive radio networks with Poissonian primary traffic, in *International Conference on Multimedia Computing and Systems, 2011, ICMCS '11*, April 7–9, 2011, pp. 1–6.

7. Mitola, J., Cognitive radio for flexible mobile multimedia communications, in *Proceedings of the IEEE International Workshop on Mobile Multimedia Communications*, 1999, pp. 3–10.

8. Chaoub, A., Ibn Elhaj, E., and El Abbadi, J., Multimedia traffic transmission over cognitive radio networks using multiple description coding, in *Proceedings of the First International Conference on Advances in Computing and Communications (ACC)*, Kochi, India, July 2011, pp. 529–543.

9. Chaoub, A. and Ibn-Elhaj, E., Multiple description coding for cognitive radio networks under secondary collision errors, in *Proceedings of the IEEE 16th Mediterranean Electrotechnical Conference (MELECON)*, Yasmine Hammamet, Tunisia, March 2012, pp. 27–30.

10. Kushwaha, H., Xing, Y., Chandramouli, R., and Subbalakshmi, K. P., Erasure tolerant coding for cognitive radios, in *Cognitive Networks: Towards Self-Aware Networks*, Q. H. Mahmoud, Ed., John Wiley & Sons, Ltd, Chichester, UK, 2007, pp. 315–331.

11. Li, H., Multiple description source coding for cognitive radio systems, in *Proceedings of the Fifth International Conference on Cognitive Radio Oriented Wireless Networks and Communications (CROWNCOM)*, Cannes, 2010, pp. 1–5.

12. Chaoub, A. and Ibn Elhaj, E., Improving reliability in cognitive radio networks using multiple description coding, in *CNC 2012, LNICST 108*, Springer-Verlag, Berlin, 2011, pp. 99–108.

13. Chaoub, A. and Ibn-Elhaj, E., Cross layer design for equal and unequal loss protection frameworks in cognitive radio networks, *Computers and Electrical Engineering*, 39(2), 571–581, 2013.

14. Albanese, A., Blmer, J., Edmonds, J., Luby, M., and Sudan, M., Priority encoding transmission, *IEEE Transactions on Information Theory*, 42, 1737–1744, 1996.

15. Mohr, A. E., Riskin, E. A., and Ladner, R. E., Unequal loss protection: Graceful degradation of image quality over packet erasure channels through forward correction, *IEEE Journal on Selected Areas in Communications*, 18(6), 819–828, 2000.

16. Mitzenmacher, M., Digital fountains: A survey and look forward, in *Proceedings of the IEEE on Information Theory Workshop*, October 2004, pp. 271–276.

17. Shokrollahi, A. and Luby, M., Raptor codes, *Foundations and Trends in Communications and Information Theory*, 6(3–4), 213–322, 2009.

18. Multimedia Broadcast/Multicast Service (MBMS): Protocols and Codecs, 3GPP TS 26.346 V7.4.0 (2007-06).

19. Transport of MPEG-2 TS-Based DVB Services over IP Based Networks, ETSI TS 102.034 V1.3.1 (2007-10).

20. Luby, M., LT codes, in *Proceedings of the 43rd Annual IEEE Symposium on Foundations of Computer Science (FOCS 2002)*, November 2002, pp. 271–280.

21. Shokrollahi, A., Raptor codes, *IEEE Transactions on Information Theory*, 52(6), 2551–2567, 2006.

22. Namjoo, E., Aghagolzadeh, A., and Museviniya, J., Robust transmission of scalable video stream using modified LT codes, *Computers and Electrical Engineering*, 37(5), 768–781, 2011.

23. Said, A. and Pearlman, W. A., A new, fast, and efficient image codec based on set partitioning in hierarchical trees, *IEEE Transactions on Circuits and Systems for Video Technology*, 6, 243–250, 1996.

24. Policy and Charging Control Architecture, 3GPP TS 23.203 V9.4.0 (2010-03).

25. IEEE 802.11a, *Part 11: Wireless LAN Medium Access Control (MAC) and Physical Layer (PHY) Specifications: High-Speed Physical Layer in the 5 GHz Band*, Technical Report, IEEE Computer Society, USA, 1999.

ARTIFICIAL INTELLIGENCE FOR MULTIMEDIA OVER CRN

Chapter 10

Bayesian Learning for Cognitive Radio Networks

Xin-Lin Huang and Jun Wu

School of Electronics and Information, Tongji University, Shanghai, China

Contents

10.1 Bayesian Inference: Preliminaries

In statistics, Bayesian inference is a method of inference in which Bayes' rule is used to update the probability estimate for a hypothesis as additional evidence is acquired. Bayesian updating is an important technique throughout statistics, especially in mathematical statistics. For some cases, exhibiting a Bayesian derivation for a statistical method automatically ensures that the method works as well as any competing method. Bayesian updating is especially important in the dynamic analysis of a sequence of data. Bayesian inference has found application in a range of fields including science, engineering, philosophy, medicine, and law. In this section, we provide a brief motivation for the Bayesian approach and establish some concepts.

The concept of exchangeability is central to many statistical approaches and may be viewed as critical in motivating Bayesian statistics. Let us assume that we are aggregating data in an attempt to make predictions about future values of the random process we are observing. If we were to make the strong assumption that the data are independent, we would treat every new data point individually without using past observations to predict future observations because

$$p(y_1, \ldots, y_n) = \prod_{i=1}^{n} p(y_i) \tag{10.1}$$

implies that

$$p(y_{n+1}, \ldots, y_m \mid y_1, \ldots, y_n) = p(y_{n+1}, \ldots, y_m) \tag{10.2}$$

A weaker assumption that often better describes the data we encounter is that of exchangeability, which states that the order in which we encounter the data is inconsequential.

Definition 1 *A sequence of random variables $y_1, y_2, \ldots, y_n$ is said to be* finitely exchangeable *if*

$$y_1, y_2, \ldots, y_n \underline{D} \, y_{\pi(1)}, y_{\pi(2)}, \ldots, y_{\pi(n)} \tag{10.3}$$

for every permutation π on $\{1, \ldots, n\}$. Here, we use the notation $\underline{D}$ to mean equality in distribution.

Definition 2 *A sequence $y_1, y_2, \ldots$ is said to* infinitely exchangeable *if every finite subsequence is finite exchangeable.*

A very important result arising from the assumption of exchangeable data is what is typically referred to as de Finetti's theorem. This theorem states that an infinite sequence of random variables $y_1, y_2, \ldots$ is exchangeable if and only if there exists a random probability measure ν with respect to which $y_1, y_2, \ldots$ are conditionally independent identically distributed (i.i.d.) with distribution ν.

Theorem 1 *If $y_1, y_2, \ldots$ is an infinitely exchangeable sequence of binary random variables with probability measure P, then there exists a distribution function Q on [0, 1] such that for all n*

$$p(y_1, \ldots, y_n) = \int_0^1 \prod_{i=1}^{n} \vartheta^{y_i} (1 - \vartheta)^{1 - y_i} \, dQ(\vartheta) \tag{10.4}$$

where $p(y_1, \ldots, y_n)$ is the joint probability mass function defined by measure P. Furthermore, Q is the distribution function of the limiting empirical frequency:

$$\theta \underline{a.s.} \lim_{n \to \infty} \frac{1}{n} \sum_{i=1}^{n} y_i, \theta \sim Q \tag{10.5}$$

From de Finetti's theorem, we see why the Bayesian perspective of the parameter yields observations i.i.d. as a random quantity with some distribution Q, rather than as a fixed and unknown quantity. We now state the more general form of the de Finetti theorem.

Theorem 2 *If y_1, y_2,... is an infinitely exchangeable sequence of real-valued random variables with probability measure P, then there exists a probability measure μ defined on the space of all probability measures $P(\mathbb{R})$ on $\mathbb{R}$ such that*

$$P\left(y_1 \in A_1, \ldots, y_n \in A_n\right) = \int_{P(\mathbb{R})} \prod_{i=1}^{n} \upsilon(A_i) \, \mu(d\upsilon) \tag{10.6}$$

Furthermore, μ is the law of a probability measure υ, where υ is almost certainly defined by the limiting empirical measure. That is,

$$\upsilon(B) \underline{a.s.} \lim_{n \to \infty} \frac{1}{n} \sum_{i=1}^{n} I_B(y_i), \upsilon \sim \mu \tag{10.7}$$

where B ranges over all elements of the Borel σ-algebra. The measure μ is often referred to as the de Finetti measure.

When we limit ourselves to the more restrictive class of finite-dimensional θ (e.g., Bernoulli, multinomial, Gaussian random variables), we can invoke the following corollaries.

Corollary 1 *Assuming the required densities exist, and assuming the conditions of Theorem 2 hold, then there exists a distribution function Q such that the joint density of $(y_1,...,y_n)$ is of the form*

$$p\left(y_1, \ldots y_n\right) = \int_{\Theta} \prod_{i=1}^{n} p\left(y_i \mid \vartheta\right) dQ(\vartheta) \tag{10.8}$$

with $p(\cdot \mid \vartheta)$ representing the density function corresponding to the finite-dimensional parameter $\vartheta \in \Theta$.

From the foregoing corollary, it is simple to see how the de Finetti theorem motivates the concept of a *prior distribution* $Q(\cdot)$ and a *likelihood function* $p(y \mid \cdot)$.

Corollary 2 *Given that the conditions of Corollary 1 hold, then the predictive density is given by*

$$p\left(y_{m+1}, \ldots y_n \mid y_1, \ldots y_m\right) = \int_{\Theta} \prod_{i=m+1}^{n} p\left(y_i \mid \vartheta\right) dQ\left(\vartheta \mid y_1, \ldots y_m\right) \tag{10.9}$$

where

$$dQ\left(\theta \mid y_1, \ldots, y_m\right) = \frac{\prod_{i=1}^{m} p\left(y_i \mid \theta\right) dQ(\theta)}{\int_{\Theta} \prod_{i=1}^{m} p\left(y_i \mid \vartheta\right) dQ(\vartheta)} \tag{10.10}$$

From the form of the predictive density, we see that our view of the existence of an underlying random parameter θ yielding the data i.i.d. has not changed. Instead, we have simply updated our *prior belief* $Q(\theta)$ into a *posterior belief* $Q(\theta \mid y_1, ... y_m)$ through an application of *Bayes rule*:

$$p(\theta \mid y) = \frac{p(y \mid \theta) p(\theta)}{\int_{\Theta} p(y \mid \vartheta) p(\vartheta) d\vartheta} = \frac{p(y \mid \theta) p(\theta)}{p(y)} \tag{10.11}$$

Here, we have written the rule in its simplest form, assuming that a density on θ exists in addition to the conditional density on *y*. Although one can view the computation of the predictive distribution as the objective in Bayesian statistics, we will often limit our discussion to the process of forming the posterior distribution from the prior by incorporating observations because this is a fundamental step in examining the predictive distribution.

From a practical perspective, we never have an infinite sequence of observations from which to characterize our prior distribution. Furthermore, even if we had such a quantity, the probability measure that the de Finetti theorem would suggest as yielding the data i.i.d. might be arbitrarily complex. Thus, we are left with two competing pragmatic choices in defining our prior:

1. Tractable inference
2. Modeling flexibility

The issue of tractable inference often motivates the use of conjugate priors. The goal of flexibility in our models leads to the study of Bayesian nonparametric methods.

In Bayesian probability theory, if the posterior distributions $p(\theta \mid x)$ are in the same family as the prior probability distribution $p(\theta)$, the prior and posterior are then called conjugate distributions, and the prior is called a conjugate prior for the likelihood function. Please refer to Ref. [11] for more details information about Bayesian inference.

10.2 Graphical Models

A graphical model is a probabilistic model for which a graph denotes the conditional dependence structure between random variables. They are commonly used in probability theory, statistics—particularly Bayesian statistics—and machine learning. Graphical models offer several advantages:

1. They provide a simple way to visualize the structure of a probabilistic model and can be used to design and motivate new models.
2. Insights into the properties of the model, including conditional independence properties, can be obtained by inspection of the graph.
3. Complex computations, required to perform inference and learning in sophisticated models, can be expressed in terms of graphical manipulations, in which underlying mathematical expressions are carried along implicitly.

A graph comprises nodes (also called vertices) connected by links (also known as edges or arcs). In a probabilistic graphical model, each node represents a random variable (or group of random variables), and the links express probabilistic relationships between these variables. The graph then captures the way in which the joint distribution over all of the random variables can be decomposed into a product of factors each depending only on a subset of the variables. We shall begin by discussing Bayesian networks, also known as directed graphical models, in which the links of the graphs have a particular directionality indicated by arrows. The other major class of graphical models is Markov random fields, also known as undirected graphical models, in which the links do not carry arrows and have no directional significance.

10.2.1 Bayesian Network

If the network structure of the model is a directed acyclic graph, the model represents a factorization of the joint probability of all random variables. More precisely, if the events are $X_1, ..., X_n$, then the joint probability satisfies

$$P[X_1, ..., X_n] = \prod_{i=1}^{n} P[X_i \mid pa_i]$$

(10.12)

where pa_i is the set of parents of node X_i. In other words, the joint distribution factors into a product of conditional distributions. For example, the graphical model in the Figure 10.1 shown above consists of the random variables *A*, *B*, *C*, and *D* with a joint probability density that factors as

$$P[A, B, C, D] = P[A]P[B]P[C \mid B, D]P[D \mid A, B, C]$$

(10.13)

Any two nodes are conditionally independent given the values of their parents. In general, any two sets of nodes are conditionally independent given a third set if a criterion called d-separation holds in the graph. Local independences and global independences are equivalent in Bayesian networks.

This type of graphical model is known as a directed graphical model, Bayesian network, or belief network. Classic machine learning models such as hidden Markov models, neural networks, and newer models such as variable-order Markov models can be considered special cases of Bayesian networks.

10.2.2 Markov Random Field

A Markov random field, also known as a Markov network, is a model over an undirected graph. A graphical model with many repeated subunits can be represented with plate notation (Figure 10.2).

Given an undirected graph $G = (V, E)$, a set of random variables $X = (X_v)_{v \in V}$ indexed by V form a Markov random field with respect to G if they satisfy the local Markov properties:

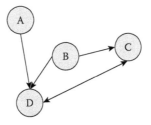

Figure 10.1   An example of a graphical model. Each arrow indicates a dependency. In this example: D depends on A, D depends on B, D depends on C, C depends on B, and C depends on D.

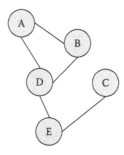

Figure 10.2   An example of a Markov random field. Each edge represents a dependency. In this example: A depends on B and D. B depends on A and D. D depends on A, B, and E. E depends on D and C. C depends on E.

Pairwise Markov property: Any two non-adjacent variables are conditionally independent given all other variables:

$$X_u \perp\!\!\!\perp X_v \mid X_{V \setminus \{u,v\}} \text{ if } \{u,v\} \notin E \tag{10.14}$$

Local Markov property: A variable is conditionally independent of all other variables given its neighbors:

$$X_v \perp\!\!\!\perp X_{V \setminus cl(v)} \mid X_{ne(v)} \tag{10.15}$$

where ne(v) is the set of neighbors of v, and cl(v) = $\{v\} \cup$ ne(v) is the closed neighborhood of v.

Global Markov property: Any two subsets of variables are conditionally independent given a separating subset

$$X_A \perp\!\!\!\perp X_B \mid X_S \tag{10.16}$$

where every path from a node in A to a node in B passes through S.

The above three Markov properties are not equivalent to each other at all. In fact, the Local Markov property is stronger than the Pairwise one, but weaker than the Global one.

10.3 System Model

We assume that there are two sets of users in cognitive radio network (CRN): (1) primary users (PUs) that have absolute priority in using licensed spectrum, and (2) secondary users (SUs) that can use the subcarriers unoccupied by PUs. We assume that there are N SUs and M subcarriers, each with a bandwidth of W Hz. We also assume the following: (1) all SUs in a cluster have the same priority to access each subcarrier; (2) each SU can detect the packets transmitted by other SUs in a cluster by detecting the packet headers or Acknowledgment from the channel; (3) A packet from an SU can be loaded on only one subcarrier, which makes it easy for the CH to detect which packet is in error; and (4) each CH maintains its member nodes, and periodically broadcasts the beacon information to guarantee synchronization in the cluster.

The utility function is determined by the subcarrier selection strategy profile of all users over each subcarrier, denoted as $S_{N \times M} = \{s_{ij}, i = 1,... N, j = 1, ... M\}$. $s_{ij} > 0$ indicates that SU$_i$ chooses

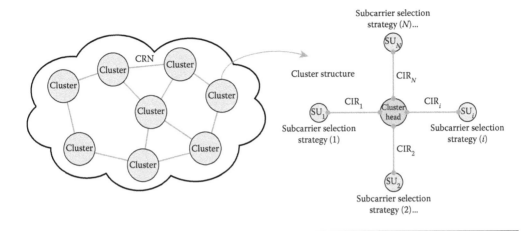

Figure 10.3   Cluster structure with dynamic subcarrier, power, and modulation allocation strategy.

subcarrier j and loads $s_{ij}(s_{ij} \leq 1)$ percentage of the required transmission rate. The traffic models of users are defined as follows.

PUs: We assume that each PU accesses the licensed channel with an ON–OFF model [1]. An ON/OFF state represents whether a PU is occupying a channel, and the channel state may alternate between state ON (active) and state OFF (inactive) [2]. We assume that each PU changes its state independently. The SUs can transmit only during the OFF time slots.

SUs: We assume that the rate requirement of multimedia applications for SU_i is R_i (bit/s) with a delay constraint. The multimedia bit stream from each SU is divided into many packets, and each packet is transmitted on only one subcarrier with suitable power and modulation schemes to maximize the utility function.

Our cognitive cross-layer scheduling uses a cluster-structure-based CRN in Figure 10.3. The CH is responsible for collecting packets from its member nodes and forwarding them to other clusters. After correctly receiving a packet, the CH will broadcast an ACK to its member node as in Ref. [3], and the Automatic Repeat Request protocol is used to guarantee that the important packets are successfully received [4]. We assume that the channel impulse response (CIR) between member nodes and CH is different from one another owing to the variable wireless environment. In this chapter, we maximize the utility function of each member node for uplink transmission (from member nodes to CH) in a distributed manner. Hence, the action strategy (i.e., dynamic subcarrier selection, power, and modulation schemes) of each member node should be aware of the action of PUs as well as other SUs in the cluster. Moreover, the uncertainty of spectrum sensing, CIR characteristics, queue congestion, interference temperature, and limited transmit power are considered in our cognitive cross-layer scheduling for multimedia transmission.

10.4 Subcarrier Selection Strategy Profile

Traffic information cannot be exchanged among SUs owing to the intolerable delay and the large bandwidth needed, which depends on the number of subcarriers M, the number of SUs N, s_{ij}, and so on. Therefore, each SU should learn the traffic information (i.e., subcarrier selection profile) of other SUs in the cluster to avoid heavy information exchange. For SU_i, other SUs' activity model

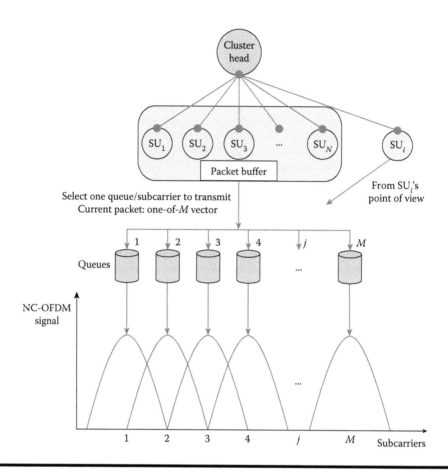

Figure 10.4 Cluster activity model in CRN (from the view point of SU$_j$).

(in the same cluster) is shown in Figure 10.4. Here, we only consider the uplink packets transmission from member nodes to the CH.

From SU$_i$'s point of view, the activity of all other SUs in the same cluster can be represented by a virtual SU in Figure 10.4. The virtual SU loads only one packet on a specific subcarrier at a time. Hence, the subcarrier selection strategy of the virtual SU can be modeled by a hidden variable $Z = (z_1, z_2,..., z_M)$, which determines the subcarrier selection action for one packet and follows the categorical distribution. In probability theory and statistics, a categorical distribution is a probability distribution that describes the result of a random event that can take on one of M possible outcomes, with the probability of each outcome separately specified. Because multiple SUs in a cluster cannot transmit packets on the same subcarrier simultaneously, we assume the SU$_i$ can detect the subcarrier selection action A (1 by M vector) of the virtual SU, which only selects one subcarrier at a time. The hidden variable Z is subject to

$$p\{z_j = 1\} = \mu_j \quad z_j \in \{0, 1\} \quad j = 1, 2,..., M \tag{10.17}$$

$$\sum_{j=1}^{M} \mu_j = 1 \tag{10.18}$$

In Equation 10.17, the parameter $\mu = \{\mu_1, \mu_2, ..., \mu_M\}$ determines the subcarrier selection strategy profile of the virtual SU. We assume the subcarrier selection $A = \{a_1, a_2, ..., a_M\}$ of the virtual SU (from SU_i's point of view) has been observed for K samples (corresponding to subcarrier selection actions for K packets), denoted as $A_{K \times M} = \{a_{k1}, a_{k2}, ..., a_{kM}\}$, $k = 1, 2, ..., K$. The likelihood function $p\{A/\mu\}$, which represents how probable the observed data set is for different settings of the parameter vector μ, can be represented as

$$p\{A/\mu\} = \mu_1^{a_1} \mu_2^{a_2} \cdots \mu_M^{a_M} = \prod_{j=1}^{M} \mu_j^{a_j} \tag{10.19}$$

And the likelihood function of K samples is

$$p\{A_{K \times M}/\mu\} = \prod_{k=1}^{K}\prod_{j=1}^{M} \mu_j^{a_{kj}} = \prod_{j=1}^{M} \mu_j^{\sum_{k=1}^{K} a_{kj}} = \prod_{j=1}^{M} \mu_j^{m_j} \tag{10.20}$$

where $m_j = \sum_{k=1}^{K} a_{kj}$ is the total time for selecting subcarrier j. We assume that prior knowledge of $\mu = \{\mu_1, \mu_2, ..., \mu_M\}$ follows the Dirichlet distribution. In probability theory and statistics, the Dirichlet distribution is a family of continuous multivariate probability distributions parameterized by a vector $\alpha = \{\alpha_1, \alpha_2, ..., \alpha_M\}$ of positive reals. It is the multivariate generalization of the beta distribution. Dirichlet distributions are often used as prior distributions in Bayesian statistics, and in fact the Dirichlet distribution is the conjugate prior probability density function (pdf) of the categorical distribution. The probability density function of the Dirichlet distribution represents the belief that the probabilities of M rival events are $\{\mu_1, \mu_2, ..., \mu_M\}$ given that each event has been observed $\alpha_j -1$ times. The pdf of μ can be represented as

$$p\{\mu_1, \mu_2, ..., \mu_M / \alpha_1, \alpha_2, ..., \alpha_M\} = \frac{1}{B(\alpha)} \prod_{j=1}^{M} \mu_j^{\alpha_j - 1} \tag{10.21}$$

where $B(\alpha) = \prod_{j=1}^{M} \Gamma(\alpha_j) / \Gamma\left(\sum_{j=1}^{M} \alpha_j\right)$, and $\Gamma(\cdot)$ is the gamma function $\left(\Gamma(x) = \int_0^{\infty} t^{x-1} e^{-t}\, dt\right)$.

The foregoing subcarrier selection strategy can be described by the probabilistic graphical model in Figure 10.5, to show the relationship between the observed information and the inside norm. In Figure 10.5, SU_i can only observe the subcarrier selection results A of the virtual SU, which is the value of the variable Z. In our system model, Z follows the categorical distribution with parameter $\mu = \{\mu_1, \mu_2, ..., \mu_M\}$. For simplicity, we assume that $\mu = \{\mu_1, \mu_2, ..., \mu_M\}$ follows the Dirichlet distribution with parameter $\alpha = \{\alpha_1, \alpha_2, ..., \alpha_M\}$, which is the conjugate pdf of the categorical distribution. Later, we will use this probabilistic graphical model to update the hyper-parameter $\alpha = \{\alpha_1, \alpha_2, ..., \alpha_M\}$ based on the observations A.

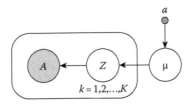

Figure 10.5 **Probabilistic graphical model of subcarrier selection strategy.**

We use the fully Bayesian model (Bayesian inference) to evaluate the uncertainty in μ after we have observed K samples $A_{K \times M}$ in the form of the posterior probability $p\{\mu/A_{K \times M}\}$.

The Bayesian theorem can be stated as follows: posterior $\propto$ (likelihood $\times$ prior), and can be extended to our problem:

$$p\{\mu/A_{K \times M}\} \propto p\{A_{K \times M}/\mu\} \cdot p\{\mu\} \propto \prod_{j=1}^{M} \mu_j^{m_j} \cdot \prod_{j=1}^{M} \mu_j^{\alpha_j - 1}$$

$$= \prod_{j=1}^{M} \mu_j^{m_j + \alpha_j - 1} \tag{10.22}$$

Hence, the pdf of μ is updated which follows the Dirichlet distribution:

$$p\{\mu/A_{K \times M}\} = \frac{1}{B(\alpha_{new})} \prod_{j=1}^{M} \mu_j^{\alpha_{j,new} - 1} \tag{10.23}$$

$$\alpha_{new} = \{m_1 + \alpha_1, m_2 + \alpha_2, ..., m_M + \alpha_M\} \tag{10.24}$$

The updating process can also be implemented after one observed action owing to

$$p\{\mu/A_{K \times M}\} \propto \prod_{j=1}^{M} \mu_j^{a_{Kj}} \left(\prod_{k=1}^{K-1} \prod_{j=1}^{M} \mu_j^{a_{kj}} \cdot \prod_{j=1}^{M} \mu_j^{\alpha_j - 1} \right) \tag{10.25}$$

In the next section, we will use the results of the subcarrier selection strategy to derive the delay-based utility function.

10.5 Multimedia Transmission Considering Quality of Service

Quality of service (QoS) is the overall performance of a telephony or computer network, particularly the performance seen by the users of the network.

To quantitatively measure the QoS, several related aspects of the network service are often considered, such as throughput, delay, and packet delay variation.

Throughput: In communication networks, such as Ethernet or packet radio, the throughput or network throughput is the rate of successful message delivery over a communication channel. These data may be delivered over a physical or logical link, or pass through a certain network node. The throughput is usually measured in bits per second (bit/s or bps), and sometimes in data packets per second or data packets per time slot.

Transmission delay: In a network based on packet switching, the transmission delay (or store-and-forward delay, also known as the packetization delay) is the time required to push all of the packet's bits into the wire. In other words, this is the delay caused by the data rate of the link. Transmission delay is a function of the packet's length and has nothing to do with the distance between the two nodes. This delay is proportional to the packet's length in bits.

Propagation delay: In computer networks, propagation delay is the amount of time it takes for the head of the signal to travel from the sender to the receiver. It can be computed as the ratio between the link length and the propagation speed over the specific medium.

Packet delay variation: In computer networking, packet delay variation is the difference in the end-to-end one-way delay between selected packets in a flow, with any lost packets being ignored. The effect is sometimes referred to as jitter, although the definition is an imprecise fit. For interactive real-time applications, for example, VoIP (Voice over Internet Protocol), packet delay variation can be a serious issue, and hence VoIP transmissions may need QoS-enabled networks to provide a high-quality channel.

10.6 Utility Function for Multimedia Transmission

We consider two important parameters (i.e., delay and throughput) to describe the QoS performance of multimedia transmissions. In this section, we will derive the closed-form results for the delay and throughput-based utility functions.

10.6.1 M/G/1 Queue Analysis

In queuing theory, a discipline within the mathematical theory of probability, an *M/G/1* queue is a queue model where arrivals are Markovian (modulated by a Poisson process), service times have a general distribution, and there is a single server [5].

A queue represented by an *M/G/1* queue is a stochastic process whose state space is the set $\{0, 1, 2, 3 ...\}$, where the value corresponds to the number of customers in the queue, including any customers being served. Transitions from state i to $i+1$ represent the arrival of a new customer: the times between such arrivals have an exponential distribution with parameter λ. Transitions from state i to $i - 1$ represent a customer who has been served, finished being served, and departing: the length of time required for serving an individual customer has a general distribution function. The lengths of times between arrivals and of service periods are random variables that are assumed to be statistically independent.

We start with the derivation for the average queue waiting time T_j for a SU over subcarrier j. We use the Pollaczek–Khinchine formula [6] as

$$T_j = N_q \cdot E(S_j) + r_j \tag{10.26}$$

where N_q is the number of packets waiting on subcarrier j during time T_j. S_j represents the subcarrier selection results of all member nodes on subcarrier j; that is, $S_j = \{s_{1j}, s_{2j}, ..., s_{Nj}\}$. Please recall that each packet is transmitted on only one subcarrier, and the subcarrier selection depends on the subcarrier selection strategy of SUs. In Section 10.4, we assumed that the subcarrier selection strategy of other SUs follows the Dirichlet distribution. $E(S_j)$ is the mean transmission time of each packet over subcarrier j, and r_j is the unfinished work in the queue (i.e., the remaining time to finish current packet). Because $r_j \ll N_q \cdot E(S_j)$, for simplicity, we represent r_j by its mean value $E(r_j) = \dfrac{X_j}{2} E(S_j^2)$ [6]. Here, $E(S_j^2)$ is the second moment of packet transmission time over subcarrier j. Hence, we can rewrite Equation 10.26 as

$$T_j = X_j T_j E(S_j) + \frac{X_j}{2} E(S_j^2) \tag{10.27}$$

where X_j is the packet arrival rate over subcarrier j, given by $\mu_j \times \lambda$, which depends on the total packet arrival rate parameter λ and subcarrier selection strategy profile $\mu = \{\mu_1, \mu_2, ..., \mu_M\}$. Hence, we have

$$T_j = \frac{\mu_j \lambda E(S_j^2)}{2[1 - \mu_j \lambda E(S_j)]}, \quad j = 1, 2, ..., M \tag{10.28}$$

Assuming the allowable delay for multimedia data transmission for packets in SU_i is d_i, the probability of packets' waiting time in the queue on subcarrier j being less than d_i is represented as

$$P_{ij}\{T_j \leq d_i\} = P_{ij}\left\{\mu_j \leq \frac{2d_i}{\lambda[E(S_j^2) + 2d_i E(S_j)]}\right\} \tag{10.29}$$

And the total packets arrival rate of other SUs is

$$\lambda = \sum_{n=1, n \neq i}^{N} \sum_{j=1}^{M} s_{nj} \lambda_n \tag{10.30}$$

In Equation 10.30, we assume the total packet arrival rate of other SUs in the cluster can be detected by SU_i. Because the transmission delay is the sum of the queue waiting time and the packet transmission time $T_{i,new}(S_j)$(which depends on the allocated modulation schemes), the probability of the transmission delay for SU_i packet on subcarrier j being less than d_i can be rewritten as

$$P_{ij}\{Delay_{ij} \leq d_i\} = P_{ij}\left\{\mu_j \leq \frac{2[d_i - T_{i,old}(S_j) - T_{i,new}(S_j)]}{\lambda\{E(S_j^2)(q_j + 1) + 2[d_i - T_{i,old}(S_j) - T_{i,new}(S_j)]E(S_j)\}}\right\} \tag{10.31}$$

where $T_{i,old}(S_j)$ is the unfinished packet transmission time of the previous packet from SU_i, which is still waiting in the queue (on the jth subcarrier), while q_j is the corresponding number of packets from SU_i waiting in the queue. In Equation 10.29, we assumed there are no packets from SU_i waiting on the jth subcarrier (i.e., $q_j = 0$). Obviously, q_j and $T_{i,old}(S_j)$ are known to SU_i. Because the marginal distribution of the Dirichlet distribution is a beta distribution, we have

$$\mu_j \sim Beta\left(m_j + \alpha_j, \sum_{i=1, i \neq j}^{M} m_i + \alpha_i\right)$$

$$P_{ij}\left\{\mu_j \leq \frac{2[d_i - T_{i,old}(S_j) - T_{i,new}(S_j)]}{\lambda\{E(S_j^2)(q_j + 1) + 2[d_i - T_{i,old}(S_j) - T_{i,new}(S_j)]E(S_j)\}}\right\}$$

$$= I_x\left(m_j + \alpha_j, \sum_{i=1, i \neq j}^{M} m_i + \alpha_i\right)\Bigg|_{x = \frac{2[d_i - T_{i,old}(S_j) - T_{i,new}(S_j)]}{\lambda\{E(S_j^2)(q_j+1) + 2[d_i - T_{i,old}(S_j) - T_{i,new}(S_j)]E(S_j)\}}} \tag{10.32}$$

where $I_x(\cdot,\cdot)$ is the regularized incomplete beta function [7]. In Equation 10.32, we integrate queuing theory with the fully Bayesian model to derive the closed-form result of the delay-based utility function for multimedia transmission.

10.6.2 Delay- and Throughput-Based Utility Function

Because each SU has opportunity to access an unused subcarrier, the utility function $u_i(S)$ of SU_i is also influenced by other SUs that select the same subcarrier and form a packet queue. The QoS of multimedia transmission depends on the delay and throughput performance that the SU can achieve on the selected subcarriers. Our utility function is oriented on the delay and throughput performance as

$$u_i(s_i, s_{-i}) = \theta_i \sum_{j=1}^{M} s_{ij} P_{ij}\{Delay_{ij} \le d_i\} + (1-\theta_i)\sum_{j=1}^{M} s_{ij}\eta_{ij}(1-p_{ij})$$

$$= \sum_{j=1}^{M} [\theta_i P_{ij}\{Delay_{ij} \le d_i\} + (1-\theta_i)(1-p_{ij})\eta_{ij}]s_{ij}$$

$$(10.33)$$

subject to

$$
\begin{cases}
(a) \; \mu_j \lambda E(S_j) + s_{ij}(R_i/L_i)\overline{T}_{i,\text{new}}(S_j) \le 1-(p_j^{\text{ON}} + p_j^{\text{OFF}} p_j^f) & \text{(subcarrier capacity constraint)} \\[2mm]
(b) \; \tilde{I}_j = P_{ij}\sigma_{ij}^2(1-\gamma_j) \le P_I & \text{(interference temperature constraint)} \\[2mm]
(c) \; \sum_j P_{ij}(t) \le P_i & \text{(power constraint)}
\end{cases}
$$

$$(10.34)$$

In Equation 10.33, the first part $\sum_{j=1}^{M} s_{ij} P_{ij}\{Delay_{ij} \le d_i\}$ represents the delay-based utility function, and the second part $\sum_{j=1}^{M} s_{ij}\eta_{ij}(1-p_{ij})$ represents the throughput-based utility function. P_{ij} is the bit error rate (BER) when SU_i transmits packets over subcarrier j with power P_{ij}, θ_i is the weight coefficient for the trade-off between the delay-based utility function and the throughput-based utility function, η_{ij} is the normalized throughput that SU_i can obtain over subcarrier j (see Equation 10.37 in Section 10.7), L_i is the average packet length, and $\overline{T}_{i,\text{new}}(S_j)$ is the average transmission time of each new packet allocated on subcarrier j. We assume that the stationary statistics of PUs' traffic patterns can be modeled by the SUs. The probability of activation (ON state) is denoted as p_j^{ON}, and the probability of inactivation (OFF state) is $p_j^{\text{OFF}} = 1 - p_j^{\text{ON}}$. We also consider the uncertainty of raw sensing information (RSI); the correct detection probability is p_j^d, and the false alarm probability is p_j^f. Hence, the probability of packet loss due to PUs' activity and false alarm is $p_j^{\text{ON}} + p_j^{\text{OFF}} p_j^f$, which also determines the amount of available time to SUs in each second and the subcarrier capacity constraint in Equation 10.34. ρ_{ij} is the frequency domain CIR between SU_i and the PU that works on subcarrier j.

Because the RSI contains noise, we use γ_j, which represents the confidence level that the SU believes the jth subcarrier is available [8], and

$$\gamma_j = \frac{p_j^{OFF} P(\Lambda_j / V_j)}{p_j^{OFF} P(\Lambda_j / V_j) + p_j^{ON} P(\Lambda_j / \overline{V}_j)}$$

$$= \frac{p_j^{OFF}(1 - p_j^f)}{p_j^{OFF}(1 - p_j^f) + p_j^{ON}(1 - p_j^d)} \qquad (10.35)$$

where V_j and $\overline{V}_j$ represent that the PU is in the OFF and ON state, respectively, and Λ_j represents the estimation result that PU's state is also OFF status based on RSI. Because the confidence level of the spectrum sensing results is not 100% owing to the presence of RSI noise, we can mitigate the interference to the PU by limiting the transmitting power [9].

In the next section, we propose an intelligent cross-layer scheduling scheme that maximizes Equation 10.33 under the constraints given in Equation 10.34 to obtain the best QoS performance for multimedia transmission.

10.7 Cognitive Cross-Layer Scheduling Scheme

The proposed cognitive cross-layer scheduling scheme considers the physical process of packet transmission, including subcarrier selection, power and modulation allocation, and the number of packets simultaneously transmitted over multiple subcarriers. We derive the optimal subcarrier selection and the power and modulation allocation schemes for each packet based on the observed traffic information of other SUs. In Equation 10.33, we cannot generate the optimal subcarrier selection strategy s_{ij} directly to maximize the utility function, without a high-complexity algorithm. In fact, we only need to select a specific subcarrier, along with the power and modulation values for each packet based on the learnt traffic profile. In Equation 10.32, $\{T_{i,old}(S_j), q_j\}$ are known to SU_i, and $\{\lambda, E(S_j), E(S_j^2)\}$ can be detected. Hence, we propose our cognitive cross-layer scheduling scheme for optimal subcarrier selection and power and modulation allocation for each packet by maximizing the utility function in Equation 10.33 under the constraint given in Equation 10.18 as

$$Maximize\{u_i(s_i, s_{-i})\} = \underset{j, P_{ij}}{Maximize}\{\theta_i P_{ij}\{Delay_{ij} \leq d_i\} + (1 - \theta_i)(1 - p_{ij})\eta_{ij}\} \qquad (10.36)$$

Maximizing Equation 10.36 for each new packet will produce the maximum utility function value in Equation 10.33. In Equation 10.36, we do not show the power consumption for obtaining the best QoS performance. Because the channel gain would vary with different subcarriers over the inter-symbol interference channel, we should consider sharing the total available power P_i among multiple subcarriers in order to maximize the utility function in Equation 10.36. Because packets may be transmitted simultaneously over different subcarriers, it is important to allocate the limited power over these multiple subcarriers to increase the throughput.

In Equation 10.34, the sum of the transmit power over multiple subcarriers should not exceed the available power P_i at any time. We assume that each packet uses only one modulation scheme during transmission. Because the transmission rate requirement of SU_i is R_i (bit/s) and

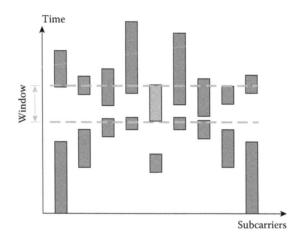

Figure 10.6   The scenario of packets transmitted over multiple subcarriers simultaneously.

the packet length is L_i (bits), while transmitting a new packet (the pink block in Figure 10.6) over a subcarrier, several other packets also share the limited power. Hence, it is important to determine the average number of packets N_i sharing the limited power when a new packet is being transmitted. We assume that the transmission rate of each subcarrier is B (bit/s), where B depends on the subcarrier bandwidth W and the modulation scheme. For example, when a new packet uses BPSK modulation, the time window size in Figure 10.6 is L_i/B (second), and thus $N_i = (L_i/B) \times (R_i/L_i) = R_i/B$. Considering the modulation schemes used in IEEE 802.11a (constellation using Gray code), we only use BPSK, QPSK, 16QAM, and 64QAM in this chapter [10], and N_i for these modulation schemes would be R_i/B, $R_i/2B$, $R_i/4B$, and $R_i/6B$ in that order.

When the normalized throughput is 1, no packet is dropped owing to insufficient power and bandwidth. However, a multimedia application may prefer a stringent delay constraint to high throughput (corresponding to a high value of the preference parameter θ_i), or the available power may not support simultaneous transmission of N_i packets. Hence, the actual number of simultaneously transmitted packets n_{ij} in a time window may be smaller than N_i; that is, $0 \le n_{ij} \le N_i$, and the normalized throughput is defined as

$$\eta_{ij} = \frac{n_{ij}}{N_i} \qquad (10.37)$$

In fact, the number of packets sharing the limited power would change from time to time, and we assume it obeys the Poisson distribution with the mean value N_i [4]. Hence, n_{ij} is determined by the selected subcarrier j (it directly determines the power consumption) and the total available power P_i, and it is the maximum number of packets that can be transmitted simultaneously. Note that the new transmitting packet shares the limited power with the previous $n_{ij} - 1$ packets, where the power assignment results for these previous $n_{ij} - 1$ packets are known to SU_i.

Moreover, under a constant BER requirement $p_i = p_{ij}$ for $j = 1,2, \ldots, M$, the required power is constant for each modulation scheme (assume the noise power in each subcarrier P_{noise} is 1), which is given in Table 10.1 [10], where $Q(x) = \int_x^{+\infty} e^{-\frac{t^2}{2}} dt$, and H_j is the frequency domain CIR on subcarrier j.

Table 10.1   Power Requirement for Each Modulation Constellation

BPSK	QPSK	16QAM	64QAM
$P_{ij} = \dfrac{1}{2\|H_j\|^2} \times \left[Q^{-1}(p_i)\right]^2$	$P_{ij} = \dfrac{1}{\|H_j\|^2}\left[Q^{-1}(p_i)\right]^2$	$P_{ij} = \dfrac{5}{\|H_j\|^2} \times \left[Q^{-1}\left(\dfrac{4}{3}p_i\right)\right]^2$	$P_{ij} = \dfrac{5}{\|H_j\|^2} \times \left[Q^{-1}\left(\dfrac{12}{7}p_i\right)\right]^2$

Table 10.2   Steps of the Proposed Cognitive Cross-Layer Scheduling Scheme

Given: $\theta_i, d_i, P_i, A_{K \times M}, p_{ij}, \sigma_{ij}^2, P_1, p_j^{ON}, p_j^d, p_j^f, T_{i,\text{old}}(S_j)$, and q_j, where $j = 1,2,\ldots,M$.

Observing: Observe the subcarrier selection results of K packets $A_{K \times M}$, the packet arrive rate λ, and the first and second moments of the packet transmission time over each subcarrier j ($E(S_j)$ and $E(S_j^2)$).

Learning: Use our strategy profile learning scheme (Section 10.4) to update the distribution of μ and the Dirichlet parameter α.

Action: On the basis of the observed information and learning results, we find out the optimal subcarrier and the power and modulation allocation according to Equation 10.36 under the constraint Equation 10.34.

Here, we fix the total transmission power of all subcarriers. However, we use the dynamic power allocation among individual subcarriers (by selecting the optimal modulation scheme for each subcarrier) to maximize the QoS of multimedia. The relationship between the selected modulation and the required power is listed in Table 10.1. For each modulation, we get the maximum utility function in Equation 10.36 and the corresponding subcarrier after we go through all subcarriers, under the constraints in Equation 10.34. Then, we can get an optimal modulation scheme and power allocation after all modulation schemes are considered. The detailed steps of our proposed scheduling scheme are given in Table 10.2.

In the Initial step, SUi has the known parameters $\left\{d_i, T_{i,\text{old}}(S_j), q_j, \theta_i, P_i, p_i, \sigma_{ij}^2, P_1, p_j^{ON}, p_j^d, p_j^f\right\}$, in which $\{d_i, T_{i,\text{old}}(S_j), q_j\}$ will be used to estimate the delay performance (see Equation 10.32), and $\left\{\theta_i, P_i, p_i, \sigma_{ij}^2, P_1, p_j^{ON}, p_j^d, p_j^f\right\}$ will be used to allocate the optimal subcarrier and the power and modulation schemes (see Equations 10.34 and 36). In the Observing step, SUi receives the ACK broadcast information from CH, and gets the subcarrier selection results $A_{K \times M}$ of the virtual SU and its statistical values of packet transmission (i.e., $E(S_j)$, $E(S_j^2)$, and λ). These parameters are essential to estimate the delay performance (see Equation 10.32). In the Learning step, we use the observations $A_{K \times M}$ to update the subcarriers selection strategy of the virtual SU, and the detailed updating process has been derived in Section 10.4. In the Action step, after updating the estimation of the distribution of μ, packet arrive rate λ, and the first and second moments of the packet transmission time over each subcarrier j (i.e., $E(S_j)$ and $E(S_j^2)$), we can derive the delay performance (i.e., the first part of Equation 10.36) based on Equation 10.32. Please note that q_j and $T_{i,\text{old}}(S_j)$ are known, and $T_{i,\text{new}}(S_j)$ depends on the modulation selected. Moreover, we can obtain the throughput performance (i.e., the second part of Equation 10.36) based on Equation 10.37, where n_{ij} is the number of simultaneously transmitted packets and can be derived by exhaustive search. (In computer science, brute-force search or exhaustive search, also known as generate and test, is a very general problem-solving technique that consists of systematically enumerating all possible candidates for the solution and checking whether each candidate satisfies the problem's statement). n_{ij} depends on the available power P_i and the selected modulation (i.e., BPSK, QPSK, 16QAM, and 64QAM). Hence, based on the foregoing delay and throughput performance, we can obtain the optimal subcarrier and the power and modulation allocation for each new packet by maximizing Equation 10.36 under constraint Equation 10.34. More detailed derivations and explanations in Sections 10.3–10.7 can be found in our recent research work Ref. [12].

References

1. K. R. Chowdhury and M. D. Felice, Search: A routing protocol for mobile cognitive radio ad-hoc networks, *Computer Communications*, Vol. 32, No. 18, pp. 1983–1997, 2009.
2. W.-Y. Lee and I. F. Akyildiz, Optimal spectrum sensing framework for cognitive radio networks, *IEEE Transaction on Wireless Communications*, Vol. 7, No. 10, pp. 3845–3857, 2008.
3. A. Dutta, D. Saha, D. Grunwald, and D. Sicker, SMACK—A SMart ACKnowledgment scheme for broadcast messages in wireless networks, *Proceedings of ACM SIGCOMM2009*, pp. 1–12, August 2009.
4. H.-P. Shiang and M. van der Schaar, Queuing-based dynamic channel selection for heterogeneous multimedia applications over cognitive radio networks, *IEEE Transactions on Multimedia*, Vol. 10, No. 5, pp. 896–909, 2008.
5. J. C. Gittins, *Multi-armed Bandit Allocation Indices*, John Wiley, Chichester, 1989, p. 77.
6. C.-S. Chang, D.-S. Lee, and C.-L. Yu, Generalization of the Pollaczek-Khinchin formula for throughput analysis of input-buffered switches, *Proceedings of IEEE INFCOM 2005*, pp. 960–970, March 2005.
7. http://en.wikipedia.org/wiki/Beta_distribution
8. R. Wang, K. Vincent, N. Lau, L. Lv, and B. Chen, Joint cross-layer scheduling and spectrum sensing for OFDMA cognitive radio systems, *IEEE Transactions on Wireless Communications*, Vol. 8, No. 5, pp. 2410–2416, 2009.
9. D. Gözüpek, S. Buhari, and F. Alagöz, A spectrum switching delay aware scheduling algorithm for centralized cognitive radio networks, *IEEE Transactions on Mobile Computing*, Vol. 12, No. 7, pp. 1270–1280, 2012.
10. X.-L. Huang, G. Wang, Y.-K. Ma, and C.-W. Zhang, An efficient bit loading algorithm for OFDM system, *Journal of Harbin Institute of Technology*, Vol. 42, No. 9, pp. 1379–1382, 2010.
11. E. B. Fox, *Bayesian Nonparametric Learning of Complex Dynamical Phenomena*, Doctoral Thesis, Massachusetts Institute of Technology, July 2009.
12. X.-L. Huang, G. Wang, F. Hu, S. Kumar, and J. Wu, Multimedia over cognitive radio networks: towards a cross-layer scheduling under Bayesian traffic learning, *Computer Communications*, Vol. 51, No. 2014, pp. 48–59, 2014.

Chapter 11

Hierarchical Dirichlet Process for Cognitive Radio Networks

Xin-Lin Huang and Jun Wu

School of Electronics and Information, Tongji University, Shanghai, China

Contents

11.1 Dirichlet Processes

In probability theory, a Dirichlet process (DP) is a random process, that is, a probability distribution whose domain is itself a set of probability distributions. Because of this hierarchical structure of distribution over distributions, we will in the following text use the terminology "draw from" a distribution, as distinct from "draw of" a distribution. In the former terminology, the distribution mentioned governs the relative chances of different outcomes; in the latter, the distribution mentioned is the random outcome.

Given a DP(H, α), where H (the base distribution or base measure) is an arbitrary distribution and α (the concentration parameter) is a positive real number, a draw from DP will return a random distribution (the output distribution) over some of the values that can be drawn from H. That is, the support of each draw of the output distribution is always a subset of the support of the base distribution. The output distribution will be discrete, meaning that individual values drawn from the output distribution will sometimes repeat themselves even if the base distribution is continuous (i.e., if two different draws from the base distribution will be distinct from the probability one). The extent to which values will repeat is determined by α, with higher values causing less repetition.

Note that the DP is a stochastic process, meaning that technically speaking it is an infinite sequence of random variables, rather than a single random distribution. The relation between the two is as follows. Consider the DP as defined earlier, as a distribution over random distributions, and call this process $DP_1(\cdot)$. We can call this the distribution-centered view of the DP. First, draw a random output distribution from this process, and then consider an infinite sequence of random variables representing values drawn from this distribution. Note that, conditioned on the output distribution, the variables are independent and identically distributed. Now, consider instead the distribution of the random variables that results from marginalizing out (integrating over) the random output distribution. (This makes all the variables dependent on each other. However, they are still exchangeable, meaning that the marginal distribution of one variable is the same as that of all other variables. That is, they are "identically distributed" but not "independent.") The resulting infinite sequence of random variables with the given marginal distributions is another view of the DP, denoted here by $DP_2(\cdot)$. We can call this the process-centered view of the DP. The conditional distribution of one variable given all the others, or given all previous variables, is defined by the Chinese Restaurant Process.

11.1.1 Formal Definition

A DP over a set S is a stochastic process whose sample path (i.e., an infinite-dimensional set of random variates drawn from the process) is a probability distribution on S. The finite dimensional distributions are from the Dirichlet distribution: If H is a finite measure on S, α is a positive real number, and X is a sample path drawn from a DP, written as

$$X \sim DP(\alpha, H) \tag{11.1}$$

then for any measureable partition of S, say $\{B_i\}_{i=1}^n$, we have

$$\left(X(B_1),...,X(B_n)\right) \sim \text{Dirichlet}\left(\alpha H(B_1),...,\alpha H(B_n)\right) \tag{11.2}$$

11.1.2 The Chinese Restaurant Process

As shown earlier, a simple distribution, the so-called Chinese Restaurant Process, considers the conditional distribution of one component assignment given all previous ones in a Dirichlet distribution mixture model with K components, and then takes the limit as K goes to infinity. It can be shown, using the above formal definition of the DP and considering the process-centered view of the process, that the conditional distribution of the component assignment of one sample from the process given in all previous samples follows a Chinese restaurant process.

Suppose that J samples, $\left\{\theta_j\right\}_{j=1}^{J}$, have already been obtained. According to the Chinese restaurant process, the $(J+1)^{\text{th}}$ sample should be drawn from

$$\theta_{J+1} \sim \frac{1}{H(S)+J}\left(H+\sum_{j=1}^{J}\delta_{\theta_j}\right) \tag{11.3}$$

where δ_θ is an atomic distribution centered on θ. Interpreting this, two properties are clear.

Even if S is a countable set, there is a finite probability that two samples will have exactly the same value. Samples from a DP are therefore discrete.

The DP exhibits a self-reinforcing property; the more often a given value has been sampled in the past, the more likely it is to be sampled again.

The name "Chinese restaurant process" is derived from the following analogy: imagine an infinitely large restaurant containing an infinite number of tables, and able to serve an infinite number of dishes. The restaurant in question operates a somewhat unusual seating policy whereby new diners are seated either at a currently occupied table with probability proportional to the number of guests already seated there, or at an empty table with probability proportional to a constant. Guests who sit at an occupied table must order the same dish as those currently seated, whereas guests allocated a new table are served a new dish at random. The distribution of dishes after J guests are served is a sample drawn as described above.

11.2 Hierarchical Dirichlet Process

There are many scenarios in which groups of data are thought to be produced by related, yet distinct, generative processes. For example, take a sensor network monitoring an environment where time-varying conditions may influence the quality of the data. Data collected under certain conditions should be grouped and described by a similar but different model from that of other data. The hierarchical Dirichlet process extends the DP to such scenarios by taking a hierarchical Bayesian approach: the group-specific distributions G_j, with

$$G_j \mid G_0, \alpha \sim \mathrm{DP}(\alpha, G_0) \tag{11.4}$$

are tied together via a global base measure G_0, which is itself given a DP prior:

$$G_0 \mid H, \gamma \sim \mathrm{DP}(\gamma, H) \tag{11.5}$$

for every $A \subset \Theta$,

$$\mathrm{E}\left[G_j(A) \mid G_0\right] = G_0(A) \tag{11.6}$$

In this sense, we can interpret G_0 as an "average" distribution across all groups. In the following text, we demonstrate that this specific choice of hierarchy implies that atoms are shared not only within groups but also between groups, as desired.

The marginal probabilities obtained from integrating over the random measures G_0 and G_j can be described in terms of a *Chinese restaurant franchise* (CRF), that is, an analog of the Chinese restaurant process. In the CRF, the metaphor of the Chinese restaurant process is extended to allow multiple restaurants that share a set of dishes.

The metaphor is as follows (see Figure 11.1). We have a restaurant franchise with a shared menu across the restaurants. At each table of each restaurant, a certain dish is ordered from the menu by the first customer who sits there, and it is shared among all customers at that table. Multiple tables in multiple restaurants can serve the same dish.

Each restaurant is represented by a rectangle. Customers (θ_{ji}'s) are seated at tables (circles) in the restaurants. At each table a dish is served. The dish is served from a global menu (ϕ_K), whereas the parameter ψ_{jt} is a table-specific indicator that serves to index items on the global menu. The customer θ_{ji} sits at the table to which it has been assigned as shown in Equation 11.7.

In this setup, the restaurants correspond to groups, and the customers correspond to the factors θ_{ji}. We also let $\phi_1,...,\phi_K$ denote K independently and identically distributed (i.i.d.) random variables distributed according to H; this is the global menu of dishes. We also introduce variables ψ_{jt} which represent the table-specific choice of dishes; in particular, ψ_{jt} is the dish served at table t in restaurant j.

Note that each θ_{ji} is associated with one ψ_{jt}, while each ψ_{jt} is associated with one ϕ_K. We introduce indicators to denote these associations. In particular, let tji be the index of the ψ_{jt} associated with θ_{ji}, and let k_{jt} be the index of ϕ_K associated with ψ_{jt}. In the CRF metaphor, customer i in restaurant j sat at table t_{ji} while table t in restaurant j serves dish k_{jt}.

We also need a notation for counts. In particular, we need to maintain counts of customers and counts of tables. We use the notation n_{jtk} to denote the number of customers in restaurant j at table t eating dish k. Marginal counts are represented with dots. Thus, $n_{jt\cdot}$ represents the number of customers in restaurant j at table t and $n_{j\cdot k}$ represents the number of customers in restaurant

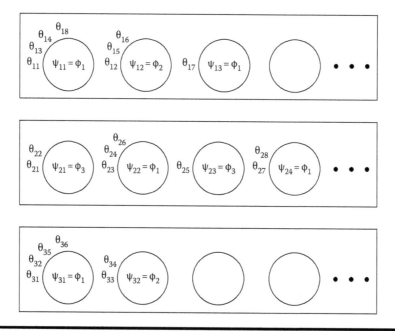

Figure 11.1   A depiction of a Chinese restaurant franchise.

j eating dish *k*. The notation m_{jk} denotes the number of tables in restaurant *j* serving dish *k*. Thus, $m_{j.}$ represents the number of tables in restaurant *j*, $m_{.k}$ represents the number of tables serving dish *k*, and *m* represents the total number of tables occupied.

Let us now compute marginals under a HDP when G_0 and G_j are integrated out. First, consider the conditional distribution for θ_{ji} give $\theta_{j1},...,\theta_{j,i-1}$ and G_0, where G_j is integrated out.

$$\theta_{ji} \mid \theta_{j1}, \ldots, \theta_{j,i-1}, \alpha_0, G_0 \sim \sum_{t=1}^{mj.} \frac{n_{jt.}}{i-1+\alpha_0} \delta_{\psi jt} + \frac{\alpha_0}{i-1+\alpha_0} G_0 \tag{11.7}$$

This is a mixture, and a draw from this mixture can be obtained by drawing from the terms on the right-hand side with probabilities given by the corresponding mixing proportions. If a term in the first summation is chosen, then we set $\theta_{ji} = \psi_{jt}$ and let $t_{ji} = t$ for the chosen *t*. If the second term is chosen, then we increment $m_{j.}$ by one, draw $\psi_{jmj.} \sim G_0$, and set $\theta_{ji} = \psi_{jmj.}$ and $t_{ji} = m_{j.}$.

Now, we proceed to integrate out G_0. Note that G_0 appears only in its role as the distribution of the variables ψ_{jt}. Because G_0 is distributed according to a DP, we can integrate it out and write the conditional distribution of ψ_{jt} as

$$\psi_{jt} \mid \psi_{11}, \psi_{12}, \ldots, \psi_{21}, \ldots, \psi_{j,t-1}, \gamma, H \sim \sum_{k=1}^{K} \frac{m_{.k}}{m_{..} + \gamma} \delta_{\phi k} + \frac{\gamma}{m_{..} + \gamma} H \tag{11.8}$$

If we draw ψ_{jt} by choosing a term in the summation on the right-hand side of this equation, we set $\psi_{jt} = \phi_k$ and let $k_{jt} = k$ for the chosen *k*. If the second term is chosen, then we increment *K* by one, draw $\phi_k \sim H$, and set $\psi_{jt} = \phi_k$ and $k_{jt} = k$.

This completes the description of the conditional distributions of the θ_{ji} variables. To use these equations to obtain samples of θ_{ji}, we proceed as follows. For each *j* and *i*, first sample θ_{ji} using Equation 11.7. If a new sample from G_0 is needed, we use Equation 11.8 to obtain a new sample ψ_{jt} and set $\theta_{ji} = \psi_{jt}$.

Note that in the HDP, the values of the factors are shared between the groups, as well as within the groups. This is a key property of HDP.

11.3 Cluster-Based CRN

We consider cluster-based CRN as shown in Figure 11.2. The SUs are deployed over a wide area, and the clusters are formed based on metrics such as location, mobility, and so on (see our previous study [1] on CRN clustering strategies). This chapter assumes the same clustering criterion as in our previous study [1]. Therefore, we can assume that all cluster members share the same spectrum map because of their close proximity. However, owing to spectrum sensing noise and errors in each cluster member, there could be minor discrepancies among their sensing results. Thus, an efficient spectrum fusion algorithm is needed in order to reach a consensus in terms of the entire cluster's spectrum patterns. In the spectrum sensing stage, all SUs in the same cluster first individually operate compressive sensing (CS) sampling in a synchronization mode based on the spectrum management commands sent from the cluster head (CH). The CH then collects the local result from each cluster member to produce a fusion result, and then broadcasts the fused information to the CHs in other clusters.

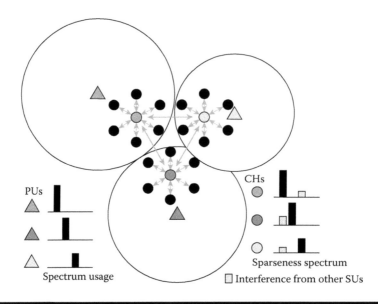

Figure 11.2   Cooperative, decentralized spectrum sensing for cluster-based CRN.

To realize our cooperative, decentralized spatiotemporal data mining, we assume the following:

1. Sampling synchronization throughout the CRN is not required, and synchronized sampling only occurs inside each cluster. When a cluster starts spectrum sensing, it may also receive the signal from neighboring clusters. Such a signal can be seen as interference from the standpoint of spectrum sensing. Hence, we need to identify which channels are occupied by the PUs, or by the SUs from the neighboring clusters.
2. The transmission power of PUs may not be high enough to cover the whole CRN area, and different clusters may have different spectrum occupancy status owing to the geography-dependent PUs. Hence, the sparseness spectrum is determined by the geography-dependent PUs as well as the distant SUs in other clusters.
3. The CS algorithm is adopted in each SU during spectrum sensing, because wideband spectrum sensing requires a high sampling rate and high cost on the analog-to-digital converter circuit according to the Nyquist sampling theorem.

11.4 Multitask Bayesian CS Modeling with Hierarchical Prior: One-Cluster Case

For simplicity, we first consider cooperative spectrum sensing in one cluster only, and then extend it to the multicluster case in the next section. In each cluster, the CH periodically broadcasts the spectrum sensing commands, and all member nodes execute signal sampling and exchange information with their CH. In the one-cluster case, we assume that all member nodes share a common sparseness spectrum. In this section, we propose to use multitask CS modeling with a hierarchical prior to detect spectrum holes in a cooperative manner. Unlike conventional cooperative spectrum sensing schemes [2–6], here we consider (1) an unknown channel impulse response (CIR) between PUs and SUs, (2) complex-valued sampling signals (including horizontal component and orthogonal component), and (3) automatic identification of PUs' and SUs' spectrum occupancy states based on spectrum assignment records from its neighboring CHs.

We assume that there are M subcarriers (also called *channels*) and N SUs in the target cluster. Through Nyquist sampling, the received signals in SU j can be represented as

$$r_j(n) = r_j^P(n) + r_j^C(n) + w_j(n), \quad n = 0,1,...,M-1 \tag{11.9}$$

where $r_j^P(n) = \sum_{i=1}^{I} h_{i,j}(n) * x_i(n)$ and $r_j^C(n) = \sum_i g_{i,j}(n) * y_i(n)$ correspond to the received signals from a total of I PUs and the interference from neighboring clusters, respectively. $h_{i,j}(n)$ is the CIR between PU i and SU j ($j = 1,2,...,N$) and (*) denotes convolution. $g_{i,j}(n)$ is the CIR between a neighboring cluster i and SU j. $x_i(n)$ and $y_i(n)$ correspond to the original transmitted signal from PU i and neighboring cluster i, respectively. $w_j(n)$ represents additive white Gaussian noise.

After applying the M-point DFT (discrete Fourier transform) to the observed signals, Equation 11.9 can be further rewritten as [2]

$$R_j(k) = \sum_{i=1}^{I} H_{i,j}(k)X_i(k) + \sum_i G_{i,j}(k)Y_i(k) + W_j(k) \tag{11.10}$$

where $H_{i,j}(k)$, $X_i(k)$, $G_{i,j}(k)$, $Y_i(k)$, and $w_j(k)$ ($k = 0,1,...,M-1$) are the complex-valued frequency-domain discrete versions of $h_{i,j}(n)$, $x_i(n)$, $g_{i,j}(n)$, $y_i(n)$, and $w_j(n)$, respectively. Owing to the sampling rate limitation of the ADC and the sparse nature of received signals in Equation 11.10, we use the CS technique in spectrum sampling as follows:

$$\upsilon_j = \Phi_j F_M F_M^{-1} R_j^P + \Phi_j F_M F_M^{-1} R_j^C + \Phi_j F_M F_M^{-1} W_j \tag{11.11}$$

where υ_j is a $m_j \times 1$ vector ($m_j << M$) of the CS observations, and $R_j^P = \sum_{i=1}^{I} H_{i,j}(k)X_i(k)$ is orthogonal to $R_j^C = \sum_i G_{i,j}(k)Y_i(k)$ (i.e., $(R_j^P)^T R_j^C = 0$). $\Phi_j F_M$ is the observation matrix, where Φ_j is a $m_j \times M$ matrix with elements constituted randomly [3], and F_M is the $M \times M$ DFT matrix [2]. In Equation 11.11, υ_j is the time-domain sampling signal and can be further rewritten as

$$\upsilon_j = \Phi_j R_j^P + \Phi_j R_j^C + \Phi_j W_j$$
$$= \Phi_j \cdot \text{Re}\{R_j^P + R_j^C + W_j\} + sqrt(-1) \cdot \Phi_j \cdot \text{Im}\{R_j^P + R_j^C + W_j\} \tag{11.12}$$

Because the *Re{•}* part is orthogonal to the *Im{•}* part and both of them have the same structure (see Equation 11.12), we only show the analysis of the *Re{•}* part here, and the *Im{•}* part can be analyzed in the same manner. Any analysis result from *Re{•}* can be symmetrically extended to the *Im{•}* case eventually. Hence, we have

$$\upsilon_j = \Phi_j \theta_j + \nu_j \tag{11.13}$$

where $\theta_j = \text{Re}\left\{R_j^P + R_j^C\right\}$ is an $M \times 1$ vector that represents the spectrum occupation states, v_j is a $m_j \times 1$ vector whose components are i.i.d. Gaussian variables. We employ a hierarchical multitask CS model for our cooperative spectrum sensing scheme based on Equation 11.13. Because v_j is *i.i.d.* draws of a zero-mean Gaussian distribution with unknown precision α_0, the likelihood function for the parameters θ_j and α_0, based on the observations v_j, can be represented as

$$p\{v_j \mid \theta_j, \alpha_0\} = (2\pi / \alpha_0)^{-m_j/2} \exp\left(-\frac{\alpha_0}{2}\left\|v_j - \Phi_j\theta_j\right\|_2^2\right) \tag{11.14}$$

where m_j is the number of measurements. It is much smaller than the Nyquist sampling rate M in Equation 11.9.

The parameters θ_j are assumed to be drawn from a product of zero-mean Gaussian distributions that are shared by the SUs in one cluster, and therefore the N tasks are statistically related to each other. Specifically, let $\theta_{j,k}$ represent the kth element of vector θ_j, we have

$$p\{\theta_j \mid \alpha, \alpha_0\} = \prod_{k=1}^{M} \text{N}\left(\theta_{j,k} \mid 0, \alpha_0^{-1}\alpha_k^{-1}\right) \tag{11.15}$$

where $\alpha = \{\alpha_1, \alpha_1, ..., \alpha_M\}$ is a hyper-parameter. To promote sparseness over θ_j, a gamma prior can be placed on the hyper-parameter α_0:

$$p\{\alpha_0 \mid a, b\} = Ga\left(\alpha_0 \mid a, b\right) = \frac{b^a}{\Gamma(a)}\alpha_0^{a-1}\exp\left(-b\alpha_0\right) \tag{11.16}$$

Thus, the posterior probability of θ_j based on the observed signals v_j and the hyper-parameter α can be represented as

$$\begin{aligned}
p\{\theta_j \mid v_j, \alpha\} &= \int p\left(\theta_j \mid v_j, \alpha, \alpha_0\right)p\left(\alpha_0 \mid a, b\right)d\alpha_0 \\
&= \int \frac{p\left(v_j \mid \theta_j, \alpha_0\right)p\left(\theta_j \mid \alpha, \alpha_0\right)}{\int p\left(v_j \mid \theta_j, \alpha_0\right)p\left(\theta_j \mid \alpha, \alpha_0\right)d\theta_j}p\left(\alpha_0 \mid a, b\right)d\alpha_0 \\
&= \frac{\Gamma(a + M / 2)\left[1 + \dfrac{1}{2b}\left(\theta_j - \mu_j\right)^T\Sigma_j^{-1}\left(\theta_j - \mu_j\right)\right]^{-(a+M/2)}}{\Gamma(a)(2\pi b)^{M/2}\left|\Sigma_j\right|^{1/2}}
\end{aligned} \tag{11.17}$$

where

$$\mu_j = \Sigma_j\Phi_j^T v_j \tag{11.18}$$

$$\Sigma_j = \left(\Phi_j^T\Phi_j + A\right)^{-1} \tag{11.19}$$

In Equation 11.19, A is a diagnose matrix ($A = diag\{\alpha_1, \alpha_1, ..., \alpha_M\}$). From Equations 11.17 to 11.19, in order to obtain the posterior probability of θ_j, we should first seek the point-value of the hyper-parameter α based on the N observations.

The maximum likelihood (ML) function of the observations from N SUs can be written as (the detailed proof is provided in the appendix):

$$\ell(\alpha) = \sum_{j=1}^{N} \log p(\upsilon_j \mid \alpha)$$

$$= \sum_{j=1}^{N} \log \int p(\upsilon_j \mid \theta_j, \alpha_0) p(\theta_j \mid \alpha, \alpha_0) p(\alpha_0 \mid a, b) d\theta_j \, d\alpha_0 \qquad (11.20)$$

$$= -\frac{1}{2} \sum_{j=1}^{N} \left[(m_j + 2a) \log(\upsilon_j^T B_j^{-1} \upsilon_j + 2b) + \log|B_j| \right] + Const$$

Here,

$$B_j = E + \Phi_j A^{-1} \Phi_j^T \qquad (11.21)$$

where E is the identity matrix. According to [4], B_j can be decomposed as

$$B_j = E + \Phi_j A^{-1} \Phi_j^T = E + \sum_{n \neq k} \alpha_n^{-1} \Phi_{j,n} \Phi_{j,n}^T + \alpha_k^{-1} \Phi_{j,k} \Phi_{j,k}^T \qquad (11.22)$$

$$= B_{j,-k} + \alpha_k^{-1} \Phi_{j,k} \Phi_{j,k}^T$$

where $B_{j,-k} \triangleq E + \sum_{n \neq k} \alpha_n^{-1} \Phi_{j,n} \Phi_{j,n}^T$ ($k = 1,2,\ldots,M$). Hence, the determinant and inversion of matrix B_j can be expressed as

$$|B_j| = |B_{j,-k}| \left| 1 + \alpha_k^{-1} \Phi_{j,k}^T B_{j,-k}^{-1} \Phi_{j,k} \right| \qquad (11.23)$$

$$B_j^{-1} = B_{j,-k}^{-1} - \frac{B_{j,-k}^{-1} \Phi_{j,k} \Phi_{j,k}^T B_{j,-k}^{-1}}{\alpha_k + \Phi_{j,k}^T B_{j,-k}^{-1} \Phi_{j,k}} \qquad (11.24)$$

Then, the contribution of the basis vector $\Phi_{j,k}$ in the likelihood function (Equation 11.20) can be separated from others, that is,

$$\ell(\alpha) = -\frac{1}{2} \sum_{j=1}^{N} \left[(m_j + 2a) \log(\upsilon_j^T B_{j,-k}^{-1} \upsilon_j + 2b) + \log|B_{j,-k}| \right] + Const$$

$$\qquad -\frac{1}{2} \sum_{j=1}^{N} \left[\log(1 + \alpha_k^{-1} s_{j,k}) + (m_j + 2a) \log\left(1 - \frac{q_{j,k}^2 / e_{j,k}}{\alpha_k + s_{j,k}} \right) \right] \qquad (11.25)$$

$$= \ell(\alpha_{-k}) + \ell(\alpha_k)$$

where $\alpha_{-k} \triangleq \{\alpha_1, \alpha_2, \ldots, \alpha_{k-1}, \alpha_{k+1}, \ldots, \alpha_M\}, s_{j,k}, q_{j,k}$, and $e_{j,k}$ are defined as

$$s_{j,k} \triangleq \Phi_{j,k}^T B_{j,-k}^{-1} \Phi_{j,k}$$

$$q_{j,k} \triangleq \Phi_{j,k}^T B_{j,-k}^{-1} \upsilon_j$$

$$e_{j,k} \triangleq \upsilon_j^T B_{j,-k}^{-1} \upsilon_j + 2b \tag{11.26}$$

To update α_k in each iteration, we fix the other hyper-parameter α_{-k} as the latest values, differentiate likelihood function $\ell(\alpha_k)$ with α_k, and set the result to zero:

$$\frac{\partial \ell(\alpha_k)}{\partial \alpha_k} = \sum_{j=1}^N \frac{s_{j,k}(s_{j,k} - q_{j,k}^2 / e_{j,k}) / \alpha_k - (m_j + 2a)q_{j,k}^2 / e_{j,k} + s_{j,k}}{2(\alpha_k + s_{j,k})(\alpha_k + s_{j,k} - q_{j,k}^2 / e_{j,k})} = 0 \tag{11.27}$$

Because α_k is the precision of the Gaussian distribution, we have $\alpha_k > 0$. We assume $\alpha_k \gg s_{j,k}$ (this is an empirical result stated in Ref. [3]), and thus $\alpha_k \gg s_{j,k} \approx s_{j,k}$ in Equation 11.27. Then, we can derive the new α_k from Equation 11.27 as

$$\text{if } \sum_{j=1}^N \frac{(m_j + 2a)q_{j,k}^2 / e_{j,k} - s_{j,k}}{s_{j,k}\left(s_{j,k} - q_{j,k}^2 / e_{j,k}\right)} > 0$$

$$\alpha_k \approx \frac{N}{\displaystyle\sum_{j=1}^N \frac{(m_j + 2a)q_{j,k}^2 / e_{j,k} - s_{j,k}}{s_{j,k}\left(s_{j,k} - q_{j,k}^2 / e_{j,k}\right)}} \tag{11.28}$$

else

$$\alpha_k = \infty \tag{11.29}$$

Hence, the SU calculates $\dfrac{(m_j + 2a)q_{j,k}^2 / e_{j,k} - s_{j,k}}{s_{j,k}\left(s_{j,k} - q_{j,k}^2 / e_{j,k}\right)}$ in each iteration and broadcasts the value to its CH. From Equations 11.28 and 11.29, we can update the hyper-parameter α_k ($k = 1, 2, \ldots, M$) after each iteration. After reaching the upper bound of the iteration times, or if the increment value of the likelihood in Equation 11.25 is less than a threshold (which means that we almost reach the maximum value of the likelihood), the CH obtains the spectrum decision for its cluster. In Equation 11.29, $\alpha_k = \infty$ means $\theta_k = 0$ ($j = 1, 2, \ldots, N$), and the subcarrier k is available to SUs.

From the foregoing analysis, one can see that: (1) the member nodes in one cluster seek a consensus spectrum map based on the multitask Bayesian CS model, and (2) the information exchanged among member nodes can be used to derive the shared hyper-parameter $\alpha = \{\alpha_1, \alpha_2, \ldots, \alpha_M\}$. An advantage of our proposed hierarchical prior (see Equations 11.15 and 11.16) is that the spatial contribution from all member nodes is collected in order to derive the common sparseness spectrum, thus removing ISI channel fading.

After several iterations, the result $\alpha = \{\alpha_1, \alpha_2, \ldots, \alpha_M\}$ will converge, and we can then make a binary spectrum decision regarding d_{PU} and d_{SU} (to be discussed in Equations 11.65 and 11.66 next), which represents the spectrum occupancy states of the PUs and SUs.

11.5 Distributed Information Exchange and Spatiotemporal Data Mining: Multicluster Case

In this section, we will extend the foregoing one-cluster spectrum sensing case to a multicluster case, that is, the entire CRN with observations from different clusters. In Section 11.4, one cluster is assumed to share a common sparseness spectrum. However, the CRN may be deployed over a large-scale area, and the sparseness spectrum decisions may vary in different positions owing to the geography-dependent PUs and signal attenuation along a path. Hence, different clusters may not be statistically interrelated with each other, and the CS observations from different clusters may not be appropriate for sharing. For example, one cluster may be located near a high TV tower (base station), which makes fewer IEEE 802.22 channels available for the SUs. In the multicluster case, we should design an efficient algorithm that first *separates* the CS observations from different clusters (multiple clusters' CS observations may belong to the same *group* as long as they obey the same spectrum statistics) and then uses the multitask Bayesian CS model (see Section 11.4) in each group to discover the common sparseness spectrum within each group. For this purpose, we introduce a DP prior to the hierarchical Bayesian CS model that has been discussed in Section 11.4. The DP prior [7] has shown the powerful capability of automatically classifying different samples into groups based on their statistical patterns. In our application, the DP prior will be used to realize both spectrum grouping and CS inversion.

In the multicluster spectrum sensing case, different clusters may have different hyperparameters—that is, $\alpha_i = \{\alpha_{i1}, \alpha_{i2}, ..., \alpha_{iM}\}$—and the cluster ID $i = 1, 2, ..., C$ (C is the total number of clusters in the CRN). We assume $\{\alpha_i, i = 1, ..., C\}$ is drawn identically from distribution G, which is a random draw from the DP:

$$\alpha_i \mid G \sim G, \ i = 1, ..., C \tag{11.30}$$

$$G \sim GP(\lambda, G_0) \tag{11.31}$$

$$E(G) = G_0 \tag{11.32}$$

Equation 11.30 is the likelihood function for G, and the hyper-parameter α_i has been derived in the multitask Bayesian CS model (see Section 11.4). Equation 11.31 is the prior knowledge of G.

When we integral out G according to Equations 11.30 and 11.31, α_i obeys the base distribution G_0. In our cluster-based CRN, when one cluster collects the hyper-parameter information $\alpha^{-i} = \{\alpha_1, \alpha_2, ..., \alpha_{i-1}, \alpha_{i+1}, ..., \alpha_c\}$ from other clusters, the base distribution G_0 is updated, and we have [8]

$$p(\alpha_i \mid \alpha^{-i}, \lambda, G_0) = \frac{\lambda}{\lambda + C - 1} G_0 + \frac{1}{\lambda + C - 1} \sum_{k=1, k \neq i}^{C} \delta_{\alpha_k} \tag{11.33}$$

where δ_{α_k} represents a mass point concentrate at α_k with probability $1/(\lambda + C - 1)$. $\{\tilde{\alpha}_k\}_{k=1}^{K}$ ($K \leq C$) represents a set of distinct hyper-parameters in $\{\alpha_k\}_{k=1}^{C}$. We assume that there are n_k^{-i} number of clusters that choose $\tilde{\alpha}_k$ in $\{\tilde{\alpha}_k\}_{k=1}^{K}$. Then Equation 11.33 can be further written as

$$p(\alpha_i \mid \alpha^{-i}, \lambda, G_0) = \frac{\lambda}{\lambda + C - 1} G_0 + \frac{1}{\lambda + C - 1} \sum_{k=1}^{K} n_k^{-i} \delta_{\tilde{\alpha}_k} \tag{11.34}$$

Equation 11.34 clearly shows the important sharing property of the DP distribution: a new sample α_i prefers to select a group $\tilde{\alpha}_k$ with a large population n_k^{-i}.

In Equation 11.30, the distribution G can be generated by the stick-breaking process, which introduces two independent random variables π_k and α_k^* ($k = 1,2,\ldots,\infty$):

$$G = \sum_{k=1}^{\infty} \omega_k \delta_{\alpha_k^*} \tag{11.35}$$

where

$$\omega_k = \pi_k \prod_{n=1}^{k-1} (1 - \pi_n) \tag{11.36}$$

$$\pi_k \mid \lambda \sim Beta(1,\lambda) \tag{11.37}$$

$$\alpha_k^* \mid G_0 \sim G_0 \tag{11.38}$$

In Equations 11.35 and 11.36, π_k and α_k^* are drawn i.i.d. from a beta distribution (Equation 11.37) and base distribution G_0 (Equation 11.38), respectively. To promote sparseness over θ_j, we assume G_0 is a multiplication of gamma distribution [3]:

$$G_0 \sim \prod_{k=1}^{M} Ga(c,d) \tag{11.39}$$

In Equation 11.35, we can see that the number of mass points is infinite. However, the total number of unique values of α_k^* is finite. Hence, we can use the finite approximation to represent DP via a modified distribution G:

$$G = \sum_{k=1}^{J} l_k \delta_{\alpha_k^*} \tag{11.40}$$

where l_k represents the weight of mass point α_k^*, and $\sum_{k=1}^{J} l_k = 1$. Hence, Equation 11.30 can be rewritten as

$$p(\alpha_i \mid G) = p\left(\alpha_i \mid \{l_k\}_{k=1,J}, \{\alpha_k^*\}_{k=1,J}\right) = \sum_{k=1}^{J} l_k \delta_{\alpha_k^*} \tag{11.41}$$

where J is the number of unique values of the hyper-parameter, obviously, $J \leq C$. Further, $\{l_1, l_2,\ldots, l_J\}$ obeys the Dirichlet distribution:

$$\{l_1, l_2,\ldots, l_J\} \sim Dir(\omega_1, \omega_2,\ldots, \omega_J) \tag{11.42}$$

11.5.1 Automatically Grouping and Distributed Information Exchange

On the basis of the foregoing DP, the hidden model shown in Equation 11.13 can be defined as

$$\upsilon_j \mid \theta_j, \alpha_0 \sim N\left(\Phi_j \theta_j, \alpha_0^{-1} E\right), \quad j = 1, 2, \ldots, C$$

$$\theta_{j,k} \mid \alpha_{z_j,k}^* \sim N\left(0, \alpha_0^{-1} \alpha_{z_j,k}^{*-1}\right), \quad k = 1, 2, \ldots, M$$

$$\alpha_0 \sim Ga(a, b)$$

$$\alpha_j \mid \{l_k\}_{k=1,J}, \{\alpha_k^*\}_{k=1,J} \sim \sum_{k=1}^{J} l_k \delta_{\alpha_k^*} \tag{11.43}$$

$$z_j \sim \text{Categorical}(l_1, l_2, \ldots, l_J)$$

$$\{l_1, l_2, \ldots, l_J\} \sim Dir(\omega_1, \omega_2, \ldots, \omega_J)$$

where z_j is an index variable to indicate which group the cluster j belongs to. In the DP model, we are interested in $\{\alpha_k^*\}_{k=1,J}$ and $\{z_j\}_{j=1,C}$, which are the required information for the spectrum decision (see Equations 11.65 and 11.66). The lower bound of the marginal log-likelihood function can be written as

$$\ell(\alpha^*, \omega) = \int\int q(z,l) \cdot \left[\log p\left(\upsilon, z, l \mid \alpha^*, \omega\right) - \log q(z, l)\right] dz\, dl$$

$$= \int\int q(z,l) \cdot \left[\log p\left(\upsilon, z, l \mid \alpha^*, \omega\right) - \log q(z, l)\right] dz\, dl$$

$$= \int\int q(l) \cdot \prod_{j=1}^{C} q(z_j) \cdot \left\{\log p(l \mid \omega) + \sum_{j=1}^{C}\left[\log p\left(z_j \mid l\right) + \log p\left(\upsilon_j \mid \alpha_{z_j}^*\right)\right]\right. \tag{11.44}$$

$$\left. -\log q(l) - \sum_{j=1}^{C} \log q(z_j)\right\} dz\, dl$$

We can use variational Bayesian (VB) inference, that is, a variational posterior distribution $q(\{z_j\}_{j=1,C}, \{l_k\}_{k=1,J}) = \prod_{j=1}^{C} q(z_j) \cdot q\left(\{l_k\}_{k=1,J}\right)$, to approximate the true posterior $p(\{z_j\}_{j=1,C}, \{l_k\}_{k=1,J} \mid \{\upsilon_j\}_{j=1,C})$ [7]. In Equation 11.44, estimates of α^* and ω can be obtained by maximizing the lower bound $\ell(\alpha^*, \omega)$ via the expectation-maximization (EM) algorithm as follows:

1. In the E-step, ω is estimated by maximizing $\ell(\alpha^*, \omega)$ given $\alpha^* = \{\alpha_k^*\}_{k=1,J}$ as the latest estimated values. Specifically, $q(l)$ and $q(z)$ are updated separately by maximizing the lower bound in Equation 11.44 given other $q(\cdot)$ values and α^*.

2. In the M-step, values of α^* are estimated by maximizing Equation 11.44 given the most current values of ω, $q(l)$ and $q(z)$. Let $\gamma_{j,k} = q(z_j = k)$, Equation 11.44 then becomes

$$\ell(\alpha^*) = \sum_{k=1}^{J} \ell_k(\alpha_k^*) \tag{11.45}$$

$$\ell_k(\alpha_k^*) = \sum_{j=1}^{C} \gamma_{j,k} \log p(\upsilon_j \mid \alpha_k^*)$$

$$= \sum_{j=1}^{C} \gamma_{j,k} \log \int p(\upsilon_j \mid \theta_j, \alpha_0) p(\theta_j \mid \alpha_k^*, \alpha_0) p(\alpha_0 \mid a, b) d\theta_j \, d\alpha_0 \tag{11.46}$$

$$= -\frac{1}{2} \sum_{j=1}^{C} \gamma_{j,k} \left[(m_j + 2a) \log(\upsilon_j^T B_{j,k}^{-1} \upsilon_j + b) + \log|B_{j,k}| \right] + const$$

where

$$B_{j,k} = E + \Phi_j A_k^{-1} \Phi_j^T \tag{11.47}$$

$$A_k = diag(\{\alpha_{k,j}^*\}_{j=1,M}) \tag{11.48}$$

From Equations 11.45 and 11.46, we can see that the elements of $\alpha^* = \{\alpha_1^*, \alpha_2^*, \ldots, \alpha_J^*\}$ are independent of each other, and can thus be obtained separately by maximizing Equation 11.46 in the M-step. To obtain the optimal values $\alpha^* = \{\alpha_1^*, \alpha_2^*, \ldots, \alpha_J^*\}$, we decompose the matrix $B_{j,k}$ in the same way as was done in Equation 11.22:

$$B_{j,k} = E + \sum_{t=1, t\neq n}^{M} \alpha_{k,t}^{*-1} \Phi_{j,t} \Phi_{j,t}^T + \alpha_{k,n}^{*-1} \Phi_{j,n} \Phi_{j,n}^T$$

$$= B_{j,k,-n} + \alpha_{k,n}^{*-1} \Phi_{j,n} \Phi_{j,n}^T \tag{11.49}$$

where $B_{j,k,-n}$ $(n = 1,2,\ldots,M)$ is used to denote the accumulated effects of $\{\alpha_{k,1}^*, \alpha_{k,2}^*, \ldots, \alpha_{k,n-1}^*, \alpha_{k,n+1}^*, \ldots, \alpha_{k,M}^*\}$. In Equation 11.49, we separate the contribution of $\alpha_{k,n}^*$ from other items. The matrix determinant and inverse identities in Equation 11.46 can be then rewritten as

$$|B_{j,k}| = |B_{j,k,-n}| |1 + \alpha_{k,n}^{*-1} \Phi_{j,n}^T B_{j,k,-n}^{-1} \Phi_{j,n}| \tag{11.50}$$

$$B_{j,k}^{-1} = B_{j,k,-n}^{-1} - \frac{B_{j,k,-n}^{-1} \Phi_{j,n} \Phi_{j,n}^T B_{j,k,-n}^{-1}}{\alpha_{k,n}^* B_{j,k,-n}^{-1} \Phi_{j,n}} \tag{11.51}$$

Substituting Equations 11.50 and 11.51 into Equation 11.46, we can further solve Equation 11.46 as

$$\ell_k\left(\alpha_k^*\right) = -\frac{1}{2}\sum_{j=1}^{C}\gamma_{j,k}\left[\left(m_j+2a\right)\log\left(\upsilon_j^T B_{j,k,-n}^{-1}\upsilon_j+b\right)+\log\left|B_{j,k,-n}\right|\right]+const$$

$$-\frac{1}{2}\sum_{j=1}^{C}\gamma_{j,k}\left[\log\left(1+\alpha_{k,n}^{*-1}s_{j,k,n}\right)+\left(m_j+2a\right)\log\left(1-\frac{q_{j,k,n}^2/e_{j,k,n}}{\alpha_{k,n}^*+s_{j,k,n}}\right)\right] \quad (11.52)$$

$$\triangleq \ell_k\left(\alpha_{k,-n}^*\right)+\ell_k\left(\alpha_{k,n}^*\right)$$

where

$$s_{j,k,n}\triangleq \Phi_{j,n}^T B_{j,k,-n}^{-1}\Phi_{j,n}$$

$$q_{j,k,n}\triangleq \Phi_{j,n}^T B_{j,k,-n}^{-1}\upsilon_j$$

$$e_{j,k,n}\triangleq \upsilon_j^T B_{j,k,-n}^{-1}\upsilon_j+2b \quad (11.53)$$

Equation 11.52 indicates the dependence of $\ell_k(\alpha_k^*)$ on the hyper-parameter $\alpha_{k,n}^*$, which can be isolated from all other parameters $\alpha_{k,-n}^*$. We assume that cluster j has N_j nodes, and the sub-Nyquist sampling rates of these N_j nodes are $\{m_{j,1},m_{j,2},\ldots,m_{j,N_j}\}$. Hence, $\ell_k(\alpha_{k,n}^*)$ in Equation 11.52 can be further rewritten as

$$\ell_k\left(\alpha_{k,n}^*\right)=-\frac{1}{2}\sum_{j=1}^{C}\gamma_{j,k}\sum_{l=1}^{N_j}\left[\log\left(1+\alpha_{k,n}^{*-1}s_{j,l,k,n}\right)+\left(m_{j,l}+2a\right)\log\left(1-\frac{q_{j,l,k,n}^2/e_{j,l,k,n}}{\alpha_{k,n}^*+s_{j,l,k,n}}\right)\right] \quad (11.54)$$

According to [7], $\lambda_{j,k}$ and $(\omega_1,\omega_2,\ldots,\omega_J)$ can be updated in the E-step for cooperative CS inversion. We extend [7] to our multicluster CRN application, and consider the contributions of all member nodes in each cluster. We assume that all member nodes in cluster j share the same membership $\lambda_{j,k}$ ($k=1,2,\ldots,J$). Then, we have [7]

$$\gamma_{j,k}=\frac{\sum_{l=1}^{N_j}e^{r_{j,l,k}}}{\sum_{m=1}^{J}\sum_{l=1}^{N_j}e^{r_{j,l,m}}} \quad (11.55)$$

$$\omega_k=\frac{1}{J}+\sum_{j=1}^{C}\gamma_{j,k} \quad (11.56)$$

where

$$r_{j,l,k} = \left[\psi(\omega_k) - \psi\left(\sum_{m=1}^{J} \omega_m\right) \right]$$

$$- \frac{1}{2}\left[(m_{j,l} + 2a)\log\left(\upsilon_{j,l}^T B_{j,l,k}^{-1} \upsilon_{j,l} + b\right) + \log\left|B_{j,l,k}\right| \right] \tag{11.57}$$

$$\psi(x) = \frac{\partial \log \Gamma(x)}{\partial x} \tag{11.58}$$

The maximization of $\ell_k\left(\alpha_{k,n}^*\right)$ in Equation 11.54 $\left(\text{i.e., } \dfrac{\partial \ell_k\left(\alpha_{k,n}^*\right)}{\partial \alpha_{k,n}^*} = 0\right)$ cannot be solved directly in a closed-loop format because the denominator of each factor is a second-order polynomial of $\alpha_{k,n}^*$. Moreover, the entire equation is the sum of $\displaystyle\sum_{j=1}^{C} N_j$ factors. Hence, we will obtain a complex equation on the order of $2\left(\displaystyle\sum_{j=1}^{C} N_j - 1\right) + 1 = 2\displaystyle\sum_{j=1}^{C} N_j - 1$, which cannot be solved in a closed loop. As with Equations 11.28 and 11.29, here too we assume that $\alpha_{k,n}^* \ll s_{j,l,k,n}$ (this is an empirical result stated in Ref. [3]). By maximizing $\ell_k(\alpha_{k,n}^*)$ in the M-step, we obtain

$$\text{if } \sum_{j=1}^{C} \gamma_{j,k} \sum_{l=1}^{N_j} \frac{(m_{j,l} + 2a)q_{j,l,k,n}^2 / e_{j,l,k,n} - s_{j,l,k,n}}{s_{j,l,k,n}\left(s_{j,l,k,n} - q_{j,l,k,n}^2 / e_{j,l,k,n}\right)} > 0$$

$$\alpha_{k,n}^* \approx \frac{\displaystyle\sum_{j=1}^{C} N_j \gamma_{j,k}}{\displaystyle\sum_{j=1}^{C} \gamma_{j,k} \sum_{l=1}^{N_j} \frac{(m_{j,l} + 2a)q_{j,l,k,n}^2 / e_{j,l,k,n} - s_{j,l,k,n}}{s_{j,l,k,n}\left(s_{j,l,k,n} - q_{j,l,k,n}^2 / e_{j,l,k,n}\right)}} \tag{11.59}$$

else

$$\alpha_{k,n}^* = \infty \tag{11.60}$$

Hence, a CH exchanges its fusion result, $\gamma_{j,k} \displaystyle\sum_{l=1}^{N_j} \frac{(m_{j,l} + 2a)q_{j,l,k,n}^2 / e_{j,l,k,n} - s_{j,l,k,n}}{s_{j,l,k,n}\left(s_{j,l,k,n} - q_{j,l,k,n}^2 / e_{j,l,k,n}\right)}$, with other CHs in each iteration. In Equation 11.60, $\alpha_{k,n}^* = \infty$ means that channel n is unoccupied by PUs and other SUs. The detailed steps of our proposed algorithm are described in Table 11.1.

Jeff Wu [9] has pointed out that if the joint distribution of hidden variables belongs to the curved exponential family, then the EM algorithm can find a stationary value of the likelihood function. In our case, $p(z,l \mid \alpha^*, \omega) = p(z \mid l)\, p(l \mid \omega)$, where $p(z \mid l)$ follows the categorical

Table 11.1   The Detailed Steps of Our Proposed Algorithm

Our Proposed DP-Based Hierarchical Bayesian CS Algorithm

1. Initialize ω_k, $\alpha_{k,n}^*$ and the corresponding $\Phi_{j,n}$ ($k = 1,2,\ldots,J$, $j = 1,2,\ldots,C$, and $n = 1,2,\ldots,M$).

2. The member node l in cluster j updates $r_{j,l,k}$ according to Equations 11.57 and 11.58, and calculates $\dfrac{(m_{j,l} + 2a)q_{j,l,k,n}^2 / e_{j,l,k,n} - s_{j,l,k,n}}{s_{j,l,k,n}\left(s_{j,l,k,n} - q_{j,l,k,n}^2 / e_{j,l,k,n}\right)}$ based on its local observations. Those two values will be collected by the CH in cluster j.

3. The CH in cluster j updates the membership $\gamma_{j,k}$ and ω_k according to Equations 11.55 and 11.48, respectively.

4. For $k = 1,2,\ldots,J$, the CH in cluster j selects a candidate basis $\Phi_{j,l,n}$ and updates $\alpha_{k,n}^*$ according to Equations 11.59 and 11.60. Here, we choose the element $\alpha_{k,n}^*$ with the maximal increment $\Delta\ell_k\left(\alpha_{k,n}^*\right)$ (see Equation 11.54) in each iteration.

5. The CH broadcasts the fusion result, $\gamma_{j,k}\displaystyle\sum_{l=1}^{N_j} \dfrac{(m_{j,l} + 2a)q_{j,l,k,n}^2 / e_{j,l,k,n} - s_{j,l,k,n}}{s_{j,l,k,n}\left(s_{j,l,k,n} - q_{j,l,k,n}^2 / e_{j,l,k,n}\right)}$, to its neighboring CHs.

6. Check the algorithm terminating criterion, which could be (a) an upper bound of the iteration times or (b) the increment of $\ell(\alpha^*, \omega)$ in each iteration being less than a threshold. (*Note:* When $\ell(\alpha^*, \omega)$ cannot be increased much, that means we have almost reached the maximum of the likelihood. Then we can stop the iterations, because our goal is to seek the maximum likelihood). If either of them meets, then stop; otherwise, go back to step (2).

distribution and $p(l|\omega)$ follows the Dirichlet distribution. Because the categorical distribution and the Dirichlet distribution belong to the exponential family, and the Dirichlet distribution is the conjugate prior of the categorical distribution, $p(z,l \,|\, \alpha^*, \omega)$ should belong to the curved exponential family. Hence, our proposed EM algorithm will finally converge to a stationary point.

After the marginal log-likelihood function (see Equation 11.44) converges to a stationary point, we obtain $\left\{\alpha_k^*, k = 1,2,\ldots,J\right\}$ as well as the membership $\{\gamma_{j,k}, j = 1,2,\ldots, C, k = 1,2,\ldots,J\}$. In our proposed DP-based hierarchical Bayesian CS model, we update α_k^* by monotonically increasing the likelihood function $\ell(\alpha^*)$ in each iteration until the convergence is achieved.

In the foregoing discussions, we have fully exploited the *spatial* relationship among the CS observations from all clusters to infer the spectrum map. To further increase the accuracy of spectrum sensing decisions, we employ the HMM (see the following text) to exploit the time-domain relevance of subcarrier states, and select the most possible candidate α_k^* for the final spectrum decision.

11.5.2 Hidden Markov Model

A hidden Markov model (HMM) is a statistical Markov model in which the system being modeled is assumed to be a Markov process with unobserved (hidden) states. An HMM can be considered the simplest dynamic Bayesian network.

In simpler Markov models (such as a Markov chain), the state is directly visible to the observer, and therefore the state transition probabilities are the only parameters. In a hidden Markov model, the state is not directly visible, but output, dependent on the state, is visible. Each state has a probability distribution over the possible output tokens. Therefore, the sequence of tokens generated by an HMM gives some information about the sequence of states. Note that the adjective "hidden" refers to the state sequence through which the model passes, not to the parameters of the model; even if the model parameters are known exactly, the model is still "hidden."

In Figure 11.3, the relationship between hidden subcarriers states and CS observations is plotted. Because the subcarriers' states should be time relevant, only a small number of subcarriers change their binary states between two consequent CS observations.

Here, we consider time-domain relevance when assigning the final hyper-parameter α_k^* to each cluster for the time tth spectrum sensing. In Figure 11.3, the previous states are considered in the HMM as well as the final spectrum decision. The probability that cluster j selects $\alpha_k^*(t)$ as its hyper-parameter can be calculated, and the hyper-parameter that leads to the maximal probability will be selected as the final hyper-parameter, that is:

$$V_{\alpha_k^*(1)}(1) = p\left(\upsilon_j(1) \mid \alpha_k^*(1)\right) p\left(\alpha_k^*(1)\right) \tag{11.61}$$

$$V_{\alpha_k^*(t)}(t) = p\left(\upsilon_j(t) \mid \alpha_k^*(t)\right) \cdot \max_{\alpha_z^*(t-1)} p\left(\alpha_k^*(t) \mid \alpha_z^*(t-1)\right) V_{\alpha_z^*(t-1)}(t-1) \tag{11.62}$$

$$\alpha^*(t) = \arg\max_{\alpha_z^*(t)} \left(V_{\alpha_z^*(t)}(t)\right) \tag{11.63}$$

The Viterbi algorithm is a dynamic programming algorithm for finding the most likely sequence of hidden states—the Viterbi path—that results in a sequence of observed events, especially in the context of Markov information sources and hidden Markov models.

Equations 11.61 to 11.63 are the standard Viterbi algorithms with the recurrence calculations. The computation complexity of the Viterbi algorithm is $O(J^2)$. Here, $V_{\alpha_k^*(t)}(t)$ is the probability of the most probable hyper-parameter sequence that is responsible for the first t observations. Further, the corresponding $\alpha_k^*(t)$ is used for the final hyper-parameter to make a decision on subcarriers' states.

In Equations 11.61 and 11.62, $p\left(\upsilon_j(t) \mid \alpha_k^*(t)\right)$ equals the membership $\gamma_{j,k}$ in Equation 11.55. Moreover, the transition probability between two adjacent hidden hyper-parameters is considered as a first-order Markov model [10]. Note that we are only interested in the binary subcarrier state $d_{k,n}^*(t)$ that corresponds to two cases: $0 < \alpha_{k,n}^*(t) < \infty$ and $\alpha_{k,n}^*(t) = \infty$, instead of the exact value of $\alpha_{k,n}^*(t)$. Hence, the transition probability $p\left(\alpha_k^*(t) \mid \alpha_z^*(t-1)\right)$ can be further rewritten as

$$p\left(\alpha_k^*(t) \mid \alpha_z^*(t-1)\right) \propto p\left\{d_k^*(t) \mid d_z^*(t-1)\right\} \tag{11.64}$$

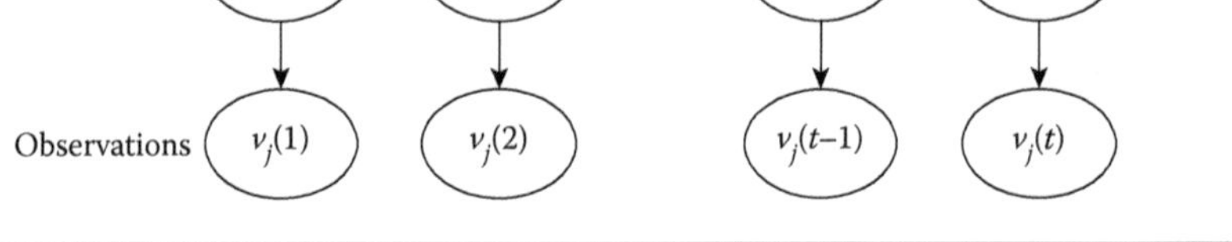

Figure 11.3 The relationship between hidden subcarriers states and CS observations.

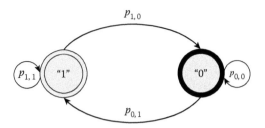

Figure 11.4 First-order Markov model for each subcarrier state.

The transition probability of the binary subcarrier state is described as a Markov model in Figure 11.4. Here, we assume the current subcarrier state (tth time) has a relationship only with the last subcarrier state (i.e., $(t-1)$th time). The nth element of $d_z^*(t-1)$ and $d_k^*(t)$, that is, $d_{z,n}^*(t-1)$ and $d_{k,n}^*(t)$, only has one state ("1" or "0"). In our first-order Markov model, we assume that the transition probabilities $p_{0,0}$, $p_{0,1}$, $p_{1,0}$, and $p_{1,1}$ are fixed [11–13] and can be detected by SUs.

After going through $\left\{\alpha_k^*(t)\right\}_{k=1,J}$ in Equation 11.62, we can obtain a candidate $\alpha_k^*(t)$ with the maximal value as the final hyper-parameter for cluster j. As we did in Section 11.4, here we again only consider the real part *Re{•}* in the mathematical analysis because the imaginary part *Im{•}* can be analyzed in the same manner. Hence, we can get two hyper-parameters $\alpha_k^*(t)$ and $\alpha_z^*(t)$ for the real part and the imaginary part, respectively. The final binary subcarriers state is determined by a threshold and those two hyper-parameters

if $\sqrt{(\alpha_{k,n}^*(t))^2 + (\alpha_{z,n}^*(t))^2} < threshold$

$$d_n^*(t) = 1 \tag{11.65}$$

else

$$d_n^*(t) = 0 \tag{11.66}$$

where $n \in \{1, 2, \ldots, M\}$.

11.5.3 Identify Spectrum Decision

After we apply the foregoing discussed spatiotemporal data mining scheme, we further employ the spectrum assignment records in neighboring CHs to identify which subcarriers are occupied by PUs, and the remaining subcarriers are regarded as the interference from the SUs in other clusters. The subcarriers temporally occupied by SUs in neighboring clusters are also spectrum opportunities, which can be accessed through negotiation or competition schemes among CHs. The binary spectrum decision $d_n^*(t)$ obtained earlier is a mixture of PUs' and SUs' occupancy state in each subcarrier. Each CH should send out the binary spectrum decision $d_n^*(t)$ and its corresponding sensing time to its neighboring CHs.

When the neighboring CHs receive $d_n^*(t)$ and the corresponding sensing time, they revise the subcarriers' state assigned for their data transmissions during the sensing time (i.e., change $d_n^*(t) = 1$ to $d_n^*(t) = 0$), and return the modified binary spectrum decision to the source CH. After exchanging the results with the neighboring CHs, a CH will obtain the final binary spectrum decision, which is determined only by the PUs' activities.

Appendix: The Detailed Steps for Equation 11.20

In Equation 11.20, the ML function can be rewritten as

$$
\ell(\alpha) = \sum_{j=1}^{N} \log p(\upsilon_j \mid \alpha)
$$

$$
= \sum_{j=1}^{N} \log \iint p(\upsilon_j \mid \theta_j, \alpha_0) p(\theta_j \mid \alpha, \alpha_0) p(\alpha_0 \mid a, b) \, d\theta_j \, d\alpha_0 \tag{11.67}
$$

$$
= \sum_{j=1}^{N} \log \int p(\alpha_0 \mid a, b) \left\{ \int p(\upsilon_j \mid \theta_j, \alpha_0) p(\theta_j \mid \alpha, \alpha_0) \, d\theta_j \right\} d\alpha_0
$$

where

$$
p(\upsilon_j \mid \alpha, \alpha_0)
$$

$$
= \int p(\upsilon_j \mid \theta_j, \alpha_0) p(\theta_j \mid \alpha, \alpha_0) \, d\theta_j
$$

$$
= \int \left(\frac{2\pi}{\alpha_0} \right)^{-\frac{m_j}{2}} \exp\left(-\frac{\alpha_0}{2} \left\| \upsilon_j - \Phi_j \theta_j \right\|_2^2 \right) \prod_{k=1}^{M} \left(\frac{2\pi}{\alpha_0 \alpha_k} \right)^{-\frac{1}{2}} \exp\left(-\frac{\alpha_0 \alpha_k}{2} \left\| \theta_{j,k} \right\|_2^2 \right) d\theta_j
$$

$$
= \left(\frac{2\pi}{\alpha_0} \right)^{-\frac{m_j}{2}} \prod_{k=1}^{M} \left(\frac{2\pi}{\alpha_0 \alpha_k} \right)^{-\frac{1}{2}} \int \exp\left[-\frac{\alpha_0}{2} \left(\upsilon_j - \Phi_j \theta_j \right)^T \left(\upsilon_j - \Phi_j \theta_j \right) \right] \cdot \exp\left(-\frac{\alpha_0}{2} \theta_j^T A \theta_j \right) d\theta_j
$$

$$
= \left(\frac{2\pi}{\alpha_0} \right)^{-\frac{m_j}{2}} \prod_{k=1}^{M} \left(\frac{2\pi}{\alpha_0 \alpha_k} \right)^{-\frac{1}{2}} \int \exp\left[-\frac{\alpha_0}{2} \left(\upsilon_j^T \upsilon_j - \upsilon_j^T \Phi_j \theta_j - \theta_j^T \Phi_j^T \upsilon_j + \theta_j^T \Phi_j^T \Phi_j \theta_j + \theta_j^T A \theta_j \right) \right] d\theta_j
$$

$$
= \left(\frac{2\pi}{\alpha_0} \right)^{-\frac{m_j}{2}} \prod_{k=1}^{M} \left(\frac{2\pi}{\alpha_0 \alpha_k} \right)^{-\frac{1}{2}} \int \exp\left\{ -\frac{\alpha_0}{2} \upsilon_j^T \upsilon_j - \frac{\alpha_0}{2} \left[\theta_j - \left(\Phi_j^T \Phi_j + A \right)^{-1} \Phi_j^T \upsilon_j \right]^T \left(\Phi_j^T \Phi_j + A \right) \right.
$$

$$
\left. \cdot \left[\theta_j - \left(\Phi_j^T \Phi_j + A \right)^{-1} \Phi_j^T \upsilon_j \right] + \frac{\alpha_0}{2} \upsilon_j^T \Phi_j \left(\Phi_j^T \Phi_j + A \right)^{-1} \Phi_j^T \upsilon_j \right\} d\theta_j
$$

$$
= \left\{ \left(\frac{2\pi}{\alpha_0} \right)^{-\frac{m_j}{2}} \prod_{k=1}^{M} \left(\frac{2\pi}{\alpha_0 \alpha_k} \right)^{-\frac{1}{2}} \int \exp\left\{ -\frac{\alpha_0}{2} \left[\theta_j - \left(\Phi_j^T \Phi_j + A \right)^{-1} \Phi_j^T \upsilon_j \right]^T \right] \left(\Phi_j^T \Phi_j + A \right) \right.
$$

$$
\left. \cdot \left[\theta_j - \left(\Phi_j^T \Phi_j + A \right)^{-1} \Phi_j^T \upsilon_j \right] \right\} d\theta_j \right\}
$$

$$
\cdot \exp\left\{ -\frac{\alpha_0}{2} \upsilon_j^T \left[E - \Phi_j \left(\Phi_j^T \Phi_j + A \right)^{-1} \Phi_j^T \right] \upsilon_j \right\} \tag{11.68}
$$

From Equation 11.63, one can see that $p(\upsilon_j \mid \alpha, \alpha_0)$ follows a zero-mean Gaussian distribution, with covariance matrix $\left\{\alpha_0\left[E - \Phi_j(\Phi_j^T\Phi_j + A)^{-1}\Phi_j^T\right]\right\}^{-1} = \dfrac{1}{\alpha_0}\left(E + \Phi_j A^{-1}\Phi_j^T\right)$. Hence, Equation 11.67 can be rewritten as

$$\ell(\alpha) = \sum_{j=1}^{N} \log \int p(\alpha_0 \mid a,b)\left\{\int p(\upsilon_j \mid \theta_j,\alpha_0)p(\theta_j \mid \alpha,\alpha_0)\,d\theta_j\right\}d\alpha_0$$

$$= \sum_{j=1}^{N} \log \int p(\alpha_0 \mid a,b)\left(\frac{2\pi}{\alpha_0}\right)^{-\frac{m_j}{2}} \frac{1}{\left|E + \Phi_j A^{-1}\Phi_j^T\right|^{\frac{1}{2}}} \exp\left[-\frac{\alpha_0}{2}\upsilon_j^T\left(E + \Phi_j A^{-1}\Phi_j^T\right)^{-1}\upsilon_j\right]d\alpha_0$$

$$= \sum_{j=1}^{N} \log\left\{(2\pi)^{-\frac{m_j}{2}} \frac{1}{\left|E + \Phi_j A^{-1}\Phi_j^T\right|^{\frac{1}{2}}} \int p(\alpha_0 \mid a,b)\alpha_0^{\frac{m_j}{2}} \exp\left[-\frac{\alpha_0}{2}\upsilon_j^T(E + \Phi_j A^{-1}\Phi_j^T)^{-1}\upsilon_j\right]d\alpha_0\right\}$$

$$= \sum_{j=1}^{N} \log\left\{(2\pi)^{-\frac{m_j}{2}} \frac{1}{\left|E + \Phi_j A^{-1}\Phi_j^T\right|^{\frac{1}{2}}} \int \frac{b^a}{\Gamma(a)}\alpha_0^{a-1+\frac{m_j}{2}} \exp\left\{-\alpha_0\left[b + \frac{1}{2}\upsilon_j^T\left(E + \Phi_j A^{-1}\Phi_j^T\right)^{-1}\upsilon_j\right]\right\}d\alpha_0\right\}$$

$$= \sum_{j=1}^{N} \log\left\{(2\pi)^{-\frac{m_j}{2}} \frac{1}{\left|E + \Phi_j A^{-1}\Phi_j^T\right|^{\frac{1}{2}}} \frac{b^a}{\Gamma(a)} \frac{1}{\left[b + \frac{1}{2}\upsilon_j^T\left(E + \Phi_j A^{-1}\Phi_j^T\right)^{-1}\upsilon_j\right]^{a+\frac{m_j}{2}}}\right\}$$

$$= \sum_{j=1}^{N} -\frac{m_j}{2}\log(2\pi) - \frac{1}{2}\log\left|E + \Phi_j A^{-1}\Phi_j^T\right| + \log\left(\frac{b^a}{\Gamma(a)}\right) - \left(a + \frac{m_j}{2}\right)\log\left[b + \frac{1}{2}\upsilon_j^T\left(E + \Phi_j A^{-1}\Phi_j^T\right)^{-1}\upsilon_j\right]$$

$$= -\frac{1}{2}\sum_{j=1}^{N}\left[(m_j + 2a)\log\left(\upsilon_j^T B_j^{-1}\upsilon_j + 2b\right) + \log\left|E + \Phi_j A^{-1}\Phi_j^T\right|\right]$$

$$+ \sum_{j=1}^{N}\left\{-\frac{m_j}{2}\log(2\pi) + \log\left(\frac{b^a}{\Gamma(a)}\right) + \left(a + \frac{m_j}{2}\right)\log 2\right\}$$

$$= -\frac{1}{2}\sum_{j=1}^{N}\left[(m_j + 2a)\log\left(\upsilon_j^T B_j^{-1}\upsilon_j + 2b\right) + \log\left|B_j\right|\right] + Const \tag{11.69}$$

where $B_j = E + \Phi_j A^{-1}\Phi_j^T$, and E is the identity matrix. We can also see the existence of a constant in the foregoing result, that is, $Const = \sum_{j=1}^{N}\left\{-\dfrac{m_j}{2}\log(2\pi) + \log\left(\dfrac{b^a}{\Gamma(a)}\right) + \left(a + \dfrac{m_j}{2}\right)\log 2\right\}$

References

1. X.-L. Huang, G. Wang, F. Hu, and S. Kumar, Stability-capacity-adaptive routing for high-mobility multihop cognitive radio networks, *IEEE Transactions on Vehicular Technology*, vol. 60, no. 6, pp. 2714–2729, 2011.
2. F. Zeng, C. Li, and Z. Tian, Distributed compressive spectrum sensing in cooperative multihop cognitive networks, *IEEE Journal of Selected Topics in Signal Processing*, vol. 5, no. 1, pp. 37–48, 2011.
3. S. Ji, D. Dunson, and L. Carin, Multitask compressive sensing, *IEEE Transactions on Signal Processing*, vol. 57, no. 1, pp. 92–106, 2009.
4. M. E. Tipping and A. Faul, Fast marginal likelihood maximisation for sparse Bayesian models, *Proceedings of AISTATS*, January, pp. 1–8, 2003.
5. S. Ji, Y. Xue, and L. Carin, Bayesian compressive sensing, *IEEE Transactions on Signal Processing*, vol. 56, no. 6, pp. 2346–2356, 2008.
6. J. A. Tropp and A. C. Gilbert, Signal recovery from random measurements via orthogonal matching pursuit, *IEEE Transactions on Information Theory*, vol. 53, no. 12, pp. 4655–4666, 2007.
7. Y. Qi, D. Liu, D. Dunson, and L. Carin, Multi-task compressive sensing with Dirichlet process priors, *Proceedings of ACM ICML*, July, pp. 768–775, 2008.
8. Y. W. Teh, M. I. Jordan, M. J. Beal, and D. M. Blei, Hierarchical Dirichlet processes, *Journal of the American Statistical Association*, vol. 101, no. 476, pp. 1566–1581, 2006.
9. C. F. Jeff Wu, On the convergence properties of the EM algorithm, *The Annals of Statistics*, vol. 11, no. 1, pp. 95–103, 1983.
10. H. Ishikawa, Transformation of general binary MRF minimization to the first-order case, *IEEE Transactions on Pattern Analysis and Machine Intelligence*, vol. 33, no. 6, pp. 1234–1249, 2011.
11. J. Borges and M. Levene, Evaluating variable-length Markov Chain models for analysis of user web navigation sessions, *IEEE Transactions on Knowledge and Data Engineering*, vol. 19, no. 4, pp. 441–452, 2007.
12. J. Wang, F. Wang, C. Zhang, H. C. Shen, and L. Quan, Linear neighborhood propagation and its applications, *IEEE Transactions on Pattern Analysis and Machine Intelligence*, vol. 31, no. 9, pp. 1600–1615, 2009.
13. H.-M. Lu, D. Zeng, and H. Chen, Prospective infectious disease outbreak detection using Markov switching models, *IEEE Transactions on Knowledge and Data Engineering*, vol. 22, no. 4, pp. 565–577, 2010.

EXPERIMENTAL DESIGN FOR MULTIMEDIA OVER CRN

Chapter 12

A Real-Time Video Transmission Test Bed Using GNU Radio and USRP

Ke Bao,[1] Fei Hu,[1] and Sunil Kumar[2]

[1]Electrical and Computer Engineering, University of Alabama, Tuscaloosa, AL, USA

[2]Electrical and Computer Engineering, San Diego State University, San Diego, CA, USA

Contents

12.1 Introduction

In this chapter, a real-time video transmission test bed based on software-defined radio (SDR) is discussed. Our ultimate goal is to develop a cognitive radio network (CRN) research platform. In this test bed, we use GNU Radio as the SDR software component, while the universal software radio peripheral (USRP) boards serve as the hardware components. GNU Radio is a free and open-source software development toolkit that provides signal processing blocks to implement software radios [2]. The USRP board is a hardware support for the GNU Radio software and is used for wireless communication applications from DC to almost 6 GHz. Various wireless communication research projects have used this SDR platform, such as public safety, spectrum monitoring, radio networking, cognitive radio satellite navigation, and amateur radio [1]. Other functions related to CRN, such as frequency hand-off and energy-based spectrum sensing, have also been implemented in this test bed.

This chapter is organized as follows. GNU Radio and USRP are introduced in Sections 12.2 and 12.3, respectively. The architecture of the real-time video transmission test bed is described in Section 12.4. Section 12.5 presents programming level details of the test bed. Other functions of the test bed related to CRN are presented in Section 12.6, including spectrum sensing, frequency hand-off, and cross-layer design. Finally, Section 12.7 provides the details of the test bed.

12.2 GNU Radio Introduction

GNU Radio provides a simple-to-use and rapid-application-development environment for the design of a real-time high-throughput radio [2]. Basically, GNU Radio code is written in both Python and C++ programming languages. C++ code is primarily in charge of implementing the critical signal processing function units, while the Python code groups them together as an entire system.

GNU Radio performs most of the signal processing tasks in an SDR. It can be used to build applications to receive data out of digital streams or to push data into digital streams, which are then transmitted using the hardware components. The GNU Radio library provides various basic wireless communication tools, such as filters, channel codes, synchronization elements, equalizers, demodulators, encoders, decoders, and many other elements (called element blocks) that are typically found in radio systems. In addition, it includes a method of connecting these blocks and manages how data are passed from one block to another. GNU Radio also allows users to develop other blocks that are not already present in the GNU Radio library.

12.2.1 How GNU Radio Works

GNU Radio can be seen as a library of blocks for accomplishing signal processing tasks. Users are able to connect these blocks together to achieve some functions of traditional communication devices. Generally, GNU Radio can be operated in two ways: GNU Radio companions (GRC) and regular code programed in Python and C++.

We explain the working principle of GNU Radio with the help of the following example (shown in Figure 12.1), which is based on the GRC graphic interface [2]. Users are able to construct a flow graph by picking up some blocks with specific functions from GRC of the GNU

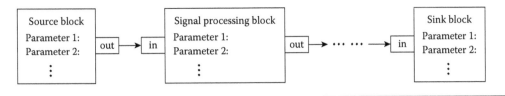

Figure 12.1 GNU Radio GRC audio example. (From http://gnuradio.org/redmine/projects/gnuradio/wiki)

Radio library and connecting them together. This system starts from the source block in the left-hand side of Figure 12.1. The source block is mainly used to generate various signals, such as audio source, random source, and noise source. The sink block (right-hand side) can be seen as a consumer of data. We can either display the processed data on screen or store it in memory of the sink block. Between the source block and the sink block, there is a series of functional processing units (i.e., blocks). GNU Radio provides plenty of functional blocks such as filter, selector, modulation, demodulation, and so on. Furthermore, a number of parameters of each block are decided. For example, the sample rate of the audio source and sink should be decided in advance. Because some blocks have multiple inputs or outputs, the system constructed by GRC can be quite complicated.

Besides the GRC interface, GNU Radio also comes with a variety of tools for a programing interface. Similar to the GRC interface, the programing interface of GNU Radio is based on the flow graph structure. In the proposed test bed, the programming interface of GNU Radio is used. The code of the test bed will be generally explained in the later sections.

12.3 USRP Introduction

GNU Radio is a pure software library. To implement SDR, hardware support is also required. Several types of hardware have been used in SDR research, such as USRP, Comedi, and Perseus. Currently, USRP is the most popular hardware among them. The family of USRP devices, produced by Ettus Research, includes different motherboards with USB or Gigabit Ethernet interfaces, with sampling rates of up to 100 MSPS and a range of front-ends for reception and transmission from 0 Hz up to 5.8 GHz. They are available as PC-bound devices or as stand-alone embedded devices [3]. All USRP devices use the same universal hardware driver (UHD), which is natively supported by GNU Radio. UHD software is developed as a separate project from GNU Radio, but all UHD devices can be easily used within GNU Radio.

Generally, a motherboard and a daughterboard are the two most critical components of a USRP hardware system. The motherboard generally includes an analog-to-digital converter (ADC), a digital-to-analog converter (DAC), and an FPGA. ETTUS Research provides a series of daughterboard products for different applications, depending on the radio frequency (RF) range and full- or half-duplex operation. Figure 12.2 shows the structure of an SDR composed of a USRP board and a host computer. First, the USRP board is connected to a host PC where GNU Radio software is installed, with a USB or Ethernet cable. Second, a proper daughterboard is plugged onto a USRP motherboard in order to send or receive RF analog signals. Third, an antenna, the front-end of the system, is connected with the daughterboard. ETTUS provides several kinds of antennas for different applications.

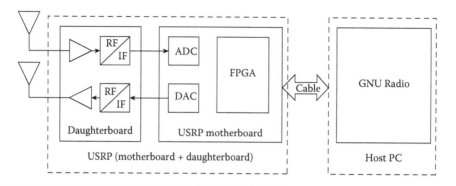

Figure 12.2   USRP board and host PC structure.

12.4 Real-Time Video Test Bed Architecture

Our test bed uses the USRP N210 motherboard and an SBX daughterboard as hardware support for real-time video transmission. An important feature of N210 is its multiple-input and multiple-output (MIMO) capability with high bandwidth and dynamic range [4]. Two N210 units may be connected together to realize a complete 2 × 2 MIMO configuration using the optional MIMO cable. Moreover, the Gigabit Ethernet interface serves as the connection between the N210 and the host computer. This enables the user to realize 50 MS/s of real-time transmission between the sender/receiver sides simultaneously (i.e., full duplex). In addition, N210 also has a faster FPGA than other USRP boards. The SBX daughterboard is a wide-bandwidth transceiver that provides up to 100 mW of output power, and a typical noise figure of 5 dB [5]. It is also MIMO capable and provides 40 MHz of bandwidth. Generally, the SBX daughterboard is able to handle most of the applications requiring access to a variety of bands in the 400 to 4400 MHz range.

The test bed uses two host computers as shown in Figure 12.3. The transmitter-side computer is connected with a web camera, which is used to capture live video. The captured video is encoded into the H.264 bit stream in the host computer and processed by GNU Radio software. In order to transmit the video data, the host computer is connected to a USRP board with an Ethernet cable. Here, the USRP board acts as an RF transmitter. Similarly, the receiving side also consists of a computer and a USRP board. This USRP board receives the video signal data from the antenna and delivers them to the host computer, which then demodulates and disassembles the received packets. Finally, an H.264 decoder on the computer decodes the packets and displays the video.

Figure 12.4 shows the system flow diagram. GNU Radio and FFmpeg software are installed on both computers. On the transmitting side, FFmpeg performs the video capture and compresses the video streaming in H.264 format. GNU Radio obtains encoded packets from FFmpeg and then performs packet header assembly and modulation before sending the packets to the USRP board. Some fields (such as the preamble, access code, CRC, etc.) are then appended to each packet. As the last step of the transmission side, the assembled packets are modulated with GMSK and sent to USRP board through the Ethernet interface. On the receiving side, the USRP board captures the analog signals from the air and passes them to GNU Radio. Through GMSK demodulation and packets disassembly by GNU Radio, the received link layer packets are dissembled into application layer packets with sequence numbers. The video packets are finally decoded and replayed by FFplay software. The communication between GNU Radio and FFmpeg/FFplay is through the UDP socket from Python.

The FFmpeg software is employed in the process of capturing real-time video by a webcam and encoding the video data in H.264. As one of the most popular media software in Linux system,

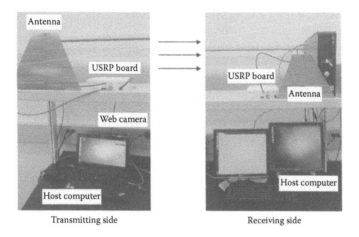

Figure 12.3 Test bed hardware setup.

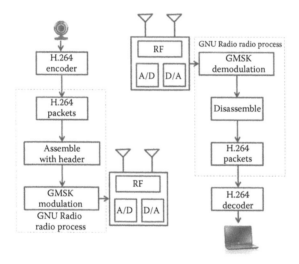

Figure 12.4 Real-time video transmission system at RF frequency.

FFmpeg is able to decode, transcode, mux, demux, stream, filter, and play most media files. H.264 is a standard for video compression that is one of the most common formats for the recording, compression, and distribution of high-definition video.

12.5 GNU Radio Python Code Explanation

12.5.1 Sender-Side Python Code

After FFmpeg exports H264 packets in real time, GNU Radio is employed to implement some data processing modules for H264 packets, including packet header assembly and modulation functions on the sender side. From the standpoint of the flow graph, the system is generally built by gluing blocks of GNU Radio library together in Python code, as discussed in the following text.

```
class my_top_block(gr.top_block):
  def __init__(self, modulator, options):
      gr. top_block._init_(self)

      if(options.tx_freq is not None):
          # Work-around to get the modulation's bits_per_symbol
          args = modulator.extract_kwargs_from_options(options)
          symbol_rate = options.bitrate / modulator(**args).bits_per_
                        symbol()

          self.sink = uhd_transmitter(options.args, symbol_rate,
                                      options.samples_per_symbol,
                                      options.tx_freq, options. tx_gain,
                                      options.spec, options.antenna,
                                      options. verbose)
        options.samples_per_symbol = self.sink._sps

      elif(options.to_file is not None):
          sys.stderr.write(("Saving samples to '%s'.\n\n" % (options.
             to_file)))
          self.sink = gr.file_sink(gr.sizeof_gr_complex, options.to_
                      file)
      else:
          sys.stderr.write("No sink defined, dumping samples to null
             sink.\n\n")
          self.sink = gr.null_sink(gr.sizeof_gr_complex)

      # do this after for any adjustments to the options that may
      # occur in the sinks (specifically the UHD sink)
      self.txpath = transmit_path(modulator, options)
      self.connect(self.txpath, self.sink)
```

First, the code defines a class named 'my_top_block' that is a container for the flow graph; it is derived from another class 'gr.top_block'. All blocks of the system are connected under this class. Generally, the sender side Python code consists of two main blocks, named 'self.txpath' and 'self. sink'. Note that 'self.' refers to the instance attribute in Python. The 'self.txpath' block defines a series of parameters and functions (including modulation, signal amplitude sent to USRP, sample frequency, bit rate, etc.) that belong to GNU Radio software. On the contrary, 'self.sink' is derived from the 'uhd_interface' class. The UHD is the driver of USRP and works as an interface between GNU Radio and USRP. All the parameters and functions of the 'self.sink' block are related to the USRP board, such as the sampling frequency in the ADC, RF signal amplitude from the antenna, and signal frequency. After these two blocks are defined, the command 'self.connect(self.txpath, self.sink)' is used to connect them together.

The 'my_top_block' class is followed by a main function, which is where the program execution starts. All functions are executed in a certain sequence in this main function. In the first dozens of lines of codes, the necessary functions and variables are defined, such as modulation options, sampling rate of USRP, and so on. Then, the class 'my_top_block' is invoked to build the flow graph in the main function, as shown in the in the following text. 'gr.enable_realtime_scheduling()' is a module from the GNU Radio library, and it is used to check if the code works in real-time scheduling.

```
# build the graph
tb = my_top_block (mods [options.modulation], options)

r = gr.enable_realtime_scheduling()
if r != gr.RT_OK:
      print "Warning: failed to enable realtime scheduling"
tb.start ()                        # start flow graph
```

After FFmpeg captures the real-time video data with a webcam and encodes it into H.264 packets, a pipeline is required to connect the Python code of GNU Radio and FFmpeg. In this case, a 'socket' module of Python enables the H264 packets to transfer from FFmpeg to GNU Radio Python code by using the UDP protocol. The 'socket' module code is given in the following text.

```
# create a socket|
s = socket.socket(socket.AF_INET, socket.SOCK_DGRAM)
host = '' # can leave this blank on the server side

try:
    s.bind((host, options.port))
except socket.error, err:
    print "Could not set up a UDP server on port %d: %s" % (options.port,
          err) raise SystemExit

while True:
    data = s.recv(4972)
    print "raw data length = %4d" % (len(data))
    if not data:
       break
```

Once GNU Radio code receives the H.264 data packets from FFmpeg, a data header assembly process is launched. In this step, we use a Python module named 'structure.pack()'. Some necessary information is added in the header, such as packet number, packet length, and so on. Each H264 packet is then sent to the USRP board. In future work, the routing information would also be added in the MAC and data link layer applications.

```
      while (len(data) > max_pkt_size): # (maximum allower packet size:
        4096)
        partial_pl = data[0:max_pkt_size]   # extract 3072 bytes
        # print "partial_pl 1 length = %4d" % (len(partial_pl))
        partial_pl = struct.pack('!H', 0x5555) + partial_pl
        # print "partial_pl 2 length = %4d" % (len(partial_pl))
        payload = struct.pack('!H', pktno & 0xffff) + partial_pl
        # print "partial_pl 3 length = %4d" % (len(payload))
        send_pkt(payload)
        n += len(payload)
        pktno += 1
        sys.stderr.write('*')   # "*" to denote split packet

        data = data[max_pkt_size:] # update "data"

     data = struct.pack('!H', 0xaaaa) + data
     payload = struct.pack('!H', pktno & 0xffff) + data
```

```
        send_pkt(payload)
        n += len(payload)
        sys.stderr.write('.')
        pktno += 1

send_pkt(eof=True)
```

12.5.2 Receiver-Side Python Code

The receiver side executes the inverse process of the sender side. After a USRP board captures live video data from the antenna, it is converted from an analog to a digital signal, and H.264 video packets are recovered. Then, the Python code based on GNU Radio software is used to separate the header from the packet and read necessary information from it. Next, the raw video data are sent to the video player and gets played in real time.

Similar to the sender side, the packets also need to go through the flow graph built by GNU Radio on the receiver side. The Python code of receiver side also begins with a 'my_top_block' class, which consists of two main blocks in the receiver side, namely 'source' and 'rxpath'. The 'source' block derives from the UHD and provides necessary parameters and functions to the receiver-side USRP board, while 'rxpath' integrates some blocks from the GNU Radio library. Generally, 'rxpath' consists of a demodulation block, an 'rx_callback' function, and an 'options' module of Python. The 'rx_callback' is an unpacking function that will be explained later. The 'options' module enables the user to input some parameters when the Python code is executed. The code of 'my_top_block' class is shown in the following text.

```
class my_top_block(gr.top_block:
    def __init__(self, demodulator, rx_callback, options):
        gr.top_block._init_(self)

        if(options.rx_freq is not None):
            args = demodulator.extract_kwargs_from_options(options)
            symbol_rate = options.bitrate / demodulator(**args).bits_
                            per_symbol()

            self.source = uhd_receiver(options.args, symbol_rate,
                                options.samples_per_Symbol,
                                options.rx_freq, options.rx_gain,
                                options.spec, options.antenna,
                                options.verbose)
            options.samples_per_symbol = self.source._sps

        elif(options.from_file is not None):
            sys.stderr.write(("Reading samples from '%s'.\n\n"
                            % (options.from_file)))
            self.source = gr.file_source(gr.sizeof_gr_complex,
                                    options.from_file)

        else:
            sys.stderr.write("No source defined, pulling samples from
                            null source.\n\n")
            self.source = gr.null_source(gr.sizeof_gr_complex)

        self.rxpath = receive_path(demodulator, rx_callback, options)

        self.connect(self.source, self.rxpath)
```

The structure of the main function of the receiver-side code is very similar to the sender-side code. The first dozens of lines of code define some parameters and simple functions. Then, the Python module 'socket' establishes a UDP protocol pipeline for sending the video data out of the GNU Radio flow graph. Next, the 'rx_callback' function, given in the following text, is used to unpack the header from raw packets.

```
def rx_callback(ok, payload):
    global n_rcvd, n_right
    print "Received packet length = %4d" % (len(payload))
    (pktno,) = struct.unpack ('!H', payload[0:2])
    n_rcvd += 1
    if ok:
        n_right += 1

    print "ok = %5s pktno = %4d n_rcvd = %4d n_right = %4d" % (ok, pktno,
                n_rcvd, n_right)

    print "Sent packet length = %4d" % (len(payload[2:]))
    cs.send(payload[2:])
```

After all the parameters and functions are ready, the flow graph of GNU Radio is built and executed using the following commands.

```
# build the graph
tb = my_top_block (demods[options.modulation], rx_callback, options)

r = gr.enable_realtime_scheduling()
if r != gr.RT_OK:
    print "Warning: Failed to enable realtime scheduling."

tb.start()     # start flow graph
tb.wait()      # wait for it to finish
```

12.6 Other Functions of Test Bed

Besides the real-time video transmission, the following functions are implemented in this test bed to realize the CRN:

1. Spectrum sensing;
2. Spectrum hand-off;
3. Cross-layer design (such as routing/MAC cross-layer design).

12.6.1 Spectrum Sensing Function

Spectrum sensing is a critical function in a CRN and is used to detect if the radio channel is occupied by the primary user. Specifically, the spectrum sensing can be seen as a binary hypothesis [6]:

1. H_0: $y[n] = w[n]$, $n = 1,...,N$
2. H_1: $y[n] = x[n] + w[n]$, $n = 1,...,N$

Here, $x[n]$ is the primary user's signal, $w[n]$ is the noise, and n presents the time factor. If the received signal $y[n]$ is in H_0 status, the channel is considered as idle, that is, available to other users. However, H_1 means the channel is occupied by the primary users. It is complicated to decide whether the channel is occupied or not in practical RF environment because $x[n]$ may be too weak to be detected, or noise signals may be very similar to the primary users and can confuse the spectrum sensing unit. Hence, several spectrum sensing techniques have been used, which can be divided into three categories [7]: energy detection, matched filter detection, and cyclostationary detection. Among them, energy detection is most widely used because it is very simple and does not require any a priori knowledge of the primary signals or complicated algorithms. In energy detection, the noise energy of the channel is sensed and compared with a predefined threshold. If the noise energy is higher than the threshold, the channel is considered as occupied; otherwise, the channel is idle.

Our test bed uses the energy detection for spectrum sensing. The Python code of spectrum sensing is explained in the following text.

12.6.2 Explanation of Code for Energy-Detection Based Spectrum Sensing

The Python code used to realize the spectrum sensing function comes from the GNU Radio library 'usrp_spectrum_sense.py'. The entire system is divided into several blocks as shown in Figure 12.5, where each block represents a basic function of the system. These blocks are glued together by the class 'gr.top_block' from the GNU Radio library.

The USRP receiving module is represented by 'self.u = uhd.usrp_source()'. This command is generally used to set necessary parameters of the USRP board and ensure that the RF signal is received properly. After streaming data are captured by the USRP board, they are sent to the fast Fourier transform (FFT) block. Between these two blocks, command 's2v = gr.stream_to_vector(…)' is introduced to take a sample of streaming data in vector format. For the next step, the output data of the FFT block goes through the 'c2mag' block (converting data from time domain to frequency domain), and it is finally accepted by a data statistics block called 'stats = gr.bin_statistics_f(…)'. Generally, this module is used to control scanning and record the frequency domain statistics. The USRP board can only detect 8 MHz bandwidth of the RF frequency channel at a time. If we want to perform sensing in a wider frequency band, the USRP board should sense 8 MHz each time and repeat this process in different frequency bands until the entire expected RF band is covered. The 'stats' block is used to record results

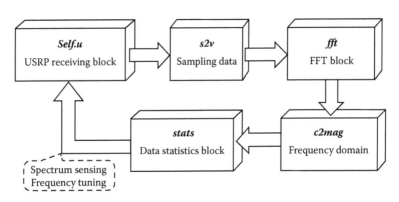

Figure 12.5 The USRP_spectrum_sense.py flowchart.

when the system gets one noise sensing process done and sets the system to the next sensing process. Moreover, an invoking function named 'tune()' is predefined to set the USRP board to the next frequency interval.

12.6.3 Spectrum Hand-Off

The spectrum hand-off is another important component of a CRN. When a primary user reoccupies the channel, the secondary users who are currently occupying this channel must vacate it immediately. In this case, the secondary user hops to another available frequency channel by using a spectrum hand-off function. A simple cross-layer application is also achieved based on frequency hand-off. The cross-layer approach allows parameter exchange between different protocol layers [8]. It improves the communication efficiency and speed through coordination, interaction, and joint optimization of protocols across different layers. The frequency hand-off is implemented in Python in our test bed. The basic idea is to realize the frequency hopping while the video is transmitting in real time. On the basis of GNU Radio and URSP hard drive software (UHD), the USRP working frequency can be simply changed by calling the frequency setting function 'set_center_freq()' from the UHD library, which is programmed in C++. The frequency setting function is already defined in the USRP hardware driver module 'uhd_interface_hopping. py'. In our test bed, we use two RF channels, and USRP can switch transmissions between them. The code is shown in the following text.

```
def set_next_freq(self):
    freq_interval = 1000000

    target_freq = self.next_freq
    if self.next_freq > self._freq + freq_interval/2:
        self.next_freq = self._freq

    elif self.next_freq < self._freq + freq_interval/2:
        self.next_freq = self._freq + freq_interval

    rn = self.u.set_center_freq(target_freq, 0)

    if rn:
        print "Succeed to set frequency to", target_freq
    else:
        print "Failed to set frequency to", target_freq

    return target_freq
```

Another problem that should be considered is how to synchronize the spectrum hand-off between the sender and the receiver sides. Currently, our USRP boards operate in the half-duplex mode. If the receiver side fails to synchronize with the sender side after the latter performs a hand-off, the packets will be lost. We use data packets to transmit a frequency hand-off command from the sender side to the receiver side. When the sender side decides to switch to another frequency channel, a command is written into the header of a control packet and sent to the receiver side. In response, the frequency-setting function is immediately executed at the receiver. The hand-off synchronization can be improved in the following two ways. A test bed with full-duplex communication between the sender and the receiver side can be used.

Alternatively, a third USRP board can be used that works as a spectrum sensing station. When a primary user is detected on the current frequency channel, it will send commands to both the sender and the receiver, and both of them will carry out the spectrum hand-off at the same time.

12.6.4 Cross-Layer Design of Test Bed

We have also implemented a cross-layer design by adjusting the video resolution with different RF channels during the real-time video transmission. The cross-layer design facilitates the information exchange between the application layer and the physical layer. Based on the aforementioned spectrum hand-off design, we can achieve this cross-layer design by producing video output in two resolutions and transmitting each of them in different FR channels. Two different resolution video outputs are achieved by using FFmpeg, as discussed in the next section. With respect to the GNU Radio flow graph code, two pipelines are employed by the socket module. Each of the two video outputs is assigned a pipeline to transmit video data from FFmpeg to the GNU Radio flow graph. Based on the different frequency channels the test bed is operating on, the GNU Radio flow graph will switch to the corresponding pipeline and receive the video streaming data.

12.7 Test Bed Demo for Real-Time Video Transmission

The following four functions are implemented in our test bed:

1. High-resolution real-time video transmission
2. Spectrum sensing
3. Frequency hopping for spectrum hand-off
4. A simple cross-layer design

Because the GNU Radio flow graph needs to invoke other modules, all the Python files of GNU Radio are put in one directory: 'gnuradio/example/digital/narrowband/'.

12.7.1 Signal Sensing and Video Transmission

The general demo steps are as follows:

1. Sense the signals on the specified spectrum bands.
2. Find out the channel with the lowest signal amplitude, and select it on both the sender and receiver sides.
3. Transmit the video in real time.

For spectrum sensing in Step 1 above, the 'hop_usrp_spectrum_sense.PY' file is executed by running a command in 'hop_usrp_spectrum_sense.py 900M 1500M'. Specifically, 1500M and 900M are the maximum and minimum values of the frequency range, respectively. Then, a spectrum sensing result is generated in seconds. Figure 12.6 shows the result of the spectrum sensing process, in which the noise amplitude is presented.

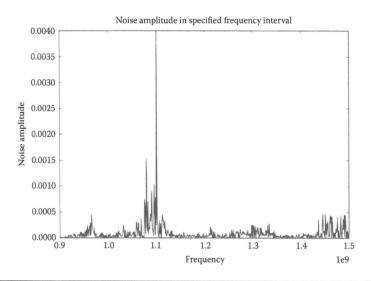

Figure 12.6 Spectrum sensing result.

In the next step, the test bed transmits the real-time video on the frequency channel that has the lowest noise amplitude.

On the sender side:

1. The FFmpeg software is executed in a new terminal window via the command 'ffmpeg– f video4linux2 –b:v 32k –s 1080*960 –r 120 –i/dev/video0 –c:v libxx264 –tune zerolatency –y –f h264 udp://127.0.0.1:12347'.
2. In a new terminal, go to the directory 'usr/local/share/gnuradio/examples/digital/narrowband'; and execute the sender side GNU Radio flow graph code with the command './video_tx.py –f 910M –port 12347'.

On the receiver side:

1. The video player ffplay is opened first in standby status using the command: `ffplay -f h264 udp://127.0.0.1:12345`
2. In a new terminal window, input the './packet_tranceiver.py' command to run the Python file. This file is used to connect the GNU Radio flow graph Python code with the ffplay software.
3. Execute the GNU Radio flow graph code in a new terminal window by using the 'video_ rx.py –f 910M –port 12347' command.

12.7.2 Frequency Hopping

The general demo steps are as follows:

1. Set two frequency channels in advance.
2. The sender side hops from one channel to the other after transmitting a certain number of packets.
3. The receiver also hops such that it operates in the same frequency channel as the sender side in order to uninterruptedly receive the real-time video packets.

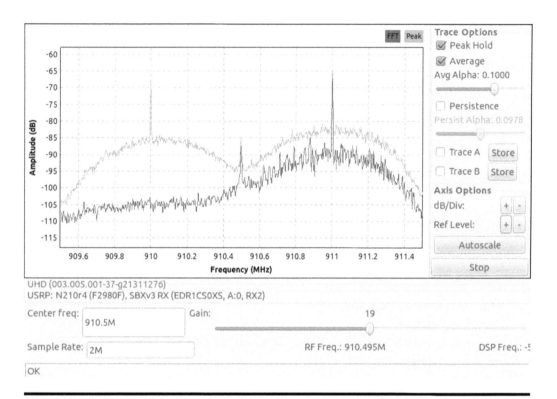

Figure 12.7   FFT result of the frequency hopping experiment.

The GNU Radio flow graphs at the sender and receiver sides are called 'freq_hopping_tx.py' and freq_hopping_rx.py', respectively. Similar to the experiment in Section 12.7.1, the FFmpeg software is run first with the same command as in Section 12.7.1. Then, the GNU Radio flow graph Python code is executed with the command './freq_hopping_tx.py'. On the receiver side, the process is also the same as in Section 12.7.1, except that 'video_tx.py' is replaced by 'freq_hopping_tx.py'. Figure 12.7 shows the FFT of the result of the frequency hopping experiment.

In Figure 12.7, the blue curve presents the real-time video transmission signal amplitude, while the green curve represents the peak amplitude for the entire transmission process. Obviously, the video test bed is transmitting on a 911 MHz channel. A moment later, the working frequency shifts to the 910 MHz channel. Note that the user is able to set two frequency channels in any RF band in the test bed. The hardware used in this test bed can handle any frequency between 850 and 4400 MHz. Owing to fast frequency switching in USRP, only a few video frames may be lost in this process, while the delay between the sender side and the receiver side is still smaller.

12.7.3 Demo of Cross-Layer Design

On the basis of the frequency hopping in Section 12.7.2, a cross-layer function is added in this part that changes the video resolution with frequency hopping. Two video outputs with different resolutions are generated from the FFmpeg software. Because all the frequency channel values and the two ports of pipelines connected with FFmpeg have been written in the GNU Radio flow graph code, the procedure here is similar to the previous sections.

Figure 12.8 **A video frame with two different resolutions (left: 196 × 90 pixels; right: 800 × 600 pixels).**

On the sender side:

1. Run the FFmpeg software with the command 'ffmpeg –f video2linux2 –b:v 32k –s 320x240 –r 64 –i/dev/video –c:v libx264 –tune zerolatency –y –f h264 udp://127.0.0.1:12347 –s 196x98 –c:v libx264 –tune zerolatency –y –f h264 udp://127.0.0.1:12346'.
2. Run the GNU Radio flow graph Python code './crosslayer_tx.py' in a new terminal window.

On the receiver side:

1. Run the ffplay software by 'ffplay –f h264 udp://127.0.0.1:12345'.
2. Input './packet_tranceiver.py' in a new terminal window.
3. Input './crosslayer_rx.py' in a new terminal window.

Figure 12.8 shows a video frame in two different resolutions at the receiving side.

References

1. L. Chunxiao, K. J. Niraj, and R. Anand, Secure reconfiguration of software-defined radio. *ACM Trans. Embed. Comput. Syst*, Vol. 11 (1), No. 10, pp. 1–22, 2012.
2. B. Eric. GNU radio: tools for exploring the radio frequency spectrum. *Linux J*, Vol. 122, p. 4, 2004.
3. B. T. Nguyen and Y. Chansu, Investigating Latency in GNU Software Radio with USRP Embedded Series SDR Platform. In Proceedings of the 2013 Eighth International Conference on Broadband and Wireless Computing, Communication and Applications (BWCCA '13). *IEEE Computer Society*, Washington, DC, USA, pp. 9–14, 2013.
4. USRP n200/n210 networked series data sheet. Available at https://www.ettus.com/content/files/07495_Ettus_N200-210_DS_Flyer_HR_1.pdf
5. USRP SBX daughter board data sheet. Available at https://www.ettus.com/product/details/SBX
6. H. Li, X. Cheng, K. Li, X. Xing, and T. Jing, Utility-based cooperative spectrum sensing scheduling in cognitive radio networks, *IEEE INFOCOM*, April 14–19, pp. 165–169, 2013.

7. M. A. Abdul Sattar and Z. A. Hussein, Energy detection technique for spectrum sensing in cognitive radio: A survey, *International Journal of Computer Networks and Communications (IJCNC)*, Vol. 4, No. 5, p. 223, 2012.

8. R. A. El-Mayet, H. M. El-Badawy, and S. H. Elramly, Modeling and analysis for effective capacity of a cross-layer optimized wireless networks, *World Academy of Science, Engineering and Technology*, Vol. 70, pp. 221–227, 2012.

Chapter 13

PR Activity Model for Multimedia Communication in NS-2

Yasir Saleem[1] and Mubashir Husain Rehmani[2]

[1]Department of Computer Science and Networked System, Sunway University, Selangor, Malaysia

[2]Department of Electrical Engineering, COMSATS Institute of Information Technology, Wah Cantt., Pakistan

Contents

13.1 Introduction

Simulation is an imitation of a real-world process or system. In telecommunication networks and networking, researchers use simulators to validate their protocols and algorithms. The most widely used simulator in the networking research community is the network simulator ns-2 [1]. The network simulator ns-2 is an open source simulator that was developed in 1989. It is also used for educational purposes, to better understand the underlying network behavior. In addition, it is also used for validating both new and existing network-related algorithms and protocols.

The network simulator ns-2 has many built-in modules; however, researchers need new simulation modules. Similarly, with the emergence of new communication networks, the need to incorporate changes in ns-2 is inevitable. The incorporation of new modules in ns-2 requires a deeper understanding of the ns-2 architecture. Thus, new building blocks and patches are required to simulate these emerging networks in ns-2.

Cognitive radio networks (CRNs) are also an emerging network, proposed to solve the problem of spectrum scarcity. In CRNs cognitive radio users use the licensed band when it is not utilized by the primary radio (PR) users. The PR users follow a different activity pattern depending on the underlying communication network. Consequently, the performance of CRNs is dependent on the PR user activity pattern. Thus, the modeling of PR user activity to support multimedia communication is very important. However, ns-2 does not support the activity of PR users. In this chapter, our goal is to discuss in detail the modeling of PR user activity modeling for multimedia communication in ns-2.

This chapter is organized as follows: In Section 13.2, we discuss PR activity models along with a few examples. We then provide the classification of PR activity models in Section 13.3. The network simulator ns-2 basics are discussed in Section 13.4. The support to simulate CRNs in ns-2 is discussed in Section 13.5. The higher-level design of PR activity models for CRNs in ns-2 is discussed in Section 13.6. Issues and challenges regarding the implementation of PR activity models in ns-2 are discussed in Section 13.7. Finally, Section 13.8 concludes the chapter.

13.2 PR Activity Models

The goal of this section is to provide a brief overview of PR activity models. First, we will discuss why PR activity modeling is important for the performance of CRNs. Second, we give an overview of the existing PR activity models used in the literature.

PR activity modeling is very important in CRNs because CR users utilize the spectrum of PR users when the latter are not utilizing it. Therefore, there is no guarantee that the spectrum will be available to CR users during the entire transmission because a PR user can arrive at any time, upon which a CR user has to vacate that spectrum band and is assigned another one. Thus, by modeling PR activity, the CR user can predict the future state of PR users by learning from the history of their spectrum utilization. In this manner, CR users can assign the best available spectrum bands for their communication. Hence, we can say that PR activity modeling plays a vital role in the performance of CRNs. The following two examples illustrate how the performance of a CRN is dependent on PR activity modeling.

Example 1: In the first example, we demonstrate how a CRN can improve the bandwidth. Figure 13.1 shows the PR occupancy over different spectrum bands. To make these spectrum holes available to CR users, one can aggregate the underutilized spectrum by combining the entire available spectrum. This concept is shown in Ref. [2] in which the aggregate spectrum occupancy is modeled as multiple ON/OFF processes that are obtained by the logical OR of the individual ON/OFF processes. In this manner, by aggregating the available spectrum, CR users can improve the bandwidth utilization.

Example 2: If PR users occupy a spectrum for long periods, this spectrum will be available to CR users for only a very short time. Thus, if a CR user still performs sensing on that spectrum,

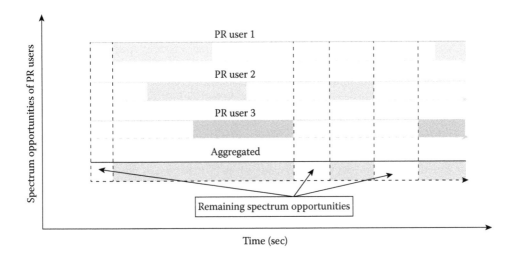

Figure 13.1 Aggregate spectrum occupancy from multiple ON/OFF processes by the logical OR of individual ON/OFF process.

it would be a waste of time and energy [3]. Therefore, PR activity modeling increases the performance of CRNs and utilizes the spectrum more effectively. In Ref. [4], different techniques have been presented for modeling PR activity, which include PR activity based on Poisson modeling, statistics, measure data, and some other techniques. Thus, by using those techniques and many other techniques that we are going to discuss in this chapter, CR users can optimize spectrum sensing and prevent wastage of time and energy.

13.3 Classification of PR Activity Models

Many PR activity models have been used in the literature. It is very important to classify them, so that researchers can use the appropriate PR activity model according to their requirements. Therefore, in this section, we will classify PR activity models based on different criteria.

We classify the PR activity models as follows:

- Markov process
  - Two-state Markov chain
  - Three-state Markov chain
  - Hidden Markov model
  - Semi-Markov model
- Queuing theory
  - M/G/1
  - Bernoulli process
- ON/OFF Periods
  - ON/OFF model
  - Block activity model
- Time series
  - ARIMA model

13.4 Network Simulator (ns-2) Basics

The network simulator ns-2 [5] is an open source simulator that was developed in 1989. ns-2 basics will be presented in this section. These basics include the main components of ns-2. It will also contain a discussion on writing simulation scripts, parsing, and post-simulation operations.

We now discuss the fundamentals of ns-2. This includes the following:

- **Basic architecture:** The basic architecture of ns-2 is simple. It consists of the TCL language, with which we can write scenario files where we describe all the simulation parameters. This TCL script then links to the C++ core code of ns-2. It has main files of protocols, and is composed of several hundreds of classes of different objects. These C++ objects are linked with OTcl language objects. This means that the core of ns-2 is composed of two languages: C++ and OTcl. The TCL language is used to write the simulation scenario and acts as an interface between the user and the core ns-2 code. After the execution of the simulation, ns-2 provides the trace file, which can be analyzed by using AWK and Perl languages. Moreover, the results can also be visualized through Trace Graph, Network Animator (NAM), and GNU plot.
- **Installation:** ns-2 is open source and is freely available on the Internet. We can run ns-2 on the Windows and Linux platforms. To run ns-2 on Windows, we need to install the Cygwin software. Cygwin is a virtual machine through which we can get the Linux environment over the Windows operating system. ns-2 can be installed into components; however, for beginners, the all-in-one package of ns-2 will be better. This all-in-one package of ns-2 contains all the related packages that the ns-2 simulator requires.
- **Running ns-2 program:** To run ns-2, we need to run the ns-2 tcl script. This script contains all the simulation scenario parameters.
- **Network object *Node*:** The *node* is a basic network object in ns-2. It acts as a router and a computer host. As a router, it forwards the packet to the connecting link based on the routing table. As a host, it delivers packets to the upper layer, that is, the transport layer agent. This transport layer agent is attached with the port as specified in the packet header. A node can be static or mobile. When nodes are mobile, the `setdest` command can be used to provide the destination of mobile nodes with the velocity information. The command `node-config` is used to configure the node parameters.
- **Packets and packet headers:** Packets are composed of packet header and data. The source address, destination address, and packet id are stored in the packet header, whereas data are stored in the data portion of the packet. In ns-2, a packet can be of several types such as UDP, TCP, CBR, audio, and video. However, one can also define their own packet types. We now discuss some packet functions used in the AODV protocol.
  - `AODV::recvAODV (Packet *p)`: This function classifies the incoming AODV packets. If the incoming packet is of type RREQ, RREP, RERR, or HELLO, It will call `recvRequest(p)`, `recvReply(p)`, `recvError(p)`, and `recvHello(p)` functions, respectively.
  - `AODV::recvRequest (Packet *p)`: When a node receives a packet of type REQUEST, it calls this function.
  - `AODV::recvReply (Packet *p)`: When a node receives a packet of type REPLY, it calls this function.
  - `AODV::recvError (Packet *p)`: This function is called when a node receives an ERROR message.

- AODV::recvHello(Packet *p): This function receives the HELLO packets and looks into the neighbor list. If the node is not present in the neighbor list, it inserts the neighbor, otherwise if the neighbor is present in the neighbor list, it sets its expiry time to: CURRENT TIME + (1.5 * ALLOWED HELLO LOSS * HELLO INTERVAL), where ALLOWED HELLO LOSS = 3 packets and HELLO INTERVAL = 1000 ms.

■ **Traffic generators:** In ns-2, different types of traffic generators are available. These include constant bit rate (CBR) traffic generator, exponential On/Off traffic generator, Pareto On/Off traffic generator, and traffic trace. From the traffic trace file, we can call a trace file and from there, we can use the traffic pattern.

■ **Helper classes in ns-2:** Timers are one of the helping classes available in ns-2. Timers are used to delay an action, or they can be used when a certain event occurs. A random number is another helping class available in ns-2. A random number generator (RNG) is employed in ns-2 for random number generation. This RNG can generate random numbers of uniform distribution, exponential distribution, normal distribution, and log-normal distribution. The error model is another helping class present in ns-2. This error model is used to incorporate errors in the packet transmission.

We now give an example of timers used in the AODV protocol.

- Broadcast Timer: This timer is responsible for purging the IDs of nodes and schedule after every BCAST ID SAVE.
- Hello Timer: It is responsible for sending Hello packets with a delay value equal to an interval = MinHelloInterval + ((MaxHelloInterval - MinHelloInterval) * Random::uniform()).
- Neighbor Timer: It purges all timed-out neighbor entries and schedule after every HELLO INTERVAL.
- RouteCache Timer: This timer is responsible for purging the route from the routing table and schedule after every FREQUENCY.
- Local Repair Timer: This timer is responsible for repairing the routes.

13.4.1 Simulation Process in ns-2

The goal of this section is to describe the simulation process in ns-2. Figure 13.2 shows the simulation process of ns-2. This includes the following:

■ **Scenario creation:** Here, the basic scenario of the simulation is created. This includes the declaration of the mobility pattern of nodes, number of nodes present, and propagation model; physical layer parameters such as the propagation model, error model, data rate, and type of channels are defined. Moreover, we can also define the MAC layer protocol, routing layer protocol, and other traffic conditions. A scenario file is created by using the TCL language; the file has the extension .tcl.

■ **Execution of simulation:** Once the simulation scenario is defined, the simulations are executed. Simulations should be run several times to obtain consistent results. For instance, a wireless simulation can run 1000 times with 95% confidence interval.

■ **Output of simulation:** The output of the simulation can be in the form of a trace file. This trace file can have a .txt or .tr extension. The name of the trace file should be declared in the scenario file. This generated trace file can be very huge, having a size of several gigabytes to terabytes. Thus, it is very difficult to manually analyze this file to obtain results.

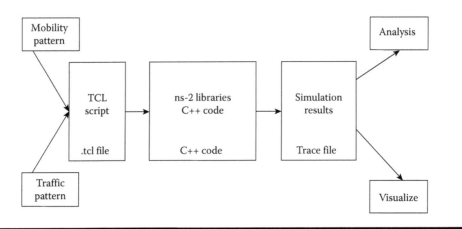

Figure 13.2   ns-2 simulation process.

- **Analysis of trace file:** The generated trace file is then analyzed. The analysis can be done through parsing of the trace file. To parse the trace file, scripting languages can be used. The well-known scripting languages are Perl and AWK.
- **Visualizing results:** After parsing the results and analyzing them, there is a need to visualize those results. Various graphical tools are available for this purpose, such as Trace Graph and GNU Plot.

13.5 Support for CRN Simulation in ns-2

In this section, we discuss how much support is available to simulate CRNs in ns-2. Several authors used ns-2 [6,7] to validate their protocols in the context of CRNs. Therefore, we will discuss some of them. We will also discuss the Cognitive Radio Cognitive Networks (CRCN) patch [8] of ns-2 available to simulate CRNs. In fact, as shown in Figure 13.3, the CRCN patch has three building blocks in order to support cognitive radio functionalities in ns-2.

Cognitive Radio Networks: Several authors have used ns-2 [6,7] to validate their protocols in the context of CRNs. Here, we discuss the CRCN patch (CRCN ns-2 Path [online]) of ns-2 available to simulate CRNs. In fact, as shown in Figure 13.3, the CRCN patch has three building blocks that support cognitive radio functionalities in ns-2. However, several basic functionalities of CRNs are still missing in this CRCN patch.

Table 13.1 shows the ns-2 existing capabilities as well as the ns-2 contributed codes.

Comparison of ns-2 with ns-3: ns-3, similar to ns-2, is also an event-based simulator. ns-3 is written in the C++ and Python programming languages. There were several problems with ns-2 such as lack of memory management, lack of coupling between different models, and so on. ns-3 is in the development stage, and it cannot replace ns-2. However, ns-3 may replace ns-2 after a few years provided that it is extensively validated and new support patches are developed. Table 13.2 compares the existing core capabilities of ns-2 and the existing capabilities of ns-3 taken from the ns-2 official website.

ns-2 and BonnMotion: BonnMotion is a Java-based software. It was developed by the University of Bonn, Germany, and is used to create and analyze mobility scenarios. These scenarios can be exported to different simulators such as ns-2, ns-3, GloMoSim, QualNet, and ONE.

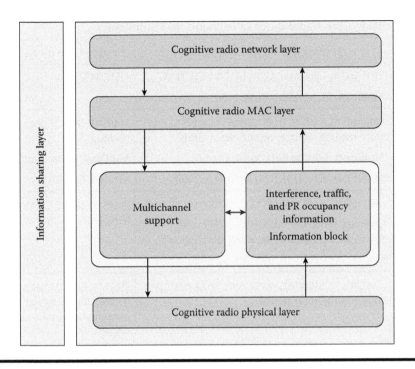

Figure 13.3 High-level PR activity design in ns-2.

Table 13.1 ns-2 Existing Capabilities and Contributed Codes

	Existing Core ns-2 Capabilities	*ns-2 Contributed Codes*
Transport Layer	TCP (many variants), UDP, SCTP, XCP, TFRC **Multicast:** PGM, SRM, RLM, PLM	TCP PEP, TCP Pacing, TCP Westwood, TCP Eifel
Network Layer	**Unicast:** IP, Mobile IP, distance vector and link state routing, source vector, Nix routing **Multicast:** SRM, generic centralized **MANET:** AODV, DSR, DSDV, TORA, IMEP	AODV+,AODV-UU, AOMDV, ZRP, DYMO, OLSR, Mobile IPv6, GPSR, RSVP, PUMA
Link Layer	ARP, HDLC, GAF, MPLS, LDP, Diffserv **Queuing:** DropTail, RED, RIO, WFQ, SRR, Semantic Packet Queue, REM, priority, VQ **MAC:** CSMA, 802.11b, 802.15.4, Satellite, ALOHA	802.16, 802.11 a multirate, TDMA DAMA, MPLS, UMTS, GPRS, Bluetooth, CSFQ
Physical Layer	Two ray, shadowing, omni antennas, energy model, satellite repeaters	ET/SNRT/BER based physical layer, IR-UWB
Support	Random number generations, tracing, NAM	emulation, BonnMotion, Tracegraph
Applications Available	Ping, vat, Telnet, multicast FTP, HTTP, web cache, traffic generators	MPEG generator, BonnTraffic, ns-2voip

Table 13.2 Existing Core Capabilities of ns-2 and ns-3 Simulators

	Existing Core ns-2 Capabilities	*Existing Core ns-3 Capabilities*
Transport Layer	TCP (many variants), UDP, SCTP, XCP, TFRC **Multicast:** PGM, SRM, RLM, PLM	UDP, TCP
Network Layer	**Unicast:** IP, Mobile IP, distance vector and link state routing, source vector, Nix routing **Multicast:** SRM, generic centralized **MANET:** AODV, DSR, DSDV, TORA, IMEP	**Unicast:** IPv4, global static routing **Multicast:** static routing **MANET:** OLSR
Link Layer	ARP, HDLC, GAF, MPLS, LDP, Diffserv **Queuing:** DropTail, RED, RIO, WFQ, SRR, Semantic Packet Queue, REM, priority, VQ **MAC:** CSMA, 802.11b, 802.15.4, Satellite, ALOHA	Point to point, CSMA, 802.11 MAC, rate control algorithms
Physical Layer	Two ray, shadowing, omni antennas, energy model, satellite repeaters	802.11a, Friis propagation model, log distance propagation loss model, basic wired
Support	Random number generations, tracing, NAM	Random number generations, tracing, mobility visualizer
Applications Available	Ping, vat, Telnet, multicast FTP, HTTP, web cache, traffic generators	ON/OFF application, asynchronous sockets API, packet sockets

BonnMotion provides several mobility models such as Random WayPoint Model, Random Walk Model, Gauss-Markov Model, Manhattan Grid Model, Reference Point Group Mobility Model, Disaster Area Model, and Random Street Model.

Adding New Modules in ns-2: The network simulator ns-2 has many built-in modules; however, researchers require new simulation modules. Similarly, with the emergence of new communication networks, the need to incorporate changes in ns-2 is inevitable. The incorporation of new modules in ns-2 requires a deeper understanding of ns-2 architecture. Thus, new building blocks and patches are required to simulate these emerging networks in ns-2. We now describe how to create a new protocol in ns-2.

Steps to Create a New Protocol in ns-2:

1. Copy and paste old protocol.
2. Change folder name from old name to new name.
3. Change all the names in the folder from old name to new name (pay attention to capital letters)

Now, also change names in the following files:

1. Makefile.in
2. Makefile
3. ns-Lib.tcl

4. ns-Packet.tcl
5. Packet.h Modify the last one in the last position and increase the total value.
6. CMU-Trace.h
7. Cmu-Trace.cc
8. Priqueue.cc
9. ns-Default.tcl
10. Modify timers in `friend` class of the header file (such as the aodv.h file)
11. Modify global variables names.

Once all these changes are done, try to recompile ns-2 as follows:

```
Make clean
Make
Sudo make install
```

Trace Format in ns-2: In ns-2, the general trace format is as follows:

```
s 0.000000000 0 RTR— 0 AODV 44 [0 0 0 0]— — - [0:255 -1:255 1 0]
    [0x11 [0 2] 4.000000] (HELLO)
s 10.000000000 0 RTR— 0 AODV 48 [0 0 0 0]— — - [0:255 -1:255 30 0]
    [0x2 1 1 [1 0] [0 4]] (REQUEST)
```

A general description of the trace format is given in Table 13.3 (ns-2 Trace File Format [Online]).

13.6 Higher-Level Design of PR Activity Model for CRNs in ns-2

In this section, we will discuss our higher-level design of the PR activity model for CRN in ns-2. We will first describe how we implement a PR activity model in ns-2.

We then proceed further and discuss the main components for the PR activity modeling in ns-2. We also discuss the PR activity block and its responsibilities.

13.6.1 ON/OFF PR Activity Model

The ON/OFF PR activity model has been widely used in the literature. Thus, we select this ON/OFF PR activity model with exponential distribution as a case study and describe how this model is implemented in ns-2. Based on this activity model, other PR activity models can be easily coded in ns-2.

13.6.2 ns-2 Modifications

We used the CRCN patch of ns-2. This patch does not model the activity of the PR nodes. Thus, we enhance the CRCN patch of ns-2 to include the PR activity model. The ns-2 code can be downloaded from here [9]. The PR activity block is responsible for generating and keeping track of PR activities in each spectrum band (spectrum utilization), that is, the sequence of ON and OFF periods by PR nodes over the simulation time. These ON and OFF periods can be modeled as a continuous-time, alternating ON/OFF Markov renewal process (MRP) [10,11].

Table 13.3   ns-2 Trace Form Description

Column Number	Event Attributes	Example Values
1	It shows the occurred event	"s" SEND, "r" RECEIVED, "D" DROPPED
2	Time at which the event occurred?	10.000000000
3	Node at which the event occurred?	Node id such as 0
4	Layer at which the event occurred?	"AGT" application layer, "RTR" routing layer, "LL" link layer, "IFQ" Interface queue, "MAC" mac layer, "PHY" physical layer
5	Show flags	—
6	Shows the sequence number of packets	0
7	Shows the packet type	"cbr" CBR packet, "DSR" DSR packet, "RTS" RTS packet generated by MAC layer, "ARP" link layer ARP packet
8	Shows size of the packet	Packet size increases when a packet moves from an upper layer to a lower layer and decreases when a packet moves from a lower layer to an upper layer
9	[....]	It shows information about packet duration, MAC address of destination, the MAC address of source, and the MAC type of the packet body
10	Show flags	—
11	[....]	It shows information about source node IP address: port number, destination node IP address (−1 means broadcast): port number, IP header TTL, and IP of the next hop (0 means node 0 or broadcast)

The ON (busy) state means the channel is occupied by the PR node, and the OFF (idle) state means the channel is unoccupied by the PR node. We consider that the channel's ON and OFF periods are both exponentially distributed, as in Refs. [11,12]. The rate parameters and λ_x and λ_y of the exponential distribution are provided as inputs in the simulation, which were measured in Ref. [12]. Then, according to this rate parameter, the channels follow the ON and OFF periods. We achieve this by binding the variables in the TCL script with the C++ code of ns-2 and by using the timers of ns-2. As soon as channels are declared in the TCL script and the simulation begins, each channel undergoes ON and OFF states throughout the simulation period.

First, we define the variables:

```
set val(channum) 8 ;# number of channels per radio
set val(acschan) 5 ;# number of available channel set (acs) channels
```

Now, we access the handle for the MAC layer in the TCL script by the following statement:

```
set Mub Mac (0) [$node (0) set mac (0)]
$ns at 0.00 "$Mub Mac (0) board $val(channum) $val(acschan)";
```

In the Command of the MAC layer, we deal the "board" as follows:

```
int MubashirMac::command(int argc, const char*const* argv) {
if (argc = = 3) {
..........
return TCL OK;
}
}
else if (argc = = 4) {
if (strcmp(argv [6], "board") = = 0) {
```

We call here the function that calls the timer:

```
return TCL OK;
}
}
return Mac::command(argc, argv);
}
```

In ns-2, timers are used to delay actions, or they can be used for the repetition of a particular action such as broadcasting of Hello packets after a fixed time interval. We used timers to simulate a channel's ON/OFF states. The timer is declared in the macmubashir.h file.

```
//Timer for channel activity – channel # 1
class ChannelOccupancy Timer : public TimerHandler {
public:
ChannelOccupancy Timer(MubashirMac *a) : TimerHandler() {a = a;}
void expire(Event *e);
protected:
MubashirMac *a ;
};
```

We further declare the Channel Occupancy Timer as a friend class of MubashirMac as follows:

```
class MubashirMac : public Mac {
friend class ChannelOccupancy Timer;
public:
ChannelOccupancy Timer ChannelOccupancy Timer;
}
```

We consider a simple MAC protocol (Maccon.cc), available with the CRCN patch of ns-2. This MAC protocol is a multiple-channel, collision- and contention-based MAC protocol. Note that in the original state, the Maccon.cc MAC protocol selects a channel randomly from the predefined set of channels, and the channel selection decision occurs at the MAC layer. We now perform channel selection at the network layer. Thus, we modify this MAC protocol and provide the capability to the network layer to make the channel selection decision.

We further add channel selection strategies RD, HD, SB, and SURF [6] to the network layer, which we describe later. On the basis of any particular channel selection strategy, the network layer takes the channel selection decision. This decision is encapsulated in the network layer packet header and passed to the MAC layer, which then switches to the channel based on the channel selection decision provided by the network layer. In the Maccon.cc MAC protocol, there are two channel states: IDLE and BUSY. These states are dependent on the channel conditions, and they are used by the MAC protocol to handle the transmission and reception activities of CR nodes. IDLE means that the no activity is going on in the channel, and the channel is free to be used for transmission by the CR node. BUSY means that the channel is occupied by a current CR transmission. In order to deal with the activities of the PR nodes, we include for each channel two more states at the MAC layer, that is, PR OCCUPIED and PR UNOCCUPIED, indicating that the channel is occupied and unoccupied by the PR node, respectively. These two states of the channel will be checked each time by the MAC protocol while performing transmission or overhearing.

13.7 Issues and Challenges Regarding the Implementation of PR Activity Models in ns-2

This section will discuss the issues and challenges regarding the implementation of PR activity model in ns-2. For instance, we will discuss how a generalized PR activity model can be implemented in ns-2 from the programming perspective. In fact, from network to network, the activity of PR users varies. Owing to this nonconformity and spatiotemporal variation in the PR activity, we cannot use a single PR activity model to capture the activity of PR users. Consequently, many activity models for PR activity have been proposed.

13.8 Conclusion

In this chapter, we have discussed the design and implementation of a PR activity model for multimedia communication in the network simulator ns-2. This chapter will help novice users and researchers to simulate CRNs in ns-2. Moreover, this chapter will help researchers to validate their protocols and algorithms for multimedia communication in CRNs.

References

1. Issariyakul, T. and Hossain, E. *Introduction to Network Simulator NS2.* Springer, US, 2009.
2. Matthias, W., Riihijarvi, J., and Mahonen, P. Modelling primary system activity in dynamic spectrum access networks by aggregated on/off processes. In *6th Annual IEEE Communications Society Conference on Sensor, Mesh and Ad Hoc Communications and Networks (SECON) Workshops,* June 22–26, 2009, pp. 1–6.
3. Berk, C., Akyildiz, I. F., and Oktug, S. Primary user activity modeling using first-difference filter clustering and correlation in cognitive radio networks. *IEEE/ACM Transactions on Networking,* Vol. 19, No. 1, pp. 170–183, 2011.
4. Masonta, M. T., Mzyece, M., and Ntlatlapa, N. Spectrum decision in cognitive radio networks: A survey. In *IEEE Communications Surveys and Tutorials,* Accepted for publication, Vol. 15, No. 3, pp. 1088–1107, 2013.
5. Network Simulator NS-2. Available from: http://www.isi.edu/nsnam/ns/
6. Rehmani, M. H., Viana, A. C., Khalife, H., and Fdida, S. SURF: A distributed channel selection strategy for data dissemination in multi-hop cognitive radio network. *Elsevier Computer Communications,* Vol. 36, pp. 1172–1185, 2013.

7. Saleem, Y., Bashir, A., Ahmed, E., Qadir, J., and Baig, A. Spectrum-aware dynamic channel assignment in cognitive radio networks. In *IEEE International Conference on Emerging Technologies (ICET)*, 2012, pp. 1–6.
8. CRCN Patch for NS-2. Available from: http://stuweb.ee.mtu.edu/ljialian/
9. Rehmani, M. H., Opportunistic data dissemination in ad-hoc cognitive radio networks, *Doctoral dissertation*, Université Pierre et Marie Curie-Paris VI, 2011. Available from http://www-npa.lip6.fr/~rehmani/NS2_Code.zip
10. Lee, W.-Y. and Akyildiz, I. Optimal spectrum sensing framework for cognitive radio networks. *IEEE Transactions on Wireless Communications*, Vol. 7, No. 10, pp. 3845–3857, 2008.
11. Min, A. W. and Shin, K. G. Exploiting multi-channel diversity in spectrum-agile networks. In *IEEE Proceedings of INFOCOM*, 2008, pp. 1921–1929.
12. Kim, H. and Shin, K. Efficient discovery of spectrum opportunities with MAC-layer sensing in cognitive radio networks. In *IEEE Transactions on Mobile Computing*, Vol. 7, No. 5, pp. 533–545, 2008.

OTHER IMPORTANT DESIGNS

Chapter 14

Multimedia Communication for Emergency Services in Cooperative Vehicular Ad Hoc Networks

Muhammad Awais Javed, Duy Trong Ngo, and Jamil Yusuf Khan

School of Electrical Engineering and Computer Science,
The University of Newcastle, Callaghan, Australia

Contents

14.1 Introduction

A vehicular ad hoc network (VANET) is a key component of future intelligent transportation systems that support safety, traffic management, and user infotainment applications. With vehicles communicating with one another via wireless links, the VANET architecture can realize cooperative collision avoidance, emergency warning notification, enhanced route guidance, and multiplayer gaming [1–3]. The enormous potential of VANETs for improving the travel safety and comfort level has attracted significant research efforts from both academia and industry in the past few years [4,5]. As a result of many recent advances, it is envisioned that the emerging VANET architecture will be widely deployed in the near future.

In a VANET, multimedia communication is employed to provide vehicles with information about ongoing road emergency events, which include accidents, buildings on fire, floods, road closure, car breakdowns, and so on. The emergency message may contain images of an accident to help ambulance vehicles reach the accident scene for rescue. It could also include a short video of a building on fire, alerting nearby vehicles and informing fire brigade units of the extent of damage. The message may take the form of an audio message that gives information about floods on the highway [6,7]. Compared to a simple text message, such multimedia messages provide more precise and informative details of the emergency situation, enabling other vehicles on the road to better decide on how to react to the event. An improved awareness of the emergency can also help rescue teams to plan in advance, and thus more effectively handle the emergency situation.

While periodic single-hop safety messages are used in VANETs for cooperative awareness applications, emergency multimedia messages propagate alert notifications to all the vehicles within a certain geographical area or even a distant rescue vehicle. This multimedia message can be initiated by either the vehicle involved in the emergency itself or a lead vehicle located nearby. Multimedia communications in a service network such as VANET is different from conventional multimedia communications in that the information is transmitted in a multicast or broadcast manner, and the communication nodes operate in an ad hoc manner. In addition, as the emergency multimedia message has to be sent to locations potentially out of a vehicle's transmission range, the communication takes place over multiple hops. Ranging from several hundred kilobytes to a few megabytes, a multimedia message is typically decomposed into many small fragments to be reliably transmitted to the destination vehicles. The destination vehicles then combine all the fragments to reconstruct the original multimedia message. The total transmission delay of a multimedia message is limited to a few seconds to ensure timely notification [8].

The efficient dissemination of emergency multimedia messages over multiple hops faces several major challenges. First, a broadcast storm occurs when broadcast messages are endlessly circulated around the network, consuming all the available bandwidth and causing network saturation [9]. In relay-assisted multihop transmissions, the excessive redundant information introduced by the large number of relays may also worsen the network congestion situation. Moreover, the emergency multimedia messages may be severely interfered with by the frequently generated safety messages, whose packet transmission requirements are generally stringent. As more redundant multihop retransmissions are required in this case, the multihop dissemination time delay will increase while the reception rate of safety messages is adversely affected [10]. The interfering situation may be further complicated by hidden nodes where packet collisions cannot be detected [11]. If a single fragment of a multimedia message (whose size is typically large) is lost, the receiver may not be able to use the message.

The concept of user cooperation has become a key performance enhancement technique for wireless networks in recent years. Here, network entities interact with each other to improve the reliability of communications and enhance the application's quality of service [12]. A good example of such cooperation is the use of spatial diversity at the physical layer that helps mitigate the effects of channel fading [13]. Similarly, cooperative multiple access control (MAC) protocols are developed to enhance the efficiency of medium access and reduce packet collisions due to hidden and exposed nodes [14,15]. At the network layer, cooperative communication is proposed to find the routing paths with the least interference and the shortest end-to-end delay [16]. Whereas user cooperation efficiently aggregates resources at the transport layer in multihoming applications [17], cooperation among different applications can support simultaneously running applications within a single communication entity [18].

In the context of VANETs, safety applications rely on the cooperation among vehicles on the road to provide neighborhood visibility. Vehicles share their location and trajectory information with their surrounding vehicles in the form of periodic safety messages at the rate of 1–10 Hz [19]. In an emergency situation, warning messages are disseminated to alert other vehicles within a geographical area [20]. Such cooperation at the application level directly translates into the exchange of communication parameters among individual vehicles, for example, to distributively control the transmit powers and packet generation rates and best utilize the available bandwidth [21]. Cooperative communications are also used to improve the medium access procedures, for example, contention window selection in carrier-sense multiple access with collision avoidance (CSMA/CA)-based MAC and time slot allocation in time-division multiple access (TDMA)-based MAC [11,14,15,22]. Similarly, cooperative techniques are proposed to enhance the efficiency of information dissemination of warning messages in VANETs [23,24].

It is noted that most of the cooperative techniques devised for multihop warning message dissemination in VANETs center around the idea of decreasing message redundancy. Usually, these solutions are applicable to multihop messages with a low packet generation rate [9,10,23,25]. On the contrary, emergency multimedia messages are large in size, and they need to be generated and transmitted at higher rates. A large transmission overhead incurred by multimedia message dissemination may as well degrade the performance of the existing safety messages. Therefore, transmission protocols designed for emergency multimedia services must be highly reliable, while ensuring a small number of transmissions and minimal effects on the reception rate of the in-place safety messages.

The aim of this chapter is to study cooperative communication techniques for emergency multimedia applications in VANETs. With this goal in mind, we organize the chapter as follows: Section 14.2 briefly introduces VANET applications and their associated data traffic requirements. Section 14.3 discusses current industrial standards in VANETs, with specific references to the data transmission procedures in inter-vehicle communications. Section 14.4 reviews the existing cooperative communication techniques that enhance the performance of emergency message dissemination in VANETs. Section 14.5 presents a new multihop protocol that efficiently disseminates multimedia messages in VANET emergency applications. Finally, Section 14.6 concludes this chapter.

14.2 VANET Applications and Data Traffic Requirements

A VANET is an ad hoc network that provides connectivity between vehicles present on a road segment. In a VANET, an onboard unit (OBU) is mounted on the vehicle and supported by a wireless transceiver and various sensors, whereas a roadside unit (RSU) is a special infrastructure deployed at strategic locations along the road [1]. Enabled by vehicle-to-vehicle (V2V) and vehicle-to-infrastructure (V2I) communications, the service applications can be divided into three main categories, namely road safety, traffic management, and user infotainment. Table 14.1 summarizes several important VANET applications along with their data traffic requirements.

14.2.1 Safety Applications

The objective of safety applications is to provide drivers with an increased awareness of the surrounding traffic, helping to prevent dangerous situations, and avoid accidents. Because the main purpose of VANETs is to improve vehicle safety, these applications impose the highest service requirements, with the maximum time delay being restricted to under 0.5 s. As can be seen from Table 14.1, forward collision, lane change, intersection collision, and left-turn collision warning applications require the periodic exchange of safety messages. In Figure 14.1, an example of a forward collision warning application is demonstrated. Using the periodic safety messages received from its neighboring vehicles, vehicle A maintains a "neighborhood" table in which the trajectory information of the surrounding vehicles (e.g., distance, speed, and lane number) is recorded. With the "neighborhood" table, each vehicle controls its own speed and thus avoids any possible forward collision. Knowing the position of vehicles in the adjacent lanes, a safe lane-change maneuver can also be achieved.

In other safety applications, warning messages are triggered when such an emergency event is detected. For instance, if vehicle braking is suddenly applied, an emergency warning is instantly generated and transmitted to alert other vehicles within immediate proximity. Depending on the application type, an event-driven (E.D.) message may also be propagated over large distances by multiple-hop communication. Such a safety mechanism assists in avoiding chain collisions, a major cause of accidents on large highways. Figure 14.2 illustrates a scenario where the warning vehicle notifies the vehicles approaching the place of accident or emergency. These vehicles may then take appropriate actions, for example, reducing the vehicle speed, detouring, and so on. Following a quick text warning notification of the emergency event, a more detailed depiction of the event in the form of images or videos may be sent using emergency multimedia, which has a maximum latency of 1–3 images per minute or frames per minute (f/m).

Table 14.1   VANET Applications and Data Traffic Requirements

Type	Application	Data	Hops	Mode	Latency (sec)
Safety	Forward collision warning	Periodic	Single	V2V	0.1
	Lane change warning	Periodic	Single	V2V	0.1
	Intersection collision avoidance	Periodic	Single	V2V/V2I	0.1
	Left-turn collision warning	Periodic	Single	V2V	0.1
	Emergency brake light	E.D.	Single	V2V	0.02
	Post-crash warning notification	E.D.	Multi	V2V/V2I	0.5
	Approaching emergency vehicle	E.D.	Multi	V2V/V2I	0.5
	Emergency multimedia	E.D.	Multi	V2V/V2I	20–60
Traffic management	Enhanced route guidance	Periodic	Single	V2I	1
	Optimal traffic signal timing	Periodic	Single	V2I	1
	Traffic signal violation warning	Periodic	Single	V2I	0.1
User infotainment	Downloading music	E.D	Single	V2I	1–5
	Multiplayer games	E.D.	Single	V2V	0.1–1
	Media streaming	E.D.	Multi	V2V/V2I	1–5
	Parking space booking	E.D.	Single	V2I	1–5
	Restaurant booking	E.D.	Single	V2I	1–5

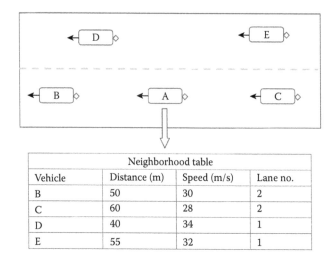

Figure 14.1   Safety applications using periodic safety messages.

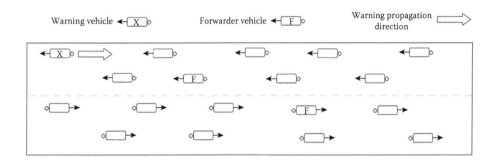

Figure 14.2   Safety applications using event-driven multihop warning messages.

14.2.2 Traffic Management Applications

In future road traffic systems, VANETs will provide intelligent trip planning and reduce road congestion [3]. The traffic management applications supported by VANETs help reduce travel time, fuel consumption, and air pollution. Compared with safety applications, these management applications have less stringent service requirements, with an acceptable latency of 1 s. As shown in Table 14.1, the enhanced route guidance employs V2I communication to determine the number of vehicles on a certain road. This piece of information is then shared with other RSUs, helping to calculate efficient routes with the shortest travel time. The same concept is applied at a local level to optimally control the traffic signal timing. Furthermore, vehicles approaching a traffic signal are informed with the current status of the traffic light (Red/Green), so as to adjust their speeds and avoid running the red light. Emergency information generated by the safety applications during an accident may also help in traffic management by avoiding congestion on the road.

14.2.3 User Infotainment Applications

Aiming to offer users information and entertainment services, the infotainment applications listed in Table 14.1 allow higher latency on the order of 1–5 s. However, the amount of data to be shared is large, particularly in music downloads and media streaming [3]. Some infotainment applications such as multiplayer games may require a low maximum latency of 0.1 to 1 s. For applications where commercial announcements are made (e.g., fuel sales, new movies, product promotions), the RSUs need to be installed at selected places along the road.

14.3 Current Standards for Vehicular Communications

Inter-vehicular communications standards have been developed by government and industrial organizations. In these standards, radio spectrum is specifically allocated for VANET applications, medium access and physical layer protocols are proposed for data dissemination, and data and management functionalities are defined for the upper layers in the protocol stack [20].

14.3.1 Spectrum Allocation for VANETs

Both the US Federal Communications Commission (FCC) and the European Conference of Postal and Telecommunications Administration (CEPT) have dedicated radio spectrum in

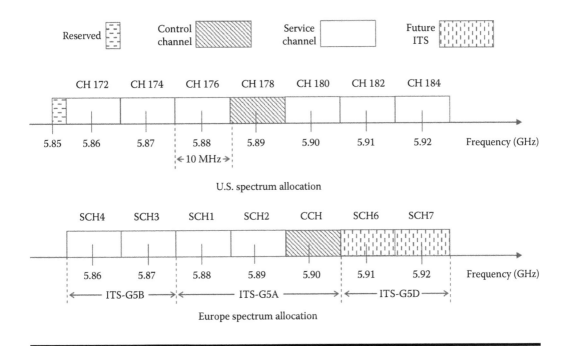

Figure 14.3   FCC and CEPT spectrum allocation for vehicular communications.

the 5.9 GHz band to vehicular communications [19,26]. The allocated spectrum consists of a single control channel (CCH) and multiple service channels (SCH), with the specific spectrum arrangements depicted in Figure 14.3.

In the United States, the dedicated short range spectrum (DSRC) has a total bandwidth of 75 MHz, ranging from 5.85 MHz to 5.925 MHz. With the initial 5 MHz reserved as guard bands, each of the remaining seven channels occupies 10 MHz. Of these, channel 172 is used for vehicle-to-vehicle safety applications (e.g., accident avoidance/mitigation, life and property safety). The next two channels, 174 and 176, are designated as service channels used for non-safety applications. Channel 178 is the control channel, whereas channels 180 and 184 are reserved as service channels. Channel 184 is assigned for both collision avoidance applications at intersections and public safety applications [19,27].

In Europe, the allocated 50 MHz is divided into the ITS-G5B (20 MHz) for non-safety applications and the ITS-G5A (30 MHz) for both safety and non-safety applications. Here, SCH 4, SCH 3, SCH 2, and SCH 1 are reserved as service channels, whereas CCH is the dedicated control channel. The remaining 20 MHz is assigned to ITS-G5D, reserved for future intelligent transportation systems (ITS) applications [26,27].

14.3.2 IEEE WAVE Standard

The standard for Wireless Access in Vehicular Environments (WAVE) was developed by two IEEE working groups: *task group p* and *group 1609* [28]. As shown in Figure 14.4, the IEEE 802.11 p standard deals with the MAC and PHY layer functionalities, whereas the IEEE 1609 standard handles the functionalities in the upper layers [19].

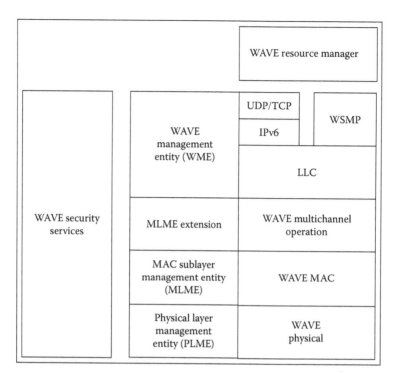

Figure 14.4   IEEE WAVE architecture.

14.3.2.1 IEEE 802.11p Standard

An amended version of the IEEE 802.11 standard, the IEEE 802.11 p standard was developed to support the WAVE architecture, where medium access control and physical layer procedures are defined in both data and management planes. The management functions at the physical and MAC layers are handled by the Physical Layer Management Entity (PLME) and the MAC Sublayer Management Entity (MLME), respectively.

The IEEE 802.11 p physical layer is almost similar to that defined in the IEEE 802.11 a standard [29]. While it is based on the orthogonal frequency-division multiplexing (OFDM) technique, the channel bandwidth is reduced from 20 MHz to 10 MHz to overcome the larger delay spread inherent to vehicular networks [19]. As a result, all the timing parameters shown in Table 14.2 have been multiplied by 2 to better guard against the inter-symbol interference (ISI). Also, the transmission data rate in the IEEE 802.11 p standard is one half of that in the IEEE 802.11 a standard.

The IEEE 802.11 p MAC is based on the Enhanced Distributed Channel Access (EDCA) procedure defined in the IEEE 802.11 e standard [29]. As shown in Figure 14.5, the CSMA-/CA-based IEEE 802.11 e MAC protocol uses the distributed coordination function (DCF) to coordinate multiple channel access. At first, the wireless node intending to transmit data performs channel sensing. If the channel has been idle for a period that is equal to the arbitrary inter frame spacing (AIFS), the node selects a random backoff value from the set of integers in a contention window (CW) to defer its access. The backoff value is decremented at each time slot. If the channel becomes busy during this process, the backoff value is suspended. After the channel becomes idle again for a duration equal to the AIFS, the backoff value continues to decrement from the previous suspended value. Finally, the data are transmitted when the backoff value is 0.

Table 14.2   IEEE 802.11 p PHY Parameters

Parameter	Value
Channel bandwidth	10 MHz
OFDM symbol duration	8 μs
Preamble duration	32 μs
Guard interval	1.6 μs
Slot duration	13 μs
Modulation scheme	BPSK, QPSK, 16-QAM, 64-QAM
Code rate	1/2, 1/3, 1/4
Data rate (in Mbps)	3, 4.5, 6, 9, 12, 15, 18, 24, 27

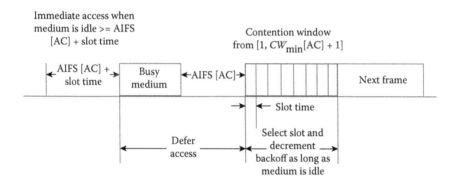

Figure 14.5   IEEE 802.11 e channel access procedure.

The contention window size is initially set to CW_{min}. After every failed transmission, the size of the contention window is increased by $2(CW + 1) - 1$ until it reaches CW_{max}. The purpose of the increase is to reduce the collision probability among the nodes that attempt to access the channel. After every successful transmission, the contention window size is reset to CW_{min}.

As in the EDCA mechanism, the IEEE 802.11 p prioritizes the traffic into four access categories (ACs), namely background traffic (BK), best effort traffic (BE), voice traffic (VO), and video traffic (VI). Because of the shorter channel sensing time, CW_{min} and CW_{max}, traffic within the highest AC, that is, video traffic as shown in Table 14.3, is given the highest priority in transmitting their packets.

The basic service set (BSS) formation in the IEEE 802.11 p standard does not require synchronization, authentication, and association steps as in the IEEE 802.11 standard. For vehicular communications where inter-vehicle connection typically lasts a few seconds, such a handshaking process would incur unacceptably long delays. Therefore, two nodes in the IEEE 802.11 p standard can immediately share data without having to join a BSS [30]. In this *WAVE mode*, communication is accomplished through a special basic service set identification (BSSID) known as the *wildcard BSSID*. A *provider* station starts the WAVE basic service set (WBSS) by sending periodic beacon messages on its control channel (CCH). Called the WAVE Service Advertisement (WSA) messages, these beacons contain information about the available service applications and parameters needed

Table 14.3　EDCA Parameters in IEEE 802.11 p

AC	CWmin	CWmax	AIFS
AC_BK	15	1023	9
AC_BE	15	1023	6
AC_VO	7	15	3
AC_VI	3	7	2

to join the WBSS (e.g., BSSID and service channel [SCH] of the WBSS). Upon scanning the WSA messages on the CCH, all the nodes can join a WBSS by switching to the SCH of that WBSS.

14.3.2.2　IEEE 1609 Standard

The upper layers of the WAVE architecture in Figure 14.4 are defined in the four standards developed by the IEEE 1609 group [31–34]. The IEEE 1609 standard manages several functions including channel coordination, channel routing, networking services, security, and remote resource access.

14.3.2.3　Channel Coordination

As discussed in Section 14.3.1, the allocated spectrum for vehicular communication comprises one CCH and multiple SCHs. The safety messages and the WSAs are transmitted on the CCH while service and business applications are run on the SCH. The periodic safety messages are named basic safety messages (BSMs) in the WAVE architecture. To allow vehicles equipped with single-channel radio to use both types of channels, the multichannel operation defined in the IEEE 1609.4 standard divides the time into a control channel interval (CCHI) and a service channel interval (SCHI) as shown in Figure 14.6. The four channel access options defined in the IEEE 1609.4 standard include continuous access to the CCH and the SCH, alternate access to the CCH and the SCH, immediate access to the SCH before the start of the SCHI, and extended SCH access without a CCHI. Recently, the US Department of Transportation (DoT) has been investigating the use of channel 172, dedicated to safety messages that are FCC compliant, while keeping the CCH reserved for management information and WSAs [27]. The channel coordination mechanism thus controls the channel intervals to select the appropriate radio channel at the right time for data exchange.

14.3.2.4　Channel Routing

The channel routing control function defined in the IEEE 1609.4 standard manages the routing of data from the logical link layer (LLC) to the appropriate channel and access category. Two types of packets are defined in the WAVE standard. The Internet Protocol (IP) packets are used to transmit non-safety messages, while the Wave Short Message (WSM) packets are used for safety messages. The WSMs are short messages whose transmission parameters such as channel number, data rate, and transmit power can be directly controlled by the higher layers so that all nodes receive the information within a certain time delay. They are suited for safety applications that have higher priority and require low transmission delay. The channel router identifies the type of packet from the EtherType field in the 802.2 header and forwards the WSM packet to the CCH or to the SCH and the IP packet to the current SCH.

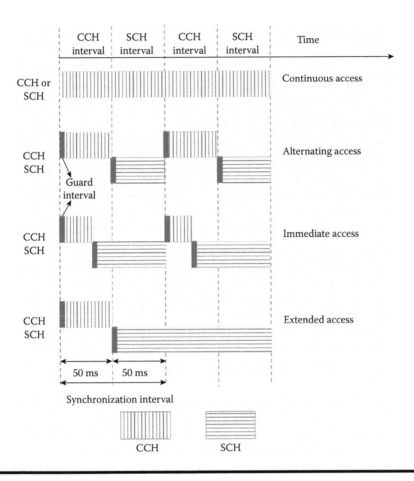

Figure 14.6   Channel access options in the IEEE 1609.4 standard.

14.3.2.5 Networking Services

The IEEE 1609.3 standard defines functions and services at the LLC, network, and transport layers of the WAVE protocol stack [32]. These services are known as the networking services, which include the data plane and Wave Management Entity (WME) as shown in Figure 14.4. The IEEE 1609.3 standard defines a new transport layer protocol known as the Wave Short Message Protocol (WSMP), which enables the transmission of WSM messages for high-priority safety applications. For non-safety information dissemination, IPv6 has been supported along with the Transmission Control Protocol (TCP) and User Datagram Protocol (UDP) transport layers. After the LLC receives a MAC service data unit, it forwards the packet to the IPv6 layer or the WSMP layer according to the EtherType field of the packet.

14.3.2.6 Security and Remote Resource Access

Security services for messages and applications are defined in the IEEE 1609.2 standard [33]. They include algorithms and mechanisms for secure data transfer between WAVE units. IEEE 1609.1 specifies the format and type of message exchange required for remote resource access [34].

To reduce the cost of hardware in the form of OBUs or RSUs for providing various vehicular services, computational intensive hardware is installed at a remote unit. The OBUs and RSUs in a VANET can request the remote unit to provide different services using the communication procedures defined in the IEEE 1609.1 standard.

14.3.3 ETSI ITS Architecture

The European standard for vehicular communication has been defined by the technical committee on Intelligent Transport Systems (TC ITS) established by the European Telecommunications Standards Institute (ETSI). Shown in Figure 14.7 is the ETSI ITS architecture, which integrates multiple communication technologies such as IEEE 802.11 p, wireless LAN, 3G, and Long-Term Evolution (LTE) at the MAC and PHY layers [26]. As can be seen from Table 14.4, the only difference between the ETSI standard and the WAVE standard at the PHY layer is their respective values of the power spectral density (PSD). The reason is that the ETSI PSD must conform to European regulations. Between the application layer and the transport layer, the ETSI standard defines a new facilities layer to collect and distribute data received from other vehicles to all applications. This new layer reduces the large overhead that would normally be incurred when individual applications gather the data separately.

In the ETSI ITS architecture, the safety messages are termed as cooperative awareness messages (CAMs) with a maximum packet generation rate of 1–10 Hz, whereas the event-driven notification messages are now called decentralized environmental notification messages (DENMs) [20].

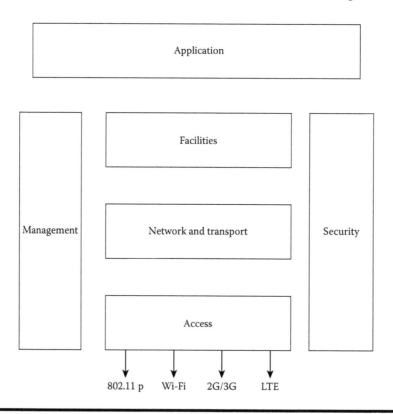

Figure 14.7 The ETSI ITS architecture.

Table 14.4   Comparison of WAVE Standard and ETSI Standard

Parameter	WAVE	ETSI
Spectrum bandwidth	75 MHz	50 MHz
Frequency range	5.855–5.925 GHz	5.855–5.905 GHz
Control channel frequency	5.89 GHz	5.9 GHz
Number of service channels	6	4
PHY/MAC layers	IEEE 802.11 p	Multiple access technologies
Higher layers	Application, transport, IEEE 1609.3, 1609.4, LLC	Applications, facilities, network, transport
Main safety messages	BSM	CAM, DENM

While the ETSI standard defines the format and message generation conditions for both CAMs and DENMs, it does not provide a channel switching option as in IEEE WAVE. Such a configuration helps improve the utilization of the control channel, but at the cost of implementing separate transceivers. Regarding the transceiver configuration, the ETSI standard offers three options [27]. In the first option, the transceiver is tuned exclusively to the CCH for safety applications. The second configuration employs multiradio transceivers where one transceiver is exclusively tuned to the CCH and the other transceiver is tuned on demand to SCH 1 and SCH 2 on the ITS-G5A. The third configuration is available for non-safety applications. Here, the two transceivers can be used on demand: The first uses SCH 1 and SCH 2 channels on the ITS-G 5 A, and the second uses SCH 3 and SCH 4 channels on the ITS-G 5 B. The ETSI ITS architecture also introduces a decentralized congestion control (DCC) that adapts the MAC and PHY parameters to resolve channel congestion. For instance, transmit power, packet generation rate, data rate, and receiver sensitivity can be dynamically adjusted according to the current network load [35]. The DCC takes into account the busy time, whose optimal lower and upper threshold values are chosen to be 10% and 40%, respectively.

14.4 Cooperative Communication Techniques in VANETs

Emergency services supported by cooperative VANETs rely on the efficient dissemination of emergency information to vehicles located within a certain geographical area. As periodic *safety messages* and *emergency messages* coexist on the same frequency bands, they can strongly interfere with each other. If not properly managed, a large number of packet losses will severely degrade the quality of communication.

14.4.1 Safety Message Dissemination

To enhance the effectiveness of safety messages, safety information should be disseminated to neighborhood vehicles on the road with a high reception rate while providing sufficient bandwidth for emergency messages. Owing to the high packet generation rate of safety messages, concurrent transmissions from the hidden vehicles may result in a large number of undetectable collisions. The communication techniques proposed in the literature aim to reduce packet collisions by adapting the transmission parameters and coming up with new medium access procedures.

14.4.1.1 Transmit Power Control

The techniques proposed in this category avoid channel congestion and maximize safety awareness by using adaptive transmit power control. To keep the load of the periodic safety messages below a predefined threshold [10], devises a cooperative transmit power control technique. On the basis of the number of vehicles within the carrier-sensing range, each vehicle finds the maximum common transmit power that satisfies the load requirement. An algorithm that achieves max–min fairness is then proposed. In Ref. [36], the transmit power is adapted to the vehicle density, which is estimated based on the vehicle's stopping time within their current trip time. Vehicle density measurement is also used by Rawat et al. [37] to control the transmit power. Reference [38] uses both vehicle density and vehicle speed to determine the optimal transmit power. Exploiting traffic context knowledge [18], devises a cooperative congestion control mechanism that adapts the transmit power in lane-change-warning applications.

14.4.1.2 Packet Generation Rate Control

Packet generation rate control techniques reduce the amount of data sent on the control channel to improve the reception rate of safety messages. The algorithm proposed in Ref. [39] employs a traffic timing structure to estimate the traffic density and thereby control the packet generation rate. In Ref. [40], adaptive beacon schemes are devised to control the packet generation rate depending on vehicle speeds. By evaluating the estimated tracking error value of its own position provided by the neighbor vehicles, Ref. [21] adapts the transmission probability of safety messages. Finally, an analytical model of packet generation probability based on vehicle density, transmission power and packet length is developed by Ye et al. [41].

14.4.1.3 Carrier Sense Threshold Control

In CSMA/CA-based medium access schemes, the carrier sense threshold controls the signal power up to which a vehicle senses a simultaneous transmission to suspend its own attempt to access the channel. A higher carrier sense threshold reduces the carrier sense range and allows a quicker medium access to the vehicles at the expense of higher interference from the distant nodes. The protocols proposed in this category find an optimal balance between the interference and medium access delay. In Ref. [42], the carrier sense threshold is adapted to the vehicle density to improve the packet reception rate at low propagation distances. On the other hand, the technique in Ref. [43] reduces the carrier sense threshold as the waiting time of messages in the transmission queue increases, thus improving the medium access delay.

14.4.1.4 Data Rate Control

Because IEEE 802.11 p supports variable modulation and data rates, the protocols in this category find their optimal combination for safety messages. The experiments performed by Jiang et al. [44] lead to the conclusion that QPSK modulation and the 6 Mbps data rate show the best performance in terms of the safety message reception rate. To reduce the effect of channel congestion [45], proposes a load-based data rate control where a higher data rate is selected when channel congestion is detected. Finally [11], proposes a combined space division multiple access and data rate control algorithm that uses different data rates for the safety messages depending on the current vehicle density to limit the number of safety message transmissions.

14.4.1.5 Medium Access Protocols

The proposals in this category suggest new medium access techniques in place of the default CSMA/CA in the IEEE 802.11p standard to improve the performance of the safety messages. By using TDMA [14,15], propose a time slot reservation mechanism with the help of cooperative communication between neighborhood vehicles. A self-organizing TDMA protocol proposed in Ref. [46] divides the synchronization interval into time slots of length equal to the safety message transmission time. After the initialization time of one synchronization interval, vehicles select an unused time slot or a time slot of the farthest vehicle to reduce interference. In Ref. [22], the Adaptive Space Division Multiplexing (ASDM) scheme is proposed for VANETs in which time slots are assigned according to a mapping function that is based on the vehicle position. By allocating empty slots on the road to the vehicles according to their measured headway, the developed protocol improves the bandwidth utilization and increases the number of transmissions per every synchronization interval.

14.4.2 Emergency Message Dissemination

The emergency message dissemination protocols aim to ensure a timely and successful delivery of emergency messages with a minimum number of required transmissions. However, hidden node collisions and broadcast storms are the main challenges in efficient propagation of emergency messages. The protocols proposed in the literature to overcome these problems can be divided into two broad categories: (1) the ones that rely on a distance-based contention mechanism to select the relay node and (2) the ones that reserve the next relay node using control packets.

14.4.2.1 Distance-Based Contention Multihop

To efficiently disseminate multihop warning messages in VANETs, the proposals in this category use a contention mechanism to select the farthest relay node. An example of this mechanism is the timer- and probability-based protocols that aim to suppress broadcast storms [9]. Based on its respective distance from the sender of a warning message, each vehicle is assigned with a wait time or a probability of transmission. Being given the shortest wait time or the highest probability of transmission, the vehicle located farthest in the range of the sender rebroadcasts the warning message. Upon receiving this duplicate message, all other vehicles withdraw their intention to rebroadcast. In Ref. [23], the proposed Distributed Vehicular Broadcast (DV-CAST) scheme employs the timer based approach to suppress broadcasts and also adopts the store-carry forward mechanism for disconnected networks. DV-CAST uses the connectivity of vehicles on the road to determine if the neighborhood is well connected, sparsely connected, or totally disconnected. It then specifies routing rules to disseminate multihop message in each of the traffic density scenarios. Taking a distance-based contention approach [47], devises the Contention-Based Forwarding protocol. Here, vehicles within a selected contention area employ a timer-based technique to disseminate multihop messages.

To consider both channel quality and distance in selecting the next relay node [24], assigns a shorter waiting time for vehicles belonging to a connected dominating set (CDS) and employs acknowledgments for every multihop message. The proposal in Ref. [48] suggests an optimized slotted 1 -persistence scheme for multihop communications. In addition, the protocol proposes the formation of vehicle clusters that include vehicles within the radio range of each other. Using the cluster front and tail, the protocol proposes a store-carry forward mechanism in sparse networks to reduce the network overhead. The work of Li et al. [25] develops the opportunistic broadcast scheme, in which the relay nodes use a long-range ACK packet to lower the redundant multihop transmissions. Moreover, the dissemination of multihop messages takes place in two phases: (1) the first one for quick propagation of the message and (2) the second one for increasing reception reliability.

14.4.2.2 Relay Reservation-Based Multihop

The protocols in this category make use of control packets to reserve the relay node. The Urban Multihop Broadcast (UMB) protocol [49] partitions the road into small segments and lets the vehicles in the farthest nonempty segment forward the received multihop message. To select the farthest node and address the hidden-node problem, the nodes transmit a black burst message of duration proportional to the distance of their segment from the source. The nodes that sense the channel as busy after their black burst is over cancel their intention to rebroadcast, and only the farthest segment forwards the message. In Ref. [50], the Smart Broadcast (SM) protocol is devised, which assigns a distance-based contention window to each road segment. A source node first sends a request to broadcast (RTB) message containing information about the contention window, segment size, and message direction. The selected relay vehicle is the one whose contention window expires first, and it sends a clear-to-broadcast (CTB) message that informs other vehicles to cancel their rebroadcast. On hearing the CTB message, the source node sends the warning message to be forwarded by the selected relay node.

14.5 A New Protocol for Emergency Multimedia Message Dissemination in VANETs

To reduce the number of transmissions required for emergency multimedia message dissemination and interference from the existing safety messages, we will present a new cooperative protocol in this section. The basic idea of the proposed cooperative time-slotted multimedia (CTSM) protocol is to employ separate time slots to transmit multimedia messages, so as to avoid interfering with the existing safety messages. Moreover, the CTSM protocol implements a segment-leader-based relay selection mechanism to reduce the number of message forwarders.

14.5.1 VANET Communication Scenario

Figure 14.8 shows the highway VANET communication scenario under consideration. In both directions of the highway, moving vehicles are connected via wireless links to form a VANET. For simplicity, we do not consider the cases with roundabouts and intersections. Assume that each vehicle is equipped with a Differential Global Positioning System (DGPS) that can accurately measure its own position on the road. An emergency multimedia message is generated by a vehicle marked as an "emergency warning vehicle" to be sent to other vehicles in the indicated

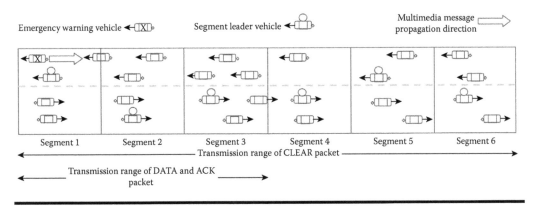

Figure 14.8   Emergency multimedia message dissemination on a highway via a VANET.

propagation direction. Because the transmission range of a vehicle is limited, we employ a multihop transmission technique to ensure that the emergency multimedia message can reach vehicles located at long distances from the emergency warning vehicle.

14.5.2 Cooperative Time-Slotted Multimedia Transmission Protocol

14.5.2.1 Road Segment Division and Segment Leader Selection

To facilitate efficient broadcast of the emergency multimedia message, we divide the highway into fixed-size road segments and designate one vehicle within each segment as the "segment leader." This is demonstrated in Figure 14.8. Using the DGPS data, each vehicle can determine the end points of its segment. Based on its own position and speed and with the known segment end points, each vehicle can also determine the remaining time that it will stay within the current segment. Information regarding the positions and speeds of other vehicles in the same segment is received via the periodic safety messages. With this information, every vehicle can also calculate the remaining time that other vehicles stay within their respective segment.

In the beginning, we select the segment leader vehicle as the one with the longest remaining time in its current segment. Note that the segment leader can be in any lane of the highway. Vehicles selected as the leaders will serve as the segment leaders for the entire duration of stay in those segments. When a segment leader realizes that it will move to another segment in a certain time, typically in five synchronization intervals (i.e., 500 ms), this vehicle marks its status as Retired for the remaining time of its presence in the current segment. This vehicle is also responsible for appointing a new segment leader, the one with the current highest remaining time in the segment. The retired segment leader notifies such an appointment to the new segment leader, and to all the current and newly arrived vehicles in the current segment.

To implement the foregoing mechanism, we propose that every vehicle add two new fields as parts of its periodic safety messages. Specifically, the MY_STATUS field indicates the current status of a vehicle, where 0, 1, and –1 represent Normal, Leader, and Retired vehicles, respectively. Further, the LEADER_ID field contains the node ID of the current segment leader. Simply put, when a vehicle receives a periodic safety message from a leader vehicle (with MY_STATUS = 1), the LEADER_ID field in that message contains the node ID of the leader.

When a segment leader retires, it finds the new segment leader and sends the safety message with the node ID of that new leader in the LEADER_ID field. Upon receiving the periodic safety message with MY_STATUS = –1, each vehicle matches its own node ID with the LEADER_ID field value. If a vehicle recognizes a match, it sets itself as the new leader and changes its MY_STATUS field to 1.

If any vehicle in a segment does not receive a periodic safety message from the leader vehicle for five consecutive synchronization intervals (i.e., 500 ms), it will find the leader using the periodic safety messages received from the neighborhood vehicles. Similarly, if a vehicle enters a segment and there is no vehicle within the segment, it would wait for five synchronization intervals before appointing itself as the segment leader. In case two vehicles in proximity consider themselves as the current segment leaders, the vehicle first receiving the periodic safety message with a MY_STATUS field equal to 1 will change its MY_STATUS to 0, thus becoming a Normal vehicle. This arrangement guarantees that there is always a single segment leader in any road segment.

14.5.2.2 Multimedia Time-Slot Reservation Mechanism

Owing to the large size of a multimedia message, we decompose the message into several fragments, each of which is to be reliably transmitted to the receiver. Upon combining all the received fragments, the receiver can recover the original message. Let S_m be the size of a multimedia message (in bytes) and S_f the size of a single fragment (in bytes). The number of packets required to completely send the multimedia message can be written as

$$P = \frac{S_m}{S_f}. \tag{14.1}$$

Let T_m be the time needed for the entire multimedia message to be transmitted. This implies that a single fragment must be transmitted within

$$T_f = \frac{T_m}{P}. \tag{14.2}$$

For emergency multimedia message transmissions, we propose to divide T_f into N time slots, each of which has the structure depicted in Figure 14.9. Let D_a be the distance between the emergency warning vehicle and the end point of the geographical area of interest. Owing to multipath fading, we assume a reliable transmission range to be one-half of the actual transmission range R_t. The number of time slots required for a single multimedia fragment transmission can be calculated as:

$$N_a = \frac{D_a}{R_t/2}, \tag{14.3}$$

which is exactly the number of hops required to disseminate the multimedia message within area D_a. To make sure that the fragment is transmitted to all the vehicles in the geographical area within the T_f duration, we propose the use of an additional number of guard time slots N_g. Altogether, the number of time slots required for a single fragment transmission is

$$N = N_a + N_g = \frac{2D_a}{R_t} + N_g \tag{14.4}$$

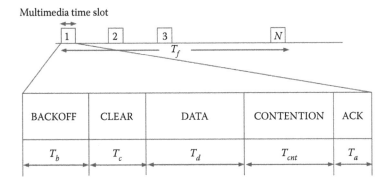

Figure 14.9 Structure of a time slot in the proposed multihop broadcast protocol.

The proposed multimedia time slot starts with a random BACKOFF phase during which vehicles with queued multimedia packets wait for a random time. It is noteworthy that the safety messages are uniformly generated within a synchronization interval (SI) of 100 ms, and that they employ the CSMA/CA mechanism. In this BACKOFF phase, safety message transmissions are also suspended.

The vehicle whose backoff timer expires first will then transmit a CLEAR packet at a transmission range R_t. The purpose of the CLEAR packet is to reserve the rest of the time slot for the multimedia packet, and also to inform the vehicles in the range R_t of the upcoming multimedia fragment transmission. The CLEAR packet also contains the additional information of T_f and N that informs all vehicles about the number of slots within T_f and the start time of each multimedia slot. Upon receiving the CLEAR packet, other vehicles suppress their transmissions until the end of the time slot. As such, all the vehicles in the range R_t that intend to send periodic safety messages during the reserved multimedia time slot will not interfere with the multimedia fragment transmission. Vehicles that do not receive a CLEAR packet may continue with their safety message transmission as soon as the backoff time is over. This arrangement allows the full utilization of the multimedia time slot in the absence of an emergency multimedia fragment.

After the CLEAR phase, the vehicle that is reserved with the multimedia time slot is allowed to send the multimedia fragment (i.e., DATA) at a transmission range of $R_t/2$. Because the CLEAR packet is transmitted over a range of R_t, all hidden nodes located within two transmission hops from the multimedia-fragment sender are made aware of the upcoming multimedia transmission. Essentially, the interference from any hidden nodes during the transmission of such a message is eliminated.

In the CONTENTION phase, each segment leader that has received the multimedia fragment computes a contention time T_{cnt}. We propose that the value of T_{cnt} is inversely proportional to the segment separation between the sender of the multimedia fragment and the segment leader (i.e., the receiver):

$$T_{cnt} = (M_s - S_d)T_s, \qquad (14.5)$$

where M_s is the maximum number of segments within the multimedia fragment transmission range, S_d is the segment separation between the sender and the receiver, and T_s is the slot time.

In the ACK phase, the segment leader with the shortest contention time T_{cnt} will transmit an ACK packet. It is worth recalling that this leader is responsible for relaying the multimedia fragment in the next time slot. After receiving the ACK of the multimedia fragment from the winning leader in the current time slot, all other segment leaders (i.e., the potential forwarders) delete the corresponding multimedia fragments in their queues. No other actions are required from these vehicles. Note that such a message cancellation policy only applies when a segment leader receives an ACK from a vehicle *located further away in the message propagation direction*. This ensures the progress of the multimedia fragment in the propagation direction. It might also happen that the vehicle who has forwarded the multimedia fragment does not receive an ACK within the current time slot. In such a case, this vehicle will resend the same multimedia fragment in one of the next time slots that are reserved for multimedia message dissemination.

14.5.3 Performance Evaluation

We evaluate the performance of the proposed cooperative time-slotted multimedia transmission (CTSM) protocol using OPNET Modeler 16.0. Specifically, we consider a highway scenario with a road length of 2 km and assume there are three lanes in each opposite direction. At medium vehicle density (120 vehicles/km), we assume an exponentially distributed inter-vehicle spacing, whereas at high vehicle densities (180 and 240 vehicles/km) we assume a normally distributed inter-vehicle spacing [51]. To represent a medium fading intensity, we assume Nakagami-m fading with $m = 3$.

We assume that every vehicle generates periodic safety messages with a transmission range of 500 m at a rate of 10 Hz. As shown in Figure 14.8, we place an emergency warning vehicle that generates and transmits a multimedia image message within the 2 km road section. Assume that the multimedia image is in JPEG format with a resolution of 320×160 pixels. If 2 bytes per pixel is required to store the red, blue, and green (RGB) light values, the image size is $S_m = 100$ kbytes. Let the fragment size be $S_f = 500$ bytes and the number of guard time slots be $N_g = 0.5N_a$. Because $T_m = 20$ s (3 images per minute), a total of 15 image messages is to be transmitted during the entire simulation time of 300 s. The practical parameters used in our simulations are listed in Table 14.5.

Referring to the time-slot structure of our proposed design in Figure 14.9, the multihop time slot size T_{ts} is computed as

$$T_{ts} = T_b + T_c + T_d + T_{cnt} + T_a ,$$ (14.6)

where T_b is the BACKOFF time, T_c is the CLEAR packet transmission time, T_d is the DATA (multimedia message) transmission time, T_{cnt} is the CONTENTION time, and T_a is the ACK packet transmission time. Here, the backoff is selected as a random integer from a contention window consisting of 7 slots, each of which is 13 μs in duration. The contention time T_{cnt} is determined by Equation 14.5, where the maximum segment size $M_s = 6$ is used for the 500 m transmission range and the 75 m road segment size.

The size of a CLEAR packet is 24 bytes (8 bytes of physical overhead, 8 bytes of T_f, and 8 bytes of N), and it has a transmission range of 1,000 m and a data rate of 6 Mbps. The DATA message is 528 bytes long (including 500 bytes of multimedia fragment and 28 bytes of MAC overhead), and its transmission range and data rate are 500 m and 12 Mbps, respectively. ACK is a short packet of 38 bytes (including 30 bytes of ACK information and 8 bytes of MAC overhead).

Table 14.5 Simulation Parameters

Parameter		Value
Highway Road	Road length	2 km
	No. of lanes	6 (3 per direction)
	Segment size	75 m
	Max. no. of seg. M_S	6
Vehicle	Medium density	120 (vehicles/km)
	High density	180, 240 (vehicles/km)
	Speed	60, 45, 30 (km/h)
Safety message	Size	344 bytes
	Data rate	6 Mbps
	Transmission range	500 m
	Generation freq.	10 Hz
Multimedia BACKOFF	Size	7
Contention Window	Slot time T_S	13 µs
CLEAR packet	Size	24 bytes
	Data rate	6 Mbps
	Transmission range	1,000 m
DATA message	Size	528 bytes
	Data rate	12 Mbps
	Transmission range	500 m
ACK packet	Size	38 bytes
	Data rate	12 Mbps
	Transmission range	500 m
Multimedia message parameters	T_m	20 s
	Image resolution	JPEG (340 × 160)
	S_m	100 kbytes
	P	205
	T_f	0.097 s
	N_a	8
	N_g	4
Fading model		Nakagami-m ($m = 3$)
Reception Rx_{th}		−91 dBm
Background noise		−99 dBm
Simulation time		300 s

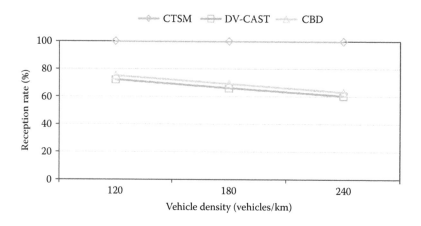

Figure 14.10   Reception rate of emergency multimedia messages within a distance of 2 km.

To calculate T_c, T_d, and T_a in Equation 14.6, we need the packet transmission time T_t, which is given by

$$T_t = \frac{L_p}{R_d}, \tag{14.7}$$

where L_p is the packet size, and R_d is the data rate. By using the packet size and data rate of CLEAR, DATA, and ACK from Table 14.5, we calculate T_c, T_d, and T_a as 0.032, 0.36, and 0.027 ms, respectively. By putting these values in Equation 14.6, T_{ts} is indeed 0.59 ms.

We compare our proposed CTSM transmission design with two existing protocols, namely DV-CAST [23] and CBD [47]. In our simulation study, the transmission range, the number of time slots, and the maximum wait time for the DV-CAST protocol are taken as 500 m, 5, and 5 ms, respectively [23]. On the contrary, the transmission range, the contention area, and the maximum contention time for the CBD protocol are set as 500 m, 300 m, and 50 ms, respectively [47]. In Figure 14.10, we show the percentage of vehicles that successfully receive all fragments of a multimedia message within the 2 km road section. As is evident from the results, our proposed solution guarantees an almost 100% delivery rate in every vehicle density scenario considered. This is a noticeable enhancement in light of the 60%–75% reception rates provided by the DV-CAST and the CBD protocols. The performance of these protocols is degraded owing to interference between the existing safety messages and the multimedia messages resulting in the loss of multimedia fragments.

Figure 14.11 shows the fragment delay, that is, the average time period required to disseminate a single fragment of a multimedia image within the 2 km road section. It is clear from the figure that the proposed CTSM protocol has a much lower dissemination time compared to both the DV-CAST and the CBD protocols. In particular, at the density of 120 vehicles/km, the CTSM fragment delay is 5 ms shorter than the DV-CAST and the CBD counterparts. As the number of vehicles per kilometer increases, the fragment delay incurred by the DV-CAST and the CBD schemes dramatically grows owing to the higher interference from the periodic safety messages. However, the effect of interference is limited in the case of our CTSM proposal because we employ separate time slots to send the multimedia message. From Figure 14.11, the resulting CTSM fragment delay only slightly increases, remaining below 60 ms for all the

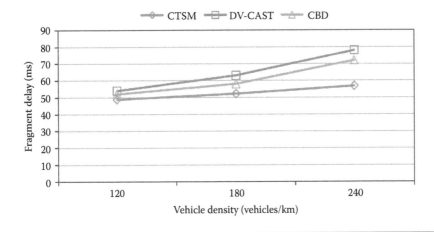

Figure 14.11 **Time required to disseminate a single fragment within a distance of 2 km.**

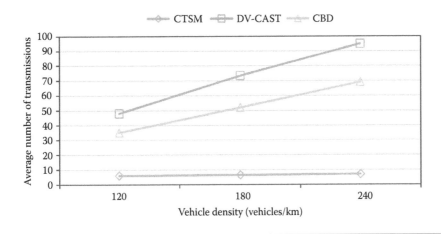

Figure 14.12 **Average number of multihop transmissions required to successfully disseminate a single multimedia fragment within a distance of 2 km.**

vehicle densities under consideration. Finally, the most pronounced advantage is observed at the vehicle density of 240 vehicles/km, where the fragment delay drops by 16–24 ms in our proposed CTSM solution.

We display in Figure 14.12 the average number of transmissions required to successfully disseminate a single fragment of an image message to the vehicles within a 2 km road section. In comparison to the DV-CAST and the CBD protocols, our proposed CTSM scheme significantly reduces the number of required transmissions. In particular, at the vehicle density of 240 vehicles/km, only seven transmissions are needed for dissemination of a single fragment. This figure represents a mere 7%–10% of the total number of transmissions required by the existing approaches. As discussed in Section 14.5.2.2, such a remarkable gain is a direct result of several advanced features of our design: (1) the suppression of broadcast storms by only allowing segment leaders to act as the potential forwarders and (2) interference avoidance and hidden node resolution by employing the time-slotted structure and the CLEAR packet.

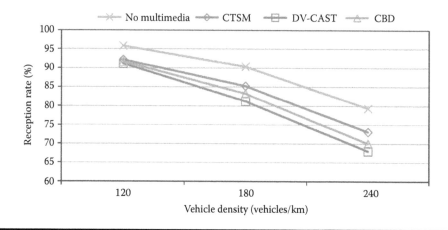

Figure 14.13 Packet success rate of the safety messages within a distance of 100 m.

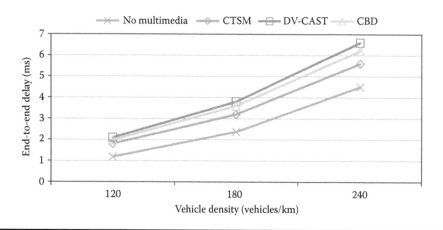

Figure 14.14 End-to-end delay of the safety messages within a distance of 100 m.

We also examine the effects of our proposed multimedia message transmission protocol on the existing periodic single-hop safety messages. It is apparent from Figure 14.13 that although the reception rate of the safety messages is degraded in the presence of a multimedia message, the effect is mild in our CTSM protocol. At a density of 240 vehicles/km, the loss of safety messages caused by the proposed CTSM multimedia transmissions is simply 6% greater than that in the case of no multimedia messages. This is a clear improvement from the 11% and 10% losses resulting from the DV-CAST and the CBD protocols. A similar trend can also be observed from Figure 14.14, albeit in terms of the end-to-end delay of safety messages. Here, our CTSM solution offers a 10%–20% reduction in the end-to-end delay, while following quite closely the delay in the case of no multimedia transmissions.

Finally, we show in Figure 14.15 the effect of the image transmission time T_m on the reception rate of the safety messages. The obtained result confirms the trade-off between a shorter image transmission time and a higher reception rate of the safety messages. In particular, the cooperative awareness of the vehicles is degraded by 3%–6% as T_m is decreased from 40 s to 10 s.

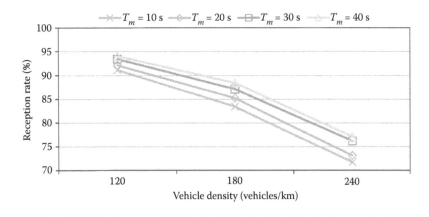

Figure 14.15   Packet success rate of the safety messages within a distance of 100 m at different values of T_m.

14.6 Conclusions

The aim of this chapter is to present research challenges and solutions for multimedia emergency services in cooperative VANETs. Specifically, we discussed VANET applications and the current industrial standards to provide insights into the data traffic requirements and transmission procedures in VANETs. We then reviewed cooperative communication techniques that improve the dissemination of emergency messages via inter-vehicle communications. We finally presented a novel cooperative protocol that reduces the transmission overhead of emergency multimedia messages without affecting the performance of the existing safety messages.

References

1. R. Chen, W.-L. Jin, and A. Regan, Broadcasting safety information in vehicular networks: Issues and approaches, *IEEE Network*, vol. 24, no. 1, pp. 20–25, 2010.
2. D. Caveney, Cooperative vehicular safety applications, *IEEE Control Systems*, vol. 30, no. 4, pp. 38–53, 2010.
3. Y. Toor, P. Muhlethaler, and A. Laouiti, Vehicle ad hoc networks: Applications and related technical issues, *IEEE Communications Surveys Tutorials*, vol. 10, no. 3, pp. 74–88, 2008.
4. NHTSA (National Highway Traffic Safety Administration), *Vehicle safety communications project, Task 3 final report: Identify intelligent vehicle safety applications enabled by DSRC*, Technical Report, NHTSA, Washington, DC, 2005.
5. ETSI (The European Telecommunications Standards Institute), *ETSI TS 102 637-2 v1.2.1—Intelligent transport systems (ITS)—Vehicular communications—Basic set of applications—Part 2: Specification of cooperative awareness basic service*, Technical Report, ETSI, Sophia Antipolis Cedex, France, 2011.
6. Z. Yang, M. Li, and W. Lou, CodePlay: Live multimedia streaming in VANETs using symbol-level network coding, *IEEE Transactions on Wireless Communications*, vol. 11, no. 8, pp. 3006–3013, 2012.
7. J.-S. Park, U. Lee, S. Y. Oh, M. Gerla, and D. S. Lun, Emergency related video streaming in VANET using network coding, in *Proceedings of the 3rd International Workshop on Vehicular Ad Hoc Networks*, September 2006, pp. 102–103.
8. F. Naeimipoor, C. Rezende, and A. Boukerche, Performance evaluation of video dissemination protocols over vehicular networks, in *Proceedings of the IEEE International Workshop on Performance and Management of Wireless and Mobile Networks*, October 2012, pp. 694–701.

9. N. Wisitpongphan, O. Tonguz, J. Parikh, P. Mudalige, F. Bai, and V. Sadekar, Broadcast storm mitigation techniques in vehicular ad hoc networks, *IEEE Wireless Communications*, vol. 14, no. 6, pp. 84–94, 2007.

10. M. Torrent-Moreno, J. Mittag, P. Santi, and H. Hartenstein, Vehicle-to-vehicle communication: Fair transmit power control for safety-critical information, *IEEE Transactions on Vehicular Technology*, vol. 58, no. 7, pp. 3684–3703, 2009.

11. J. Sahoo, E.-K. Wu, P. Sahu, and M. Gerla, Congestion-controlled-coordinator-based MAC for safety-critical message transmission in VANETs, *IEEE Transactions on Intelligent Transportation Systems*, vol. 14, no. 3, pp. 1423–1437, 2013.

12. W. Zhuang and M. Ismail, Cooperation in wireless communication networks, *IEEE Wireless Communications*, vol. 19, no. 2, pp. 10–20, 2012.

13. J. Laneman, D. Tse, and G. W. Wornell, Cooperative diversity in wireless networks: Efficient protocols and outage behavior, *IEEE Transactions on Information Theory*, vol. 50, no. 12, pp. 3062–3080, 2004.

14. H. Omar, W. Zhuang, and L. Li, VeMAC: A TDMA-based MAC protocol for reliable broadcast in VANETs, *IEEE Transactions on Mobile Computing*, vol. 12, no. 9, pp. 1724–1736, 2013.

15. S. Bharati and W. Zhuang, CAH-MAC: Cooperative ad hoc MAC for vehicular networks, *IEEE Journal on Selected Areas in Communications*, vol. 31, no. 9, pp. 470–479, 2013.

16. A. Ibrahim, Z. Han, and K. Liu, Distributed energy-efficient cooperative routing in wireless networks, *IEEE Transactions on Wireless Communications*, vol. 7, no. 10, pp. 3930–3941, 2008.

17. H.-Y. Hsieh and R. Sivakumar, A transport layer approach for achieving aggregate bandwidths on multi-homed mobile hosts, *Wireless Networks*, vol. 11, no. 1–2, pp. 99–114, 2005.

18. M. Sepulcre, J. Gozalvez, J. Harri, and H. Hartenstein, Contextual communications congestion control for cooperative vehicular networks, *IEEE Transactions on Wireless Communications*, vol. 10, no. 2, pp. 385–389, 2011.

19. J. Kenney, Dedicated short-range communications (DSRC) standards in the United States, *Proceedings of the IEEE*, vol. 99, no. 7, pp. 1162–1182, 2011.

20. R. Stanica, E. Chaput, and A.-L. Beylot, Properties of the MAC layer in safety vehicular ad hoc networks, *IEEE Communications Magazine*, vol. 50, no. 5, pp. 192–200, 2012.

21. C.-L. Huang, Y. Fallah, R. Sengupta, and H. Krishnan, Adaptive intervehicle communication control for cooperative safety systems, *IEEE Network*, vol. 24, no. 1, pp. 6–13, 2010.

22. J. J. Blum and A. Eskandarian, A reliable link-layer protocol for robust and scalable intervehicle communications, *IEEE Transactions on Intelligent Transportation Systems*, vol. 8, no. 1, pp. 4–13, 2007.

23. O. Tonguz, N. Wisitpongphan, and F. Bai, DV-CAST: A distributed vehicular broadcast protocol for vehicular ad hoc networks, *IEEE Wireless Communications*, vol. 17, no. 2, pp. 47–57, 2010.

24. M. Slavik and I. Mahgoub, Spatial distribution and channel quality adaptive protocol for multi-hop wireless broadcast routing in VANET, *IEEE Transactions on Mobile Computing*, vol. 12, no. 4, pp. 722–734, 2013.

25. M. Li, K. Zeng, and W. Lou, Opportunistic broadcast of event-driven warning messages in vehicular ad hoc networks with lossy links, *Computer Networks*, vol. 55, no. 10, pp. 2443–2464, 2011.

26. ETSI (The European Telecommunications Standards Institute), *ETSI ES 202 663 v1.1.0—Intelligent transport systems (ITS)—European profile standard for the physical and medium access control layer of intelligent transport systems operating in the 5 GHz frequency band*, Technical Report, ETSI, Sophia Antipolis Cedex, France, 2010.

27. C. Campolo and A. Molinaro, Multichannel communications in vehicular ad hoc networks: A survey, *IEEE Communications Magazine*, vol. 51, no. 5, pp. 158–169, 2013.

28. R. Uzcategui and G. Acosta-Marum, Wave: A tutorial, *IEEE Communications Magazine*, vol. 47, no. 5, pp. 126–133, 2009.

29. IEEE (The Institute of Electrical and Electronics Engineers), *IEEE Std 802.11-2007 (Revision of IEEE Std 802.11-1999), IEEE Standard for information technology—Telecommunications and information exchange between systems—Local and metropolitan area networks—Specific requirements—Part 11: Wireless LAN medium access control (MAC) and physical layer (PHY) specifications*, Technical Report, IEEE, New York, USA, 2007.

30. D. Jiang and L. Delgrossi, IEEE 802.11p: Towards an international standard for wireless access in vehicular environments, in *Proceedings of the IEEE Vehicular Technology Conference (VTC-Spring)*, May 2008, pp. 2036–2040.

31. IEEE (The Institute of Electrical and Electronics Engineers), *IEEE Std 1609.4-2011, IEEE standard for wireless access in vehicular environments (WAVE)—Multi-channel operation*, Technical Report, IEEE, New York, USA, 2011.

32. IEEE (The Institute of Electrical and Electronics Engineers), *IEEE Std 1609.3-2010, IEEE standard for wireless access in vehicular environments (WAVE)—Networking services*, Technical Report, IEEE, New York, USA, 2010.

33. IEEE (The Institute of Electrical and Electronics Engineers), *IEEE Std 1609.2-2012, IEEE standard for wireless access in vehicular environments (WAVE)—Security services for applications and management messages*, Technical Report, IEEE, New York, USA, 2012.

34. IEEE (The Institute of Electrical and Electronics Engineers), *IEEE Std 1609.1-2006, IEEE standard for wireless access in vehicular environments (WAVE)—Resource Manager*, Technical Report, IEEE, New York, USA, 2006.

35. ETSI (The European Telecommunications Standards Institute), *ETSI ES 102 687 v1.1.1—Intelligent transport systems (ITS)—Decentralized congestion control mechanisms for intelligent transport systems operating in the 5 GHz range; Access layer part*, Technical Report, ETSI, Sophia Antipolis Cedex, France, 2010.

36. M. Artimy, Local density estimation and dynamic transmission-range assignment in vehicular ad hoc networks, *IEEE Transactions on Intelligent Transportation Systems*, vol. 8, no. 3, pp. 400–412, 2007.

37. D. Rawat, D. Popescu, G. Yan, and S. Olariu, Enhancing VANET performance by joint adaptation of transmission power and contention window size, *IEEE Transactions on Parallel and Distributed Systems*, vol. 22, no. 9, pp. 1528–1535, 2011.

38. K. Hafeez, L. Zhao, B. Ma, and J. Mark, Performance analysis and enhancement of the DSRC for VANET's safety applications, *IEEE Transactions on Vehicular Technology*, vol. 62, no. 7, pp. 3069–3083, 2013.

39. Y. Park and H. Kim, Application-level frequency control of periodic safety messages in the IEEE WAVE, *IEEE Transactions on Vehicular Technology*, vol. 61, no. 4, pp. 1854–1862, 2012.

40. R. Schmidt, T. Leinmuller, E. Schoch, F. Kargl, and G. Schafer, Exploration of adaptive beaconing for efficient intervehicle safety communication, *IEEE Network*, vol. 24, no. 1, pp. 14–19, 2010.

41. F. Ye, R. Yim, S. Roy, and J. Zhang, Efficiency and reliability of one-hop broadcasting in vehicular ad hoc networks, *IEEE Journal on Selected Areas in Communications*, vol. 29, no. 1, pp. 151–160, 2011.

42. R. Stanica, E. Chaput, and A.-L. Beylot, Physical carrier sense in vehicular ad-hoc networks, in *Proceedings of the IEEE Conference on Mobile Adhoc and Sensor Systems*, October 2011, pp. 580–589.

43. R. K. Schmidt, A. Brakemeier, T. Leinmüller, F. Kargl, and G. Schäfer, Advanced carrier sensing to resolve local channel congestion, in *Proceedings if the ACM International Workshop on Vehicular Internetworking*, September 2011, pp. 11–20.

44. D. Jiang, Q. Chen, and L. Delgrossi, Optimal data rate selection for vehicle safety communications, in *Proceedings of the ACM International Workshop on Vehicular Internetworking*, September 2008, pp. 30–38.

45. Y. Mertens, M. Wellens, and P. Mahonen, Simulation based performance evaluation of enhanced broadcast schemes for IEEE 802.11-based vehicular networks, in *Proceedings of the IEEE Vehicular Technology Conference (VTC-Spring)*, May 2008, pp. 3042–3046.

46. K. Sjoberg, E. Uhlemann, and E. Strom, Delay and interference comparison of CSMA and self-organizing TDMA when used in VANETs, in *Proceedings of the IEEE Wireless Communications and Mobile Computing Conference*, July 2011, pp. 1488–1493.

47. M. Torrent-Moreno, Inter-vehicle communications: Assessing information dissemination under safety constraints, in *Proceedings of the Conference on Wireless on Demand Network Systems and Services*, January 2007, pp. 59–64.

48. R. S. Schwartz, R. R. R. Barbosa, N. Meratnia, G. Heijenk, and H. Scholten, A directional data dissemination protocol for vehicular environments, *Computer Communications*, vol. 34, no. 17, pp. 2057–2071, 2011.

49. G. Korkmaz, E. Ekici, F. Özgüner, and U. Özgüner, Urban multi-hop broadcast protocol for inter-vehicle communication systems, in *Proceedings of the ACM International Workshop on Vehicular Adhoc Networks*, October 2004, pp. 76–85.

50. E. Fasolo, A. Zanella, and M. Zorzi, An effective broadcast scheme for alert message propagation in vehicular ad hoc networks, in *Proceedings of the IEEE International Conference on Communications*, June 2006, pp. 3960–3965.

51. K. Abboud and W. Zhuang, Modeling and analysis for emergency messaging delay in vehicular ad hoc networks, in *Proceedings of the IEEE Global Telecommunications Conference*, December 2009, pp. 1–6.

Chapter 15

Opportunistic Spectrum Access in Multichannel Cognitive Radio Networks

Ning Zhang,[1] Nan Cheng,[1] Ning Lu,[1] Haibo Zhou,[2]
Jon W. Mark,[1] and Xuemin (Sherman) Shen[1]

[1]Department of Electrical and Computer Engineering, University of Waterloo, Waterloo, ON, Canada

[2]Department of Electrical Engineering, Shanghai Jiao Tong University, Shanghai, China

Contents

15.1 Introduction

With the surging popularity of multimedia applications, the demand for spectrum is rising dramatically. Owing to the spectrum scarcity problem, the limited available spectrum bands have been considered as the major bottleneck for high-quality multimedia transmission. Currently, spectrum resources are managed by a static assignment policy that allocates spectrum to licensed users on a long-term basis to avoid interference among different wireless systems. However, it has been discovered that some assigned spectrum bands are underutilized [1,2]. To facilitate ever-increasing multimedia applications, the unused spectrum can be exploited for multimedia transmission. Recently, cognitive radio has been envisioned as a promising paradigm to make better use of spectrum resources, because it allows users to access the unused spectrum bands opportunistically [3–9]. In cognitive radio networking, licensed users and unlicensed users are referred to as primary users (PUs) and secondary users (SUs), respectively.

Thanks to the merits of cognitive radio, great attention has been devoted to multimedia transmission over cognitive radio networks (CRNs), which enable multimedia users to utilize additional spectrum opportunistically [10–12]. In Ref. [10], multimedia transmission over CRNs is studied, where digital fountain code is leveraged to distribute multimedia content over unused subchannels. The issue of how to select channels to support the quality of service (QoS) requirement of multimedia applications is further investigated. In Ref. [11], considering the various requirements of multimedia users in terms of the transmission rate and delay, a queuing-based dynamic channel selection scheme is proposed. The priority virtual queue is leveraged to calculate the expected delays. Based on the expected delay, a dynamic learning algorithm is proposed with the objective of maximizing the user's utility dynamically. In Ref. [12], multimedia streaming over multihop CRNs is studied, considering the heterogeneous channel availabilities and the dynamics of network conditions. The goal of this study is to maximize the quality of the received video and achieve fairness among different video sessions, with the constraint of the collision rate to PUs due to spectrum sensing errors. The problem is formulated as a mixed integer nonlinear programming (MINLP) problem, and different algorithms are devised to achieve the objective.

All the foregoing studies [10–12] investigate multimedia transmission over CRNs, where spectrum sensing is adopted. In spectrum sensing, SUs need to detect idle spectrum bands (i.e., spectrum holes) before transmission and have to vacate once the PUs reclaim spectrum [13,14]. However, spectrum sensing is energy consuming and may not be accurate owing to channel fading or shadowing [15]. To address the limitations of spectrum sensing, two forms of cooperation have been introduced in CRNs: cooperative spectrum sensing and cooperative networking. In cooperative spectrum sensing [16,17], SUs cooperate with each other to better exploit the spectrum access opportunities when PUs are inactive. In cooperative cognitive radio networking (CCRN) [18–23], SUs cooperate with PUs to improve the latter's transmission performance, and in return gain spectrum access opportunities when PUs are active. Note that both PUs and SUs can benefit from cooperation in CCRN, and a win–win situation can be created. Cooperative sensing has been extensively studied in the literature, while CCRN has just emerged recently and needs to be further investigated. We will focus on CCRN in this chapter.

In CCRN, the cooperation schemes can be classified into two categories: three-phase [18,19] and two-phase cooperation schemes [20,21,23]. In the three-phase scheme, the PU first transmits a message to the cooperating SUs, and then SUs relay the PU's message to the destination in the second phase. In the last phase, SUs access the reward period granted by the PU for their own transmission. In the two-phase scheme, the last two phases can be combined by applying advanced physical layer techniques, for example, superimposed coding. In Ref. [18], a set of SUs cooperate with a PU to increase the transmission rate of the PU, and in return the PU grants a period of access time to the SUs as a reward. In this reward time, the SUs transmit simultaneously by selecting a suitable transmission power. In Ref. [19], the PU aims to maximize the transmission rate and the revenue obtained from SUs through cooperation. The cooperating SUs share the rewarding time via a payment mechanism. In Ref. [21], a two-phase cooperation scheme is proposed whereby the SU decodes the received signal from the first phase and superimposes it with its own signal to broadcast in the second phase, using different power levels. In Ref. [20], an FDMA-based two-phase cooperation is proposed whereby SUs cooperate with the PU to enhance its physical layer security in one portion of the frequency band, while transmitting their own data in another portion of the frequency band. In Ref. [23], a two-phase cooperation scheme based on space division multiple access is proposed whereby each SU simultaneously relays the primary message and transmits its own message by exploiting multiple antennas. In Ref. [24], different cooperation schemes are proposed whereby the PU can cooperate with trustworthy SUs to enhance its security level, and SUs can gain transmission opportunities. In summary, all the foregoing work only considers cooperation over a single channel. In practice, a system usually consists of multiple channels, allowing users to communicate simultaneously. Therefore, cooperation among multiple PUs and multiple SUs could be performed over different channels simultaneously. Therefore, cooperation in multichannel CRNs needs to be investigated.

In this chapter, we focus on multi-channel CCRN, where SUs gain reward access time for transmission by cooperating with PUs operating over different channels to improve the PUs' throughput. To study cooperation over multiple channels, a single case is studied first using the Stackelberg game. By analyzing this game, the strategies for a PU–SU pair can be obtained, that is, the access time allocation coefficient of the PU and the optimal transmission power of the SU. Based on the strategies obtained, cooperation over multiple channels is studied, with the objective of maximizing the total network utility, which is defined as the aggregate rewarding access time of different channels. To this end, a cluster-based cooperation scheme (CBC) is proposed whereby SUs first form a cluster based on geographic locations, cooperate with PUs to obtain transmission opportunities over different channels using the strategies for the single-channel case, and then share the resource obtained as a reward. In order to achieve the maximum aggregate access time, the best SUs are determined to cooperate with PUs over different channels by using maximum weight matching. For fair sharing of the obtained reward, the SU follows an approach using congestion game and quadrature signaling. Specifically, *active SUs*, which participate in cooperation with PUs as relays, stay in the current operating channels and employ the in-phase component of quadrature amplitude modulation (QAM) for transmission, whereas *inactive SUs*, that is, the rest of the SUs not selected as relays, choose access channels for their own interests following the Nash equilibrium (NE) of the congestion game and employ the quadrature component of QAM for transmission. By employing quadrature signaling, active and inactive SUs can access channels simultaneously without interference to each other [25]. With the congestion game, each inactive SU can gain certain transmission opportunities in a fair way.

The rest of the chapter is organized as follows. The detailed description of the system model is given in Section 15.3.1. Cooperation over single channel and multiple channels are studied in Sections 15.3 and 15.4, respectively. Concluding remarks are provided in Section 15.5.

15.2 Cognitive Radio Networks and Cooperative Communication

15.2.1 Cognitive Radio Networks

As the enabling technology for CRNs, cognitive radio (CR) is a radio that can sense the surrounding wireless environments and adjust its transmission parameters accordingly. For a more precise definition, FCC defines it as follows: *Cognitive radio: A radio or system that senses its operational electromagnetic environment and can dynamically and autonomously adjust its radio operating parameters to modify system operation, such as maximize throughput, mitigate interference, facilitate interoperability, access secondary markets* [26].

The two key characteristics of CR are cognitive capability and reconfigurability. The cognitive capability corresponds to the awareness of CR regarding the transmitted waveform, channel availability, the communication network, user need, security policy, and so on. The reconfigurability represents capability of adaption to the wireless environments according to the obtained information [2].

With CR technology, SUs can coexist with PUs and opportunistically utilize the temporarily unused spectrum bands owned by PUs. Therefore, CRN architecture comprises two components: a primary network and a secondary network, both of which can be deployed in either a centralized mode or an ad hoc mode. For the centralized mode, communications are coordinated by central nodes such as base stations. For the ad hoc mode, communications are performed in a peer-to-peer fashion.

The primary network can be any existing network that holds a license for operation in a certain spectrum band and has the exclusive privilege to access it. When the primary network is infrastructure based, PUs can be coordinated through the base station to access the spectrum bands. When the primary network is ad hoc, PUs communicate with each other without any infrastructure. The transmissions of PUs in the primary network should be protected from the interference caused by secondary networks. Generally speaking, because PUs and primary base stations are not equipped with CR functions, SUs have to sense the spectrum bands to protect PUs.

The secondary network, composed of a set of SUs, does not have the license to operate in any licensed spectrum bands. Similar to the primary network, the secondary network can also be classified into two types: infrastructure based and ad hoc [27]. For an infrastructure-based secondary network, there exists a central controller, for example, a secondary base station, which can coordinate the opportunistic spectrum access of SUs. For an ad hoc secondary network, SUs communicate with each other through multihop wireless links in either the licensed or unlicensed spectrum bands. Typically, both SUs and secondary base stations are equipped with CR technology.

15.2.2 Cooperative Communication

Recently, cooperative communications have attracted much attention in the literature, which can be leveraged to improve the transmission performance in terms of increasing the transmission rate, saving energy, enhancing the reliability, and so on. The main idea is that when the source node sends a message to the destination node, the intermediate nodes can also receive it owing to the broadcast nature of the wireless media. Those intermediate nodes can perform some processing and then retransmit the message to the destination node. Therefore, the destination node can receive multiple copies of the message and utilize them to create spatial diversity so as to improve the reception performance. In what follows, we will give a brief introduction to the two basic cooperative communication modes: amplify-and-forward (AF) and decode-and-forward (DF).

For the AF relaying mode, the source node first sends the signal to the destination node, which is also overheard by the relaying node. After receiving the signal from the source node, the relaying node just scales what it received by a factor, and then forwards the amplified version to the destination node. After receiving those copies, the destination node combines them using maximal ratio combining (MRC) to achieve the optimal reception. The overall signal-to-noise ratio (SNR) at the destination is equal to the sum of the received SNRs from both links. The main advantage of the AF mode is implementation simplicity, while the disadvantage is that the noise at the relaying node is also amplified and forwarded to the destination node.

For the DF relaying mode, the relaying node first decodes the received signal, then re-encodes it, and finally forwards it to the destination node. If the signal is decoded successfully, the noise component from the source node can be removed perfectly at the relaying node. If the signal is decoded incorrectly, the relayed signal is meaningless to the destination node. Therefore, the overall performance of the DF mode is determined by the worst link between the link from the source node to the relaying node, and the link from the relaying node to the destination node. In addition, to further improve the performance, the adaptive DF mode can be employed, which allows the relaying node to forward the message only when the message is decoded successfully. Compared with the AF mode, the DF mode has the advantage of reducing the adverse effects of noise at the relaying node, though it is more complex in terms of implementation.

15.3 Cooperation over Single Channel

In this section, cooperation between a PU and an SU over a single channel is investigated. Owing to the poor channel condition, the PU wishes to improve its throughput. Toward this end, the PU chooses an SU as a relay to increase the throughput, and in return grants a period of access time to the selected SU. As a result, the PU increases its throughput, and the SU gains spectrum access opportunities. Therefore, cooperation is performed on a mutual benefits basis. The procedure for cooperation is modeled by the Stackelberg game, where the PU acts as a leader, and the SU acts as a follower. By analyzing the game, the players' best strategies are derived, which constitute the Stackelberg equilibrium.

15.3.1 System Model

As shown in Figure 15.1, the PU communicates with the base station (BS) over a single channel in a time slot with length T. To improve the throughput, the PU (e.g., PU_j) chooses an SU (e.g., SU_i)

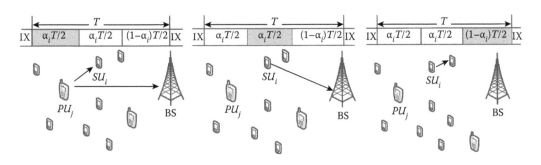

Figure 15.1   Cooperation over the single channel.

for cooperation, and the cooperation between PU_j and SU_i is carried out in the following way. A fraction α_i of T $(0 < \alpha_i \leq 1)$ is for cooperative communication. In the first duration of $\dfrac{\alpha_i T}{2}$, PU_j sends a message to SU_i, and in the subsequent duration of $\dfrac{\alpha_i T}{2}$, SU_i relays the received message to the BS by adopting the AF mode [28]. In the last duration of $(1 - \alpha_i)T$, that is, the reward access time, the cooperating SU_i transmits its own message to the corresponding secondary receiver.

Rayleigh block-fading channels are considered, which remain constant within each slot and vary over different slots. h_{pb}, h_{ps}^i, h_{sb}^i, and h_s^i, are channel gains from PU_j to BS, from PU_j to SU_i, from SU_i to the base station, and from SU_i to its corresponding receiver, respectively. As in Refs. [18,19,21], the channel state information (CSI), which is assumed to be known in the system, can be estimated by periodic pilots. The bandwidth for each channel is W. During cooperation, PU_j devotes power P_c^i for the transmission from PU_j to SU_i, while SU_i is constrained to spend the same power P_s^i for both cooperation and its own transmission in order to ensure that SU_i spends at least the same power for cooperation as for its own transmission. The one-sided power spectral density of white Gaussian noise is N_0.

15.3.2 Stackelberg Game between PU and SU

Assume that PUs and SUs are selfish and rational; that is, they are only interested in maximizing their own utilities. Therefore, game theory is suitable for modeling the interactions between them. In addition, considering PUs and SUs have different priorities on spectrum usage, the Stackelberg game is the most suitable for modeling the cooperation procedure, where the PU acts as the leader, and the SU acts as the follower. As the leader, the PU selects the best strategies, being aware of the effect of its decision on the strategies of the follower (the SU), whereas the SU can just choose its strategies given the selected strategies of the PU. In the Stackelberg game, the utility functions for both the PU and the SU are defined as follows. By analyzing the game, the best cooperating SU and the optimal cooperation parameters can be determined.

15.3.2.1 Primary User

For the PU, increasing throughput is equivalent to increasing the average transmission rate, given a fixed time duration T. Toward this end, the PU selects the best SU from the set $\mathbf{S}_p$ of its one-hop neighbors. Suppose that SU_i is selected for cooperation, the PU decides the time allocation coefficient α_i and the transmission power P_c^i to maximize the utility.

For the case without cooperation, the transmission rate of direct communication can be given by

$$R_d = W \log_2 \left(1 + \frac{P|h_{pb}|^2}{N_0} \right). \tag{15.1}$$

The utility function U_p^i of the PU is defined as the transmission rate through AF cooperative communication when cooperating with SU_i, which can be given as follows:

$$U_p^i = \frac{\alpha_i W}{2} \log_2 \left[1 + \frac{P|h_{pb}|^2}{N_0} + f\left(P|h_{ps}^i|^2, P_s^i|h_{sb}^i|^2 \right) \right], \tag{15.2}$$

where

$$f\left(P\left|h_{ps}^{i}\right|^{2}, P_{s}^{i}\left|h_{sb}^{i}\right|^{2}\right) = \frac{1}{N_{0}} \frac{P\left|h_{ps}^{i}\right|^{2} P_{s}^{i}\left|h_{sb}^{i}\right|^{2}}{P\left|h_{ps}^{i}\right|^{2} + P_{s}^{i}\left|h_{sb}^{i}\right|^{2} + N_{0}}.$$

The factor $\dfrac{\alpha_{i}}{2}$ accounts for the fact that $\alpha_{i}T$ is used for cooperative communication, which is further split into two equal phases. The PU chooses cooperation only when the transmission rate via cooperation is greater than that of direct communication. The objective of the PU is to maximize its utility, and the strategy is to choose the best SU, the time allocation coefficient α_{i}, and the transmission power P_{c}^{i} for cooperation.

15.3.2.2 Secondary User

The SU can acquire transmission opportunities through cooperation with the PU. Assuming cooperation with the PU, SU_{i} decides its transmission power, pertaining to the given α, which is determined by the PU. The objective of the SU is to maximize throughput without expending too much energy. Following the cooperation agreement, SU_{i} devotes the same power P_{s}^{i} for both cooperative and secondary transmissions. The transmission rate R_{s}^{i} for secondary transmission from SU_{i} to its corresponding receiver is given by

$$R_{s}^{i}(\alpha_{i}) = (1 - \alpha_{i}) W \log_{2}\left(1 + \frac{P_{s}^{i}\left|h_{s}^{i}\right|^{2}}{N_{0}}\right). \tag{15.3}$$

Considering the energy consumption $P_{s}^{i}\left(1 - \dfrac{\alpha_{i}}{2}\right)T$, the utility function of SU_{i} can be defined as $R_{s}^{i}(\alpha_{i})T - c \cdot P_{s}^{i}\left(1 - \dfrac{\alpha_{i}}{2}\right)T$. Note that c $(0 < c < 1)$ is the weight of energy consumption in the overall utility. A smaller c means that the SU values throughput more than energy consumption, and vice versa. Over the duration T, the utility function of SU_{i} is given by

$$U_{s}^{i}(\alpha_{i}) = W \log_{2}\left(1 + \frac{P_{s}^{i}\left|h_{s}^{i}\right|^{2}}{N_{0}}\right)(1 - \alpha_{i}) - c\left(1 - \frac{\alpha_{i}}{2}\right)P_{s}^{i}. \tag{15.4}$$

The objective of SU_{i} in the game is to maximize its utility, and the strategy is to choose the optimal transmission power P_{s}^{i}.

15.3.3 Game Analysis

The backward induction method can be used to analyze the Stackelberg game, which consists of two main steps. For the first step, the optimal strategy of the follower (i.e., the SU) is analyzed, assuming that the strategy of the leader (i.e., the PU) is given. For the second step, the PU determines the optimal strategy to maximize its utility, being aware of the results of the previous step. By doing so, the best response functions of both the PU and the SU can be derived, with which the corresponding utilities are maximized. Then, the Stackelberg equilibrium can be achieved based on the best response functions.

15.3.3.1 Best Response Function of the SU

Assuming that the PU chooses α_i for cooperation, SU_i aims to maximize its utility by selecting the optimal transmission power, which can be formulated as the following optimization problem:

$$\max_{P_s^i} U_s^i(\alpha_i) = (1-\alpha_i)W \log_2\left(1+\frac{P_s^i|h_s^i|^2}{N_0}\right) - c\left(1-\frac{\alpha_i}{2}\right)P_s^i$$

$$\text{s.t. } 0 \leq P_s^i \leq P_{max},$$

where P_{max} is the power constraint for SU_i. The best response function of the SU, that is, the optimal transmission power can be determined by solving the foregoing problem.

Definition 1: Let $P_s^{*i}(\alpha_i)$ be the best response function of SU_i if the utility of SU_i can achieve the maximum value by selecting $P_s^{*i}(\alpha_i)$, given α_i, that is, $\forall\ 0 < \alpha_i < 1, U_s^i\left(P_s^{*i}(\alpha_i),\alpha_i\right) \geq U_s^i\left(P_s^i(\alpha_i),\alpha_i\right)$.

Theorem 1 *The best response function of the secondary user* $P_s^{*i}(\alpha_i)$ *is given by* $P_s^{*i}(\alpha_i) =$

$$\min\left\{\frac{(1-\alpha_i)W}{c\left(1-\frac{\alpha_i}{2}\right)\ln 2}-\frac{N_0}{|h_s^i|^2}, P_{max}\right\}, \textit{ when the primary user selects } \alpha_i \textit{ for cooperation.}$$

Proof: Given the time allocation coefficient α_i, the utility function of SU_i is given by

$$U_s^i(\alpha_i) = (1-\alpha_i)W \log_2\left(1+\frac{P_s^i|h_s^i|^2}{N_0}\right) - c\left(1-\frac{\alpha_i}{2}\right)P_s^i. \tag{15.5}$$

Without considering the power constraint, it is easy to prove that the utility function first increases and then decreases as P_s^i increases, based on the foregoing equation. Therefore, there exists an optimal transmission power such that U_s^i can be maximized. Taking the first-order partial derivative of the utility function with respect to P_s^i yields

$$\frac{\partial U_s^i}{\partial P_s^i} = \frac{(1-\alpha_i)W|h_s^i|^2}{\left(1+\frac{P_s^i h_s^{i2}}{N_0}\right)N_0 \ln 2} - c\left(1-\frac{\alpha_i}{2}\right). \tag{15.6}$$

Setting $\dfrac{\partial\left(U_s^i\right)}{\partial\left(P_s^i\right)} = 0$ yields the optimal transmission power, which is given by

$$\frac{(1-\alpha_i)W}{c\left(1-\frac{\alpha_i}{2}\right)\ln 2}-\frac{N_0}{|h_s^i|^2}. \tag{15.7}$$

Taking the power constraint into consideration, we have the best response function $P_s^{*i}(\alpha_i)$ as follows:

$$P_s^{*i}(\alpha_i) = \min\left\{\frac{(1-\alpha_i)W}{c\left(1-\dfrac{\alpha_i}{2}\right)\ln 2} - \frac{N_0}{\left|h_s^i\right|^2}, P_{max}\right\}.$$ (15.8)

This completes the proof.

The first-order derivative of the best response function with respect to α_i is given by $\dfrac{-\alpha_i W}{(-2+a)^2 c \ln 2}$, which is negative. Thus, the best transmission power of SU_i is a decreasing function of α_i. This corresponds to the fact that the SU is willing to spend more transmission power for cooperation if the PU allocates more time for the SU's transmission.

15.3.3.2 Best Response Function of the PU

Aware of the best response function of the SU, the PU selects its own best strategy to maximize the utility. The best response function of the PU can be derived by solving the following optimization problem:

$$\max_{\alpha_i, P_c^i, i} \frac{\alpha_i W}{2} \log_2\left[1 + \frac{P_c^i \left|h_{pb}\right|^2}{N_0} + f\left(P_c^i \left|h_{ps}^i\right|^2, P_s^i \left|h_{sb}^i\right|^2\right)\right]$$

$$\text{s.t.} \ \ 0 < P_c^i \le P_{max}, 0 < \alpha_i \le 1, SU_i \subseteq \mathbf{S}_p.$$

Definition 2: α^*, P_c^{*i}, i^* are the optimal parameters associated with the best response function of the PU such that the utility can be maximized when this strategy is selected.

Theorem 2 *The best response function of the PU, α^*, P_c^{*i}, i^*, is given by $\left(\alpha^*, P_c^{*i}, i^*\right) = \arg\max_{\alpha_i, P_c^i, i} U_p^i$. In particular, $i^* = \arg\max U_p^i\left(P_c^{*i}, \alpha_i^*\right)$, where*

$$P_c^{*i} = P_{max}$$

$$\alpha_i^* = \begin{cases} 1 - \sqrt{1 - \dfrac{2W\left|h_{sb}^i\right|^2}{P_{max}\left|h_{ps}^i\right|^2 c + 2W\left|h_{sb}^i\right|^2 + N_0 c}}, & \text{if } \dfrac{W}{c\ln 2} - \dfrac{N_0}{\left|h_s^i\right|^2} < P_{max} \\[4ex] \max\left\{2 + \dfrac{2}{\dfrac{c\ln 2}{W}\left(P_{max} + \dfrac{N_0}{\left|h_s^i\right|^2}\right) - 2}, 1 - \sqrt{1 - \dfrac{2W\left|h_{sb}^i\right|^2}{P_{max}\left|h_{ps}^i\right|^2 c + 2W\left|h_{sb}^i\right|^2 + N_0 c}}\right\}, & \text{otherwise} \end{cases}$$

(15.9)

α_i^* and P_c^{*i} are optimal time allocation coefficients and the optimal transmission power, respectively, assuming cooperation with SU_i. The optimal P_c^{*i} and α_i^* correspond to the selected i^*.

Proof: The utility function U_p is a monotonically increasing function of P_c^i, because the first-order derivative of U_p with respect to P_c^i is always positive. In addition, considering that the parameters P_c^i and α_i are independent, P_c^i should be the maximum power such that the utility can be maximized. Therefore, solving the optimization problem is equivalent to optimizing the utility function when $P_c^i = P_{max}$ and SU_i chooses the best response $P_s^{*i}(\alpha_i)$. Because the first term in Equation 15.8 monotonically decreases with respect to α_i, its maximum value is $\dfrac{W}{c\ln 2} - \dfrac{N_0}{|h_s^i|^2}$.

When $\dfrac{W}{c\ln 2} - \dfrac{N_0}{|h_s^i|^2} < P_{max}$, $P_s^{*i}(\alpha_i)$ always takes the value of the first term in Equation 15.8.

Substituting $P_c^i = P_{max}$ and $P_s^{*i}(\alpha_i) = \dfrac{(1-\alpha_i)W}{c\left(1 - \dfrac{\alpha_i}{2}\right)\ln 2} - \dfrac{N_0}{|h_s^i|^2}$ into the utility function of PU, the utility can be rewritten as

$$U_p^i = \frac{\alpha_i W}{2}\log_2\left[1 + \frac{P_{max}|h_{pb}^i|^2}{N_0} + f\left(P_{max}|h_{ps}^i|^2, P_s^{*i}(\alpha_i)|h_{sb}^i|^2\right)\right], \tag{15.10}$$

which is a function of α_i. The first-order derivative of Equation 15.10 can be derived as follows:

$$\frac{\partial U_p^i}{\partial \alpha_i} = A\cdot\alpha_i^2 + B\cdot\alpha_i + C, \tag{15.11}$$

where

$$A = P_{max}|h_{ps}^i|^2 c + 2W|h_{sb}^i|^2 + N_0 c$$

$$B = -2P_{max}|h_{ps}^i|^2 c - 4W|h_{sb}^i|^2 - 2N_0 c = -2\cdot A$$

$$C = 2W|h_{sb}^i|^2.$$

To find the optimal α_i^* such that U_p can achieve the maximum value, set the first-order derivative of Equation 15.10 to 0. Because $C < A$, we have $B^2 - 4AC > 0$. Therefore, there exist real roots for the foregoing quadratic function. Considering the range α_i ($0 < \alpha_i < 1$), there exists one and only one root α_r. Then, the optimal α_i^* can be determined as follows:

$$\alpha_i^* = \alpha_r = 1 - \sqrt{1 - \frac{C}{A}}$$

$$= 1 - \sqrt{1 - \frac{2W|h_{sb}^i|^2}{P_{max}|h_{ps}^i|^2 c + 2W|h_{sb}^i|^2 + N_0 c}} \tag{15.12}$$

When $\dfrac{W}{c\ln 2} - \dfrac{N_0}{\left|h_s^i\right|^2} \geq P_{max}$, there exists α_0 in the range 0 to 1, such that $P_s^*(\alpha_0) = P_{max}$.

Specifically, $\alpha_0 = 2 + \dfrac{2}{D-2}$, where $D = \dfrac{c\ln 2}{W}\left(P_{max} + \dfrac{N_0}{\left|h_s^i\right|^2}\right)$. This is because the range of D is from

0 to 1 owing to the assumption that $\dfrac{W}{c\ln 2} - \dfrac{N_0}{\left|h_s^i\right|^2} \geq P_{max}$. For $\alpha_i \leq \alpha_0$, $P_s^{*i}(\alpha_i)$ always takes the value

P_{max}. Hence, U_p^i can achieve the maximum value in that range by selecting α_0. For $\alpha_0 < \alpha_i \leq 1$, there exists one and only one root α_r for the foregoing quadratic function, which lies in the range

0 to 1. If $\alpha_r < \alpha_0$, then $\dfrac{\partial U_p}{\partial \alpha_i} < 0$ when $\alpha_0 < \alpha_i \leq 1$. The derivative of U_p with respect to α_i is mono-

tonically decreasing. Therefore, the optimal $\alpha_i^* = \alpha_0$. Otherwise, the optimal $\alpha_i^* = \alpha_r$.

On the basis of the foregoing analysis, the optimal α_i^* can be achieved as given by Equation 15.9 in Theorem 2.

This completes the proof.

Using the backward induction method, the PU can determine the optimal strategies α_i^* and P_c^{i*} when it cooperates with SU_i. Substituting Equation 15.9 into Equation 15.8, the SU_i can also decide its best strategy $P_s^{*i}(\alpha_i^*)$. Note that, with α_i^* and P_c^{i*}, the PU can achieve the maximum utility, and the utility of the SU_i can also be maximized with its best response function $P_s^{*i}(\alpha_i^*)$. Hence, a Stackelberg equilibrium is reached. Moreover, the best SU_{i^*} will be selected so that the PU will have the highest expected throughput when it cooperates with this SU.

15.3.4 Numerical Results

In this part, we provide numerical results to evaluate the proposed cooperation scheme for the single-channel case. Similar to [18], we normalize the distance between the PU and the BS. Then, the SU is approximately placed at a distance $d \in (0,1)$ from the PU and $1 - d$ from the BS. Adopting the path loss model, the average power gains from the PU to the SU, and from the SU to BS, are

$\left|h_{ps}^i\right|^2 = \dfrac{1}{d^\zeta}$ and $\left|h_{sb}^i\right|^2 = \dfrac{1}{(1-d)^\zeta}$, respectively, where $\zeta = 3.5$ is the path loss exponent. The maximum

secondary transmission power P_{max} is normalized to 1, and we choose $P_{max}/N_0 = 0$ dB.

Figure 15.2 shows the PU's throughput on a certain channel as a function of α, with the parameter of weights c as 0.4 and 0.6 for the normalized distance $d = 0.5$ and $d = 0.7$, respectively. The normalized distance d_s between the SU and its corresponding receiver is set to 0.5. It can be seen that there exists an optimal α^* that can maximize the throughput via cooperation. With a smaller weight, the throughput of the PU is higher. The reason is that with a smaller weight, the SU values throughput more than energy consumption, and hence the SU is willing to spend more power for cooperation. Moreover, it is also seen that different distances lead to different through-puts and optimal α^*.

Figure 15.3 shows the throughput of the PU, averaged over fading, with respect to the normalized distance d, for $c = 0.2$ and 0.5. It can be seen that there exists a cooperation range in which the PU can cooperate with the SU to achieve a higher throughput than that of direct transmission. Furthermore, a smaller weight c leads to a larger cooperation range for a similar reason as before.

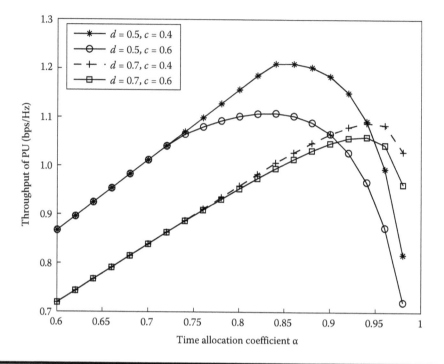

Figure 15.2   Throughput of PU versus time allocation coefficient α.

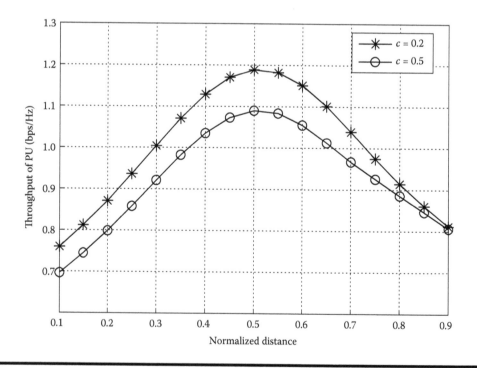

Figure 15.3   Throughput of PU averaged over fading, versus the normalized distance *d*.

15.4 Cooperation over Multiple Channels

In this section, we extend the cooperation scheme from the single-channel case to the multichannel case, where each SU can only select one channel each time to cooperate over that channel. The single-channel approach cannot bring the maximum benefit to the whole network because it only focuses on the interest of individual users. It is possible that multiple SUs compete with each other over some channels for transmission opportunities, while no SUs exploit the transmission opportunities over the other channels. Therefore, from the perspective of the whole secondary network, the transmission opportunities are not efficiently used; from the perspective of the individual SU, it is not guaranteed that it can obtain a chance to access the channel because it also depends on other SUs selecting the same channel. To maximize the total utility of the secondary network, the CBC scheme is proposed for multichannel CRNs in the section.

15.4.1 System Model

As shown in Figure 15.4, there exist K channels in the primary network, which allow K PUs to transmit simultaneously. Each PU communicates with the base station over one channel in a time slot with length T. The PU operating over a certain channel can be indicated by the channel index; for example, PU_j denotes the PU operating over channel j, where $j \in \{1,2,...,K\}$. In the secondary network, SUs transmit data to the corresponding receivers, and each SU can only select one channel at one time to exploit transmission opportunities by performing cooperation with the PU over that channel. For cooperation, each PU selects one SU, which acts as a relay to forward the PU's message to improve the throughput. Suppose PU_j selects SU_i for cooperation; it adopts the strategy obtained from the single-channel game during cooperation. Specifically, for channel j, the cooperation between PU_j and SU_i is carried out in the following way. A fraction $\alpha_i(j)$ of the time slot duration T ($0 < \alpha_i(j) \le 1$) is used for cooperative communication. Note that for $\alpha_i(j)$, i corresponds to SU_i and j corresponds to channel j or PU_j. In the first duration of $\frac{\alpha_i(j)T}{2}$, PU_j transmits data to SU_i, and in the subsequent duration of $\frac{\alpha_i(j)T}{2}$, SU_i relays the received data to the BS. In the last period of $(1 - \alpha_i(j))T$, which is the reward time, the cooperating SU_i transmits its own data to the corresponding secondary receiver.

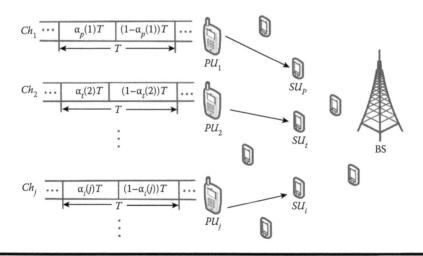

Figure 15.4   Cooperative cognitive radio network with multiple channels.

15.4.2 CBC Scheme

To exploit the spectrum access opportunities efficiently, the CBC scheme is proposed with the objective of maximizing the total utility of the secondary network, which is defined as the aggregate rewarding access time of all channels. For a given channel, the rewarding access time can be obtained when the SU and the PU use the Stackelberg equilibrium strategy for cooperation over that channel. Considering that the secondary network is formed in ad hoc mode, to maximize the total network utility and the average access time per user, SUs first form a cluster, cooperate with PUs to gain transmission opportunities, and then share the obtained resource fairly.

Specifically, based on the geographic location, SUs form a cluster N with size N to share the CSI. Then, the best SUs can be selected for each channel, which cooperate with PUs in order to achieve the maximum aggregate rewarding access time of different channels. This problem corresponds to a maximum weight matching problem, which can be represented by the bipartite graph in Figure 15.5. In particular, the vertices correspond to the SUs and PUs, and the weight on each edge represents the rewarding access time $1 - \alpha_i(j)T$ when SU_i and PU_j cooperate with each other. Finding the best SUs for cooperation to maximize the aggregate rewarding time is equivalent to finding the maximum weight matching in Figure 15.5. The well-known Hungarian algorithm can be performed to find the matching such that the sum of the weights can be maximized [29].

Then, the selected SUs cooperate with the corresponding PUs over different channels using the Stackelberg equilibrium strategy to obtain the rewarding access time. After that, the SUs in

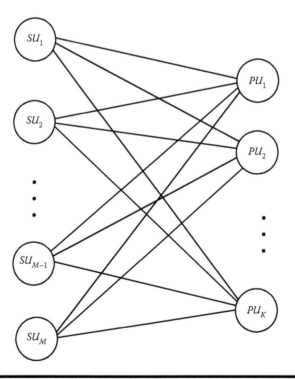

Figure 15.5   Maximum weight matching.

the cluster start to share the obtained rewarding time fairly. For this purpose, SUs are divided into two classes: active SUs (i.e., the selected SUs to cooperate with PUs as relays) and inactive SUs (the rest of the SUs with the size of M). Considering that the active SUs spend the transmission power during cooperation, they should have a larger share of the rewarding time. To this end, two classes of users first share the channels using quadrature signaling; that is, the active SUs stay in the current operating channels and use the in-phase component of QAM for transmission, while each inactive SU selects one channel to access and transmit by employing the quadrature component of QAM. By leveraging quadrature signaling, the active and inactive SUs can transmit concurrently without interference with each other. For each inactive SU, it also has to decide which channel to access to maximize their own utilities, that is, the shares of rewarding time for accessing the channels. The decision-making process is modeled by a congestion game and the Nash equilibrium (NE) strategy can be found. The share that each inactive SU can obtain is determined by the NE. With the CBC scheme, each SU can be guaranteed to gain a certain access time. Moreover, the average access time obtained using the CBC scheme is longer than that using the random channel access approach, which will be shown in the numerical results.

Each inactive SU selects an access channel among the multiple channels with different reward times, aiming to maximize its own utility. The congestion game is leveraged to model this process, which is defined by the tuple $\left\{ M, K, \left(\sum i \right)_{i \in M}, \left(U_j^i \right)_{i \in M, j \in K} \right\}$, where $M = \{1, 2, \ldots, M\}$ is the set of inactive SUs, $K = \{1, 2, \ldots, K\}$ denotes the set of channels, $\sum i$ represents the strategy space of SU_i, and U_i^j is the utility function of SU_i when selecting channel j. Note that U_i^j is a decreasing function of the total number of SUs selecting the channel j, because of the competition or congestion. In other words, when more SUs select the same channel, each SU can obtain a lower share. Each SU tries to maximize its utility by deciding which channel to access. The utility function of SU_i can be defined as follows:

$$U_i^j = \Psi_j \zeta(n_j). \tag{15.13}$$

where Ψ_j is the length of the rewarding access time of channel j, $\zeta(n_j)$ is the share of the rewarding access time that SU_i obtains over channel j, and n_j is the total number of inactive SUs selecting channel j. Therefore, U_j^i represents the access time that SU_i can have. For simplicity, the inactive SUs selecting the same channel share the rewarding time equally using TDMA, and then $\zeta(n_j) = 1/n_j$.

In the congestion game, if each player has chosen a strategy and no player can increase its utility by changing the strategy unilaterally, the current strategy profile constitutes an NE.

Definition 3: A strategy profile $S^* = (s_1^*, s_2^*, \ldots, s_M^*)$ is an NE if and only if

$$U_i(s_i^*, s_{-i}^*) \geq U_i(s_{i'}, s_{-i}^*), \forall i \in M, s_{i'} \in S_i \tag{15.14}$$

where s_i and s_{-i} are the strategies selected by SU_i and the other SUs, respectively. NE means no one can increase its utility unilaterally.

It is known that the congestion game always exists as a pure NE. The condition for NE in the congestion game is given as follows:

$$n_i = \left\lceil \frac{\Psi_i M - \sum_{j \neq i, j \in K} \Psi_j}{\sum_{j \in K} \Psi_j} \right\rceil + n', \tag{15.15}$$

where $n' \in \left\{ 0,1,2,\ldots, \left\lceil \frac{\Psi_i M + \Psi_i(K-1)}{\sum_{k \in K} \Psi_k} \right\rceil - \left\lceil \frac{\Psi_i M - \sum_{k \neq i, k \in K} \Psi_k}{\sum_{k \in K} \Psi_k} \right\rceil - 1 \right\}$. The detailed proof

can be found in the appendix. Because any strategy profile that satisfies the foregoing condition in Equation 15.15 will constitute an NE, there exist multiple NEs in the proposed congestion game. In order for the SUs to select an NE strategy, procedure 2 in algorithm 1 can be used for SUs to determine which channel to access.

The whole procedure of the CBC scheme is presented in Algorithm 1, which consists of two main parts: the selection of the best SUs and rewarding access time sharing.

Algorithm 1

1: //**Initialization**: Form the cluster based on geographic locations
2: //**Procedure 1**: Selection of Best SUs
3: **for** each $SU_i \in \mathcal{N}$ **do**
4:     **for** PU_j on channel j, $j \in$ K **do**
5:         Calculate access time allocation $\alpha_{i,j}$ using Equation 15.9
6:         Calculate reward periods $\Psi_{i,j} = 1 - \alpha_{i,j}$.
7:     **end for**
8: **end for**
9: Run Hungarian algorithm to find the best SUs for cooperation
10: //**Procedure 2**: Rewarding Access Time Sharing
11: Set congestion vector $n(S) = (n_1,\ldots,n_K) = (0,0,\ldots,0)$.
12: Order the rewarding periods on each channel $[\Psi_1,\Psi_2,\ldots,\Psi_K]$ decreasingly according to the length.
13: **for** each $SU_i \subseteq \mathcal{N}$ **do**
14:     **if** SU_i is active SU **then**
15:         SU_i stays in the current operating channel.
16:         SU_i employs the in-phase component for transmission.
17:     **else**
18:         **for** each Ψ_j, where $j \subseteq$ K **do**
19:             Calculate $\Psi_j \zeta(n_j + 1)$.
20:         **end for**
21:         SU_i selects the channel with maximum $\Psi_j \zeta(n_j + 1)$.
22:         SU_i employs the quadrature component for transmission.
23:         $n_j = n_j + 1$.
24:     **end if**
25: **end for**
26: **return**

15.4.3 Numerical Results

To evaluate the performance of the CBC scheme, similar to [30], we set up the simulation scenario as follows: The base station is placed at the origin (0, 0), and PUs are randomly located between $(0, d_{p,min})$ and $(0, d_{p,max})$, while SUs are randomly located between $(0, d_{s,min})$ and $(0, d_{s,max})$. The number of PUs is set to 5. The distances between nodes are normalized by $d_{p,max}$, and the previous path loss model is used to calculate average power gains.

We define the channel selection indicator of channel i as the number of inactive SUs selecting channel i divided by the total number of inactive SUs, that is, n_i/M, which can reflect the popularity of channel i. Figure 15.6 shows the impact of the number of inactive SUs (M) on the NE of the congestion game. When M is small, some channels may not be chosen by any SUs. For example, when $M = 8$, there is no inactive SU choosing channel 1. As M increases, all the channels are selected by at least one SU, and the selection indicator of each channel also varies owing to the changes in the NE.

Figure 15.7 shows the average access time per user, averaged over fading, versus the size of the cluster. We compare the proposed scheme with the random channel access approach. It can be seen that each SU can obtain a longer access time using the proposed scheme, compared with the random channel access approach. The reason is that the best SUs are selected to obtain the maximum aggregate time, which is fairly shared by all the SUs.

Similar to Ref. [31], we define fairness as $\dfrac{\left(\sum_i U_i\right)^2}{N\sum_i U_i^2}$, where U_i is the access time obtained by SU_i. Figure 15.8 shows the fairness among SUs. It can be seen that the fairness of the CBC scheme outperforms the random access approach. This is because each SU can obtain a certain share of access time using the CBC scheme, while only a few SUs can exclusively access the channel using the random channel access approach.

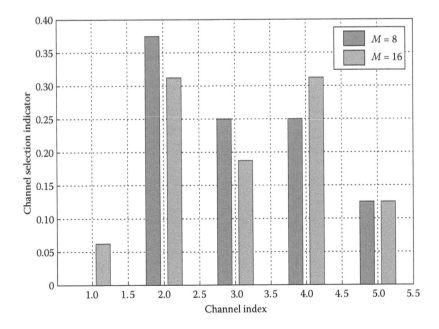

Figure 15.6 Impact of the number of inactive SUs on Nash equilibrium.

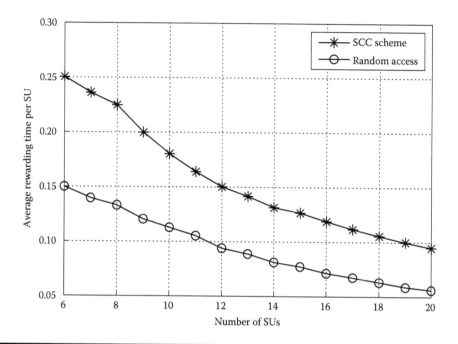

Figure 15.7 Average access time per SU averaged over fading for CBC scheme and random channel access.

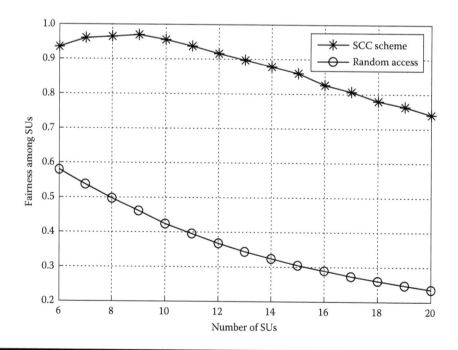

Figure 15.8 Fairness among SUs versus the number of SUs.

15.5 Conclusions

In this chapter, we have studied cooperative cognitive radio networking to yield additional spectrum for multimedia transmission. We have investigated cooperation over a single channel with the Stackelberg game. Based on the results of the single-channel scenario, we have proposed a CBC scheme for cooperation over multiple channels. In the CBC scheme, SUs form a cluster to maximize the total utility of the secondary network and share the obtained resources based on the congestion game and quadrature signaling. Numerical results have demonstrated that, with the proposed schemes, the PUs can achieve a higher throughput, while the SUs can obtain a longer average access time, compared with the random channel access approach.

To better facilitate multimedia transmission over CRNs, further research is required in the following aspects: (1) various quality of service (QoS) requirements of heterogeneous multimedia users need to be considered; (2) the effect of imperfect CSI on cooperation has to be studied because the channel estimation cannot be perfect; (3) security issues should be considered because cooperation may involve malicious users.

Appendix: NE Condition

A strategy profile is a set of strategies of all inactive SUs and is denoted by $S = s_1, s_2, \ldots, s_M$, where s_i is the strategy of SU_i. Denote by $n(S) = (n_1, \ldots, n_K)$ the congestion vector corresponding to the strategy profile S, where n_i represents the total number of SUs choosing channel i. For an NE, according to the definition of NE, it holds that

$$\Psi_i \zeta(n_i) \geq \Psi_k \zeta(n_k + 1), \quad \forall k \in \mathcal{K}, k \neq i. \tag{15.16}$$

For simplicity, the inactive SUs selecting the same channel share the rewarding time equally using TDMA, and then $\zeta(n_i) = 1/n_i$. To constitute an NE, for any two arbitrary channels i and j, according to Equation 15.16, we have

$$\frac{\Psi_i}{n_i} \geq \frac{\Psi_j}{n_j + 1} \text{ and } \frac{\Psi_j}{n_j} \geq \frac{\Psi_i}{n_i + 1},$$

which can be further written as

$$\frac{\Psi_j}{\Psi_i} n_i - 1 \leq n_j \leq \frac{\Psi_j}{\Psi_i} n_i + \frac{\Psi_j}{\Psi_i}, \quad j \neq i. \tag{15.17}$$

For any channel $k \in K$, $k \neq i, j$, it also holds that

$$\frac{\Psi_k}{\Psi_i} n_i - 1 \leq n_k \leq \frac{\Psi_k}{\Psi_i} n_i + \frac{\Psi_k}{\Psi_i}, \quad k \neq i, j. \tag{15.18}$$

Combining Equations 15.17 and 15.18, we have

$$\sum_{j \neq i, j \in \mathcal{K}} \left(\frac{\Psi_j}{\Psi_i} n_i - 1 \right) \leq \sum_{j \neq i, j \in \mathcal{K}} n_j \leq \sum_{j \neq i, j \in \mathcal{K}} \left(\frac{\Psi_j}{\Psi_i} n_i + \frac{\Psi_j}{\Psi_i} \right), \tag{15.19}$$

which can be further written as

$$\frac{\sum_{j\neq i, j\in\mathcal{K}}\Psi_j}{\Psi_i}n_i - (K-1) \leq \sum_{j\neq i, j\in\mathcal{K}} n_j \leq \frac{\sum_{j\neq i, j\in\mathcal{K}}\Psi_j}{\Psi_i}n_i + \frac{\sum_{j\neq i, j\in\mathcal{K}}\Psi_j}{\Psi_i}. \tag{15.20}$$

Since $\sum_{j\neq i, j\in\mathcal{K}} n_j = M - n_i$, we have

$$\frac{\Psi_i M - \sum_{j\neq i, j\in\mathcal{K}}\Psi_j}{\sum_{j\in\mathcal{K}}\Psi_j} \leq n_i \leq \frac{\Psi_i M + \Psi_i(K-1)}{\sum_{j\in\mathcal{K}}\Psi_j}. \tag{15.21}$$

The difference between the right side and the left side of Equation 15.21 can be written as

$$\frac{\Psi_i M + \Psi_i(K-1)}{\sum_{j\in\mathcal{K}}\Psi_j} - \frac{\Psi_i M - \sum_{j\neq i, j\in\mathcal{K}}\Psi_j}{\sum_{j\in\mathcal{K}}\Psi_j} > 1.$$

Moreover, for the left side of Equation 15.21, the following holds:

$$-1 < \frac{\Psi_i M - \sum_{k\neq i, k\in\mathcal{K}}\Psi_k}{\sum_{k\in\mathcal{K}}\Psi_k} < M$$

Therefore, the proposed congestion game has the following solution:

$$n_i = \left\lceil \frac{\Psi_i M - \sum_{j\neq i, j\in\mathcal{K}}\Psi_j}{\sum_{j\in\mathcal{K}}\Psi_j} \right\rceil + n', \tag{15.22}$$

where $n' \in \left\{0, 1, 2, \ldots, \left\lceil \frac{\Psi_i M + \Psi_i(\mathcal{K}-1)}{\sum_{k\in\mathcal{K}}\Psi_k} \right\rceil - \left\lceil \frac{\Psi_i M - \sum_{k\neq i, k\in\mathcal{K}}\Psi_k}{\sum_{k\in\mathcal{K}}\Psi_k} \right\rceil - 1 \right\}.$

References

1. FCC, Spectrum policy task force, *ET Docket*, vol. 2, p. 135, 2002.
2. S. Haykin, Cognitive radio: Brain-empowered wireless communications, *IEEE Journal on Selected Areas in Communications*, vol. 23, no. 2, pp. 201–220, 2005.
3. I. Akyildiz, W. Lee, M. Vuran, and S. Mohanty, Next generation/dynamic spectrum access/cognitive radio wireless networks: A survey, *Computer Networks*, vol. 50, no. 13, pp. 2127–2159, 2006.
4. N. Zhang, N. Cheng, N. Lu, H. Zhou, J. W. Mark, and X. Shen, Cooperative cognitive radio networking for opportunistic channel access, in *Proceedings of IEEE GLOBECOM*, December 9–13, 2013.
5. S. Gunawardena and W. Zhuang, Service response time of elastic data traffic in cognitive radio networks, *IEEE Journal on Selected Areas in Communications*, vol. 31, no. 3, pp. 559–570, 2013.

6. T. Han, T. Xing, N. Zhang, K. Liu, B. Tang, and Y. Liu, Wireless spectrum sharing via waiting-line auction, in *Proceedings of 11th IEEE Singapore International Conference on Communication Systems*, November 19–21, 2008.

7. F. Zeng, C. Li, and Z. Tian, Distributed compressive spectrum sensing in cooperative multihop cognitive networks, *IEEE Journal of Selected Topics in Signal Processing*, vol. 5, no. 1, pp. 37–48, 2011.

8. Y. Liu, L. X. Cai, and X. Shen, Spectrum-aware opportunistic routing in multi-hop cognitive radio networks, *IEEE Journal on Selected Areas in Communications*, vol. 30, no. 10, pp. 1958–1968, 2012.

9. N. Zhang, N. Cheng, N. Lu, H. Zhou, J. W. Mark, and X. Shen, Risk-aware cooperative spectrum access for multi-channel cognitive radio networks, *IEEE Journal on Selected Areas in Communications*, vol. 32, no. 3, pp. 516–527, 2014.

10. H. Kushwaha, Y. Xing, R. Chandramouli, and H. Heffes, Reliable multimedia transmission over cognitive radio networks using fountain codes, *Proceedings of the IEEE*, vol. 96, no. 1, pp. 155–165, 2008.

11. H.-P. Shiang and M. van der Schaar, Queuing-based dynamic channel selection for heterogeneous multimedia applications over cognitive radio networks, *IEEE Transactions on Multimedia*, vol. 10, no. 5, pp. 896–909, 2008.

12. D. Hu and S. Mao, Streaming scalable videos over multi-hop cognitive radio networks, *IEEE Transactions on Wireless Communications*, vol. 9, no. 11, pp. 3501–3511, 2010.

13. T. Yucek and H. Arslan, A survey of spectrum sensing algorithms for cognitive radio applications, *IEEE Communications Surveys and Tutorials*, vol. 11, no. 1, pp. 116–130, 2009.

14. H. T. Cheng and W. Zhuang, Simple channel sensing order in cognitive radio networks, *IEEE Journal on Selected Areas in Communications*, vol. 29, no. 4, pp. 676–688, 2011.

15. I. F. Akyildiz, B. F. Lo, and R. Balakrishnan, Cooperative spectrum sensing in cognitive radio networks: A survey, *Physical Communication*, vol. 4, no. 1, pp. 40–62, 2011.

16. A. S. Cacciapuoti, I. F. Akyildiz, and L. Paura, Correlation-aware user selection for cooperative spectrum sensing in cognitive radio ad hoc networks, *IEEE Journal on Selected Areas in Communications*, vol. 30, no. 2, pp. 297–306, 2012.

17. K. M. Thilina, K. W. Choi, N. Saquib, and E. Hossain, Machine learning techniques for cooperative spectrum sensing in cognitive radio networks, *IEEE Journal on Selected Areas in Communications*, vol. 31, no. 11, pp. 2209–2221, 2013.

18. O. Simeone, I. Stanojev, S. Savazzi, Y. Bar-Ness, U. Spagnolini, and R. Pickholtz, Spectrum leasing to cooperating secondary ad hoc networks, *IEEE Journal on Selected Areas in Communications*, vol. 26, no. 1, pp. 203–213, 2008.

19. J. Zhang and Q. Zhang, Stackelberg game for utility-based cooperative cognitive radio networks, in *Proceedings of the Tenth ACM International Symposium on Mobile Ad Hoc Networking and Computing*, May 18–21, 2009.

20. N. Zhang, N. Lu, N. Cheng, J. W. Mark, and X. Shen, Towards secure communications in cooperative cognitive radio networks, in *Proceedings of IEEE ICCC*, August 12–14, 2013.

21. Y. Han, A. Pandharipande, and S. Ting, Cooperative decode-and-forward relaying for secondary spectrum access, *IEEE Transactions on Wireless Communications*, vol. 8, no. 10, pp. 4945–4950, 2009.

22. N. Zhang, N. Lu, R. Lu, J. W. Mark, and X. Shen, Energy-efficient and trust-aware cooperation in cognitive radio networks, in *Proceedings of IEEE ICC*, June 10–15, 2012.

23. S. Hua, H. Liu, M. Wu, and S. Panwar, Exploiting MIMO antennas in cooperative cognitive radio networks, in *Proceedings of IEEE INFOCOM*, April 10–15, 2011.

24. N. Zhang, N. Lu, N. Cheng, J. W. Mark, and X. Shen, Cooperative spectrum access towards secure information transfer for CRNS, *IEEE Journal on Selected Areas in Communications*, vol. 31, no. 11, pp. 2453–2464, 2013.

25. V. Mahinthan, J. W. Mark, and X. Shen, A cooperative diversity scheme based on quadrature signaling, *IEEE Transactions on Wireless Communications*, vol. 6, no. 1, pp. 41–45, 2007.

26. FCC (Federal Communications Commission), Notice of proposed rulemaking and order: Facilitating opportunities for flexible, efficient, and reliable spectrum use employing cognitive radio technologies, ET Docket no. 03-108, p. 73, 2005.

27. I. Akyildiz, W. Lee, and K. Chowdhury, Crahns: Cognitive radio ad hoc networks, *Ad Hoc Networks*, vol. 7, no. 5, pp. 810–836, 2009.

28. J. Laneman, D. Tse, and G. Wornell, Cooperative diversity in wireless networks: Efficient protocols and outage behavior, *IEEE Transactions on Information Theory*, vol. 50, no. 12, pp. 3062–3080, 2004.
29. D. B. West, *Introduction to Graph Theory*, vol. 2, Prentice Hall, Englewood Cliffs, 2001.
30. L. Dong, Z. Han, A. P. Petropulu, and H. V. Poor, Improving wireless physical layer security via cooperating relays, *IEEE Transactions on Signal Processing*, vol. 58, no. 3, pp. 1875–1888, 2010.
31. R. Jain, D.-M. Chiu, and W. R. Hawe, *A Quantitative Measure of Fairness and Discrimination for Resource Allocation in Shared Computer System*, Eastern Research Laboratory, Digital Equipment Corporation, Hudson, MA, 1984.

Chapter 16

A Policy-Based Framework for Cognitive Radio Networks*

Gianmarco Baldini,[1] Ricardo Neisse,[1]
Abdur Rahim Biswas,[2] and Alberto Trombetta[3]

[1]*European Commission, Joint Research Centre (JRC), Institute for the Protection and Security of the Citizen (IPSC), Nuclear Security Unit, Via Enrico Fermi, Ispra VA, Italy*

[2]*CREATE-NET, Trento, Italy*

[3]*DiSTA Insubria University, Varese, Italy*

Contents

16.1 Introduction

Cognitive radio (CR) technology allows efficient radio frequency spectrum utilization by enabling adaptive configuration of the terminal based on information collected from the environment [1]. The ever-increasing number of new applications that require broadband wireless connectivity is the key motivation for investigating novel approaches for improved spectrum utilization. CR can be used for

---

* This work was funded by the EC through the FP7 projects iCore (287708).

innovative spectrum management approaches such as dynamic spectrum access, which can include various scenarios. A valid taxonomy of dynamic spectrum access scenarios is presented in Ref. [2], which identifies models based on dynamic use of the spectrum where frequency bands can be reassigned to different users in different dimensions: space (e.g., different locations), time (e.g., different time cases), or frequency or opportunistic use of the spectrum with primary users (PUs), which can coexist with other secondary users (SUs). One example of opportunistic spectrum access is the white spaces concept [3], which is a term indicating the spectrum available for a radio communication network (i.e., SUs) at a given time in a given geographical area on a noninterfering/nonprotected basis with regard to primary services such as digital TV broadcasters. Secondary cognitive radio networks can access the white spaces, but they should not generate harmful interference to primary services.

In dynamic spectrum access models, the rules to access and use the spectrum are defined by spectrum regulations and standards, which define various parameters including the frequency bands to be used by each service, the maximum emission levels, the activity factor of the signal, the modulations, the density of the transmitting nodes, and so on. All these parameters can contribute to the level of wireless interference to PUs in the radio frequency spectrum. For example, to mitigate the risk of harmful interference to primary services, CR systems and devices should conform to spectrum emission masks defined by spectrum regulations. In a scenario based on dynamic spectrum access, conformance to spectrum regulations can be expressed and applied through interference-limiting policies.

Although the rules may be well defined, the deployment of CR systems in the real world may suffer from elements that are not easy to predict. For example, the characteristics of electronic components may diverge from the model on which the regulations or standards were specified. In another example, the landscape where the CR systems must operate could be different from the recorded settings. Finally, CR systems may malfunction, intentionally or unintentionally. Intentional malfunction is due to attacks by CR systems either to cause basic denial of service (DoS) or for selfish purposes to use spectrum bands that are rightfully assigned to another CR system (which will also cause DoS). Unintentional malfunction may be due to a failure of the CR system or device, or it could be related to an incorrect model of the physical environment.

CR systems can be based on a cooperative approach, where CR nodes exchange messages to coordinate the use of spectrum resources, or on a noncooperative approach, where each CR system tries to use the spectrum resources independently of the other CR systems. As described in Ref. [4], cooperative approaches generally outperform noncooperative approaches, as well as closely approximate the optimum use of spectrum resources, even if the CR must use some communication resources to exchange information. On the basis of this consideration, in this chapter, we will focus only on CR systems based on a cooperative approach.

All the previous considerations indicate the need for a framework to control the spectrum resources in the system. This framework should be able to represent and enforce the rules defined in regulations and standards and to prevent misuse of the resources either intentionally or unintentionally. In addition, the framework should be able to support a change in the context. This change in context may be due to events such as a malicious attack, an emergency crisis that requires a reallocation of spectrum resources, or the need to change some policies owing to incorrect deployment models.

In addition, the framework should support different types of users with different levels of authorities for organizational reasons (e.g., public safety, military or commercial organizations) or business reasons (e.g., different types of network contracts or service level agreements).

This chapter will present a policy-based framework, which provides the capabilities described in the previous paragraph.

The design of CR networks based on a policy management framework is not new. The concept of policy-controlled CR has been suggested, among others, by the DARPA's neXt Generation (XG) project, which defined a policy language called CoRaL presented in Ref. [5]. CoRaL supports both permissive policies and restrictive policies. Permissive policies describe conditions under which transmission is allowed, and restrictive policies describe conditions under which transmission is not allowed. Section II provides a survey on the background research for the application of policy frameworks to CR networks.

The policy management mechanism can play a crucial role in dynamic spectrum access in order to ensure conformance to regulations and prevent harmful interference to the surrounding incumbent and SUs. It is important that the spectrum policies and rules are clearly defined and validated by all policy-controlled CR nodes in the area and that they are distributed in a secure way. In the rest of the chapter, we identify the policy-controlled CR node simply as the CR node.

On the top of the possibility of intentional and unintentional malfunctions of CR systems, policy-based CR networks can be vulnerable to various security threats as well. An extensive survey on security threats to CR networks is presented in Ref. [6]. Malicious devices can jeopardize the operation of the CR system by intercepting and spoofing policies or by compromising the exchange of policies. Malicious CRs can act as eavesdroppers and execute forgery of the input sensing data needed for the policy derivation process. This can lead to derivation of wrong policies and misbehavior of the CR users. Finally, the malicious devices can affect the policy reasoning and enforcement processes by performing PU emulation attacks.

Therefore, there is a need for a security mechanism to enable secure distribution and enforcement of the derived policies in CR systems. Usually, a centralized authority, given an a priori fixed set of rules describing the resources and conditions under which they can be accessed, enforces the conventional wireless medium access approach. Wireless terminals are usually identified through a standard login-and-password online mechanism. Other more sophisticated off-line authentication mechanisms require the presence of heavyweight, centralized, and rather rigid infrastructures, such as Public Key Infrastructures (PKIs). Thus, a more flexible and agile security mechanism is required in order to effectively manage the dynamic spectrum access in a distributed policy-based CR system.

Trust negotiation [7] is an example of such an approach. Trust negotiations allow two—initially mutually untrusting—parties wishing to exchange information to establish a mutual trust relationship. The trust is established through an exchange of digital credentials. The digital credentials represent digital statements of the relevant properties of the parties, and may be recommended by trusted entities (i.e., certification authorities [CAs]), or other entities that are trusted by the negotiating parties. The required credentials for the negotiation are defined on the fly according to a given negotiation's goals. During the negotiation, each entity chooses the credentials that it is prepared to disclose to the counterpart and under what conditions. Such conditions are expressed by rules called disclosure policies. Trust negotiation was originally designed for open distributed computing environments [8], where the goal is to allow unknown parties to gain access to services and resources. Trust negotiation is based on the iterative requests and disclosures for credentials among the parties to achieve an adequate level of trust, which permits access to the resources.

As indicated in Ref. [8], the practical deployment of trust negotiation techniques can be different, depending on the context and factors such as the diversity of the computing devices, the link used to transmit the credentials, and the computing power of the devices. As described in Ref. [9], trust negotiation in mobile ad hoc networks requires intensive public key cryptographic calculation and extensive checking and exchange of credentials, which can be excessively onerous on mobile devices and wireless links. For this reason, an essential criterion for the choice of the trust negotiation technique is the optimization of computing and communication resources.

Resource-constrained trust negotiation (RCTN) is a trust negotiation technique that has been specifically designed to operate on mobile ad hoc networks where mobile devices and links have resource constraints [9]. RCTN is based on a hierarchical trust model composed of two phases: basic trust and application trust. Digital credentials in RCTN are similar to digital credentials in traditional trust negotiation (e.g., implemented in X.509 certificates), but the credential attributes are encrypted with symmetric keys. Digital credentials are signed by a credential issuer using its private key and are verified by the credential issuer's public key. In the basic phase, RCTN exchanges the credential with encrypted attributes to reach a basic level of trust. After this phase, RCTN implements application trust model that is specific to the context. RCTN needs less memory space than that needed by traditional trust negotiation, because in the RCTN, two parties save only one credential, while every party saves many credentials in the tradition trust negotiation.

Finally, Trust-X has already been proposed in Ref. [10] for its application to CR networks. In comparison to other negotiation techniques, Trust-X supports the recovery of crashed negotiations. Briefly, the framework provides a way to save the state of the ongoing trust negotiation from time to time. If the negotiation is interrupted because of a communication failure—for example, a loss of connectivity of one of the negotiating parties—then it will possible, once the communication has been restored, to recover the interrupted negotiation. This is an important feature in wireless networks, and it improves the overall robustness of the CR network.

The proposed framework is based on elements of RCTN (for improved efficiency) and Trust-X (for the robustness in case of loss of connectivity).

This chapter is organized as follows: Section 16.2 describes the overall architecture of a policy-based system CR network. Section 16.3 presents the application of the proposed policy framework and the trust negotiation technique to a specific operational scenario. Finally, Section 16.4 concludes this chapter and describes future developments.

16.2 Architecture of Policy-Controlled CR Networks

Policy frameworks for CR are usually associated with two main functions in terms of the policy-based decision making process: policy reasoning and policy enforcing. The policy reasoning process refers to the identification of a possible solution space (i.e., one or a set of policies) based on the used and available policies defined at a specific moment in the CR system or device. In the rest of the chapter, we will use the term CR node, a CR system, or device, which is policy-enabled, which means that it is equipped with the capability to interact with the policy management framework. Instead, the process of policy enforcing has the objective to enforce the policy in the specific CR system or device. The two processes are associated with the key components in the policy framework: the policy enforcing point (PEP) and the policy decision point (PDP). Whereas the PEP can be implemented with a very simple logic, the PDP must be more sophisticated because it implements the reasoning for the selection of the best policy to be applied in answer to a specific request. To perform its duties, the PDP may need information about the context and must have access to the database of all the applicable policies. In the CR network, these processes and components are described in Figure 16.1.

The PEP is placed inside every CR node in a policy-enforced CR network. The PEP in the CR system or device performs the following essential tasks.

The first task is to receive and process events, which may request a change in the applicable policies. These events could be a request for a new connection, change in traffic demand or change in the context, which may require a new policy.

The second task is the request for a new policy to the PDP and the processing of the answer. The request of the new policy may also include some specific information such as the description of the processed event, the capabilities and status of the CR node, and so on.

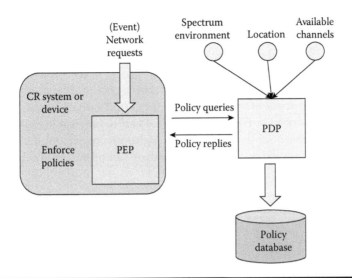

Figure 16.1 PDP and PEP in CR systems.

The third task is the enforcement of the policy in the CR node. If the PEP and PDP are designed and operate correctly, the enforcement of the policy should always work. In case the enforcement of the policy is not executed correctly because the CR node cannot implement it (e.g., the CR node does not have a specific feature or capability as requested by the policy), the PEP notifies the PDP and requires a new policy. The PDP also records the failure to update its internal database.

The PEP can also perform another function: the collection of data on the environment using the CR node functions if they are available. For example, the PEP can record the position of the CR node if the CR node is equipped with a GNSS receiver or it can record a snapshot of the spectrum environment if the CR node is equipped with spectrum sensing capability.

On the basis of the policy replies received from the PDP, the PEP is able to control and enforce various parameters of the radio transmissions and communications: communication channels, emission power, user data flow, and so on.

The PDP can be placed both in a CR node or in an entity that which does not have specific CR capabilities but is anyway connected to the CR network. The PDP can use various types of data to perform the task of identifying the optimal policy or set of policies, which must be transmitted to the PEP. The PDP may need to be aware of the spectrum environment, the location and the quantity of the CR devices in the network, the list of available channels, and information on wireless services distinct from the CR nodes but participating in the dynamic spectrum access scenario. For example, the PDP may also need the information on the location and features of the PUs, which are provided in a geolocation database. The concept of geolocation databases is described in Ref. [11]. Basically, the geolocation database stores information on the known PUs' transmitters (e.g., base stations, DTV broadcasters) and the spectral bands used by them in a known area. This information is used to define protection areas, which must be respected by SUs when they access the shared radio frequency spectrum. The information on the PUs can change in place or time, and the geolocation database must update this information and communicate it to the SUs.

Another database used in the idea proposed in this chapter is the policy database. The policy database includes the policies defined by the spectrum regulators for specific set of frequencies and geopolitical areas of competence (e.g., the United States, Europe, or the European member states).

While it is recommended that the PEP be implemented or hosted in the CR node to facilitate the enforcement of the policies in the radio communication devices, there are different options for the deployment of the PDPs and the policy database can also have different options.

In one option, described in Figure 16.2, the PDPs are present in the CR nodes themselves. Not all CR nodes need to have a PDP, and we can envision the presence of CR nodes only with the PEP. The policy database can be in one of the CR nodes or it can be distributed across the CR nodes.

In a second option described in Figure 16.3, the PDPs are not present in the CR nodes but are hosted in a separate component of the network with higher processing and storing capacity than the generic CR node. In a similar way, the policy database can be hosted in this specific component of the network.

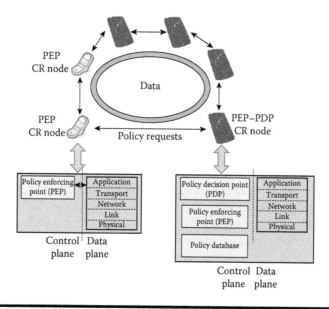

Figure 16.2   First option with distributed architecture.

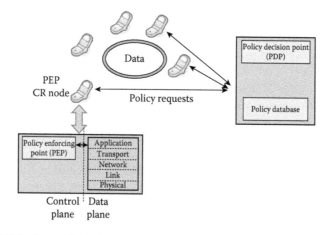

Figure 16.3   Second option with centralized architecture.

Another way to classify the two options is (1) centralized and (2) distributed. The first option is more appropriate to support a distributed architecture where PDPs can be distributed in the CR network. The second option is more appropriate in a centralized architecture, where the CR nodes must interrogate the same PDP in the network. The trade-off is well known and documented in literature: a distributed approach requires the synchronization of the PDPs and policy database, which can create excessive overhead in the CR network. In addition, a distributed approach in a CR network can create misalignments, especially when the CR nodes are not able to maintain full connectivity among themselves. The trade-off for a centralized approach is based on the advantage of a simpler architecture and less complex mechanisms for synchronization, with the disadvantage that the central PDP and policy database can become a bottleneck in the CR network, especially when a change in the context may cause numerous PEP to PDP requests at the same time.

The CR network proposed in this book chapter is based on a distributed architecture where the definition and activation of the policies is implemented through a model-based security toolkit that is presented in the following section. In the toolkit, the term *usage control* is used to define access to resource through policies. The exchange of the policies is instead regulated through trust negotiation.

16.2.1 Model-Based Security Toolkit (SecKit)

Usage control extends access control and policy management notions with the concept of obligations, which specifies future constraints in the behavior of a system. The main application of usage control is data protection, for example, to control the use of data after access is granted. Usage control can also be applied to control the usage of any type of resource, which in the work described in this chapter consists of the spectrum usage in a CR network.

In our solution, we apply a usage control framework previously developed by one of the authors in Ref. [12] that supports the specification of policies using mechanisms according to an event–condition–action (ECA) rule structure. These mechanisms have their *Action* part executed when an *Event* pattern is observed and the *Condition* expression evaluates to true.

These rules use as a reference a set of interrelated design models representing different aspects of the CR system, and are used as input for the runtime components in the framework. This solution to enable monitoring of ECA rules and execution of security enforcement behavior is called the Model-Based Security Toolkit, or just SecKit.

SecKit consists of a collection of metamodels for the specification of a computer system structure, information, behavior, context, identities, organizational roles, and security rules. These metamodels provide the foundation for security engineering tooling add-ons and metamodel extensions to address requirements of governance, security, privacy, and resource usage. SecKit adopts a generic design language to represent the architecture of a distributed system across application domains and levels of abstraction, including refinement relations support inspired in the Interaction System Design Language (ISDL) [13].

Figure 16.4 gives a high-level overview of the model, components, and tools supported by SecKit. The design models represent the system structure, behavior, information, organizational roles, and security rules. These design models can be used to support a model-based engineering approach or can be obtained from a running system by a manual or automatic reverse-engineering approach. The objective of the design models is to support analysis tools and the specification of runtime models.

The runtime models are representations of the running system that are used as input by runtime components such as PDPs and PEPs. Examples of runtime models are the current existing entities, data items, and events representing the execution of the system behavior. PEPs are essentially technology-specific components that are embedded in the system platform and applications,

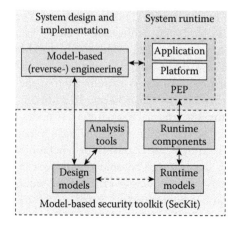

Figure 16.4   SecKit overview.

and are capable of enforcing the security and resource usage rules defined in the design models and monitored by the runtime components. The ECA rules representing the policies specified using SecKit are signaled by the PEP components.

In order to support both preventive and detective enforcement, two types of events are specified: tentative and actual events. Tentative events indicate that an activity (action or interaction) is ready to be started in the monitored system but has not yet been started, and actual events indicate that an activity has been completed.

For tentative events, the Action part of a mechanism specifies an authorization action that allows the activity to be executed, blocked, delayed, or modified. For actual events, the Action part simply triggers the execution of additional actions.

The Condition expression of a mechanism supports propositional, temporal, cardinality, and event operators. Our language for the specification of conditions is based on linear temporal logic (LTL) and evaluates event traces considering a discrete time step with a predefined granularity. The condition part of a mechanism can be parameterized with variables, which allows re-use and modularity in the specification of usage control policies.

From a usage control perspective, our mechanisms can be used to specify authorization and obligations without any changes. Authorizations are essentially mechanisms specified by domain administrators for enforcement on their own domains. Obligations are mechanisms specified by domain administrators that are delegated to other domains when interactions that require a sensitive resource (e.g., data or network bandwidth) occur.

In contrast to existing frameworks for access and usage control, our toolkit is more expressive and can express complex authorizations and obligations. For example, a policy stating that access should be denied to users after three unsuccessful logins cannot be expressed using existing access control languages such as XACML [14] or CoRaL [15] because these languages do not support cardinality or temporal operators.

16.2.2 Trust Negotiation

The main reason for exchanging policy-related information among the PCDs is that each CR node in the network may have specific information on the spectrum environment that can be beneficial to all policy reasoners in the network.

In a CR distributed network, it is important that the CR nodes can trust each other to distribute and enforce the policies. The establishment of trust is particularly important when a new CR node joins the CR networks, because it needs to have access to new policies or compare its policies with those of the rest of the network.

Different functions may also have different levels of trust. The request for context (e.g., spectrum sensing data) by a CR node to another may require a low level of trust, because the CR node can always decide to ignore the received information if it is clearly mismatched from its own information. On the contrary, the distribution of policies from a PCD to an OCD requires a high level of trust when the policies are supposed to be enforced in the OCD. As a consequence, the CR network must use different levels of trust for different functions and operations. In addition, CR nodes may not be present in the CR network with the same level of trust. A new CR node that joins the CR network from another area or network may start with a low level of trust in comparison to the other CR nodes already present.

The proposed policy framework is presented in Figure 16.5. The CR nodes (either with PDP or PEP) must authenticate through the trust negotiation of the system in order to be allowed to exchange the policies and transmit/receive data or the credentials used in the trust negotiation algorithms as described in Section 16.3. The policy defines the allowed usage of available channels or spectrum bands and other transmission parameters such as the maximum emission power, modulation scheme, and so on. It is recommended that the channel used to distribute the policies be based on a well-defined and identified spectrum band (e.g., by spectrum regulators) and the transmission characteristics (e.g., modulations, protocol) be well known by all the CR nodes a priori in contrast to the data on white spaces channels, which are dynamically defined on the basis of the context. An example of this channel is the cognitive control channel (CCC) described in Ref. [16], which is defined as the channel used to exchange information between network elements and terminals in a CR system. In Ref. [17], the exchanged information can include available

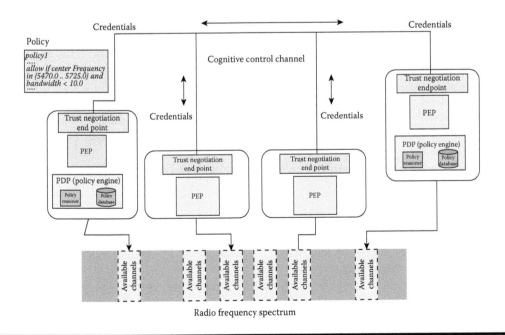

Figure 16.5 Architecture of policy-controlled CR radios.

frequency bands, detected PUs and SUs, and spectrum usage policies, and it can be easily extended to the exchange of the trust negotiation credentials as well. As a consequence, the application of CCC is suitable for the framework described in this chapter as well.

16.3 Application of Trust Negotiation to Policy-Controlled CR Networks in the Opportunistic Spectrum Access Scenario

16.3.1 Operational Scenario Based on Opportunistic Spectrum Access

We now apply the policy framework model and the trust negotiation technique to the use case of opportunistic spectrum access in the digital TV band, which is usually known as TV White Spaces (TVWS). These are fallow bands in the range of frequencies usually licensed to digital TV (DTV) providers [18]. CR devices, which are the SUs, can use the white spaces to transmit and communicate if they do not create harmful interference to the DTV providers, which are the PUs in this scenario. Various design solutions and functions have been proposed to mitigate the risk of wireless interference by SUs to PUs. They include (1) geolocation spectrum databases, which store information on the position and features of the PUs and define the protection areas with a specific emission mask and (2) spectrum sensing functions of the CR node to sense the presence of PUs in the area. This scenario comprises two separate sub-scenarios. The basic scenario is where public/ open users or business users can access the white spaces with two different levels of access (basic for open users and higher for business users), and an emergency scenario is where government users (e.g., public safety users) can access the white spaces with a higher level of access than open users and business users for the duration of the emergency situation.

Figure 16.6 shows a screenshot of the SecKit graphical user interface (GUI) implementation for the specification of design models. More specifically, this figure shows the structure design model of the CR system. We model the system as a collection of devices (CR nodes) that interact with the network using an interface.

Figure 16.7 shows the data design model that represents the CR parameters. In the example, the CR parameters used in the policy are

- At the level of CR parameters: The maximum allowed emission level, the maximum level of the activity factor (between 0 and 1), the type of modulation, and the maximum allowed density of CR nodes in a geographical area to mitigate the impact of the aggregated wireless interference.
- At the level of allowed frequency bands for transmission and reception: The upper and lower limit of the frequency bands.

Figure 16.8 shows the behavior design model of the system. We model only the device behavior with one possible action type: "activateVideoStreamingService." This action is executed with the parameters of the CR network and results in a tentative event every time a device in the network tries to execute this activity.

Figure 16.9 shows the context design model. In this model, we represent the types of context information and context situations. Context information is a simple type of information about an entity that is acquired at a particular moment in time, and context situations are a complex type that models a specific condition that begins and finishes at specific points in time. For example, the GPS location is an example of a context information type, which is used to limit the density

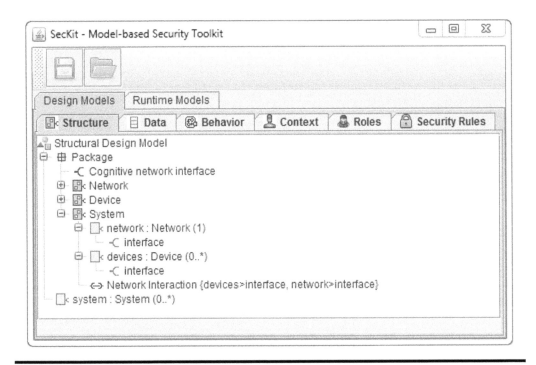

Figure 16.6   Structure design model.

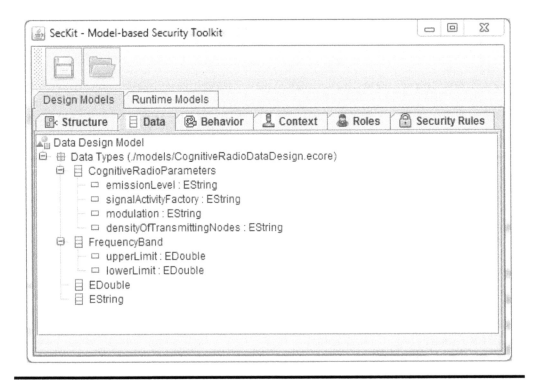

Figure 16.7   Data design model.

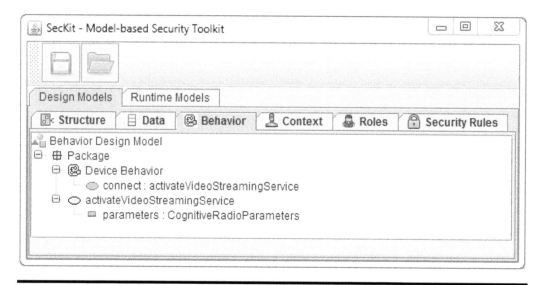

Figure 16.8   Behavior design model.

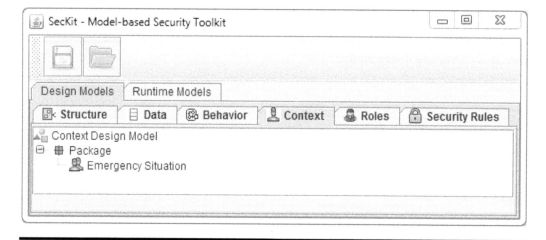

Figure 16.9   Context design model.

of CR nodes in a specific geographical area. In this figure, only one context situation is defined called emergency situation, which differs from a basic context where public/open users are the only users accessing the white spaces. In the emergency situation, the government users need to access the white spaces for the duration of the emergency, and they are allowed to use such spectrum resources through the predefined policies.

Figure 16.10 shows the role design model. We represent in this model three types of users/ roles: government, business, and public/open. The objective of the role, context, behavior, data, and structure design models is to support the specification of security policy rules that govern the usage of CR resources (e.g., white spaces).

Figure 16.11 shows the tab with the security rules design model. The security rule template "Deny video streaming service by default" specifies that when the action type "activateVideo-StreamingService" is about to be executed (tentative event or *try*), it should be denied. This rule

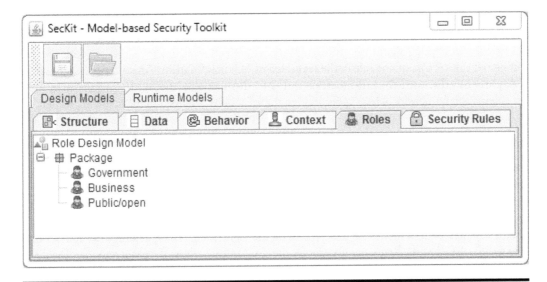

Figure 16.10　Role design model.

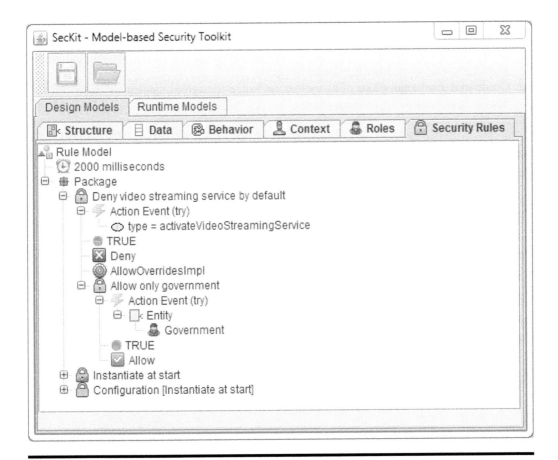

Figure 16.11　Security rules design.

template contains a nested rule that allows the activity to be executed if the user performing this action has the role "Government." The conditional part of these rules is simply TRUE. These nested rules follow a combining algorithm, "Allows Overrides," which evaluates as "allow" the result for a set of nested rules if at least one rule in the set evaluates to "allow," and any "deny" result is ignored.

As previously discussed, the authentication and "safe" distribution of the policies is regulated by the trust negotiation framework, whose application to the scenario is described in the following section.

16.3.2 Proposed Trust Negotiation Framework

Trust negotiation represents a collaborative process between two entities, Requester and Controller, with the goal of establishing mutual trust in order to access or exchange certain resources. It is assumed that the resource description is encoded into a credential, which represents a list of the relevant attributes and their values of the given information. Moreover, it is also assumed that the resource is protected by a disclosure policy (held by the Controller), which defines what conditions ought to be satisfied by the Requester in order for the Controller to release the given resource. The disclosure policy should not be confused with the spectrum policy or the policies defined in the toolkit. Typically, the Requester's conditions are encoded into predicates about credentials, which are to be disclosed themselves to the Controller. The negotiation process is divided into three distinct phases:

1. **Introductory phase**: The entities identify the resource R (e.g., a white space channel or a policy) required by the Requester (e.g., a CR node).
2. **Policy Evaluation phase**: The entities sequentially exchange disclosure policies in order to agree upon a set of credentials to be exchanged.
3. **Credential Exchange phase**: The entities exchange the agreed credentials based on the disclosure policies in the previous phase.

The credentials can be based on attributes of the CR device and its owner, such as type of contract, transmission features, or type of model. As suggested in Section 16.2, this basic negotiation process has been extended in several ways to adapt it to CR networks and to provide a variety of features. One of the extensions is to support the possibility of renegotiation of the requested resources, as described in the Trust-X system [11]. In many specific cases, the initially requested resource R can be defined as a composite resource. If the disclosure policy associated with such a resource is nonsatisfactory for the Controller (e.g., the CR node does not have information on its owner or the telecom provider), then the Requester can re-request the resource from the Controller with a less demanding disclosure policy (e.g., CR node model). In turn, the Controller may offer a subset of resources that comprise the requested composite resource R. The Requester may choose to accept or refuse the proposal or offer a more/less demanding disclosure policy in order to obtain the desired resource. If a given negotiation is partially completed (for some parts of the given composite resource), the protocol allows the remaining parts to be obtained by means of another trust negotiation. This negotiation process will evaluate the disclosure policies protecting the original resource, taking into consideration the fact that such policies have partially satisfied the previous negotiation.

Both the policy evaluation and the credential exchange are supported with a recovery phase if a negotiation crashes owing to a communication failure due to a loss of connectivity between one or more of the CR nodes. This approach provides a method for recording the state of the ongoing trust negotiation. If the negotiation is interrupted, it will be possible to recover the interrupted negotiation once the communication has been restored. The recording rate is not agreed to by the entities, and thus each entity may define its own time interval between the creation of a negotiation state according to its preferences. Obviously, operational requirements of the CR network can impose time constraints on the completion of the negotiations, but the trust framework is flexible enough to support the requirements.

The Trust-X system will take care of the reconciliation of the saved states when the negotiation is restored. This feature has been further improved to allow the negotiating entities to suspend a negotiation on a voluntary basis (also presented in Ref. [19]). This is useful when one of the negotiating parties is required to provide a certain credential that is not currently available, but will be in the future (e.g., it can be provided by the owner of the CR node). This feature is achieved by creating a saving state, called a negotiation tree, at a time instant agreed to by both entities. It enables a resumption of the negotiations from the previously recorded state.

We now provide an example of a simple negotiation exchange between a new CR node entering the network (Requester) and a CR node already present in the network (Controller). As described before, in the exchange, we use disclosure policies, which consist of a credential or set of credentials, terms, and the Boolean connectives {AND(), OR()}.

The workflow of the trust-negotiation-based authentication described in Figure 16.12 is as follows:

1. A new CR node (called the Requester) with only the PEP enters the network where CR nodes are already present with both PEP and PDP. The initial step for the Requester CR node is to require some basic information on the spectrum environment from an existing CR node (called the Controller) in the CR network and to collect the basic policy information for the PEP.
2. The Controller responds with a disclosure policy based on credentials where the attributes are encrypted.
3. The Requester provides the requested credentials.
4. The Controller verifies the credentials and provides the requested information on the spectrum environment and the basic information for the PEP. At this step, a basic level of trust is reached between the Requester and the Controller. At this stage, a negotiation tree (see Figure 16.13) has been created to represent the information that an initial level of trust with a specific entity (i.e., the new CR node or Requester) has been reached.
5. The user associated with the Requester would like to activate a Video streaming service. The Requester requests the Controller the permission to use spectrum resources to activate a video streaming session, which requires a high data rate and considerable spectrum resources. The Requester provides the Controller the characteristics of the required video streaming service: QoS, data rate, time of the communication, and the number of the CR nodes involved in the communication. These characteristics or service requirements correspond to the required amount of spectrum, maximum emission power, and the modulation schemes that are allowed and needed to guarantee the video streaming session. Note that this type of information cannot be provided by the Controller at the basic level of trust established in the previous step.
6. The Controller transmits to the Requester the disclosure policy. For example, the disclosure policy can be described as follows:

```
(device(CR node identifier) = "true")∧(contract(IDnetworkprovider)
    = "true")∧(MaxEmissionPower(-70dBm)) = "true"
```

This means that the CR node must have a valid identifier (e.g., belonging to a set of registered identifiers), have a maximum emission power of –70 dBm, and the Identifier (Id) of the network provider must be equal to a specific value.

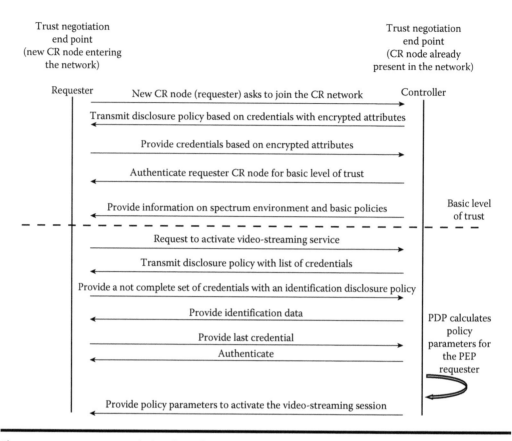

Figure 16.12 Trust-negotiation-based authentication example.

7. The Requester can provide the ID of the network provider only to specific Controllers. As a consequence, the Requester responds with almost the complete required set credentials but also with a disclosure policy asking for the ID of the Controller.

8. The Controller sends its ID.

9. The Requester verifies the ID of the Controller and sends the ID of the network provider to the Controller.

10. The Controller authenticates the Requester to a higher level of trust and sends this notification to the Requester. The negotiation tree is updated (see Figure 16.13).

11. At this point, the Requester and the Controller have reached an adequate level of trust, and the Controller can send the Requester the parameters decided by the PDP on the basis of the requested video streaming services and the current network context, which includes the available spectrum resources (e.g., available white spaces), the density of the CR nodes (to mitigate the risk of aggregate interference), and potential propagation errors (e.g., owing to physical obstacles), and so on.

12. The PEP component in the Requester receives the policy information from the Controller and enforces them to activate the video streaming service. This policy information may include the white spaces channels to be used, the modulation schemes, maximum emission power, and so on. This step concludes the overall process. A pictorial description of the overall scenario is presented in Figure 16.13.

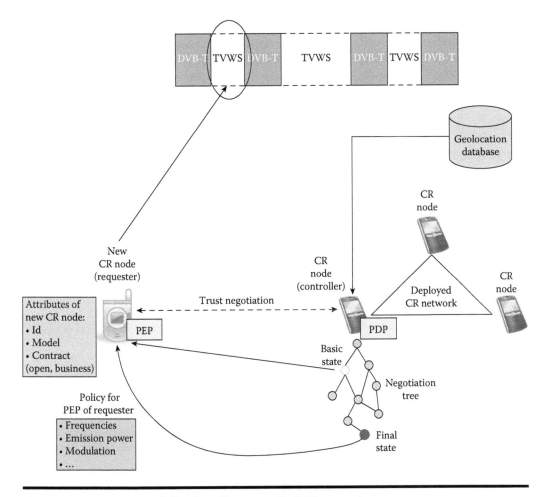

Figure 16.13   Trust negotiation in policy-controlled CR networks.

16.4 Conclusion and Future Developments

The nature of policy-controlled CR networks leaves them vulnerable to a variety of security threats that can decrease the system performance of the CR networks as well as the surrounding wireless networks in the vicinity. It is important that CR nodes in the network are able to trust each other to exchange policies. This chapter proposes a security approach that combines the concepts of trust negotiation with the concept of policy-controlled CR nodes implemented with a toolkit.

The combination of the two concepts provides a higher degree of flexibility in comparison to approaches based only on policy management. First of all, trust negotiation is fully distributed, and it does not require a centralized entity in the operational phase in the field. Second, the trust negotiation framework adopted in this chapter supports the stop and restart of a negotiation in case of a communication failure.

The potential disadvantage of the proposed approach is the slowness in the convergence of the trust negotiation technique in CR networks. Although this may not be an issue in Internet domains, where the trust negotiation technique was initially applied, it could degrade the overall performance of CR networks. This chapter is a preliminary description of the proposed approach, and future work will provide a quantitative model based on realistic operations scenarios identified in standardization bodies or spectrum regulatory technical reports. In particular, the performance of the CR network will be analyzed in relation to (1) the admission of a new CR node in an existing CR network, (2) the performance of a CR network for different type of services, and (3) communication of CR node failure, which interrupts the trust negotiation.

References

1. J. Mitola and G. Q. Maguire, Cognitive radio: Making software radios more personal, *IEEE Personal Communications*, vol. 6, pp. 13–18, 1999.
2. M. M. Buddhikot, Understanding dynamic spectrum access: Models, taxonomy and challenges, in *2nd IEEE International Symposium on New Frontiers in Dynamic Spectrum Access Networks, 2007, DySPAN 2007*, pp. 649–663, April 17–20, 2007.
3. A. Rabbachin, G. Baldini, and T. Q. S. Quek, Aggregate interference in white spaces, in *7th International Symposium on Wireless Communication Systems (ISWCS), 2010*, pp. 751–755, September 19–22, 2010.
4. C. Peng, H. Zheng, and B. Y. Zhao, Utilization and fairness in spectrum assignment for opportunistic spectrum access, *Mobile Networks and Applications*, vol. 11, no. 4, pp. 555–576, 2006.
5. D. Wilkins, G. Denker, M.-O. Stehr, D. Elenius, R. Senanayake, and C. Talcott, Policy-based cognitive radios, *IEEE Wireless Communications*, vol. 14, no. 4, pp. 41–46, 2007.
6. G. Baldini, T. Sturman, A. R. Biswas, R. Leschhorn, G. Godor, and M. Street, Security aspects in software defined radio and cognitive radio networks: A survey and a way ahead, *IEEE Communications Surveys and Tutorials*, vol. 14, no. 2, pp. 355–379, 2012.
7. A. C. Squicciarini, E. Bertino, A. Trombetta, and S. Braghin, A flexible approach to multisession trust negotiations, *IEEE Transactions on Dependable and Secure Computing*, vol. 9, pp. 16–29, 2012.
8. A. Lee, K. Seamons, M. Winslett, and T. Yu, Automated trust negotiation in open systems secure data management in decentralized systems, *Advances in Information Security*, vol. 33, pp. 217–258, 2007.
9. G. Yajun and W. Yulin, Establishing trust relationship in mobile ad-hoc network, in *International Conference on Wireless Communications, Networking and Mobile Computing, 2007, WiCom 2007*, pp. 1562–1564, September 21–25, 2007.
10. G. Baldini, V. Rakovic, V. Atanasovski, and L. Gavrilovska, Security aspects of policy controlled cognitive radio, in *IFIP International Conference on New Technologies, Mobility and Security (NTMS'12)*, Istanbul, Turkey, May 7–10, 2012.
11. D. Gurney, G. Buchwald, L. Ecklund, S. L. Kuffner, and J. Grosspietsch, Geo-location database techniques for incumbent protection in the TV white space, in *IEEE Symposium on New Frontiers in Dynamic Spectrum Access Networks, 2008 (DySPAN 2008)*, vol. 3, pp. 1–9, October 14–17, 2008.
12. N. Ricardo, A. Pretschner, and V. D. Giacomo, A trustworthy usage control enforcement framework, *International Journal of Mobile Computing and Multimedia Communications (IJMCMC)*, vol. 5, no. 3, pp. 34–49, 2013. doi: 10.4018/jmcmc.2013070103.
13. D. Quartel, *Action relations—Basic design concepts for behaviour modelling and refinement*, PhD Thesis, University of Twente, 1998.
14. E. Rissanen, eXtensible Access Control Markup Language (XACML) Version 3.0, OASIS Standard, Jan-2013. Available at: http://docs.oasis-open.org/xacml/3.0/xacml-3.0-core-spec-os-en.pdf
15. D. Elenius, G. Denker, M.-O. Stehr, R. Senanayake, C. Talcott, and D. Wilkins. CoRaL—Policy language and reasoning techniques for spectrum policies, in *2007 IEEE Workshop on Policies for Distributed Systems and Networks*, Bologna, Italy, June 13–15, 2007.
16. V. Stavroulaki, K. Tsagkaris, P. Demestichas, J. Gebert, M. Mueck, A. Schmidt, R. Ferrus, et al., Cognitive control channels: From concept to identification of implementation options, *IEEE Communications Magazine*, vol. 50, no. 7, pp. 96–108, 2012.

17. M. Winslett, T. Yu, K. E. Seamons, A. Hess, J. Jacobson, R. Jarvis, B. Smith, and L. Yu, Negotiating trust in the Web, *IEEE Internet Computing*, vol. 6, no. 6, pp. 30–37, 2002.
18. Unlicensed Operations in the TV Broadcast Bands, Second Memorandum Opinion and Order, FCC 10-174, September 23, 2010.
19. S. Braghin, I. N. Fovino, and A. Trombetta, Advanced trust negotiation in critical infrastructures, in *International Conference on Infrastructure Systems 2008*, pp. 10–12, November 2008.

Chapter 17

Context-Aware Wideband Localization Using Cooperative Relays of Cognitive Radio Network

Homayoun Nikookar

Faculty of Electrical Engineering, Mathematics and Computer Science,
Delft University of Technology, Delft, The Netherlands

Contents

The author thanks Dr. Hao Lu for his contributions to this chapter.

389

17.1 Introduction

Wireless locating technology and wireless location-based services have been receiving increasing attention over the past decades. Remarkable commercial, public, and military potentials underlie various wireless localization applications, which encompass cellular system management, security, monitoring, tracking, logistics, radio planning, and so on.

Although Global Positioning System (GPS) is the most popular localization solution currently, it is well known that GPS is not always suitable for all scenarios and locations [1]. For those so-called GPS-denied scenarios, such as indoor, underground, urban canyons, and tree canopies, the GPS's requirement of line of sight (LOS) to multiple satellites becomes difficult, if not impossible. Hence, an alternative positioning and navigation method has become a need, which can be either used in locations that are unreachable by satellites or be adopted as a backup. Beacon localization is an option; it relies on a terrestrial fixed infrastructure, such as GSM base stations or WiFi access points (APs).

Furthermore, remote servers, such as war-driving activities, WiFi clubs, or some huge workstations, are required to provide a database of location information. In order to offer up-to-the-second location information, those powerful remote servers should be constantly updated. However, in sparse wireless network coverage areas, where the population density is low, the lack of base stations or WiFi APs might cause insufficient location estimation, and consequently increase localization errors [2]. Conventionally, a higher density base station deployment or the utilization of a higher-power base station enables higher accuracy in a localization system. However, both these approaches are impractical in real life owing to the cost reason. Therefore, a localization system that achieves high accuracy in rigorous scenarios with limited infrastructure and rigid power constraint will attract considerable attention among researchers. A basic method to solve this issue is employing cooperative localization as well as wideband transmission.

Cooperation between wireless nodes is a simple way to address the need for precise localization in power-limited, low-density base station deployment. The cooperative diversity introduced by the mobile devices helping each other, helps estimate the location and further improves the localization accuracy. The benefit of cooperative localization based on time domain measurements is illustrated in Figure 17.1, where mobile device 1 is not in the transmission range of base stations 3 and 4, while mobile device 2 cannot communicate with base stations 1 and 2. Both mobile devices cannot trilaterate their location merely on the basis of information from their neighboring base stations, whereas if communication between mobile devices 1 and 2 is established—that is, mobile devices 1 and 2 cooperate with each other—the localization for both mobile devices is enabled.

Ultra-wideband (UWB) radio has an absolute bandwidth of more than 500 MHz, or its relative (fractional) bandwidth is larger than 20%. Such a wide bandwidth is advantageous in multiple ways for both wireless communications and wireless localization. For wireless communications, both large relative and large absolute bandwidths alleviate small-scale fading. For wireless localization, the UWB signal is especially appropriate owing to its centimeter accuracy in ranging.

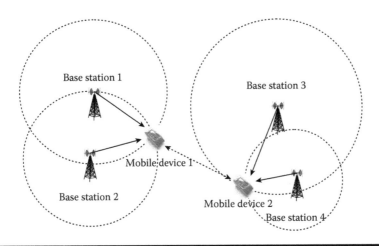

Figure 17.1 Cooperation between nodes benefits conventional localization.

Furthermore, a large bandwidth improves reliability for both communication and localization, because a large number of frequency components increases the probability that at least some of them can propagate, go around, or penetrate obstacles. By using the UWB signal, cooperative wireless networks enhance the communication speed and refine location estimation.

In this chapter we focus on cooperative wideband localization and review and analyze different cooperative relaying techniques for wideband localization. The rest of the chapter is organized as follows. In Section 17.2, we review the general cooperative localization procedure. Because a wideband signal is favorable for time-measurement-based localization, we narrow down the topic into cooperative wideband time-difference-of-arrival (TDOA) estimation in Section 17.3. In Section 17.4, we compare different cooperative relaying techniques with respect to the localization resolution, system complexity, and bandwidth usage. In Section 17.5, simulation results are shown to justify the theoretical claims. Finally, Section 17.6 concludes the chapter.

17.2 Cooperative Localization Procedure

The cooperative localization procedure is typically divided into two phases: measurement phase and position update phase [3]. In the measurement phase, position-related information, including the received signal strength (RSS), angle of arrival (AOA), or propagation time, should be measured. In the position update phase, the measurements are collected and used as inputs for executing the localization algorithm, to determine the location of the mobile device. For instance, when a mobile device obtains a ranging estimation with respect to three base stations, the mobile device can calculate its position by trilateration, given that it obtains the locations of the base stations. In order to achieve precise cooperative localization, the accuracy in both the measurement phase and the position update phase should be enhanced.

17.2.1 Measurement Phase

17.2.1.1 Signal Strength Measurement

The RSS is measured by the received signal power. Because the signal fades during the transmission, this approach is based on the relationship between power loss and the distance from the transmitter to the receiver, and determines the distance between the transceivers. The RSS measurement is simple and attractive owing to its low cost for the system designer. However, an RSS measurement always includes a large error, and only provides accurate localization with dense nodes in the wireless networks.

17.2.1.2 Time Measurement

Estimating the wireless signal propagation time usually provides a finer resolution in distance measurements, and it is more useful with low-density nodes in wireless networks. Propagation time measurement techniques include the time of arrival (TOA), round trip time of arrival (RTOA), and TDOA, which are all calculated from the signal cross-correlation. The RTOA is a practical distributed TOA scheme, because a common time reference between nodes is not necessary.

The TDOA refers to calculating the time difference between two signal propagations. The signal can be received at two synchronized receivers or at one common receiver. Therefore, for the TDOA estimation, there must be either two transmitters sending the same signal simultaneously to one receiver or two spatially separated synchronized receivers acquiring the signal from the same transmitter.

17.2.1.3 Angle Measurement

When the receiver is equipped with antenna arrays, AOA estimation determines the orientation of the received signal and, accordingly, the direction of the transmitter. Naturally, the AOA positioning approach with antenna arrays incurs more system costs.

17.2.1.4 Measurement in Wideband Localization

In wideband or UWB localization, a very large number of paths can be seen. This is due to the very wide bandwidth, especially in indoor scenarios. Consequently, accurate AOA estimation becomes a challenge, because there is significant scattering from objects in the multipath transmission. The very wide bandwidth leads to a very high temporal resolution, which makes the wideband system ideal for high-accuracy time-based localization applications. Furthermore, the aforementioned extraordinary characteristics of UWB, that is, the very wide bandwidth, cannot be used to improve the resolution for the RSS localization technique. The power emission limitations of the UWB protocols handicap RSS measurements as well. Although the RSS approach is easier than the time-measurement-based approach, the ranging information acquired from the RSS measurement is coarser compared to that obtained from the time measurement. Thus, the time-based approaches are better suited for wideband cooperative localization than the more costly AOA-based technique and the less accurate RSS-based technique.

17.2.2 Position Update Phase

In the position update phase, measurements are collected and exploited to calculate the position of each mobile device and update its location information. Localization and navigation technique development has a long history. In the cooperative localization scenarios, some new characteristics for the cooperative localization algorithm are worth noting here.

17.2.2.1 Hybrid Centralized-Distributed Algorithm

This cooperative localization algorithm is a hybrid algorithm that combines the advantages of centralized and distributed features. The centralized approaches collect measured information in the central processor before calculating the position; the position information is then transmitted by the central processor to the corresponding mobile devices. Distributed algorithms require the mobile device to locate its position iteratively. Mobile devices therefore need to recalculate their locations many times until convergence. There are two reasons for hybrid algorithms. First, centralized algorithms, unlike distributed algorithms, have no need to address the convergence issue, and are likely to lower the energy cost. Second, for some practical scenarios, the central processor will not be available to process the calculations for all mobile devices. Further, when many mobile devices simultaneously forward their measurement data to a central processor, a communication bottleneck appears in the data transmission.

Therefore, it is wise to adopt hybrid centralized–distributed algorithms for cooperative localization. For those mobile devices near the base station, the centralized algorithms are adopted; in this case, the base station plays the role of the central processor, collects the measurements, and calculates the position for each mobile device. Generally, when the number of hops for the mobile device to the base station exceeds the necessary number of iterations for convergence at the mobile device, the distributed algorithms are adopted; otherwise, the centralized algorithms are used.

17.2.2.2 Cooperative and Relative Algorithm

Unlike the conventional localization algorithms, in cooperative localization, mobile devices need not be within the transmission range of base stations to communicate with each other. Therefore, a high density of base stations or a large base station transmission range is no longer necessary. Because the mobile devices can receive data from both base stations and other mobile devices, cooperation between mobile devices can offer a higher localization accuracy and can provide positioning information to the remote mobile devices.

Cooperative localization algorithms can also exploit the range or angle measurement data between different mobile devices at undetected locations. A mobile device in that neighborhood or local environment can then obtain the relative positions of the mobile devices. By introducing the position information of the base station through communication between the base station and the mobile devices, the relative position of the mobile device can be transferred into an absolute position. The challenge of cooperative localization is to allow mobile devices that are not in the range of any detected location device to be localized and, further, to perform the location estimation for all mobile devices. Many cooperative location estimation algorithms have been proposed to determine the position of a device. A detailed review of these algorithms can be found in Ref. [4].

17.3 Cooperative Wideband TDOA Estimation

17.3.1 Why Cooperative Wideband TDOA?

Wideband or UWB systems are inherently suitable for time-measurement-based cooperative localization. This is because of the extremely wide bandwidths for transmission, which result in a fine temporal resolution and accurate ranging. Further, UWB provides multiple access communications capability. Moreover, UWB signals have the capability of passing around or penetrating obstacles.

In modern wideband wireless systems, orthogonal frequency division multiplexing (OFDM) technology is widely used. Owing to low spikes in the power spectrum density together with low transmitted power, zero-padding (ZP)-OFDM has been proposed in Ref. [5] for the IEEE UWB standard. Furthermore, ZP-OFDM provides advantages over its conventional counterpart, cyclic-prefix (CP)-OFDM, in terms of accurate blind time synchronization, better blind channel estimation, and better transmission performance. Thus, ZP-OFDM will play an increasingly visible role in wideband wireless systems.

In time-measurement-based localization, TDOA possesses many advantages that rely on its immunity to the clock bias of the transmitters. Therefore, TDOA approaches have been used for positioning system with asynchronous transmitters for decades; there are many applications in GPS and cellular localization. In the following sections of this chapter, we will focus on cooperative wideband TDOA estimation based on ZP-OFDM.

17.3.2 Multipath as Major Error Source in Cooperative TDOA

Multipath transmission means that transmitted signals reach the receiver through multiple paths. This is either radio reflection or the scattering phenomenon. This practical issue is considered deleterious for localization because the overlapped paths in multipath channels introduce interference in received signals. Multipath interference is a major source of the time-based range error. In the multipath channel, the late-coming multipath signals may dramatically degrade the signal-to-noise-and-interference ratio (SINR) of the desired LOS signal. In particular, some multipath components can arrive just behind the LOS signal and the so-called early-coming multipath components. They contribute to their cross-correlations and obscure the peak location related to the direct-path-transmitted signal. It can also happen that the direct-path-transmitted signal is severely attenuated, and even weaker than the late-coming multipath signals. This phenomenon leads to the direct-path-transmitted signal being totally lost, and causes unacceptable errors in the TDOA estimation.

In multipath and NLOS (non-line-of-sight) scenarios, the receiver should extract the first peak of the cross-correlation, while not tracking the highest peak. Because the first peak stands for the LOS signal, when the direct-path-transmitted signal is more attenuated, the first peak detection can be fulfilled by employing the thresholding technique for cross-correlation peak finding or the leading edge template matching approach [6].

However, scattering objects can be regarded as virtual cooperative mobile devices that can forward copies of the transmitted signals to the destination and introduce multipath diversity. Thus, with proper design of the transceiver, the signal detection at the receiver side can be improved by the gain from multipath diversity. In this chapter, we will illustrate how to gain from cooperative-multipath diversity and contribute to cooperative ZP-OFDM localization.

17.4 Relaying Techniques in TDOA-Based Cooperative Localization

17.4.1 Conventional Relaying for Cooperative Localization

We consider a cooperative wideband localization system in this chapter, and assume that the primary receiver has already recorded the positions of the base station and the relay. Scatterers spread between the primary receiver (mobile device), base station, and relays (mobile devices), and form a multipath transmission scenario. ZP-OFDM, a widely used paradigm in UWB systems, is employed as the modulation scheme here. In general, in order to obtain the location of the primary receiver in q dimensions, we need to perform $(q + 1)$ TDOA estimations with hyperbolic cross-point finding.

We begin by briefly reviewing the conventional TDOA estimation based on the amplify-and-forward (AF) and the decode-and-forward (DF) relay protocols. In the AF relay protocol, the relay amplifies the received signal first, and then sends it to the destination. In the DF relay protocol, the received signal is decoded by the relay first. If the full decoding is guaranteed, the decoded information will be re-encoded by the relay and retransmitted to the destination. If the decoding is not successful, this relay will be discarded. As shown in Figure 17.2, the relay receives the broadcast signal from the base station, and then employs the AF protocol or the DF protocol to transmit the signal to the primary receiver.

For cooperative localization using the AF relay, the signal received at the relay is amplified and forwarded, and the whole process is simple, but the noise and interference during the transmission from the base station to the relay is amplified as well. For cooperative localization using the DF relay, by decoding the signal, the OFDM symbol boundary can be detected. Some statistical features of each OFDM symbol (block) can consequently be used to estimate the TDOA, leading to bandwidth-efficient TDOA. The block features tested in Ref. [7] include the mean, average symbol's phase, variance, peak power, peak-to-average power ratio (PAPR), skewness, and kurtosis.

By exploiting the classical cross-correlation of two received signals transmitted from the base station and the relay to the primary receiver, respectively, we can calculate TD_{br}, that is, the TDOA between the base station to the primary receiver link and the relay to the primary receiver link. Then, multiplying TD_{br} by the speed of light easily translates the time difference into the distance difference. Based on the property of hyperbola, the primary receiver must appear on the hyperbola with the positions of the base station and the relay as the two foci. Combined with another two hyperbolas delineated from another base station and relay sets, the primary receiver can be located in the 2D plane.

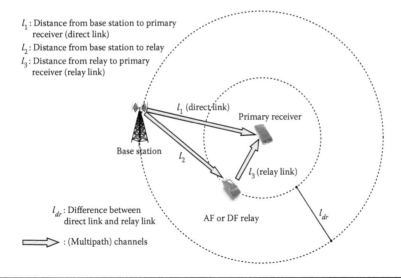

Figure 17.2 Cooperative TDOA based on conventional AF relay or DF relay.

17.4.2 Trigger Relaying for Cooperative Localization

The merit of the AF relay is its simple manipulation, while the advantage of the DF relay is that it eliminates the channel interference and the noise effect. The DF relay with a block of OFDM symbols can further reduce the transmitted data and improve the bandwidth efficiency. Thus, we propose a trigger relay mechanism that possesses both the aforementioned strong points, and leads to a cooperative TDOA estimation with high resolution, low complexity, and bandwidth efficiency. As shown in Figure 17.3, we assume that the locations of the base station and relays are both known by the primary receiver. After receiving the signal from the base station, each relay immediately sends a short predefined signal to the primary receiver, which is the so-called trigger relay.

The signal transmission diagram of the cooperative TDOA estimation using trigger relays is depicted in Figure 17.4. Relay 1 and relay 2 are triggered by receiving signals from the base station at t_1 and t_2. Then, the two relays send a predetermined short pilot or preamble to the primary receiver at t_1' and t_2', respectively. The processing times of the two relays are assumed to be identical; that is, $t_1' - t_1 = t_2' - t_2$. Finally, the signals from relay 1 and relay 2 reach the primary receiver at

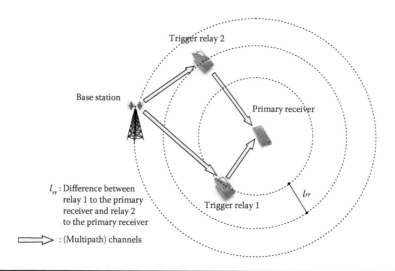

Figure 17.3 Cooperative TDOA mechanism based on trigger relays.

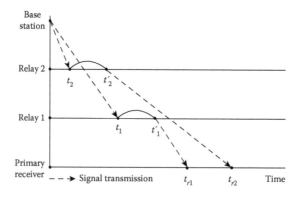

Figure 17.4 Signal transmission diagram of cooperative TDOA estimation using trigger relays.

t_{r1} and t_{r2}, respectively. By detecting the leading edge of the received signal, we can determine all the time points t_1, t_2, t_{r1}, and t_{r2}.

Subsequently, the TDOA between the link from relay 1 to the primary receiver and the link from relay 2 to the primary receiver, (TD$_{rr}$), can be denoted as

$$TD_{rr} = (t_{r2}-t_{r1}) + (t_1-t_2) \tag{17.1}$$

where $(t_{r2}-t_{r1})$ can be detected from the cross-correlation result of signals transmitted from the two trigger relays. For the AF and DF relay cases, (t_1-t_2) can be obtained from the positions of the base station and relays, which have been recorded by the primary receiver.

Next, we compare and analyze the performance of the trigger relay, AF relay, and DF relay with the block feature in the context of cooperative wideband TDOA.

In the conventional cooperative TDOA with the AF relay, the relay needs to forward the complete received signal, and requires a large bandwidth to transmit the data. On the contrary, the DF relay with the block feature needs to extract one statistical feature from each OFDM symbol, and only forwards the feature data. This scheme significantly reduces (by a factor of about 160 according to the MB-OFDM standard [5]) the transmitting data compared to the conventional TDOA estimation technique, which calculates the cross-correlation of the original received signals, and therefore leads to a bandwidth-efficient TDOA estimation performance. In the trigger relay case, only short pilots need to be transmitted in order to estimate the TDOA, and accordingly this significantly reduces the data transmission because resending of the complete signal is not required. Therefore, one of the major advantages of the trigger relay scheme is its bandwidth efficiency.

As the name implies, there should be an appropriate power amplifier in the AF relay to compensate for the attenuation the signal suffered during the transmission from the base station to the relay [8]. The DF relay should not only decode the received signal but also abstract the block features. Thus, both of these conventional relays increase the complexity of the cooperative localization system. On the contrary, the trigger relay technique does not need to process the received data, that is, the received data need not be decoded or amplified. It only needs to be switched on by the incoming signal, and a simple pilot sent to the primary receiver. Naturally, the proposed trigger relay enjoys the lowest system complexity among the three aforementioned relays.

For the estimation of the TDOA, the DF relay with the block feature scenario can only attain a resolution up to the block level. For example, when the MB-OFDM standard [5] is adopted, the feature-based TDOA estimation resolution is lower-bounded by its block interval, that is, 312.5 ns. This resolution can be enhanced by using a smaller FFT size multicarrier modulation and consequently reducing the block (OFDM symbol) interval. By using the trigger relay or AF relay, the cross-correlation of the sampled signals can have the sample-level resolution, which is 1.894 ns. It is noteworthy that the AF relay always suffers from noise and interference during the transmission from the base station to the relay, while the trigger relay is triggered by the leading edge of the incoming signal and is more robust to the noise and interference effect. Therefore, compared to the AF and DF relays, the trigger relay achieves the highest TDOA resolution.

17.4.3 Cooperative Multipath Diversity of Cooperative Localization with Trigger Relay

When a ZP-OFDM signal is transmitted through the multipath channels, cooperative TDOA estimation with trigger relay can further benefit from cooperative multipath diversity. As shown in Figure 17.5, we consider that each relay cluster has one core relay, represented by the black

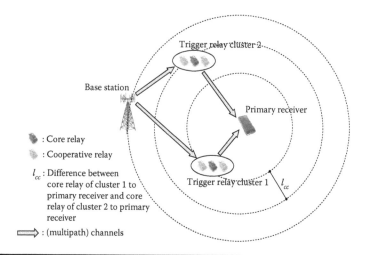

Figure 17.5   Cooperative multipath diversity for a trigger-relay-based cooperative TDOA system.

devices in the relay cluster of this figure. We assume that the primary receiver already knows the locations of the base station and the core relays. The neighboring cooperative relays, which are shown as gray devices, are temporally synchronized with the core relay and are grouped into one cluster by the core relay. The cooperative relays transmit the same signal as the core relay does, and help the core relay gain from cooperative-multipath diversity. The diversity gain combats the effect of noise and interference, and enhances signal detection. Improving signal detection based on the cooperative multipath diversity scheme will improve the resolution of the TDOA estimation.

17.5 Simulation Results

To evaluate the performance of cooperative TDOA estimation using the proposed trigger relay, we simulate a ZP-OFDM system using BPSK modulation. Zero-padding equals 25% of the OFDM symbol interval, and the sample period T_s = 1.894 ns. Different situations are simulated for a number of subcarriers N = 8, 16. We also consider different multipath diversity order scenarios, multipath tap number L = 1, 2, 3, and 4, and the delay between two neighboring channel tap equals one sample period, that is, 1.894 ns.

Example 1 (Comparison of three relaying techniques): In this case, we consider the number of subcarriers N = 8. For the DF relay protocol, we calculate the PAPR feature based on the first seven samples of the OFDM symbol. For AF relay and trigger relay protocols, which correlate the raw signals at the primary receiver, we do not use channel equalization or decoding. For the trigger relay with multipath diversity case, we use the linear transceiver proposed in Ref. [9]. In Figure 17.6, a three-path channel is simulated, that is, L = 3. It is shown that the DF relay protocol using the block feature only reaches the block-level resolution, that is, 18.94 ns. Because the AF relay always amplifies the noise and interference before forwarding, whereas the trigger relay does not, we can see from the figure that in the middle single-to-noise ratio (SNR) region, the trigger relay slightly outperforms the AF relay and provides a better TDOA estimation error. When the multipath diversity scheme is adopted, the TDOA estimation performance of the trigger relay can be further improved.

Context-Aware Wideband Localization ■ 399

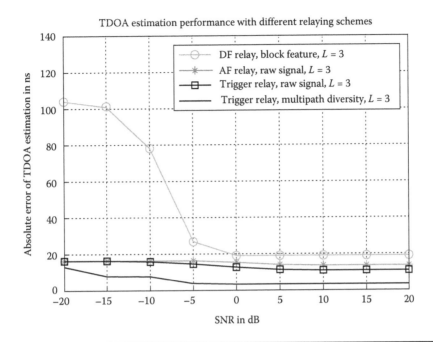

Figure 17.6   Comparison of TDOA estimations in different relaying schemes.

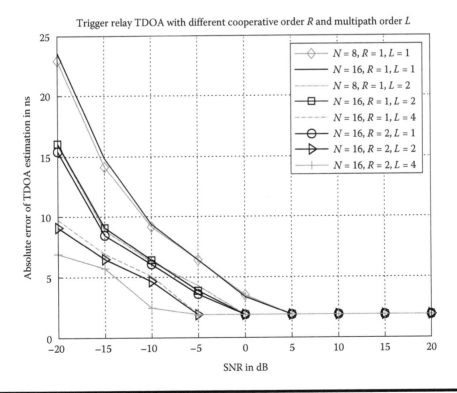

Figure 17.7   Cooperative multipath diversity for cooperative TDOA estimation using trigger relay.

Example 2 (Trigger relay TDOA estimation gaining from cooperative multipath diversity): In this case, we consider a two-cluster trigger relay scheme as shown in Figure 17.5: we assume the number of relays in one cluster (i.e., *R*) as 1 or 2, and the multipath diversity order *L* = 1, 2, and 4. Furthermore, we assume that all relays undergo the same order multipath channel. By adopting a linear transceiver proposed in Ref. [9], which holds a tall Toeplitz structure of the channel, we can guarantee channel matrix invertibility, signal detection, and the full cooperative multipath diversity gain for the ZP-OFDM system [10]. As shown in Figure 17.7, cooperative multipath diversity gains enhance the TDOA estimation accuracy, regardless of the change in the number of subcarriers. In the low-SNR region, diversity gains can be used to eliminate the noise effect. Thus, in this region, as the cooperative multipath diversity order increases, the TDOA estimation error decreases; while in the high-SNR region, the absolute error of TDOA estimation can reach its lower bound of 1.894 ns.

17.6 Conclusions

The era of extraordinarily accurate ubiquitous location awareness is on the horizon, boosted by cooperation in the wireless network. The research on cooperative localization will keep growing as larger cognitive radio networks appear and more applications requiring precise position information are deployed. The combination of wideband localization and cooperative relay is a practical way to address the issue of how to attain high resolution in a rigorous scenario with limited infrastructure and rigid power constraints. This chapter provided an overview of wideband localization using cooperative relays. First, different cooperative localization techniques were introduced briefly, and their characteristics, challenges, and future trends for development were unveiled. Then, cooperative relaying techniques in the context of wideband wireless localization were analyzed and discussed. Our discussions will hopefully inspire future efforts to develop a novel relaying paradigm for cognitive and cooperative localization.

References

1. E. Kaplan, Ed., *Understanding GPS: Principles and Applications*, Reading, MA: Artech House, 1996.
2. M. Z. Win, A. Conti, S. Mazuelas, Y. Shen, W. M. Gifford, D. Dardari, and M. Chiani, Network localization and navigation via cooperation, *IEEE Communications Magazine*, vol. 49, no. 5, pp. 56–62, 2011.
3. F. Gustafsson and F. Gunnarsson, Mobile positioning using wireless networks: Possibilities and fundamental limitations based on available wireless network measurements, *IEEE Signal Processing Magazine*, vol. 22, pp. 41–53, 2005.
4. H. Wymeersch, J. Lien, and M. Z. Win, Cooperative localization in wireless networks, *Proceedings of the IEEE*, vol. 97, no. 2, pp. 427–450, 2009.
5. A. Batra et al., Multi-band OFDM physical layer proposal for IEEE 802.15 Task Group 3a, IEEE P802.15-04/0493r1, September 2004. Online resource: http://www.ieee802.org/15/pub/2003/Jul03/03268r2P802-15_TG3a-Multi-band-CFP-Document.pdf
6. N. Patwari, J. N. Ash, S. Kyperountas, A. O. Hero, III, R. L. Moses, and N. S. Correal, Locating the nodes: Cooperative localization in wireless sensor networks, *IEEE Signal Processing Magazine*, vol. 22, pp. 54–69, 2005.
7. H. Lu, P. Martinez, and H. Nikookar, Cooperative TDOA estimation with trigger relay, in *Proceedings of the IEEE International Symposium on Personal, Indoor and Mobile Radio Communications (PIMRC 2011)*, Toronto, Canada, pp. 1–5, September 2011.

8. J. N. Laneman, D. N. C. Tse, and G. W. Wornell, Cooperative diversity in wireless networks: Efficient protocols and outage behavior, *IEEE Transactions on Information Theory*, vol. 50, no. 12, pp. 3062–3080, 2004.

9. B. Muquet, Z. Wang, G. B. Giannakis, M. Courville, and P. Duhamel, Cyclic prefixing or zero padding for wireless multicarrier transmissions, *IEEE Transaction on Communications*, vol. 50, no. 12, pp. 2136–2148, 2002.

10. H. Lu, T. Xu, H. Nikookar, and L. P. Ligthart, Performance analysis of the cooperative ZP-OFDM: Diversity, capacity and complexity, *Springer International Journal on Wireless Personal Communications*, vol. 68, no. 3, pp. 587–608, February 2013.

Chapter 18

Throughput Improvement in Mobile Ad Hoc Networks Using Cognitive Methods

Barbaros Preveze

Electronics and Communication Engineering, Cankaya University, Ankara, Turkey

Contents

18.1 Introduction

In this chapter, we consider a network in which all the nodes transmit data, video, and voice packets to random destination points through a multihop path. All the nodes in the network share the entire spectrum. It is known that any network in all likelihood uses one of the two well-known spectrum sharing techniques: OFDMA and TDMA [1–3]. In this chapter, we will focus on the systems combining these two techniques, such as 802.16.

If any performance criterion is to be improved, first a novel method must be proposed, and then a simulation program must be developed for implementation of the proposed method. This simulation program must be able to run for any N mobile relay nodes in a structure. Finally, the simulation results must also be compared with the results of mathematical calculations for confirmation.

In a cognitive network, all the nodes in the network can have any kind of information about other nodes and can be aware of the experiences of others [4]. This information can be used to improve the end-to-end performance of the entire network [4] by decreasing the delay amount or the number of lost packets.

In this chapter, a 802.16j network simulation [5] will be investigated that implements some cognitive methods on a network and improves the system performance by employing cognitivity.

The most important performance criterion for a network that an end user can sense is the throughput. The cognitive methods proposed for improving the system throughput performance must either decrease the packet loss rate of the network or successfully transmit the packets faster. The most important factors affecting these parameters are the packet loss rate, delay, and efficient spectrum usage. In this chapter, some cognitive methods to overcome these problems are explained, and the improvement in the system throughput performance is demonstrated.

18.2 Parameters Affecting Throughput Performance

18.2.1 Packet Losses

In a network, increasing the number of packets being transmitted in a second may increase the number of arrived packets and the throughput, but it may also cause some packet drops because of the congestion on the intermediate nodes; some packet losses may occur. To adjust the transmission rate carefully, it can adaptively be changed according to conditions of the network at that time, and the packet dropping rate can be decreased to increase the number of successfully arrived packets.

On the contrary, the buffers must also be managed perfectly to decide which packet should be forwarded for a lower probability of being dropped. Further, the optimum buffer size must also be calculated to avoid long waits for packets in the larger buffers and to avoid dropping and losing of packets with smaller buffers. The sizes of the buffers in which the incoming packets are stored are used as they are calculated in Section 18.4.4, and the control of the packets selected for transmission is according to the buffer management (BM) method described in Section 18.5.1.3. These methods cognitively help to reduce the packet loss rate of the overall network.

18.2.2 Delay

If the considered system is a packet-switched network, the nodes will be forwarded from one hop to another, and each packet will follow its own route with a different delay amount during the transmission. However, the important point here is that the positions of the mobile nodes will always change according to the used mobility model. So the route has to be determined each time to select the best route, as mentioned in Section 18.5.1.1. In this way, the route is shortened, the hop count is decreased, and the delay amount is decreased. It must be noted that the routing algorithm therefore also has a great influence on the network delay.

18.2.3 Efficient Spectral Usage

Efficient spectral usage is the spectral usage amount used by the packets that successfully arrive at the destination. If a packet gets lost at a node, the spectrum used by that packet up to that node will be wasted, and this will cause a decrease in efficient spectral usage. If an extra hop is used in the transmission of an arrival packet, this causes spectral wastage and decreases the effective spectral usage. The cognitive methods must also help the system to use the spectrum efficiently by using correct paths and by selecting correct packets to transmit. Therefore, the investigated cognitive methods in this chapter also tackle these issues.

18.3 Maximum Theoretical Data Rate of 802.16j Network

Before proposing and implementing the cognitive methods, the system model must be simulated. Fortunately, the common system parameters are defined by the 802.16j network standards [1,3,5] as follows:

Bandwidth (B) = 10 MHz, total used frequency (FT) = 1/TST = 9.718 kHz, total symbol time (TST) = 102.8 μs, sampling guard interval (TSG) = 11.4 μs, data symbols per frame (DSPF) = 40, guard rate (GR) = 1/8, oversampling factor (OSF) = 8/7, frame length (FL) = 5 ms, sampling period (TS) = 1/FS = 91.4 μs, forward error correction (FEC) = ¾, subchannel capacity (SCC) = 0.95 Mbps, number of subchannels (NOS) = 30, frequency spacing (FS) = 10.94 MHz, QAM level = 6.

The given parameter values and the formulations in Equations 18.1 through 18.12 are taken from calculations in Ref. [1] and the standards [1,3].

$$\text{Frames per second (FPS)} = 1/\text{FL} \tag{18.1}$$

$$\text{Bits per symbol (BPS)} = \log_2(\text{QAM}) \times \text{NODS} \tag{18.2}$$

$$\text{Symbols per frame (SPF)} = \text{FL}/\text{TST} \tag{18.3}$$

$$\text{Subchannel data rate in a frame with FEC (SCDRFWF)} = \text{DSPF} \times \text{MAU} \tag{18.4}$$

$$\text{Subchannel data rate per second with FEC (SCDRSWF)} = \text{SCDRFWF} \times \text{FPS} \tag{18.5}$$

$$\text{Maximum allocatable unit (MAU)} = \text{BPSWF}/\text{NOS} \tag{18.6}$$

$$\text{Sampling rate (SR)} = \text{B} \times \text{OSF} \tag{18.7}$$

$$\text{Data rate per second with FEC (DRPSWF)} = \text{FEC} \times \text{DRPS} \tag{18.8}$$

$$\text{Data rate per sec (DRPS)} = \text{SRPS} \times \text{BPS} \tag{18.9}$$

$$\text{Bits per symbol with FEC (BPSWF)} = \text{FEC} \times \text{BPS} \tag{18.10}$$

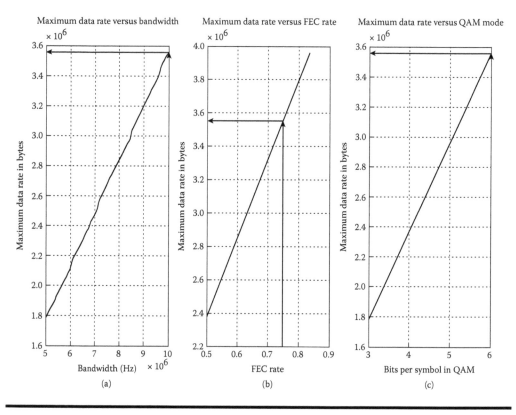

Figure 18.1   Data rates versus (a) bandwidth, (b) FEC rate, and (c) QAM mode.

$$\text{Symbol rate per sec (SRPS)} = \text{FPS} \times \text{DSPF} \qquad (18.11)$$

$$\text{Subchannel data rate in a second with FEC (SCDRSWF)} = \text{SCDRRFWF} \times \text{FPS} \qquad (18.12)$$

where NODS in Equation 18.2 stands for the number of data subcarriers.

The maximum spectrum usage by all the nodes in the network is calculated by multiplying the subchannel capacity (SCC) by the number of subchannels (NOS) as follows:

$$\text{Max. spectrum usage (MSU)} = \text{SCC} \times \text{NOS} \qquad (18.13)$$

as in Refs. [1,3].

Figure 18.1a–c represent the maximum data rates using the foregoing parameter values for various input parameters, where the result of the given parameter set is highlighted in all graphs (B = 10 MHz, FEC = 3/4, QAM level = 6).

18.4 Packet Size and Buffer Size Adjustments

18.4.1 Real-Time Multimedia Packet Size

If the nodes in the system are considered to generate 50 picture frames per second for video conversation and 200 OFDM frames are considered in a second, 50 of the 200 OFDM frames will be used to send video packets in a second within a subchannel [1]. By multiplication of a given SCC value and the video packet sending rate (VPSR) of 50/200, the video rate can be evaluated as in Equation 18.14:

$$\text{video rate} = \text{SCC} \Big/ \left(\frac{1}{\text{VPSR}} \right) \tag{18.14}$$

By using the number of subchannels in each frame, the maximum spectrum usage, and OFDMA frame length defined by the standards [1,3], the video packet size can also be calculated as

$$\text{VDPS} = \text{SCDRFWF}$$

$$= \frac{\dfrac{\text{MSU}}{\text{NOS}}}{\dfrac{1}{\text{FL}}} \tag{18.15}$$

where MSU stands for the maximum spectral usage.

On the contrary, for worst-case conditions, if we consider each node to have voice conversation and video conversation at the same time, a 16 kbps voice stream can be handled for low latency and the voice packet size can be considered as [1]

$$\text{VCPS} = \frac{16 \text{ kbits}}{200} = 10 \text{ bytes} \tag{18.16}$$

18.4.2 Non-Real-Time Packet Size

The nodes in the considered system may use a different number of subchannels for each non-real-time data type. The number of subchannels allocated for non-real-time data transmission in the current frame can be calculated by the rest of the subchannels that are not used by real-time video and voice packets as [5]

$$\text{TSFDT} = \left(\text{SC} - (\text{TSCFVD} + \text{TSCFVC}) \right) \tag{18.17}$$

In Equation 18.17,
TSFDT stands for the total number of subchannels allocated for data packets,
TSCFVD stands for the total number of subchannels allocated for video packets, and
TSCFVC stands for the total number of subchannels allocated for voice packets.

Because only 1 out of 4 subchannels is used by each node for video transmission, totally 1 or 2 subchannels can be used in a frame for video packet transmission and N subchannels will be used by voice packet transmission of N packets. Hence, the nodes will transmit their non-real-time data packets using NOS – (N + 1), NOS – (N + 2) subchannels.

On the contrary, the optimum data packet size for minimum bandwidth wastage nearest to the chosen packet size can be calculated as [5]:

$$\text{DTPS} = \frac{(\text{SCDRFWF})}{\text{floor} \left(\dfrac{\text{SCDRFWF}}{\text{DTPS}_{\text{ref}}} \right)} \tag{18.18}$$

where DTPS_{ref} stands for the DTPS that we want to use in our system, and *DTPS* stands for the optimum packet size value nearest to DTPS_{ref} that exactly fits the subchannel for a given *N*.

In this way, multiples of the packets will exactly fit the size of a subchannel, and no bandwidth wastage will occur.

18.4.3 Maximum Bandwidth Usage

According to the previous section, N subchannels among NOS subchannels will be used for voice packet transmissions for N nodes, and 1 or 2 subchannels will be required for video packet transmissions. Then the remaining NOS – $(N + 2)$ or NOS – $(N + 1)$ of NOS subchannels (from Equation 18.17) will be used by data packet transmissions. Considering all these, the wasted bandwidth by data transmission (WBDT) is calculated as [5]:

$$\text{WBDT} = \left(\text{SCC} - \left(\text{DTPS} \; x \; \text{int}\left(\frac{\text{SCC}}{\text{DTPS}} \right) \right) \right) \times \left(\text{NOS} - \left(\text{N} + (1 \text{ or } 2) \right) \right) \qquad (18.19)$$

The calculation in Equation 18.19 is made considering the last packet getting lost when the channel capacity is not exactly equal to multiples of the size of a data packet. $(N + 1)$ in Equation 18.19 is the sum of the number of subchannels used by real-time video and voice packets and not used by data packets. If DTPS is selected as in Equation 18.18, no bandwidth wastage will occur because of data packets.

On the contrary, because the number of hop counts cannot exceed $N - 1$ hops, each packet can remain in the network for a maximum duration of $N - 1$ frames. As a result, a total of N voice packets in the current frame plus $(N - 1) \times (N)$ voice packets in the previous $(N - 1)$ frames, that is, N^2 voice packets, will be sent through the subchannels in each frame for the worst case.

Then a total of $N^2 \times 10$ bytes of voice packets will be transmitted if a voice packet size (VCPS) of 10 bytes is used as calculated in Section 18.4.1. The wastage by voice packets in a second (WVCPIS) is calculated in Ref. [5] as

$$\text{WVCPIS} = \left(\left(N \times \text{SCC} \right) - \left(\text{VCPS} \times N^2 \right) \right) \times \frac{1}{\text{FL}} \qquad (18.20)$$

However, if the nodes are considered to use only one subchannel for video conversation for 2 of every 4 frames [1], half of 2 packet sizes (for the last packet not fitting the subchannel) will be wasted. If we choose the size of video packets to equal the SCC calculated in Equation 18.15, multiples of it will also fit the subchannel and no wastage would occur. As a result, the maximum possible successful bandwidth usage (MPSBU) under the given conditions can be evaluated as in Equation 18.21 [5]:

$$\text{MPSBU} = \left(\text{SCC} \times \text{NOS} \right) - \left(\left(N \times \text{SCC} \right) - \left(\text{VCPS} \times N^2 \right) \right)$$

$$- \left(\text{VPSR} \times \text{VDPS} \right) \times \left(\frac{1}{\text{FL}} \right) \text{ Bytes} \qquad (18.21)$$

18.4.4 Calculation of Optimum Buffer Size

The buffer size of the nodes in a system is another critical point for not requiring extra memory on each node and not allowing more packets to wait in the buffers. For calculating the optimal buffer size, we consider a total of $(N) \times (\text{AHC} + 1) \times \text{PPF}$ (packets per frame) packets that will be transmitted by all nodes during each frame on average, one for the current transmission and AHC packets for forwarding the previous packets during the AHC hops.

Formulation of the optimum data packet buffer size (ODBS) is given in Equation 18.22 and an example calculation is given in Equation 18.23 [5], where AHC is taken as 3, and other parameter values are taken as they are defined by the standards [1,3]:

$$\text{ODBS} = \left(\frac{\log_2(\text{QAM}) \times \text{NODS} \times \text{FEC} \times \text{DSPF}}{8 \times \text{NOS}} \right) \times (\text{NOS} - (N + (1 \text{ or } 2)))$$

$$\times (\text{AHC} + 1) \text{ bytes} \tag{18.22}$$

$$\text{ODBS} = \left(\frac{\log_2(64) \times 720 \times \frac{3}{4} \times 44}{8 \times 30} \right) \times (30 - 6 - 2) \times (3 + 1) = 52272 \text{ bytes} \tag{18.23}$$

From this equation, it is clearly seen that the ODBS value is directly related to the AHC value. Because when smaller AHC value is provided the required buffer size for not losing the packets will be smaller. Here, the optimum number of slots for data packets (ONSDT) is calculated using:

$$\text{ONSDT} = \text{ODBS}/\text{ODPS Slots} \tag{18.24}$$

The optimum buffer sizes for voice and video buffers can also be generalized and formulated as in Equation 18.25 [5]:

optimum video or voice buffer size = (video or voice packet size) $\times$ (VPSR) $\times$ (N)
$$\times (\text{AHC} + 1) \text{ bytes with } (\text{VPSR}) \times (N) \times (\text{AHC} + 1) \text{ slots} \tag{18.25}$$

As an example of this calculation, selecting the voice packet size as 10 bytes, video packet size as 594 bytes, video sending rate as ¼, voice sending rate as 1 at each frame (1/1), and using AHC as 3 and $N = 6$, the optimum buffer sizes for voice and video packets can be calculated using Equation 18.25 as [5]:

optimum voice buffer size = $(10) \times (1) \times (6) \times (3 \times 1) = 240$ bytes $(\text{VPSR} = 1)$ \hfill (18.26)

optimum video buffer size = $(594) \times \left(\frac{1}{4} \right) \times (6) \times (3 + 1) = 3564$ bytes

$$= 3.48 \text{kb with } (\text{VPSR}) \times (N) \times (\text{AHC} + 1) \text{ slots} \left(\text{VPSR} = \frac{1}{4} \right) \tag{18.27}$$

where VPSR is the video/voice packet sending rate.

The 3D graph of the results of this calculation is given in Figure 18.2 for changing node counts and AHC values on the network [5].

For the foregoing example, the total buffer size needed to manage the traffic for various AHC and N values is graphically presented in Figure 18.3.

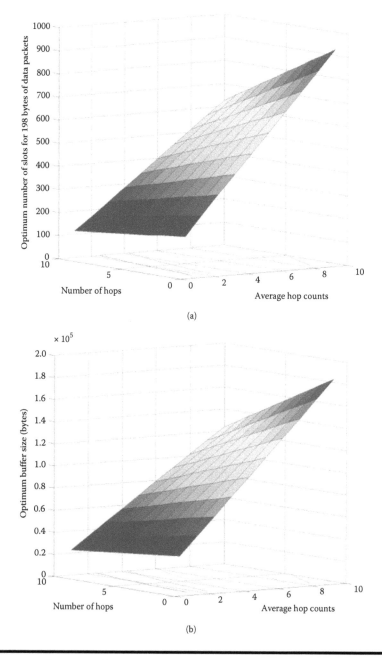

(a)

(b)

Figure 18.2 **(a) Optimum number of slots and (b) optimum buffer size for different numbers of average hop counts and nodes sharing the spectrum. (From B. Preveze and A. Safak,** *International Journal of Wireless and Mobile Networks,* **vol. 2, pp. 120–140, 2010.)**

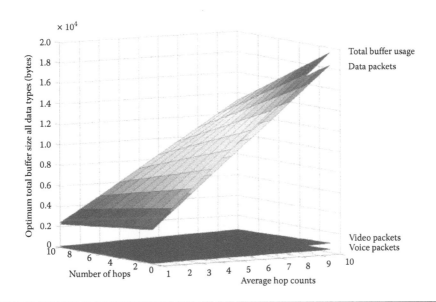

Figure 18.3 Optimum buffer sizes for data, voice, and video packets and their sum for different AHC values and different numbers of nodes sharing the spectrum where other parameter values are fixed at their default values. (From B. Preveze and A. Safak, *International Journal of Wireless and Mobile Networks***, vol. 2, pp. 120–140, 2010.)**

18.5 Cognitive Approaches to Improving Network Throughput

In a cognitive system, all the nodes work for the common benefits of the overall network, and they collect information about the entire network. They manage the packets in their buffers using this information. If they meet with an abnormal condition, they announce it to the others and adaptively make some changes in the algorithm they use.

In a mobile network, all the nodes move with random speeds toward random destination points, and their positions continuously change. Therefore, every calculation must frequently be repeated at each node for an optimum solution of packet losses. The packet loss ratio in the system is calculated as in Equation 18.28:

$$\text{Packet loss ratio} = \text{lost packets}/(\text{lost packets} + \text{successfully sent packets}) \qquad (18.28)$$

In a conventional OFDMA system, the nodes can access any number of subchannels whenever they need to. In a network with a limited number of subchannels, assigning separate subchannels for every node at the same time may not be possible. As a result, the node using the spectrum may be unsuccessful in transmission, and in this way it may waste the bandwidth. In order to solve this problem, the entire spectrum can be used by the selected nodes at a time. These nodes can be selected as the nodes having the greatest number of full slots in their buffers, as creating more free slots on a congested node will avoid packet drops at this node in the future. After making this selection, the selected transmitter nodes should also select the correct packets in their buffers to be transmitted such that their routes toward their destination include the nodes with more free slots. This selection is also done for lower probability of dropping packets on the route. After this point, the transmitter may also use the whole allocated spectrum and may free its buffer at the current OFDMA frame.

As a result, the number of lost packets can be decreased, and the throughput of the network can be increased.

These methods are applied in an OFDMA system in Ref. [5], and the resulting improvements are illustrated in Figure 18.7 [5].

In the case of congestion, more packets may be dropped and get lost at the intermediate nodes. Therefore, the number of packets transmitted in a second by the nodes can be decreased or increased adaptively. This is called the adaptive rate (AR) algorithm. By using the AR algorithm [6], the nodes will be able to produce and transmit fewer data packets, but because of less packet losses, an increased throughput performance will be obtained.

18.5.1 Cognitivity in Wireless Multihop Mobile Networks

18.5.1.1 Route Reconstruction Algorithm

If the nodes in the network are always in motion, the positions of the nodes always change. Therefore, each node has to recalculate the path of each packet before forwarding it to the next hop. A route reconstruction algorithm is therefore used in all nodes before forwarding each packet.

18.5.1.2 Maximum Data Rate

When the network becomes congested and packet losses start to occur, the transmission rate is decreased using the AR data transmission algorithm. This method adds learning and experience features to the network, where the nodes adapt to changing network conditions. It is also shown in Refs. [5,6] that the adaptive rate has a positive effect on the system throughput. For a network with heavy packet traffic, the maximum data packet generation/transmission rate in a frame in the case of zero packet loss using the AR algorithm can be calculated as in Equation 18.29 [5]:

$$MDRPN = MSU\,(bytes/s) \times \frac{TSFDT}{NOS} \times \frac{1}{N} \times (FL) \times Packet\,Loss\,Ratio \qquad (18.29)$$

18.5.1.3 Buffer Management (BM) by Queue Optimization

Once a node takes the spectrum usage right and starts to send its packets, it first determines the routes of its packets. It re-sorts the packets in its buffer such that the packets with the same destination are grouped to be sent together. According to the dynamic spectral aids algorithm, the packets whose next node has the buffer with most free memory slots will be sent first among these.

18.5.1.4 Dynamic Spectral Aids (SA) Algorithm

According to this algorithm, the nodes with the fullest buffers are selected to transmit their packets first, so that the buffers of the congested nodes are emptied first to prevent packet losses. For a system in which N (number of nodes) of NOS (number of subchannels) subchannels are allocated for voice packets of N nodes (one for each), and two of every four subchannels are allocated for video packets of the nodes, and the rest of the subchannels are all allocated to the node with the most full buffer and that therefore needs the spectrum usage most. When a node starts to transmit its packets, it performs buffer optimization and transmits the packets whose calculated next node's buffer has the largest free memory according to the BM algorithm, but if the buffer of the freest node is also full, the transmitter loses its first trial packet of the chosen destination and yields its turn to the node that alerted that it needs the spectrum most other than itself. At the end of the frame, after transmitting each packet, the spectrum usage turn is again allocated to the node that needs the spectrum the most.

18.6 Performance Evaluation

The throughput improvement of a network first requires improvements in efficient bandwidth usage and data packet loss rate under different conditions. Therefore, any kind of cognitive method can be proposed and implemented that increases the spectrum usage efficiency and decreases the packet loss rate for systems. After this point, the effect of each cognitive method on the system performance can be highlighted by rerunning the simulation with and without applying the novel method to the system and observing the system performance for each case. However, all the results must be simultaneously evaluated for exactly the same conditions.

Figure 18.5 illustrates the results of such a study [5] in which these cognitive methods are implemented. The pure system is simulated, and the simulation results are generated under normal conditions. Then the simulation is rerun, and the results are evaluated for each case of excluding each method for six nodes ($N = 6$). The numeric values of sent/lost packets, efficient spectral usage amounts, and throughput amounts taken from the simulation results of that study for all cases are listed in Tables 18.1 and 18.2 [5]. It is seen from Table 18.2 that both the efficient spectrum usage efficiency and packet loss rate are improved, so the throughput is also improved, as seen in Figure 18.5.

It is normal to think that the packet loss ratio decreases if larger buffer sizes are used and that it increases if smaller buffer sizes are used. Similarly, it is normal to think that using larger buffers has a positive effect on the throughput. It is shown in Figure 18.6 that the packet loss ratio is really decreased owing to a larger buffer size relative to the optimum calculated buffer size, but Figure 18.5 indicates that the throughput of the system is decreased by using a larger buffer size relative to the calculated optimum buffer size. This is because larger buffers store and hold more packets inside, and they cause the packets to wait for longer durations in the buffers. This extra delay decreases the throughput of the system.

However, it is observable from the results in Figure 18.5 that deactivating the buffer management algorithm allows more congestion on the buffers of the nodes and increases the packet losses.

It is also clear from Figure 18.4 that because a separated subchannel is allocated for each real-time packet of each node at each OFDMA frame, the number of sent voice/video packets are approximately equal for each case, and because their buffers are not overloaded by the system as buffers of data packets, no voice or video packet losses occur during the transmission.

As a result, because the numbers of sent/lost video/voice packets are the same for all cases, the number of sent/lost packets of high traffic loaded non-real-time data packets shown in Figure 18.4 determines the system throughput performance of each algorithm.

As shown in Figure 18.5 [5], the throughput is maximized up to 1,556,458 bytes, enabling all the novel cognitive methods at the same time, and the system average throughput performance is decreased most during the absence case of SA algorithm, which means the improvement is provided most by the SA algorithm.

However, in order to confirm the correctness of the evaluated results, the same results must also be evaluated mathematically.

Table 18.1 Number of Sent/Lost Packets Results of Simulations for Application Conditions of the Methods during 1000 OFDMA Frames (5 s)

Simulation Output Data (N = 6)	End-to-End Successful Average Transmit Time of a Packet (ms)			Number of Video Packets		Number of Voice Packets		Number of Data Packets		Packet Loss Ratio	
	Video	Voice	Data	Sent	Lost	Sent	Lost	Sent	Lost	Video/ Voice	Data
No methods	10,12	2,09	0,17	494	0	2389	0	29344	4456	0.00%	13.18%
All – adaptive rate	9,86	1,94	0,14	507	0	2576	0	35953	653	0.00%	1.78%
All + larger buffer	9,71	1,97	0,14	515	0	2532	0	36691	58	0.00%	0, 16%
All – buffer management	9,67	1,94	0,15	517	0	2582	0	33978	654	0.00%	1.89%
All – spectral Aid	9,96	1,94	0,16	502	0	2575	0	31419	786	0.00%	2,44%
All + calc. buffer	9,71	1,97	0,13	515	0	2544	0	37631	182	0.00%	0.48%

Source: B. Preveze and A. Safak, *International Journal of Wireless and Mobile Networks*, vol. 2, pp. 120–140, 2010.

Table 18.2   The Results of Simulations for Application Conditions of the Methods during 1000 OFDMA Frames

Simulation Output Data (N = 6)	Average Data Packet Loss Ratio	Average Effective Bandwidth Usage in 2,913,300 byte/s	Average Effective Bandwidth Usage Percentage in 2,913,300 byte/s	Average Overall Throughput Provided byte/s	Improvement (%)	Improvement Loss of the Throughput by Not Applying the Algorithm (i.e., Improvement of the Algorithm)
No methods	13.18%	2642358	90%	1225488	0	27%
All – adaptive rate	1.78%	2790893	95%	1489122	22	5%
All – spectral aid	2.44%	2782438	95%	1308980	7	20%
All – buffer management	1.89%	2751435	94%	1412112	15	12%
All + larger buffer	0.16%	2808181	96%	1519210	24	3%
All + opt. buffer	0.48%	2801605	96%	1556458	27	0%

Source: B. Preveze and A. Safak, *International Journal of Wireless and Mobile Networks*, vol. 2, pp. 120–140, 2010.

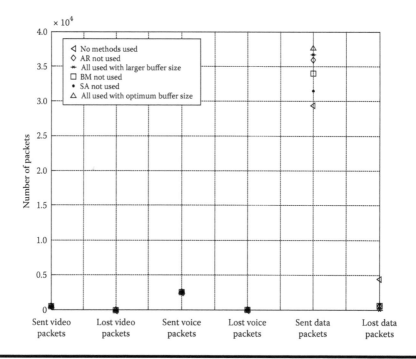

Figure 18.4 The improvement amounts of cognitive methods for the number of video, voice, and data packets successfully transmitted to the final destination in 1000 frames. (From B. Preveze and A. Safak, *International Journal of Wireless and Mobile Networks*, vol. 2, pp. 120–140, 2010.)

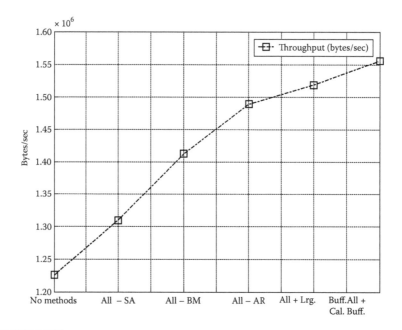

Figure 18.5 Throughput improvement (bytes) provided by the cognitive methods. (From B. Preveze and A. Safak, *International Journal of Wireless and Mobile Networks*, vol. 2, pp. 120–140, 2010.)

If we assume that all the packets are distributed to the nodes in the network in proportion to their waiting durations, the average loss probability of the packet at each of the other nodes can be calculated using Equations 18.30 and 18.31.

If we let the packet distribution rates of the nodes be

$$\text{Node}_1 \rightarrow 1, \text{Node}_2 \rightarrow 2, ..., \text{Node}_{N-1} \rightarrow N-1, \text{Node}_N \rightarrow N$$

Then, the probability of losing a packet in Node$_n$ can be calculated as

$$P_{lost}(n) = \frac{\dfrac{\text{Packet Dist. rate of the Node}_n}{\text{Sum of Dist. rates of all nodes}} \times \text{Total Packet count}}{\text{Buffer size}}$$

$$= \frac{\dfrac{(n)}{\dfrac{N \times (N+1)}{2}} \times \text{Total Packet count}}{\text{Buffer size}}$$

$$P_{lost}(n) = \frac{2 \times (n) \times \text{Total Packet count}}{\text{Buffer size} \times N \times (N+1)} \tag{18.30}$$

The probability of a packet getting lost in any of the nodes can now be evaluated by calculating the probability of selecting a packet for transmitting to a node (with a transmitting probability of $1/(N-1)$ excluding the sender itself) and losing the packet there, as [5]

$$P_{loss} = \frac{1}{N-1} \times P_{lost}(1) + \frac{1}{N-1} \times P_{lost}(2) + ... + \frac{1}{N-1} \times P_{lost}(N-1)$$

$$P_{loss} = \frac{1}{(N-1)} \left(\sum_{n=1}^{N-1} P_{lost}(n) \right) \tag{18.31}$$

Using the parameter values used in the simulation in Ref. [5], for $N = 6$, buffer size = 264, and total packet count = 264, it results in:

$$P_{loss} = \frac{1}{N-1} \sum_{n=1}^{N-1} \times P_{lost}(n)$$

$$= \sum_{n=1}^{N-1} \frac{1}{N-1} \times \frac{2 \times (n) \times \text{Total Packet count}}{\text{Buffer size} \times N \times (N+1)}$$

$$= \sum_{n=1}^{N-1} \frac{1}{6-1} \times \frac{2 \times (n) \times 264}{264 \times 6 \times (6+1)}$$

$$= \frac{1}{5} \times \left[\frac{1}{21} + \frac{2}{21} + \frac{3}{21} + \frac{4}{21} + \frac{5}{21} \right] \tag{18.32}$$

$$= 0.142 = \%14.2$$

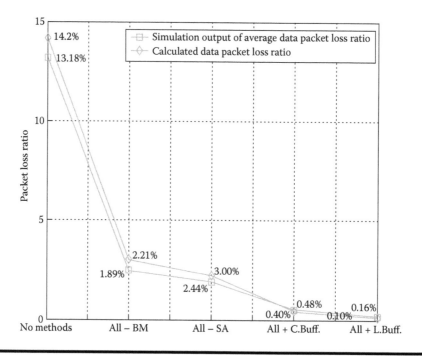

Figure 18.6 Average data packet loss ratio of the system with different methods and confirmation of the simulation results with the results evaluated using Equations 18.33 through 18.36.

The calculated result also matches the simulation results presented in Table 18.2 and Figure 18.6 [5].

If the proposed BM technique is employed, packet loss occurs if and only if all the packets in the transmitting node's buffer will be transmitted to a node with a full buffer. That is, sum of the probability of losing a packet at any of the nodes to which one of the other nodes will make transmission.

$$
P_{loss\_BM} = \frac{\binom{N-1}{1}p^1 + \binom{N-1}{2}p^2 + \binom{N-1}{3}p^3 + \dots + \binom{N-1}{N-1}p^{N-1}}{\sum\limits_{n=1}^{N-1}\binom{N-1}{n}}
$$

$$
P_{loss\_BM} = \frac{\sum\limits_{n=1}^{N-1}\left\{\binom{N-1}{n}\times p^n\right\}}{\sum\limits_{n=1}^{N-1}\binom{N-1}{n}}
$$

(18.33)

Again for $N = 6$, we have [5],

$$P_{loss\_BM} = \frac{\binom{5}{1}p^1 + \binom{5}{2}p^2 + \binom{5}{3}p^3 + \binom{5}{4}p^4 + \binom{5}{5}p^5}{\binom{5}{1} + \binom{5}{2} + \binom{5}{3} + \binom{5}{4} + \binom{5}{5}}$$

$$P_{loss\_BM} = \frac{5p^1 + 10p^2 + 10p^3 + 5p^4 + p^5}{5 + 10 + 10 + 5 + 1}, \quad \text{for } p = 0.142 \tag{18.34}$$

$$P_{loss\_BM} = 0.03 = 3.00\%$$

which also corresponds and matches with the result of All-SA in Figure 18.6.

We can calculate the packet loss ratio for the case of applying all algorithms without the BM algorithm ("All – Buffer Management" in Table 18.2) by multiplying the probability of having packets with n (from 1 to $N - 1$) different types of destinations by the probability of not losing the packet at the rest of the nodes and adding the results of all calculations evaluated from all combinations as in Equation 18.35 [5]:

$$P_{loss\_SA} = \left(p^1 \times (1-p)^{(N-1)-1} \times \frac{1}{N-1} \right) + \left(p^2 \times (1-p)^{(N-1)-2} \times \frac{2}{N-1} \right) + \cdots$$

$$+ \left(p^n \times (1-p)^{(N-1)-(n)} \times \frac{n}{N-1} \right) \tag{18.35}$$

$$P_{loss\_SA} = \sum_{n=1}^{N-1} P^n \times (1-p)^{((N-1)-n)} \times \frac{n}{N-1}$$

The p values in Equations 18.33 through 18.35 are used for P_{loss}, as shown in Equation 18.31.

For $N = 6$ and $p = 14.2\%$, we have $P_{loss\_SA} = 2.21\%$ from Equation 18.35, a result that also corresponds to and agrees with the result of "All – Buffer Management" in Figure 18.6.

When we apply both the BM and SA algorithms and combine Equations 18.33 and 18. 35, we obtain Equation 18.36 [5] as follows:

$$P_{loss\_rate\_ALL} = \sum_{n=1}^{N-1} \left(\frac{\binom{N-1}{n} \times p^n}{\sum_{r=1}^{N-1} \binom{N-1}{r}} \times (1-p)^{((N-1)-n)} \times \frac{n}{N-1} \right) \tag{18.36}$$

Again, for $N = 6$ and using 264 slots in the buffer, we have $P_{loss\_ALL} = 0.40\%$ from Equation 18.36, and we have $P_{loss\_ALL} = 0.10\%$ using 500 slots (a larger buffer size), which again corresponds to and agrees with the results of "All + L. Buff" in Figure 18.6.

As was mentioned at the beginning of the chapter, throughput is improved by increasing the spectrum usage efficiency and decreasing the packet loss ratio. As shown in Tables 18.1 and 18.2, both these objectives are achieved with cognitive methods. After implementing all the cognitive methods in the simulation, it is also shown in Ref. [5] that the used cognitive methods that increase the efficient spectral usage efficiency and decrease the packet loss rate also improve the throughput, as seen on Figure 18.7 [5].

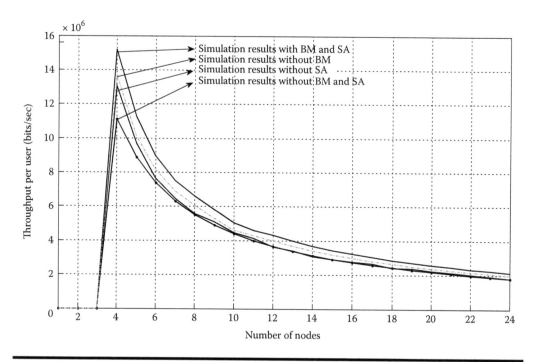

Figure 18.7 The simulation and calculation results of cognitive methods for throughput per node performance and throughput improvement amounts when applying AR, AR+BM, and AR+BM+SA.

18.7 Conclusion

In this chapter, cognition was defined, the parameters affect the system throughput performance were explored, and the improvement provided by each of these methods was analyzed separately. Then some cognitive methods implemented on a mobile ad hoc network, the effects on throughput improvement were investigated.

Finally, it was shown that the proposed cognitive methods can be implemented in a network without any extra hardware requirement to increase the spectrum usage efficiency, to decrease the packet loss rate and to improve the throughput of the network.

References

1. A. Kumar, *Mobile Broadcasting with WiMAX: Principles, Technology, and Applications* (Focal Press Media Technology Professional Series), Focal Press, April 2008.
2. WiMAX Forum, *Mobile WIMAX—Part II: A Comparative Analysis*, 2006, available at: http://www.wimaxforum.org/.
3. WiMax Forum, *Mobile WiMAX—Part I: A Technical Overview and Performance Evaluation*, August 2006, available at: http://www.wimaxforum.org/news/downloads/Mobile_WiMAX_Part1_Overview_and_Performance.pdf
4. K. C. Chen, Y. J. Peng, N. Prasad, Y. C. Liang, and S. Sun, Cognitive radio network architecture: Part I—general structure. In Proceedings of the 2nd international conference on Ubiquitous information management and communication (ICUIMC 2008). *ACM*, New York, USA, 114–119.

5. B. Preveze and A. Safak, Throughput improvement of mobile multi-hop wireless networks, *International Journal of Wireless and Mobile Networks*, vol. 2, pp. 120–140, 2010.
6. L. Iannone and S. Fdida, Can multi-rate radios reduce end-to-end delay in mesh network? A simulation case study, in *International Symposium on Mobile Ad Hoc Networking and Computing*, Florence, Italy, May 22–25, pp. 15–22, 2006.

Chapter 19

Network Formation Games in Wireless Multihop Networks

Walid Saad[1] and Tamer Başar[2]

[1]*Wireless@VT, Bradley Department of Electrical and Computer Engineering, Virginia Tech, Blacksburg, VA, USA*

[2]*Coordinated Science Laboratory, University of Illinois at Urbana-Champaign, IL, USA*

Contents

19.1 Introduction

The rapid proliferation of smartphones, tablets, and bandwidth-intensive wireless applications such as multimedia services has strained the resources of current wireless systems, thus warranting major technical advances in modern-day wireless systems [1]. Indeed, novel techniques such as cognitive radio, dynamic spectrum access, and device-to-device (D2D) communications are expected to lie at the heart of next-generation wireless systems [2,3]. In particular, D2D over multiple portions of the wireless spectrum, which includes cellular and Wi-Fi bands, is seen as a major enabler for new application services such as public safety communications, and for improved wireless performance via network offload and cooperative multihop relaying.

Deploying D2D communications requires overcoming various technical challenges, in terms of resource allocation, mode selection, and interference management [4–12]. In particular,

This research is supported by the National Science Foundation under Grants CCF 11-11342, CNS-1253731, CNS-1443914 and CNS-1406947.

performance analysis and resource allocation issues in cellular systems with underlaid D2D communications have been one of the major research foci in D2D systems, such as in Refs. [4–10]. Optimal power control strategies for D2D communications have been studied and analyzed in Ref. [4] for a cellular network with a single D2D link. Link discovery and interference management in D2D systems have been studied in Ref. [5]. A game-theoretic model for network selection in backhaul-constrained heterogeneous systems with underlaid D2D communication has been studied in Ref. [6]. In Ref. [9], D2D communications in wireless heterogeneous networks has been proposed for the purpose of caching data at the mobile devices, thus improving the performance of multimedia services and reducing the overall network load. The use of cooperative game theory for frequency selection and resource sharing in a single-cell D2D network has been discussed in Ref. [10]. The notion of mode selection between D2D and cellular has been studied in Ref. [11]. A survey of design challenges in D2D systems can be found in Ref. [12].

Naturally, the possibility of D2D over cellular bands will expedite the potential deployment of advanced communication techniques such as cooperative transmission. Indeed, cooperation has recently emerged as an important communication technique to improve the performance of wireless communication networks at different layers. One prominent example is the use of relaying in order to mitigate the fading effects of the wireless channel. In essence, a number of relay nodes can use D2D links among them to cooperate with a source node in the transmission of its data to a distant destination. Such a "cooperative transmission" provides spatial diversity gains for the source node without the burden of having several antennas physically present on the source [13]. Indeed, cooperative transmission and relaying has been shown [13–16] to yield a significant performance improvement in terms of throughput, bit error rate, capacity, or other metrics. These prospective gains coupled with the possibility of D2D has led to the inclusion of cooperative communications as an important feature of emerging standards such as 3GPP's Long-Term Evolution (LTE-Advanced) [17].

The efficient deployment of D2D-enabled cooperative transmission and relaying in next-generation networks has been extensively studied in the literature [18–30]. However, most existing work focuses on performance analysis or resource allocation in networks where the D2D links are predetermined and already established. In other words, the existing body of work on D2D or relaying, such as Refs. [4–12,18–30], typically assumes that the network architecture that interconnects the devices (sources and relays) is known. However, in practice, it is desirable that the network devices engage in a *wireless network formation process* using which they can, in a self-organizing manner, decide on the formation of the D2D links that will interconnect them. This wireless network formation will eventually yield the network architecture that will govern the network, on top of which D2D communications and relaying will operate. Here, we note that, although some studies such as Refs. [28,31–38] have investigated the relay selection problem, most of these works are based on simple architecture (such as a line network) and rely on centralized approaches that may be unsuitable in large-scale D2D networks.

The main goal of this chapter is to study the wireless network formation process in next-generation wireless networks from a game-theoretic point of view. In particular, we focus on developing algorithms suitable for the distributed formation of the network architecture that will connect the wireless devices, using D2D links, for the purpose of cooperative transmission or multihop relaying in next-generation systems. The chapter will introduce the basics of *network formation games* [39], an analytical framework from game theory and social networks that allows the study of the interactions between individuals or devices that seek to form a network graph. This is done by providing a step-by-step introduction on how to apply network formation games in two representative applications pertaining to uplink transmission in multihop wireless systems. We conclude the chapter by shedding some light on future work in this emerging area.

The rest of this chapter is organized as follows: Section 19.2 presents the basic network formation game model. In Section 19.3, we introduce a network formation algorithm for finding a solution of the game. Numerical results are presented and analyzed in Section 19.4. Finally, conclusions are drawn in Section 19.5.

19.2 Basic Network Formation Game Model for Uplink Cooperative Transmission

Consider a network of M devices that are acting as relay nodes (RNs) that transmit data received from source nodes to a central wireless base station (BS) via multihop links. In this network, a number of source mobile stations (MSs) can connect to any of the RNs via a direct D2D link so as to transmit their data to the BS over the RNs' network. Hereinafter, the term MS is used exclusively to refer to those devices that have data to transmit, whereas the term RN refers to devices that act as relays only. Once an RN receives data from the MS, it can then transmit them to the BS via cooperative transmission with the assistance of other RNs.

Cooperative transmission can be performed via a number of techniques that include amplify-and-forward and decode-and-forward relaying. For the model considered in this section, we restrict our attention to a decoded relaying multihop diversity channel, such as in Ref. [15], in which each intermediate RN on the path between a transmitting RN and the BS combines, encodes, and re-encodes the received signal from all preceding terminals before relaying using decode-and-forward. We assume that each MS k has a data traffic arrival that follows a Poisson distribution with an arrival rate λ_k and that the RNs use a first-in-first-out queuing approach. Consequently, assuming that the Kleinrock independence approximation [40, chapter 3] holds in the network, then, each RN can be viewed as an M/D/1 queuing system with an average total arrival rate of $\Lambda_i = \sum_{l \in \mathcal{L}_i} \lambda_l$ where $\mathcal{L}_i$ is the set of source nodes served by an RN i of cardinality $|\mathcal{L}_i| = L_i$. Beyond the source data, each RN i also receives packets from other RNs that are connected to it with a total rate Δ_i. An illustrative example of this model is shown in Figure 19.1.

Under this network model, one key goal for the RNs is to decide on how to form an uplink tree structure for relaying the source data to the central base station. To this end, we use the game-theoretic framework of network formation that has recently attracted attention in the game theory and social networking literatures [41–47]. In general, network formation games constitute a class of problems that involve a number of independent decision makers referred to as players that must interact with one another in order to decide on the formation of a communication graph that will interconnect them. This graph can represent various factors such as friendship relationships or the frequency of interaction. Essentially, the final network graph G that results from a network formation game is highly dependent on the goals, objectives, and incentives of every player in the game. Analogous to a strategic noncooperative game, a network formation game is defined by three components: (1) the players, (2) the strategies or actions, and (3) the utilities.

Consequently, we model the uplink multihop transmission problem as a network formation game in which the RNs are the players. In this game, the RNs will eventually form a *directed* graph $G(\mathcal{V}, \varepsilon)$ with $\mathcal{V} = \{1,\ldots, M + 1\}$ denoting the set of all graph vertices (M RNs and the BS) and ε denoting the set of all edges (links) that connect different pairs of RNs. Each directed link between two RNs i and j, denoted as $(i, j) \in \varepsilon$, corresponds to an uplink traffic flow from RN i to RN j. A *path* between two nodes $i \in \mathcal{V}$ and $j \in \mathcal{V}$ is thus defined as a sequence of nodes $i_1,\ldots,i_K$ (in $\mathcal{V}$) such that $i_1 = i$, $i_K = j$, and each directed link $(i_k, i_{k+1}) \in G$ for each $k \in \{1,\ldots, K-1\}$. As the

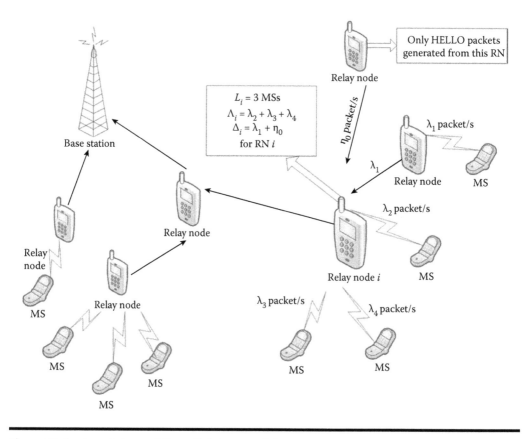

Figure 19.1   A prototype of the uplink tree model.

focus of our model is on uplink tree structures, we further assume that each RN i is connected to the BS via at most one path, denoted by q_i.

Apart from the players, the next key component of a network formation game is the set of possible strategies or actions that these players can take. In the proposed model, the action space of each RN i will simply consist of the RNs (or the BS) that node i wants to use as its next hop. Hence, the strategy of any RN i is to select the link that it wants to form from its available strategy space. Naturally, an RN i cannot connect to another RN j that is already connected to I, that is, if $(j, i) \in G$, then $(i, j) \notin G$. For a given network graph G, we let $\mathcal{A}_i = \{j \in \mathcal{V}, \{i\} | (j, i) \in G$ be the set of RNs from which RN i accepted a link (j, i), and $\mathcal{S}_i = \{(i, j) | j \in \mathcal{V}/(\{i\} \cup \mathcal{A}_i)\}$ be the set of links corresponding to the nodes (RNs or the BS) with which RN i wants to connect. Accordingly, the strategy of an RN i is to select the link $s_i \in \mathcal{S}_i$ that it wants to form. This translates into choosing the RN that it will be connected to. Note that for each selection, s_i, by an RN i corresponds a path q_i to the BS (if $s_i = \varnothing$, then the RN chooses to be disconnected from the network).

Our next step is to define a utility function that can properly capture the performance of multihop cooperative transmission in the system under investigation. In particular, we can introduce a cross-layer utility function that captures the trade-off between the packet success rate (PSR) and the delay induced by multihop transmission. Here, every packet transmitted by any RN is subject to a bit error rate (BER) due to the communication over the wireless channel using one or more hops. For any data transmission between an RN $V_1 \in \mathcal{V}$ to the BS, denoted by V_{n+1}, going through $n - 1$ intermediate RNs $\{V_2,...,V_n\} \subset \mathcal{V}$, let N_r be the set of all receiving terminals; that is, let $N_r = \{V_2...V_{n+1}\}$ and $N_{r(i)}$ be the set of terminals that transmit a signal received by a node V_i. Hence, for an RN V_i on the path from the source V_1 to the destination V_{n+1}, we have $N_{r(i)} = \{V_1,...,V_{i-1}\}$. With this notation, the BER achieved at the BS V_{n+1} between an RN $V_1 \in \mathcal{V}$ that is relaying data to the BS via a path $q_{V_1} = \{V_1,...,V_{n+1}\}$ can be calculated using the bound given in Ref. [15, equation 10] for a channel with Rayleigh fading and BPSK modulation:

$$P_{q_{V_1}}^e \leq \sum_{N_i \in N_r} \frac{1}{2} \left(\sum_{N_k \in N_{r(i)}} \left[\prod_{\substack{N_j \in N_{r(i)} \\ N_j \neq N_k}} \frac{\gamma_{k,i}}{\gamma_{k,i} - \gamma_{j,i}} \times \left(1 - \sqrt{\frac{\gamma_{k,i}}{\gamma_{k,i}+1}} \right) \right] \right). \tag{19.1}$$

Here, $\gamma_{i,j} = \dfrac{P_i \cdot h_{i,j}}{\sigma^2}$ is the average received SNR at node j from node i, with P_i being the transmit power of i, σ^2 the noise variance, and $h_{i,j} = \dfrac{1}{d_{i,j}^\mu}$ the path loss (where $d_{i,j}$ is the distance between i and j, and μ is the path loss exponent). For an RN i that is connected to the BS via a direct path q_i^d with no intermediate hops, the BER is $P_{q_i^d}^e = \dfrac{1}{2} \left(1 - \sqrt{\dfrac{\gamma_{i,BS}}{1+\gamma_{i,BS}}} \right)$ [14,15], where $\gamma_{i,BS}$ is the average received SNR at the BS from RN i. The PSR $\rho_{i,qi}$ perceived by an RS i over any path q_i is thus given by

$$\rho_{i,q_i}(G) = \left(1 - P_{q_i}^e \right)^B, \tag{19.2}$$

where B is the number of bits per packet. The PSR is a function of the network graph G as the path q_i varies depending on how RN i is connected to the BS in the formed network tree structure.

Communication over multihop wireless links involves a significant delay due to multihop transmission as well as buffering. Therefore, any utility function that is to be defined must properly capture this trade-off between improved capacity, due to cooperative transmission, and increased latency, due to multihop delay. One suitable criterion for characterizing such a trade-off is via the concept of *system power*, defined as the ratio of some power of the throughput and the delay [48]. With this in mind, for any RN i with L_i connected MSs, we can define the following utility:

$$u_i(G) = \begin{cases} \dfrac{\left(\Lambda_i \cdot \rho_{i,q_i}(G) \right)^{\beta_i}}{\tau_{i,q_i}(G)^{(1-\beta_i)}}, & \text{if } L_i > 0, \\[4mm] \dfrac{\left(\eta_0 \cdot \rho_{i,q_i}(G) \right)^{\beta_i}}{\tau_{i,q_i}(G)^{(1-\beta_i)}}, & \text{if } L_i = 0, \end{cases} \tag{19.3}$$

where $\tau_{i,qi}(G)$ is the delay, $\Lambda_i \cdot \rho_{i,qi}(G)$ represents the effective throughput of RN i, and $\beta_i \in (0,1)$ is a trade-off parameter. Here, the calculation of the delay $\tau_{i,qi}(G)$ depends on various factors such as the queuing model and the network state. In this chapter, we assume that this delay is computed via the approach that we developed in Ref. [49]. Naturally, any other approach can also be adopted with minor changes to the proposed game solution. In Equation 19.3, η_0 represents the rate of "HELLO" packets transmitted by any RN i when no source data are available at this RN, that is, $\mathcal{L}_i = \varnothing$, $\Lambda_i = 0$, and $\Delta_i = 0$. These packets are used to keep the overall network active.

The utility in Equation 19.3 can model a general class of services, with each class of service having a different β_i that can be chosen individually by the RN. As β_i increases, the service becomes more delay tolerant and more throughput demanding. For an RN i, the parameter β_i can depend on the requirements of its served MSs. Note that unless stated otherwise, hereinafter, the term *power* will be used to refer to the ratio of throughput to delay and not to the transmit power of the nodes.

19.3 Network Formation Game Solution

Given the network formation game formulated earlier, our next step is to develop an algorithm that can be adopted by the RNs to adaptively form the network tree structure for uplink transmission. Essentially, such an algorithm must reach a stable point of the game, as defined later in this section. In the proposed model, if any RN is unable to connect to other suitable RNs to form a link, this RN will connect to the BS using direct transmission. Therefore, as an initial starting point for our solution, we consider that the game starts with a star topology in which all the RNs are connected directly to the BS, prior to interacting for further network formation decisions.

From a notation perspective, when an RN i chooses a strategy $s_i \in S_i$ while all the remaining RNs maintain a vector of strategies $\boldsymbol{s}_{-i}$, we let $G_{s_i,s_{-i}}$ be the resulting network graph. By inspecting Equation 19.3, we can see that whenever an RN j accepts a link, its utility may decrease as the delay increases from the received traffic. Hence, even though each RN $i \in \mathcal{N}$ can pick any action from S_i, there might exist some link $s_i = (i, j) \in S_i$ where the receiving RN j does not accept the formation of s_i if this leads to a significant decrease in its utility. In this regard, denoting by $G + s_i$ the graph G modified when an RN i deletes its current link in G and adds the link $s_i = (i, j)$, we introduce the concept of a *feasible* strategy as follows.

Definition 1: A strategy $s_i \in S_i$, that is, a link $s_i = (i, j)$, is a *feasible strategy* for an RN $i \in V$ if and only if $u_j\left(G_{s_i,s_{-i}} + s_i\right) \geq u_j\left(G_{s_i,s_{-i}}\right) - \varepsilon$, where ε is a small positive constant. $\hat{S}_i \subseteq S_i$ denotes the set of all feasible strategies of an RN $i \in V$.

In other words, for an RN i, the feasible strategy is a link $s_i = (i, j)$ that the receiving RN j is willing to form with i. Here, we stress that any RN $j \in V$ is willing to accept a connection from any other RN $i \in V$ as long as the formation of the link (i, j) does not decrease the utility of j by more than ε. For any RN $i \in V$, given the set of feasible strategies $\hat{S}_i$, a strategy $s_i^* \in \hat{S}_i$ is said to be a *best response* for an RN $i \in V$ if $u_i\left(G_{s_i^*,s_{-i}}\right) \geq u_i\left(G_{s_i,s_{-i}}\right)$, $\forall s_i \in \hat{S}_i$. Thus, an RN i's best response is to select the feasible link that maximizes its utility given that the other RNs maintain their vector of feasible strategies $\boldsymbol{s}_{-i}$.

For solving the network formation game among the RNs, we generate a myopic algorithm for network formation inspired by those applied in economics (e.g., in Refs. [44,45]), but modified to accommodate the specifics of the wireless model. A myopic algorithm is one in which players take their decisions based on the current state of the game, rather than on the prospective evolution of this state. Such an algorithm is an iterative process with each round consisting of two phases: a network formation phase and a multihop transmission phase. In the network formation phase, the RNs interact over pairwise control links in order to engage in the formation of the D2D links among them. Here, the RNs are assumed to make their decisions sequentially in an arbitrary order. Each RN i can select a certain feasible strategy that allows it to improve its current utility. During every iteration t of the studied algorithm, each RN i chooses to play its best response $s_i^* \in S_i$ in order to maximize its utility at each round given the current network graph resulting from the strategies of the other RNs. The feasible best response of each RN can be seen as a *replace* operation using which an RN will replace its current link to the BS with another link that maximizes its utility (if available). Upon convergence, as discussed in Ref. [49], this algorithm reaches a stable outcome in which no RN has an incentive to change its current network selection. Such a stable outcome of a network formation game is known as a Nash network, which is formally defined as follows.

Definition 2: A network graph $G(\mathcal{V}, \varepsilon)$ in which no RN i can improve its utility by a unilateral change in its feasible strategy $s_i \in \hat{S}_i$ is a *Nash network* in the feasible strategy space $\hat{S}_i, \forall i \in \mathcal{V}$.

A Nash network is simply the game-theoretic notion of a Nash equilibrium applied to a network formation game. In the game considered here, a Nash network would be a stable uplink multihop transmission tree.

From a convergence point of view, it is known that, for network formation games in which utilities do not have a special structure and in which discrete network formation strategies are used, proving the convergence of any algorithm analytically can be challenging [44,50]. Nonetheless, the algorithm here, as discussed in Ref. [49], can either converge to a Nash network or cycle between a number of networks in the case of nonconvergence [44,50]. The undesirable cycling behavior can be avoided by introducing additional constraints on the strategies of the RNs. One such constraint is to let the RNs make their decisions based, not only on the current network graph but also on the history of moves or strategies taken by the other RNs. Such an approach mimics those used in dynamic games such as repeated games [50]. Alternatively, in the case of nonconvergence, the RNs may be instructed by the network operator to seek a mixed-strategy Nash network, which is guaranteed to exist [50]. As explained in Ref. [49], using such approaches, it can be shown that the proposed algorithm can be guaranteed to converge to a suitable outcome. The nature and properties of this outcome naturally depend on whether history is used or not, and on whether the RNs are seeking a mixed Nash instead of a traditional pure-strategy Nash network.

Once the network formation phase converges, the RNs will be connected through a tree structure G_T, and the second phase of the algorithm begins. During this phase, the actual multihop network operation occurs as the RNs transmit the data over the existing tree architecture G_T. A summary of the proposed algorithm is given in Table 19.1.

The proposed algorithm can be implemented in a self-organizing manner within any next-generation wireless multihop network, with a little reliance on a centralized controller such as the BS. Indeed, the only role of the BS in this network formation algorithm is to inform the RNs, over a control channel, of the graphs reached during past iterations, if history is needed. Beyond this, the algorithm relies on distributed decisions taken individually by the RNs. Within every iteration t, during its turn, each RN can engage in pairwise negotiations with the surrounding RNs in order to find its best response, among the set of feasible strategies and given the graphs that were reached in previous iterations. With the advent of the D2D discovery and communication protocol, these pairwise interactions can be easily deployed in practical wireless systems, with little corresponding overhead.

TABLE 19.1 Proposed Network Formation Algorithm

Initial State
The starting network is a star network graph.
The proposed algorithm consists of two phases.
Phase I -- Network formation:
repeat
In an arbitrary but sequential order, the RNs engage in a network formation game.
a) In every iteration t of Phase I, each RN i plays its best response.
b) The best response s_i^* of each RN is a *replace* operation through which an RN i splits from its current parent RN and replaces it with a new RN that maximizes its utility, given that this new RN *accepts* the formation of the link.
until convergence to a final tree G_T after T iterations.
Phase II -- Multihop transmission:
During this phase, data transmission from the MSs occurs using the formed network tree structure G_T.

19.4 Numerical Results and Analysis

We assess the performance of the foregoing network formation game using numerical simulations. We set up a wireless network confined within a square area of 3 km × 3 km at the center of which the BS is placed. RNs and traffic generating MSs are deployed in this area with the transmit power of each RN and MS set to 50 mW. The noise level is set to –100 dBm, the bandwidth per RN is set to $W = 100$ kHz, and the propagation loss is set to $\mu = 3$. We consider a traffic of 64 kbps, divided into packets of length $B = 256$ bits with an arrival rate of 250 packet/s. We let $\beta_i = \beta = 0.7$ to represent services that are slightly delay tolerant. Further, the parameter ε is selected to be equal to 1% of any RN's current utility, that is, an RS accepts the formation of a link if its utility does not decrease by more than 1% of its current value. Statistical results are averaged over a large number of independent simulation runs.

Figure 19.2 shows the network resulting from the network formation game process that occurs between $M = 10$ randomly deployed RNs. Prior to the presence of traffic-generating MSs, the RNs engage in the proposed network formation algorithm based on "HELLO" packets only, and this process converges to the final Nash network structure shown by the solid lines in Figure 19.2. Subsequently, 30 randomly located traffic-generating MSs are introduced. The dashed lines in Figure 19.2 show how the RNs self-organize and adapt the network's topology to this newly introduced traffic. For example, RS 9 can improve its utility by disconnecting from RN 8 and connecting to RN 6 instead. This improvement stems from the fact that, although connecting to RN 8 provides a better BER for RN 9, in the presence of the traffic, choosing a shorter path via RN 6 can reduce the delay perceived by RN 9, hence improving the overall utility. The incoming traffic coupled with the change of strategy of RN 9 leads RN 8 to disconnect from RN 6 and connect directly to the BS. This allows RN 8 to avoid the extra delay that exists at RN 6 when

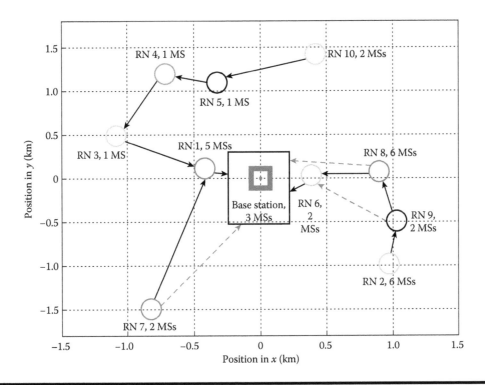

Figure 19.2 Snapshot of a tree topology formed using the proposed network algorithm with M = 10 RNs before (solid line) and after (dashed line) the random deployment of 30 traffic-generating MSs.

MSs are deployed. Further, RN 7 finds it beneficial to replace its current link with the congested RN 1 with a direct link to the BS. Figure 19.2 summarizes the operation of the proposed adaptive network formation algorithm with and without the presence of external traffic from the MSs.

In Figure 19.3, we study the impact of mobility on the network formation game. Here, given the network of Figure 19.2 *prior to the deployment of the traffic*, we let RN 9 move horizontally in the direction of the negative x-axis, while the other RNs remain static. The variations in the utilities of the main concerned RNs during this mobile scenario are shown in Figure 19.3. As soon as RN 9 starts to move, its utility increases because its distance to RN 8 decreases. Similarly, the utility of RN 2, which is connected to RN 9, also increases. As RN 9 moves about 0.2 km, it decides to replace its current link with RN 8 and connect to RN 6 instead. Here, RN 6 will accept the incoming connection from RN 9 because this does not affect its utility negatively, as shown in Figure 19.3 at 0.2 km. As RN 9 continues to move, its utility improves as it draws closer to RN 6, while the utility of RN 2 decreases because RN 9 is distancing itself from it. After moving for about 0.5 km, RN 9 approaches quite close to the BS, and, thus, its best response is to directly connect to the BS. This action taken by RN 9 at 0.5 km also improves the utility of RN 2. Meanwhile, RN 9 continues its movement, and its utility as well as that of RN 2 begin to drop as RN 9 distances itself from the BS. Once RS 9 moves for a total of 1.3 km, RN 2 decides to disconnect from it and connect directly to the BS. In summary, the results shown in Figure 19.3 clearly illustrate how the RNs can self-organize via the use of a network formation game, in the presence of low mobility.

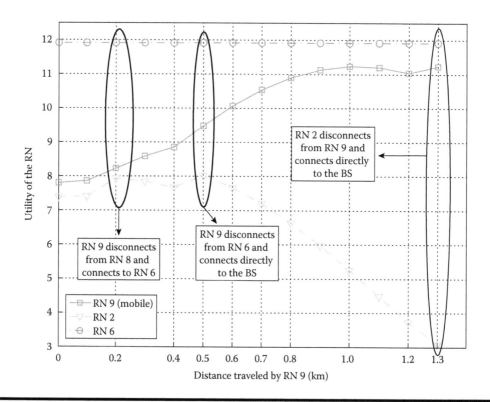

Figure 19.3 Adaptation of the network's tree structure to mobility of the RNs shown through the changes in the utility of RN 9 of Figure 19.2 as it moves on the x-axis in the negative direction prior to any MS presence.

From a performance point of view, Figure 19.4 shows the average utility achieved per traffic-generating source MS for a network with $M = 10$ RNs as the number of source nodes increases. The performance of the proposed network formation algorithm is compared against the direct transmission performance where no relaying is used and with a nearest neighbor algorithm in which each node selects the closest partner to connect to. Figure 19.4 clearly shows that as the number of MSs in the network increases, the performances of both the network formation game approach and that of the nearest neighbor algorithm decrease. This is due to the fact that as more traffic is deployed, the delay from multihop transmission increases, and, thus, the average utility decreases. In contrast, in the case of no RNs, the performance is unaffected by the increase in traffic owing to the absence of multihop delay. It is also interesting to note that, owing to the increased number of MSs, the performance of the nearest neighbor algorithm drops below that of direct transmission at around 20 MSs. Finally, Figure 19.4 shows that at all network sizes, the network formation scheme has a significant advantage over both the nearest neighbor algorithm and the direct transmission case. This advantage is at least 17.1% compared to the direct transmission case (for 50 MSs), and it reaches up to 40.3% improvement relative to the nearest neighbor algorithm at 50 MSs.

Figure 19.5 shows the average utility per MS as the number of RNs M varies, for a network with 40 MSs. Figure 19.5 shows that, as M increases, the performance of the network formation scheme as well as that of the nearest neighbor algorithm increase. This is because as the number of RNs increases, the possibility of benefiting from cooperative transmission gains increases, thus improving the overall utilities and network quality of service. In contrast, for the direct

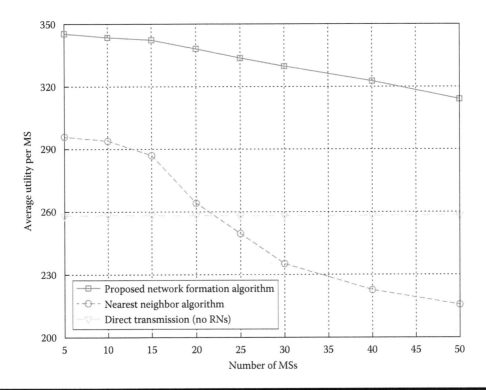

Figure 19.4 **Performance assessment of the proposed network formation algorithm in terms of average utility per traffic-generating source MS for a network having** M **= 10 RNs as the number of MSs varies.**

transmission case, the performance is constant as M varies, because direct transmission does not use relaying via the RNs. Figure 19.5 demonstrates that at all network sizes, the network formation algorithm presents a significant performance gain, reaching up to 52.8% and 38.5%, respectively, relative to the nearest neighbor algorithm and the direct transmission case.

In Figure 19.6, we show the average number of hops and the average maximum number of hops in the resulting network structure as the number of RNs increases for a network with 40 traffic-generating mobiles. The number of hops shown in Figure 19.6 does not account for the extra MS-RN hop. Figure 19.6 shows that as the number of RNs increases, both the average number of hops and the average maximum number of hops in the tree structure increase. The average number of hops and the average maximum number of hops vary, respectively, from 1.85 and 2.5 at M = 5 RNs, up to around 3 and 5 at M = 25 RNs. Figure 19.6 consequently shows that, due to the delay cost for multihop transmission, both the average number of hops and the average maximum number of hops increase very slowly with the network size M. For instance, one can observe that up to 20 additional RNs are needed in order to increase the average number of hops of by around 1 hop and the average maximum number of hops by only around 2 hops.

Figure 19.7 shows the average number of iterations and the maximum number of iterations needed until convergence of the algorithm in a network with only HELLO packets and no external source traffic. In Figure 19.7, we can see that as the number of RNs increases, the total number of iterations required for the convergence of the algorithm increases. This result is because as M increases, the relaying options for every RN increase, and, thus, more actions are required

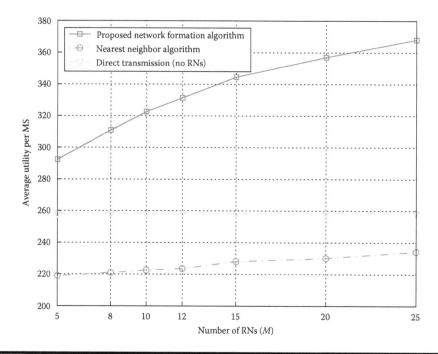

Figure 19.5 Performance assessment of the proposed network formation algorithm in terms of average utility per MS for a network having 40 MSs as the number of RNs M varies.

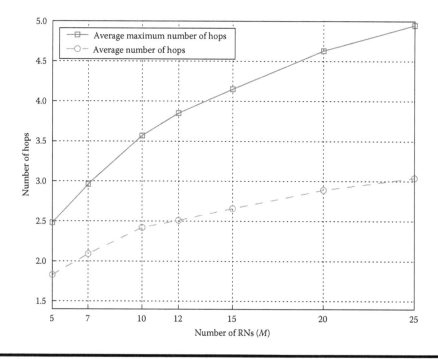

Figure 19.6 Average number of hops and the average maximum number of hops in the final tree structure for a network with 40 MSs versus number of RSs M in the network.

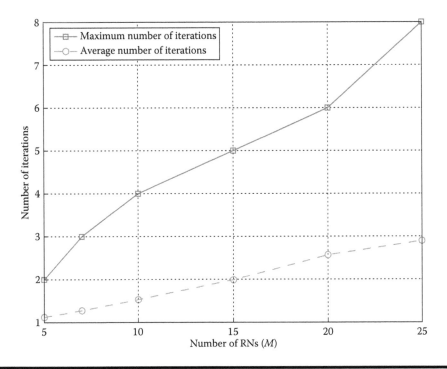

Figure 19.7 Average number of iterations and the maximum number of iterations until convergence versus number of RSs *M* in the network.

prior to convergence. Figure 19.7 shows that the average number of iterations and the maximum number of iterations vary, respectively, from 1.12 and 2 at $M = 5$ RNs up to 2.9 and 8 at $M = 25$ RNs. Hence, this result demonstrates that, on average, the speed of convergence of the proposed algorithm is quite reasonable even for relatively large networks. Similar results can be seen for the convergence of the algorithm when MSs are deployed or when the RNs are mobile.

19.5 Conclusions and Future Work

In this chapter, we have explored the potential of using the analytical framework of network formation games for studying an illustrative multihop transmission scenario in the uplink of a single-cell wireless system. The results indicate a promising outlook for this framework. On the one hand, the framework of network formation games allows one to clearly capture the heterogeneous quality-of-service goals of the nodes by incorporating node-specific utilities. On the other hand, the algorithmic aspect of network formation games enables the development of low-complexity, self-organizing algorithms for performing network formation in a wireless environment. The simulation results have shown that by using a game-theoretic approach for network formation, a wireless system can reap the promised benefits of advanced D2D and relaying techniques such as cooperative transmission. With the massive deployment of D2D, the development of network formation algorithms for various D2D-based applications will become a central theme in next-generation networks. To this end, this chapter can provide the starting point for developing such algorithms.

Several future studies can be envisioned. First, this chapter was restricted to studying an uplink, single-cell model. Naturally, this network formation model can be extended to multiple cells, which is particularly interesting within a heterogeneous cellular network consisting of a dense deployment of low-cost, low-power small cell base stations. In such an environment, cell selection and D2D network formation present a challenging problem for which a self-organizing, game-theoretic approach can prove suitable. Second, in the studied model, it is assumed that the D2D tier has its own dedicated spectrum, and no interference between D2D transmissions occurs. In D2D over cellular, interference management during network formation becomes an important issue. Studying an interference-aware network formation game model is thus an important future investigation. Such a study can build on earlier approaches such as those in Refs. [51–53]. Studying the convergence of the network formation game algorithms in a wireless setting is one promising future avenue for exploration. How to ensure convergence within a reasonable time given the dynamic nature of the wireless environment is a key question for future work on convergence properties. In addition, developing predictive network formation algorithms that take into account not only the current network setting but also the future evolution of the network is yet another important direction for future work. Finally, introducing network formation game models for novel application scenarios such as the coexistence of ad hoc cognitive networks and the deployment of D2D over cellular systems is another interesting challenge for future work.

References

1. Cisco, *Cisco Visual Networking Index: Global Mobile Data Traffic Forecast Update, 2010–2015*, white paper, Cisco, 2011.
2. T. Q. S. Quek, G. de la Roche, I. Guvenc, and M. Kountouris, *Small Cell Networks: Deployment, PHY Techniques, and Resource Management*, Cambridge University Press, Cambridge, UK, 2013.
3. D. Niyato, E. Hossain, and Z. Han, *Dynamic Spectrum Access in Cognitive Radio Networks*, Cambridge, UK: Cambridge University Press, 2009.
4. C.-H. Yu, K. Doppler, C. B. Ribeiroa, and O. Tirkkonen, Resource sharing optimization for device-to-device communication underlaying cellular networks, *IEEE Transactions on Wireless Communications*, vol. 10, no. 8, pp. 2752–2763, 2011.
5. B. Kaufman, B. Aazhang, and J. Lilleberg, Interference aware link discovery for device to device communication, in *Proceedings of Asilomar Conference on Signals Systems and Computers*, Pacific Grove, CA, November 2009.
6. C. Diaz, W. Saad, B. Maham, D. Niyato, and A. S. Madhukumar, Strategic device-to-device communications in backhaul-constrained wireless small cell networks, in *Proceedings of the IEEE Wireless Communications and Networking Conference*, Istanbul, Turkey, April 2014.
7. C.-H. Yu, O. Tirkkonen, K. Doppler, and C. B. Ribeiro, Power optimization of device-to-device communication underlaying cellular communication, in *Proceedings of the International Conference on Communications*, Dresden, Germany, June 2009.
8. P. Janis, C.-H. Yu, K. Doppler, C. B. Ribeiro, C. Wijting, K. Hugl, O. Tirkkonen, and V. Koivunen, Device-to-device communication underlaying cellular communications systems, *IEEE Transactions on Wireless Communications*, vol. 10, no. 8, pp. 2752–2763, 2011.
9. N. Golrezaei, A. Dimakis, and A. Molisch, Wireless device-to-device communications with distributed caching, in *Proceedings of the IEEE International Symposium on Information Theory (ISIT)*, Cambridge, MA, July 2012.
10. K. Akkarajitsakul, P. Phunchongharn, E. Hossain, and V. K. Bhargava, Mode selection for energy-efficient D2D communications in LTE-advanced networks: A coalitional game approach, in *Proceedings of the IEEE International Conference on Communication Systems*, Singapore, November 2012.

11. S. Hakola, T. Chen, J. Lehtomaki, and T. Koskela, Device-to-device (D2D) communication in cellular network performance analysis of optimum and practical communication mode selection, in *Proceedings of the IEEE Wireless Communications and Networking Conference*, Sydney, Australia, April 2010.

12. G. Fodor, E. Dahlman, G. Mildh, S. Parkvall, N. Reider, G. Miklos, and Z. Turanyi, Design aspects of network assisted device-to-device communications, *IEEE Communications Magazine*, vol. 50, no. 3, pp. 170–177, 2012.

13. J. Laneman, D. Tse, and G. Wornell, Cooperative diversity in wireless networks: Efficient protocols and outage behavior, *IEEE Transactions on Information Theory*, vol. 50, pp. 3062–3080, 2004.

14. A. Sadek, W. Su, and K. R. Liu, Multinode cooperative communications in wireless networks, *IEEE Transactions on Signal Processing*, vol. 55, pp. 341–355, 2007.

15. J. Boyer, D. Falconer, and H. Yanikomeroglu, Multihop diversity in wireless relaying channels, *IEEE Transactions on Communications*, vol. 52, pp. 1820–1830, 2004.

16. Z. Han and K. J. Liu, *Resource Allocation for Wireless Networks: Basics, Techniques, and Applications*, Cambridge University Press, Cambridge, UK, 2008.

17. 3GPP TR 36. 814 Technical Specification Group Radio Access Network, Further Advancements for E-UTRA, Physical Layer Aspects, Technical Report.

18. B. Lin, P. Ho, L. Xie, and X. Shen, Optimal relay station placement in IEEE 802.16j networks, in *Proceedings of the International Conference on Communications and Mobile Computing*, Honolulu, HI, August 2007.

19. R. Schoenen, R. Halfmann, and B. H. Walke, MAC performance of a 3GPP-LTE multihop cellular network, in *Proceedings of the International Conference on Communications*, Beijing, China, pp. 4819–4824, May 2008.

20. S. W. Peters, A. Panah, K. Truong, and R. W. Heath, Relay architectures for 3GPP LTE-advanced, *EURASIP Journal on Wireless Communication and Networking*, vol. 2009, 2009.

21. B. Lin, P. Ho, L. Xie, and X. Shen, Relay station placement in IEEE 802.16j dual-relay MMR networks, in *Proceedings of the International Conference on Communications*, Beijing, China, May 2008.

22. D. Niyato, E. Hossain, D. I. Kim, and Z. Han, Relay-centric radio resource management and network planning in IEEE 802.16j mobile multihop relay networks, *IEEE Transactions Wireless Communications*, vol. 8, no. 12, pp. 6115–6125, 2009.

23. T. Wirth, V. Venkatkumar, T. Haustein, E. Schulz, and R. Halfmann, LTE-advanced relaying for outdoor range extension, in *Proceedings of the Vehicular Technology Conference (VTC-Fall)*, Anchorage, AK, September 2009.

24. O. Teyeb, V. V. Phan, B. Raaf, and S. Redana, Dynamic relaying in 3GPP LTE-advanced networks, *EURASIP Journal on Wireless Communication and Networking*, vol. 2009, 2009.

25. O. Teyeb, V. V. Phan, B. Raaf, and S. Redana, Handover framework for relay enhanced LTE networks, in *Proceedings of the International Conference on Communications*, Dresden, Germany, June 2009.

26. E. Visotsky, J. Bae, R. Peterson, R. Berryl, and M. L. Honig, On the uplink capacity of an 802.16j system, in *Proceedings of the IEEE Wireless Communications and Networking Conference*, Las Vegas, NV, April 2008.

27. Y. Yu, S. Murphy, and L. Murphy, Planning base station and relay station locations in IEEE 802.16j multi-hop relay networks, in *Proceedings of the IEEE Consumer Communications and Networking Conference*, Las Vegas, NV, January 2008.

28. H. Lee, H. Park, Y. Choi, Y. Chung, and S. Rhee, Link adaptive multi-hop path management for IEEE 802.16j, in *IEEE C802/16j-07/1053*, January 2007.

29. W. Saad, Z. Han, M. Debbah, and A. Hjørungnes, Network formation games for distributed uplink tree construction in IEEE 802.16j networks, in *Proceedings of the IEEE Global Communication Conference*, New Orleans, LA, December 2008.

30. W. Saad, Z. Han, M. Debbah, A. Hjørungnes, and T. Başar, A game-based self-organizing uplink tree for VoIP services in IEEE 802.16j networks, in *Proceedings of the International Conference on Communications*, Dresden, Germany, June 2009.

31. S. Sharma, Y. Shi, T. Y. Hou, H. D. Sherali, and S. Kompella, Cooperative communications in multi-hop wireless networks: Joint flow routing and relay node assignment, in *Proceedings of the IEEE International Conference on Computer Communications (INFOCOM)*, San Diego, CA, March 2010.

32. Y. Jing and H. Jafarkhani, Single and multiple relay selection schemes and their achievable diversity orders, *IEEE Transactions on Wireless Communications*, vol. 8, no. 3, pp. 1414–1423, 2009.

33. Y. Li, P. Wang, D. Niyato, and W. Zhuang, A dynamic relay selection scheme for mobile users in wireless relay networks, in *Proceedings of the IEEE International Conference on Computer Communications (INFOCOM)*, Shanghai, China, March 2010.

34. C. K. Lo, R. W. Heath, and S. Vishwanath, The impact of channel feedback on opportunistic relay selection for hybrid-ARQ in wireless networks, *IEEE Transactions Vehicular Technology*, vol. 58, no. 3, pp. 1255–1268, 2009.

35. R. Madan, N. Mehta, A. Molisch, and J. Zhang, Energy-efficient cooperative relaying over fading channels with simple relay selection, *IEEE Transactions on Wireless Communications*, vol. 7, no. 8, pp. 3013–3025, 2008.

36. Y. Chen, G. Yu, P. Qiu, and Z. Zhang, Power-aware cooperative relay selection strategies in wireless ad hoc networks, in *Proceedings of the IEEE International Conference on Personal, Indoor and Mobile Radio Communications*, Helsinki, Finland, September 2006.

37. L. Sun and M. R. McKay, Opportunistic relaying for MIMO wireless communication: Relay selection and capacity scaling laws, *IEEE Transactions on Wireless Communications*, vol. 10, no. 6, pp. 1786–1797, 2011.

38. S. Luo, H. Godrich, A. Petropulu, and H. V. Poor, A knapsack problem formulation for relay selection in secure cooperative wireless communication, in *Proceedings of the IEEE International Conference on Acoustics, Speech, and Signal Processing*, Prague, Czech Republic, May 2011.

39. M. O. Jackson and A. Watts, The evolution of social and economic networks, *Journal of Economic Theory*, vol. 106, pp. 265–295, 2002.

40. D. Bertsekas and R. Gallager, *Data Networks*, NJ: Upper Saddle River, Prentice Hall, 1992.

41. M. O. Jackson, *Social and Economic Networks*, Princeton, NJ: Princeton University Press, 2010.

42. M. O. Jackson and A. van den Nouweland, Strongly stable networks, *Games and Economic Behavior*, vol. 51, no. 2, pp. 420–444, 2005.

43. G. C. Chasparis and J. S. Shamma, Distributed dynamic reinforcement of efficient outcomes in multi-agent coordination and network formation, *Dynamic Games and Applications*, vol. 2, no. 1, pp. 18–50, 2012.

44. G. Demange and M. Wooders, *Group Formation in Economics: Networks, Clubs and Coalitions*, Cambridge, UK: Cambridge University Press, 2005.

45. R. Johari, S. Mannor, and J. Tsitsiklis, A contract based model for directed network formation, *Games and Economic Behavior*, vol. 56, pp. 201–224, 2006.

46. J. Derks, J. Kuipers, M. Tennekes, and F. Thuijsman, *Local Dynamics in Network Formation*, Maastricht, The Netherlands: Department of Mathematics, Maastricht University, 2007.

47. W. Saad, Z. Han, M. Debbah, A. Hjørungnes, and T. Başar, Coalition game theory for communication networks: A tutorial, *IEEE Signal Processing Magazine, Special Issue on Game Theory in Signal Processing and Communications*, vol. 26, no. 5, pp. 77–97, 2009.

48. L. Kleinrock, Power and deterministic rules of thumb for probabilistic problems in computer communications, in *Proceedings of the International Conference on Communications*, Boston, MA, June 1979.

49. W. Saad, Z. Han, T. Başar, M. Debbah, and A. Hjørungnes, Network formation games among relay stations in next generation wireless networks, *IEEE Transactions on Communications*, vol. 59, no. 9, pp. 2528–2542, 2011.

50. T. Başar and G. J. Olsder, *Dynamic Noncooperative Game Theory*, Philadelphia, PA: SIAM Series in Classics in Applied Mathematics, 1999.

51. W. Saad, Q. Zhu, T. Başar, Z. Han, and A. Hjørungnes, Hierarchical network formation games in the uplink of multi-hop wireless networks, in *Proceedings of the IEEE Global Communication Conference*, Honolulu, HI, December 2009.

52. Q. Zhu, Z. Yuan, J. B. Song, Z. Han, and T. Başar, Interference aware routing game for cognitive radio multi-hop networks, *IEEE Journal on Selected Areas in Communications*, vol. 30, no. 10, pp. 2006–2015, 2012.

53. Q. Zhu, J. B. Song, and T. Başar, Dynamic secure routing game in distributed cognitive radio networks, in *Proceedings of the IEEE Global Communication Conference*, Houston, TX, December 2011.

Chapter 20

Rapid Prototyping for Video Coding over Flexible Radio Links

Matthieu Gautier,[1] Emmanuel Casseau,[1]
Hervé Yviquel,[2] Ganda Stéphane Ouedraogo,[1]
Mickaël Raulet,[2] and Olivier Sentieys[1]

[1]University of Rennes 1, IRISA, INRIA, France

[2]INSA of Rennes, IETR, France

Contents

20.1 Introduction

When new signal processing applications are being developed, rapid prototyping makes it possible to experience some of the features of the applications beforehand. At present time, software-based prototyping approaches are user convenient, and they provide flexibility in a processor-based running environment (i.e., general purpose processor [GPP] or digital signal processor [DSP]). However, these approaches reach their limits for dynamic multimedia streams when data-rate increases.

To this end, FPGA (field programmable gate array) technology is expected to play a key role in the development of platforms for multimedia over cognitive radio networks. Though the technology is continuously evolving, FPGA-based design methodology has not changed over the decades and is still based on low-level descriptions of the applications (register transfer level [RTL]). However, designing an application at a low level is a time-consuming process and does not meet the time-to-market factor. In our context, it is crucial to rapidly implement new video codecs and radio waveforms to achieve the flexibility paradigm [1]. This chapter describes two rapid prototyping approaches dedicated to video coding over flexible radio links applications. The design flow starts with a high-level description of the application, and an RTL description that can be easily synthesized for hardware implementation is generated.

The video coding design flow is based on the Reconfigurable Video Coding (RVC) standard [2]. This standard uses dataflow programming and the CAL Actor Language (CAL) to benefit from explicit concurrency and to simplify multimedia development. The radio waveform design flow relies on high-level synthesis (HLS) principles and leverages the emerging HLS tools [3]. It is also based on dataflow programming, but its entry point is a domain-specific language (DSL) that partly handles the complexity of programming an FPGA and integrates reconfigurable features.

An MPEG-4 decoder and an IEEE 802.11g transceiver have been explored using these approaches and the results have been highlighted in this chapter.

20.2 High-Data-Rate Flexible Multimedia Prototyping

20.2.1 System Model and Prototyping Methodology

Implementing signal processing applications in embedded systems is a complex and error-prone task that involves many issues such as handling complex algorithms, design methodologies, complex execution platforms, exploiting parallelism, user-defined constraints, time-to-market pressure, and so on.

Expressing an algorithm so as to exploit all its parallelism in order to meet high data rates is already a first challenging task. In this chapter, we start from a block-based model of the application, namely a dataflow program that offers a flexible development approach to building complex applications while expressing parallelism explicitly. A dataflow program is described as a directed graph composed of a set of functional blocks (vertices) that communicate through a set of communication channels represented by the directed edges (Figure 20.1). The communication

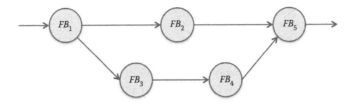

Figure 20.1   A dataflow graph with five interconnected functional blocks (FB).

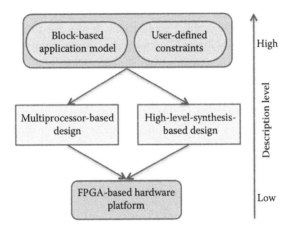

Figure 20.2   From high-level specification to hardware implementation.

corresponds to a stream of atomic data objects, called *tokens*, that follows the FIFO (first in first out) strategy. The functional blocks first read some tokens from their input channels, then process some internal computations, and finally write some tokens to their output channels.

Dataflow programming can be easily used for the development of signal processing applications owing to its consistency with the natural representation of digital signal processing. Moreover, describing the application using a set of interconnected functional blocks enhances its ability to be executed on various platforms with different concurrency capabilities, ranging from platforms with a single processor to platforms with a set of hardware processing units and processors.

From the high-level specification of the application, the design process performs the mapping of the functional blocks to the available computing resources of the executing platform according to user-defined constraints (Figure 20.2).

In this chapter, we describe two design approaches that ease multimedia development for hardware platforms/embedded systems thanks to a dataflow modeling of the application. They aim to bridge the gap between a flexible application specification and efficient execution on the hardware platform. The first approach attempts to address both hardware efficiency and software flexibility based on a multiprocessor platform, whereas the second approach targets high-performance dedicated platforms leveraging the high-level synthesis (HLS) method.

With both approaches, an register transfer level (RTL) description of the platform is generated that can be easily synthesized for rapid prototyping.

20.2.2 Introduction to Reconfigurable Video Coding

Digital video is currently used for a wide range of applications thanks to advances in computing and communication technologies, as well as in video compression techniques. A few years ago, MPEG introduced an innovative framework called RVC (Reconfigurable Video Coding) that is dedicated to the development of video coding tools in a modular fashion [2]. RVC was introduced to improve the standardization process of video compression standards. It is built upon a dataflow-based Domain-Specific Language (DSL) known as CAL Actor Language [4]. A DSL is a programming language or executable specification language that offers, through appropriate notations and abstractions, expressive power focused on, and usually restricted to, a particular problem domain [5]. A DSL can thus provide knowledge about the application domain that is lacking in general-purpose languages [6]. For example, in our case, the language used must be suitable for describing the behavior of video applications. It also has to allow the parallelization process to be simplified so as to provide high-performance implementations of video codecs that are required by the market.

In practice, an RVC-based description is made of two levels: the functional blocks, called actors, and the interconnection network [7]. A specific programming language, called Functional Unit Network Language (FNL), is used to specify the interconnection network between all the actors. A subset of CAL, known as RVC-CAL, is used to describe the actor's behavior. The type system is one of the major differences between the original CAL and the one standardized within the RVC framework. Whereas CAL keeps an abstract type system authorizing untyped data, RVC-CAL defines a practical type system dedicated to the development of signal processing algorithms, such as logical data type, bit-accurate integer data types, and floating-point types coded with 16, 32, and 64 bits, and so on.

Due to the data-dependent behavior of some functional blocks in a video codec, RVC is based on dynamic dataflow programming. The model of computation used to specify how computations progress is known as Dataflow Process Network (DPN) [8]. This model is composed of vertices (i.e., actors) and unidirectional edges representing unbounded communication channels based on the FIFO principle. DPN introduces the notion of firing. An actor firing, a so-called action, is an indivisible quantum of computation that corresponds to a mapping function of input tokens to output tokens applied repeatedly and sequentially on one or more data streams. As stated earlier, this mapping is composed of three ordered and indivisible steps: data reading, then the computational procedure, and finally data writing. These functions are guarded by a set of firing rules that specify when an actor can be fired, that is, the number and the values of the tokens that have to be available on the input ports to fire the actor. When several firing rules are satisfied at the same time, priorities or FSM-based behaviors define a partial order relation on the firing rules.

Among the available RVC video decoders, the MPEG-RVC working group has standardized three video decoders: MPEG-4 Part 2, also known as MPEG-4 visual; MPEG-4 Part 10, also known as H.264/AVC; and HEVC also known as H.265.

Our goal is to provide an entire design flow for the rapid prototyping of RVC applications, from the dataflow program specification up to the execution platform. Application-specific integrated circuits (ASICs) or dedicated hardware accelerators provide high performance, but they are dedicated to a single application and their design is time consuming. On the contrary, software processors, for example, general purpose processor (GPP), are highly flexible thanks to the compiler that translates the application to machine code, but this flexibility is achieved at the cost of performance and power consumption. We target a multiprocessor platform comprising low-complexity processors in order to provide performance and flexibility at the same time.

20.2.3 Toward Field Programmable Gate Array (FPGA)-Based Software-Defined Radio

In the field of telecommunications, the application flexibility has been mainly addressed with the introduction of the software-defined radio (SDR) concept. SDR refers to the capability to change the features of a radio transceiver in order to adapt it to various air interfaces with a unique hardware. It was first introduced by Mitola [9,10] and was later used to introduce the flexibility in the cognitive radio (CR) domain. Outstanding work has been carried out on SDR solutions, and an interesting survey of SDR platforms is given in Ref. [11].

One of the mainstream approaches to specifying an SDR application consists in implementing the processing part on a Digital Signal Processor (DSP) coupled to a hardware accelerator such as an FPGA or ASIC fabrics [10]. The reason why such heterogeneous DSP-centric SDR platforms were popularized is mainly related to the fact that current DSP fabrics offer important reconfiguration capabilities. In these platforms, the critical computations are most often performed by specialized hardware fabrics. DSPs also suffer from relatively higher power consumption and limited data rate than specialized hardware fabrics. FPGAs are an interesting alternative. They are specialized hardware fabrics with high computation performance, and they show some improvements with regard to energy consumption and the operating frequency aspects as compared to DSP architectures.

The FPGA programming model makes it possible to leverage an important dataflow parallelism thanks to its native parallel computation capability, as opposed to the sequential nature of DSPs. Thus, FPGA-based SDR is quite an old paradigm [12,13], and most of the attempts rely on intellectual property (IP) described at the RTL and getting the SDR concept far away from its initial ideal idea, that is to say, a software-only platform. One of the drawbacks of this technology is especially the low-level interface language used to program applications. These hardware languages such as Verilog or VHSIC Hardware Description Language (VHDL) require a strong background in order to use them efficiently. To tackle this issue, a software language is proposed in this chapter to target FPGAs based on the design flow shown in Figure 20.2. Indeed, radio link applications can be modeled as a dataflow graph (e.g., Figure 20.1) using a subset of dataflow graphs named synchronous dataflow (SDF) graph, where channels are buffered memories that model infinite FIFOs [14]. Behavior does not depend on the value of the data, as for dynamic dataflow programming.

The implementation of an SDF application is twofold. The first part consists in specifying the system in an abstracted way using high-level languages (HLLs). This representation highlights the sub-elements of the system and their interactions. It often considers data as floating points, which require a refinement to fixed points when it comes to implementing the system for a hardware platform.

The second part consists in a hardware description of the system. Hardware description languages (HDLs) such as Verilog or VHDL are used to implement the specification of the system written in a HLL. The implementation gives a functional description of the system and takes into account the computing and communication resources available on the platform.

A mainstream trend is to merge these two designing steps into a single one. The HLS bridges the gap between the specification and the implementation. It takes as an input a function specification written in a HLL and generates its RTL description for a specific target platform (e.g., an FPGA). The existing HLS tools emphasize the functional aspects of the application, and they propose many design optimization techniques to fulfill the performance requirements. Nevertheless, the control aspect is not well addressed by those tools, and the users have to manually perform this step.

Our approach is based on a DSL that describes the SDF model such that the transmitted frame structure is used to describe the radio waveform and to generate the RTL control.

20.3 Design Flow for Video Coding and Radio Waveform

20.3.1 From Dataflow Programming to Video Coding Implementation

As stated previously, dataflow programming is a high-level programming method that describes parallel applications inherently. Starting from the dataflow network, the design process has to map the actors onto the available processing units of the execution platform and the FIFO channels onto the communication and memory resources.

20.3.1.1 Flexible Implementation of Dataflow Programs

The proposed approach targets multiprocessor platforms, that is to say, architectures made of interconnected processor cores. Usually, the number of available processors on the platform is less than the number of actors of the dataflow graph. Several actors thus have to share the same processor. The mapping process defines the set of actors a processor executes. For example, in Figure 20.3, only three processor cores are available to execute the five actors composing the application. Actors A_1 and A_3 are mapped on processor P_1, A_4 and A_5 are mapped on P_2, and A_2 is mapped on P_3.

Executing programs efficiently on such platforms requires balancing the computational load over the processors. Furthermore, knowing that the communications and synchronizations may also become a bottleneck, the connectivity between actors, as well as the communication rates, are also key factors of the actor mapping. The mapping process can be translated to a graph partitioning problem. It is an NP-complete problem, but some algorithms find high-quality partitions in a short time using a multilevel scheme. In practice, we use the Metis tool [15] for this step. The application is previously profiled to find out the computation workload of each actor and the communication rates such that the actors will be balanced onto the processors while minimizing data exchange between processors.

In theory, the dataflow model of computation assumes an ideal execution model offering unlimited computation resources and unbounded communication channels, which enables the execution

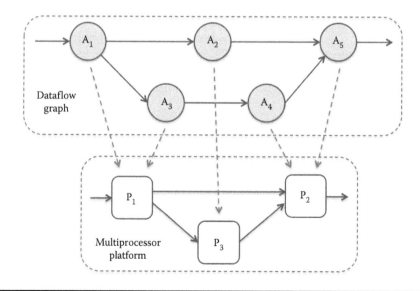

Figure 20.3 Dataflow graph mapping onto processor cores.

of all actors in parallel. In practice, owing to the limited number of processing units, a user-level scheduler is required [8]. This scheduler can sequentially test the firing rules from several actors and fire an actor if a firing rule is valid. An efficient scheduling for dataflow programs thus consists in finding an order of actor firings such that the use of all the processing units of the execution platform is maximized. Because actors in a DPN may have data-dependent behaviors, determining an optimal schedule of a program is not possible at compile time, that is, in the general case, the scheduling can only be done at run time. Actually, we use a simple scheduling strategy called round-robin strategy [16], based on a compile-time ordering of the actor execution. This strategy is easy to implement and usually provides high data rates, thanks to spatial/temporal locality. The scheduler continuously goes over a static list of actors: it evaluates the firing rules of an actor, fires the actor if a rule is met, and continues to evaluate the same actor until it no longer meets a firing rule. Then the scheduler switches to the next actor in the list. This scheduling policy guarantees to each actor an equal chance of being executed, and avoids deadlock and starvation. Contrary to the classical round-robin scheduling, there is no notion of time slice, so the timing is performed at run time: an actor is executed until it cannot fire in order to minimize the frequency of actor switching and consequently the scheduling overhead. The reason for this actor switching is that, in practice, the FIFO channels will finally be full or empty because of their bounded sizes. The scheduling strategy is applied locally to avoid global synchronizations that require excessive time for high-performance applications. A local scheduler is thus assigned to every processor. A local scheduler is in charge of scheduling the actors mapped on the processor it is attached to, and it implements the round-robin strategy on these actors only.

20.3.1.2 Toward Efficient Implementations of Video Coding Applications

The execution platform we target is made of homogeneous low-complexity processor cores. Using processors that can be configured by software code enables flexibility. Low-complexity processors provide low power consumption that embedded systems require, and the number of processor cores makes it possible to meet performance.

The architectural model of the platform is presented in Figure 20.4. Both shared and local memories are used to limit the traditional memory bottleneck. For example, processors (P1, ..., Pi) use their own local memories (LM$_1$, ..., LM$_i$) for executing their actors, but the processors are also connected, through an interconnection network, to a set of shared memories (SM$_1$, ..., SM$_j$) devoted to inter-processor communications. A local memory stores the internal FIFO channels (i.e., circular buffers and the read/write pointers) that let actors assigned to the same processor communicate together. The local memory also may have to store the current states of these actors and store the heap and the stack used during the execution of the actions, similar to traditional programs.

The shared memories store the FIFO channels that connect together two actors mapped onto two different processors. Knowing a single shared memory can contain multiple channels, the design flow has thus to assign not only actors to processors but also FIFO channels to shared memory components. Actually, FIFO channels can be freely mapped to memory components because they are not dependent on each other. However, some architectural constraints may have to be considered, such as the topology of the interconnection network or the size of the memory components.

The multiprocessor platform is based on the transport-triggered architecture (TTA) processor, for example, a very long instruction word-style processor that executes multiple instructions at each clock cycle [17]. TTA processors are extremely configurable thanks to the open source TTA Co-design Environment (TCE) [18], which offers a toolset for their design and programming. The TCE toolset enables the design of custom processors and their realization into VHDL files and memory images that can be easily synthesized on an FPGA platform. For example,

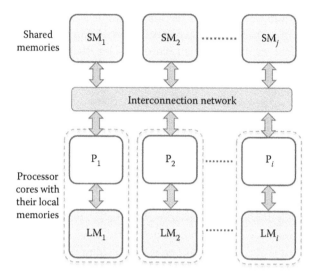

Figure 20.4 Architecture model of the platform.

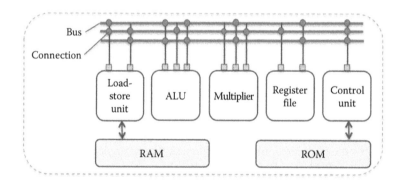

Figure 20.5 A simple TTA-based processor.

the designer can make the processor tiny and energy efficient or, on the contrary, he/she can increase the instruction-level parallelism of the processor by adding more execution units and buses. Figure 20.5 presents a simple TTA-based processor composed of three buses, one arithmetic and logic unit (ALU); one multiplier, one register file, one load-store unit to manage data accesses in the RAM memory and input/output communication; and one control unit connected to the ROM memory containing the instructions. The hardware description of each processor is generated from its high-level description by the TCE using a preexisting database of standard hardware components.

The design flow is presented in Figure 20.6. It is implemented around two open-source tool-sets: Orcc (Open RVC-CAL Compiler) [19] and TCE [18]. Orcc can be considered as an RVC-CAL front-end for TCE and TCE as a TTA back-end for Orcc. The inputs of the design flow are the application description (i.e., the RVC network and the actor specifications), the platform configuration (i.e., the model of the TTA processor [functional units, register files, number of buses, etc.] and the number of processors), and the mapping specification (i.e., the mapping of the actors

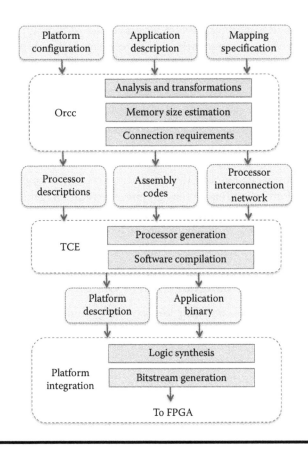

Figure 20.6 **Design flow based on Orcc and TCE toolsets.**

onto the processors) previously performed with Metis tool as stated earlier. Orcc generates a high-level description of the processors, an intermediate representation of the software code associated with each actor, and the processor interconnection requirements. Then TCE uses this information to generate a complete multiprocessor platform design: the model of the processors enables the generation of their VHDL descriptions using the preexisting database of hardware components (HDL view of the functional units, register files, etc.), and the software code is compiled into an executable binary code that will execute the actors on the processors.

Two processors have to be interconnected together through a shared memory if they execute, respectively, actors A_i and A_j that communicate with each other, that is, if A_i is connected to A_j through a FIFO channel in the dataflow graph, but A_i and A_j are not mapped onto the same processor. If A_i and A_j are assigned to the same processor, a local memory is used. In order to instantiate correctly the corresponding RAM blocks (BRAM) that will be required on the FPGA platform, the sizes of the memories (local memories and shared memories) are estimated by Orcc according to the FIFO channels they will implement.

With the proposed design flow, dataflow-based programs are thus compiled into instruction codes executable on the multiprocessor platform in two successive steps. The first step translates the whole dataflow description into a procedural intermediate representation (IR) that is low level but still independent of the targeted processor, which has been developed for the LLVM project [20]. This step is performed by Orcc. Several transformations are performed. Among them, the LLVM IR allows the designer to express bit-accurately the word length of each variable and communication channel. When a computation has to be performed with two variables of different word lengths, the correct result is ensured by the use of an explicit cast instruction. Moreover, when an actor is selected to be scheduled, an action scheduler evaluates the firing rules of this actor so as to determine the action to fire. The firing rules are evaluated successively according to the partial order defined within the actor (priorities and FSM). The action scheduler evaluates three conditions to determine the fireability of an action: (1) the amount of tokens required in the input FIFO channels, (2) the potential condition on the values of the tokens and state variables, and, due to the restricted size of the memories used to implement the FIFO channels in practice, (3) the number of available rooms in the output FIFO channels. Actually, Orcc partitions the dataflow application over the platform according to the mapping specification, generating a separate program for each processor. Then, a second step compiles successively the program of each processor from the LLVM IR into processor instructions thanks to the processor description, making the whole application executable on the multiprocessor platform. This step is performed by TCE.

20.3.2 A Domain-Specific Language for FPGA-SDR

The radio standards evolve rapidly, and their low-level programming is still written and maintained manually. In addition, the growth of the platform's complexity highlights the limitations of the current programming languages. To tackle these issues, our approach encompasses both a DSL that formalizes the application structure, behavior, and requirements in a declarative way, and an associated DSL-waveform compiler to generate multiple artifacts such as source code. The proposed approach is similar to the model-driven engineering (MDE) concept [21]. The MDE approach implies a "correct-by-construction" development of the final product.

Thus, the objective of the SDR-FPGA approach is the automatic generation of the RTL description of a radio waveform. An example of this description is shown in Figure 20.7, which is the RTL implementation model of the dataflow graph of Figure 20.1. Toward this end, we have defined a DSL that provides the primitives to model an SDF waveform together with its data frame structure. This framework is featured with the HLS tools and will be detailed later in this section. It is also a library-based framework because all the functional blocks will be loaded from imported HLS and RTL libraries. The combination of the waveform specification and the data

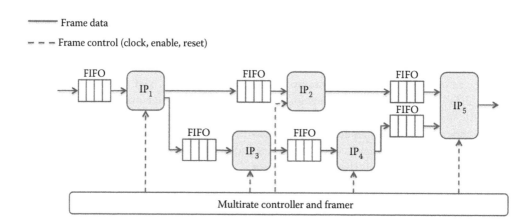

Figure 20.7 RTL implementation of the SDF representation.

frame structure provides sufficient information to build the processing unit (data path) and the control unit (Multirate Controller and Framer [MRCF]) of Figure 20.7.

Methods for automatic hardware synthesis from data frame specifications are not often encountered in the signal processing community. As an example, in Ref. [22], an XML-based waveform specification flow, including frame definition, is discussed. It is a proposal for the SDR physical layer's specification and uses XML as an entry point, in contrast to our approach based on a DSL. XML hardware-specific interpreters are used in that case to convert the XML physical layer description for a specific running platform. Our solution is quite similar and is based on a DSL. Moreover, we target the multirate architectures and accordingly include primitives to specify such behavior. An algorithm is also proposed to build the control unit from the DSL specification.

20.3.2.1 Frame-Based Waveform Description

Data framing is used in most of the telecommunication standards. A data frame is composed of a set of fields conveying meaningful data. Each field of a given data frame is mainly characterized by its duration together with the nature of the data it transports, which can be frame-specific information (e.g., modulation schemes or coding rate), a field appended for synchronization purpose (e.g., preamble), or useful data (e.g., data payload) originating from the upper layers of the application. An example of a generic data frame is given in Figure 20.8.

A data frame F is defined as a collection of fields, that is,

$$F = \cup_{i=1}^{N} F_i,$$ (20.1)

where F_i denotes the ith field. Each field F_i is characterized by its duration T_i and its *constant* or *variable* nature *State* of its transported data *Payload*:

$$F_i = \{T_i, State, Payload\}.$$ (20.2)

The duration T_F of the overall frame F is computed as

$$T_F = \sum_{i=1}^{N} T_i.$$ (20.3)

The data path, on both the transmitter and receiver side, is a set of interconnected functional blocks as described in Figure 20.1 characterized by their latency (L), throughput (TP), and input

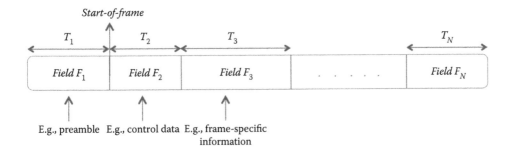

Figure 20.8   Generic data frame *F*.

and output data rates (f_{in} and f_{out}). For each block, the input's and output's rates are known, and they can be managed using enable signals.

Let FB_j be the jth functional block:

$$FB_j = \left\{ f_{in_j}, f_{out_j}, L_j, TP_j \right\}. \tag{20.4}$$

In the proposed DSL, while specifying the data path, the instantiation of each functional block includes a listing of the fields that it will process and their respective sources. For example, the fact that FB_1 operates on fields F_1 and F_3 is a piece of information that will be specified up-front in the DSL.

This information is used to build the appropriate architecture. On the one hand, the frame specification gives some specific information such as the duration of each field, which helps to generate the read and write clock signals during the appropriate period of time. The oversampling or down sampling mechanisms do not affect the fields' duration but they imply some rate changes that also have to be constantly tracked during the processing. On the other hand, the data path specification gives information on the input and output rates of each functional block and their relation with the frame specification. Combining these pieces of information (duration, sources, and rates within the DSL), a suitable control unit is inferred for the specified multirate signal processing waveform. This control is named *Multirate Controller and Framer* (MRCF) as introduced in Figure 20.7.

In order to achieve such a workflow, the DSL provides a set of language primitives to characterize a physical layer within the SDR context. A typical PHY DSL-based specification starts with generalities, namely libraries inclusion, data rate requirements, and constant definition. These are performed by using the key words *#include, frequency*, and *#define*, respectively. After specifying the generalities, a description of each field of the data frame has to be completed. A field is specified in a structure-like manner. First, the keywords *#fieldC* or *#fieldV* highlight the constant or the variable nature of the field. Then, the field is assigned a name followed by curly braces. Within the braces, different attributes of the field are enumerated. These attributes are its duration, its data payload, or a potential data redundancy in the payload. A resulting frame is built out of the fields and a *Start-Of-Frame* flag is designated among the fields for synchronization purpose.

The data-path is mainly made out of the functional block imported from the aforementioned libraries. These blocks originate from diverse HLS tools because the workflow leverages the HLS concept for rapid prototyping. Traditional RTL (VHDL and Verilog) blocks are also being considered. Dedicated key words such as *#catapultc, #vivadohls*, and *#rtl* are used to stamp each block with a specific synthesizing tool. It is important to recall that while instantiating each block, a listing of fields is appended to restrict blocks to process only those fields of the frame. The attributes of an instantiated block consist of input and output ports mapping and also of diverse architectural constraints. Associated with each port mapping is a data rate that enables to generate the reading and writing of clock signals. The architectural constraints are used to explore the design via the HLS capabilities.

The resulting PHY operates in two modes. In transmitter (TX) mode, the main task consists of data framing in order to comply with the desired standard prior to transmission. The MRCF handles the data framing by activating the FBs at the required instants. It also controls the insertion of constant data within the frame. That is, constant fields are one-time computed and inserted into the frame at run time. In some modern standards, pilot signal are inserted to estimate and compensate for the channel distortion over the transmitted signal. Pilot insertion is also controlled by the MRCF. Because the transmitter has a feed-forward architecture, its overall control is not a complex task to achieve.

In receiver (Rx) mode, the detection of a specific signal called *Start-Of-Frame* triggers the frame decoding scheduled by the controller. Here again, the MRCF is responsible for the activation and deactivation of the processing IPs at the appropriate time. This activation is done under Boolean conditions which depend on the structure of the data frame and the given specification of the data path. A block intended to process a specific field of the frame will be activated during the duration of that field. The rates of reading and writing the data are given in the specification, and they help to generate the appropriate *enable* signal over this entire duration.

20.3.2.2 Waveform Compiler and MRCF Controller

The process of describing an SDR application through the proposed framework is threefold [23]. Figure 20.9 illustrates the complete flow, which is a top–down flow. The early step of the process consists in describing the application using the syntax offered by the DSL. At this stage, the user is given the freedom to model an application by considering constrained HLS blocks or RTL IPs from the available libraries or to design his or her own IPs. The overall waveform description is based on the frame structure, as we explain later. The system control logic is inferred from this specification, and the scripting is interfaced with different libraries of components. At compile time, the required functional blocks are loaded from the libraries and synthesized into RTL using *.tcl* scripts generated as one of the DSL artifacts. The *Waveform compiler* performs the interconnection of the different blocks according to the description made. This interconnection is preceded by data type consistency checks and FIFO selection as the interface between two blocks. At the block level, the constraints can be inferred using annotations *#pragma* within the code or made explicit in each *.tcl* script.

Finally, a classical *Platform Integration* step is performed to synthesize the generated RTL description into bitstream and integrate the design on a dedicated FPGA platform.

In addition to the data path, the MRCF control logic is built up from the frame specification. This control decides the signal roadmap throughout the design by providing enable signals to

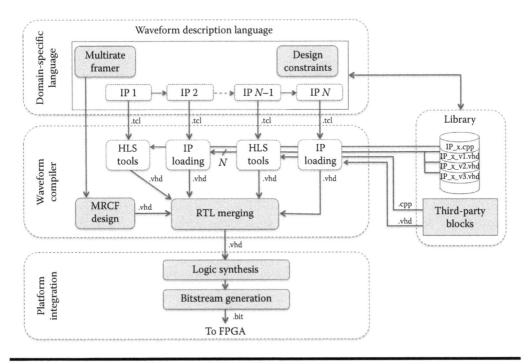

Figure 20.9   Top-down FPGA-SDR design flow.

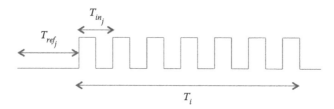

Figure 20.10   Enable distribution for the activation of *FB_j* operating on *F_i*.

the processing blocks. The proposed language automatically generates such a controller based on an abstracted specification of the dataflow waveform. In Rx mode, the *Start-Of-Frame* notion is introduced to point out the moment the receiver is correctly synchronized. As mentioned earlier, most of the modern standards stream some synchronization fields within their data frame. After the receiver is woken up, those synchronization fields are, in most of the cases, processed before the data payload. Thus, prior to the *Start-Of-Frame* detection, all the blocks processing the pre-start-of frame fields are held active. In other words, these blocks keep on computing the incoming data until a *Start-Of-Frame* is detected. After a *Start-Of-Frame* detection, the *MRCF* undertakes the control of the rest of the architecture to sketch the decoding process of the data payload.

An underlying algorithm is used to perform the computation of the enable signals. Given a functional block FB_j computing the field F_i at rate f_{in_j}, the block has to be enabled a number of times equal to

$$k_{i,j} = T_i f_{in_j} \tag{20.5}$$

The enable distribution is detailed in Figure 20.10.

Each remaining block is stamped with a distance as compared to a reference time. This reference time is initialized with the detection of the *Start-Of-Frame* signal. Then, this distance is used to activate each block, and it reflects the combined latency of the preceding blocks. Let T_{ref_j} be the distance between the reference block (following *Start-Of-Frame* detection) and the block FB_j. The heuristic used to estimate this distance is the maximum of the sum of the latencies of the blocks operating between FB_{ref} and FB_j.

20.4 Experiments and Rapid Prototyping Validation

The proposed approaches allow the rapid prototyping for video coding over flexible radio links. As a study case, an MPEG-4 decoder and an IEEE 802.11g receiver have been implemented. We focus on the decoding part of the targeted applications as their data path is more complex than their encoding part.

The high-level description of these applications combined with the proposed design flows allows architecture design flexibility. Design exploration results illustrating this flexibility are given in this section.

20.4.1 MPEG-4 Decoder over IEEE 802.11g System

The MPEG-4 Part 2 standard, also known as MPEG-4 visual, was released in 1999 by the joint ISO/ITU consortium. The popular DivX and Xvid codecs, which have largely contributed to the development of video sharing over the Internet, implement this standard. In fact, The Simple Profile (SP) of MPEG-4 Part 2 decoder was the first application standardized by the RVC working group. Figure 20.11 presents the normative version of the description. The graph is partitioned into four parts, each one corresponding to a dedicated processing. The parser extracts the values needed by the next processing from the compressed data stream. The stream is decompressed thanks to entropy decoding, after which the syntax elements composing the stream are extracted in order to be transmitted to the concerned actors. The texture decoding decodes the error resulting from the image predication using the IDCT inverse transform. The third part is dedicated to motion compensation. To increase the parallelism exposed within the decoder, the parser separates the processing of each image components, luma and chroma, in three parallel paths (Y, U, and V). The image components are then merged back at the end of the processing. MPEG-4 Part 2 decoder is designed with 41 actors and 143 FIFO channels. Computation and communication workloads of each actor, which are used for the mapping of the actors onto the processors, have been profiled. Computation workload ranges between 0.5% and 17% of the total computation workload of the MPEG-4 decoder, depending on the granularity level and the complexity of the actor. Two categories of communications are identified: video streams characterized by a large amount of data such as for actors within the motion compensation part, and control communications characterized by a small amount of data, as is the case for most of the actors within the parser part.

IEEE 802.11g is part of the IEEE 802.11 standard used in the popular WiFi technology. The DSL proposed in Section 20.2.2 relies on the frame description of the radio waveform and the SDF model of the system. The IEEE 802.11g frame of Figure 20.12 has been described using the proposed language by setting the duration and the state of each field. The DSL description also specifies the *Start-Of-Frame* position that is set after the synchronization field named *Long Preamble*. Then the SDF model of the receiver is constructed by connecting the different functional blocks and by specifying which block operates on which field.

The IEEE 802.11g receiver architecture is detailed in Figure 20.13. The physical layer is mainly based on OFDM (orthogonal frequency division multiplexing) modulation. Our design mainly implements the Hamming windowing and the forward fast Fourier transform (FFT), required to decode an OFDM-based signal, The time synchronization has also been implemented to generate the *Start-Of-Frame* signal. In our design, the FFT computes 256 points, and the mapping is a 16-QAM modulation. The FFT is implemented with a straightforward in-place Radix4 algorithm.

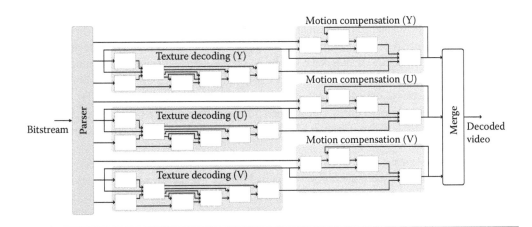

Figure 20.11   **RVC-based description of the MPEG-4 Part 2 SP decoder.**

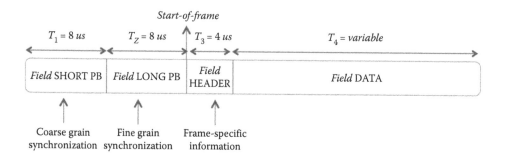

Figure 20.12 IEEE 802.11g frame structure.

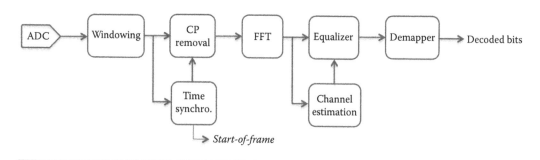

Figure 20.13 SDF model of the IEEE 802.11g receiver.

The synchronization is performed using a matched-filter-based correlation. All theses functional blocks have been programmed using the CatapultC HLS tool [23], and the generated IPs are merged and controlled using the generated MRCF.

20.4.2 Rapid Prototyping Implementation Results

20.4.2.1 MPEG-4 Decoder

Using the proposed design flow based on Orcc and TCE toolsets, we can evaluate in a short time the complexity and the performance of an implementation of an RVC application, and then generate both the RTL description of the platform and the application binary to prototype the application. As stated previously, the actors are mapped onto the processors based on graph partitioning using the Metis tool, and scheduled on the processor they are assigned according to the round-robin strategy.

The TTA processor we used for the multiprocessor platform is based on Ref. [24]. It mainly includes 3 arithmetic and logic units, 1 multiplier, 3 register files, and 18 buses. Table 20.1 shows the synthesis results for a Xilinx Virtex 6 FPGA platform (XC6VLX240T). The Xilinx ISE tool was used for synthesis.

During the experiments, in order to ease the design and not impact the performance, the size of the FIFO channels is kept the same for every channel (8192). The performance evaluation is made thanks to the instruction-set simulator included in the TCE. The decoding is simulated on the TTA-based multiprocessor platform clocked at 100 MHz. Figure 20.14 shows the influence of the number of processors on the frame rate (in frames per second [FPS]) of the MPEG-4 Part 2 Simple Profile decoder. We consider the decoding of a video sequence with a QCIF resolution (176 × 144 pixels).

Table 20.1 Synthesis Result for One TTA Processor

Slices	FF	LUT	DSP	BRAM 18 Kbits	BRAM 36 Kbits
1570	2577	4132	3	0	20

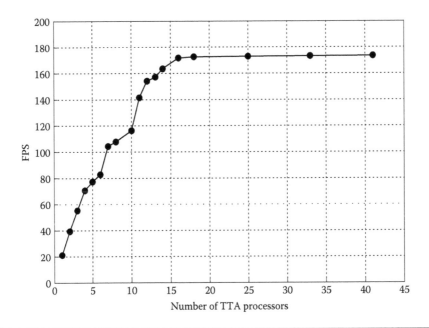

Figure 20.14 Influence of the number of TTA processors on the performance of MPEG-4 Part 2 SP decoder.

Table 20.2 Synthesis Result for MPEG-4 Part 2 SP Decoder with 5 TTA Processors

Slices	FF	LUT	DSP	BRAM 18 Kbits	BRAM 36 Kbits
13578	12274	33629	15	20	302

The maximum decoding frame rate of our MPEG-4 Part 2 SP decoder is reached with 16 TTA processors, and the speed is 8.1 times higher in comparison with MPEG-4 decoding on a single TTA processor. Increasing the number of processors further does not provide a higher decoding frame rate. Thirty-two shared memories are used in this case. The complexity of the decoder is shown in Table 20.2 using five TTA processors as an example.

20.4.2.2 IEEE 802.11g Receiver

The design of the IEEE 802.11g receiver was synthesized on a Virtex 6 FPGA platform (XC6VLX240T), and Xilinx ISE tool was used for synthesis. The clock frequency was set to 100 MHz.

First, an 8 Mbit/s IEEE 802.11g receiver design was generated by the DSL. The DSL waveform compiler output being a VHDL file, the design is then synthesized using ISE tools.

Table 20.3 Synthesis Result for DSL-Based Design of IEEE 802.11g Receiver

Slices	FF	LUT	DSP	BRAM 18 Kbits	BRAM 36 Kbits
961	803	2832	0	5	0

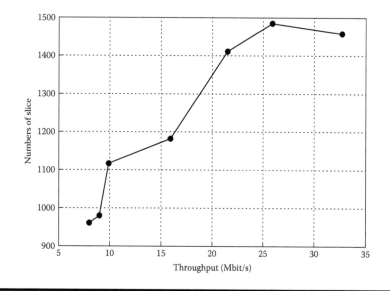

Figure 20.15 Resource estimation versus data rate of the IEEE 802.11g receiver.

Synthesis results after place and route are detailed in Table 20.3. In order to have a fair comparison between the different architectures, the number of DSP blocks has been forced to zero. Thus, only slices are used, and the number of slices will be used for comparison.

The data rate of the receiver mainly depends on the FFT block. Without optimization, the FFT FB takes 9996 cycles to compute a 256-point FFT, which sets the symbol duration to 99.96 μs with 100 MHz clock frequency. In an OFDM symbol, 200 carriers are used to transmit 16-QAM mapped symbols (4 bits per symbol), leading to 800 bits per 99.96 μ s. Finally, the data rate is 8 Mbit/s.

Several optimizations are made available on a typical HLS tool [23]. They make it possible to optimize the design at the area or throughput level. The two design optimizations that are used for the exploration with CatapultC are loop pipelining and loop unrolling. Loop pipelining provides a way to increase the throughput of a loop (or decrease its overall latency) by initiating the (*i*+1)th iteration of the loop before the *i*th iteration has completed. Loop unrolling reduces the total loop iterations by duplicating the loop bodies. The number of loop iterations is then reduced, but care must be taken regarding the data dependencies when using this technique. Loop unrolling impacts the block latency and consequently the throughput of the application.

Using the proposed design flow, we explored in a short time some designs that could be achieved by optimizing some IPs in the proposed design flow. The FFT block is the main degree of freedom of the exploration. Exploration results are shown in Figure 20.15 with the number of slices as a function of the data rate. The slice number varies from 1 up to 1.6 while increasing the throughput from 1 up to 4.

20.5 Conclusion

In this chapter, we have presented two approaches to the implementation of dataflow programs.

The first approach makes it possible to implement dynamic dataflow programs as for video coding and targets a network of low-complexity processors. Processors offer flexibility thanks to the compiler that translates the application to machine code, and performance depends on the number of processors the platform embeds.

The second approach is dedicated to the synchronous dataflow description of radio waveforms. It is based on high-level synthesis and a DSL that includes a data frame description. It allows the exploration of the design space according to the user-defined data rate constraints.

Both approaches aim at simplifying the design process, from the application level up to the RTL description of the hardware platform, and enable the rapid prototyping of video and SDR applications.

References

1. D. Amos, A. Lesea, and R. Richter. *FPGA-Based Prototyping Methodology Manual: Best Practices in Design-for-Prototyping*. Synopsys Press, 2011.
2. M. Mattavelli, M. Raulet, and J. W. Janneck. MPEG reconfigurable video coding. In S. S. Bhattacharyya, E. F. Deprettere, R. Leupers, and J. Takala, Eds, *Handbook of Signal Processing Systems*, pp. 281–314. Springer, New York, 2013.
3. T. Bollaert. Catapult synthesis: A practical introduction to interactive C synthesis. In *High-Level Synthesis: From Algorithm to Digital Circuit*, pp. 29–52. Springer, New York, 2008.
4. J. Eker and J. W. Janneck. *CAL Language Report: Specification of the CAL Actor Language*. Technical report, University of California, Berkeley, 2003.
5. A. Van Deursen, P. Klint, and J. Visser. Domain-specific languages: An annotated bibliography. *ACM Sigplan Notices*, 35: 26–36, 2000.
6. A. Pasha, S. Derrien, and O. Sentieys. System level synthesis for wireless sensor node controllers: A complete design flow. *ACM Transactions on Design Automation of Electronic Systems (TODAES)*, 17(1): 2.1–2.24, 2011.
7. International Standard ISO/IEC FDIS 23001-4: MPEG systems technologies—Part 4: Codec Configuration Representation.
8. E. A. Lee and T. Parks. Dataflow process networks. *Proceedings of the IEEE*, 83(5): 773–801, 1995.
9. J. Mitola, III. Software radios: Survey, critical evaluation and future directions. *IEEE Aerospace and Electronic systems Magazine*, 8, 25–36, 1993.
10. J. F. Jondral. Software-defined radio: Basics and evolution to cognitive radio. *EURASIP Journal on Wireless Communications and Networking*, 3: 275–283, 2005.
11. M. Dardaillon, K. Marquet, T. Risset, and A. Scherrer. Software defined radio architecture survey for cognitive testbeds. In *2012 8th International Wireless Communications and Mobile Computing Conference (IWCMC)*, pp. 189–194, August 2012.
12. M. Cummings and S. Haruyama. FPGA in the software radio. *IEEE Communications Magazine*, 37(2): 108–112, 1999.
13. A. Di Stefano, G. Fiscelli, and C. G. Gianconia. An FPGA-based software defined radio platform for the 2.4GHz ISM band. In *IEEE PRIME 2006. IEEE Ph.D. Research in Microelectronics and Electronics*, vol. 12, pp. 73–76, 2006.
14. E. A. Lee and D. G. Messerschmitt. Synchronous data flow. *Proceedings of the IEEE*, 36: 24–35, 1987.
15. G. Karypis and V. Kumar. A fast and high quality multilevel scheme for partitioning irregular graphs. *SIAM Journal on Scientific Computing*, 20(1): 359–392, 1998.
16. H. Yviquel, E. Casseau, M. Wipliez, and M. Raulet. Efficient multicore scheduling of dataflow process networks. In *2011 IEEE Workshop on Signal Processing Systems (SiPS)*, pp. 198–203, October 2011.

17. H. Corporaal. *Microprocessor Architectures: From VLIW to TTA*. John Wiley, Chichester, UK, 1997.

18. TCE. The TTA-based Co-design Environment. http://tce.cs.tut.fi/

19. Orcc. The Open RVC-CAL Compiler: A development framework for dataflow programs. http://orcc.sourceforge.net

20. C. Lattner and V. Adve. LLVM: A compilation framework for lifelong program analysis and transformation. In *International Symposium on Code Generation and Optimization, 2004. CGO 2004*, pp. 75–86, Palo Alto, California, 2004.

21. D. C. Schmidt. Guest editor's introduction: Model-driven engineering. *Computer*, 39(2): 25–31, 2006.

22. E. Grayver, H. S. Green, and J. L. Roberson. SDRPHY—XML description for SDR physical layer. In *Military Communications Conference, (MILCOM 2010)*, pp. 1140–1146, October 2010.

23. V. Bahtnagar, G. S. Ouedraogo, M. Gautier, A. Carer, and O. Sentieys. An FPGA software defined radio with a high-level synthesis flow. In *IEEE Vehicular Technology Conference (VTC-Spring13)*, June 2013.

24. O. Esko, P. Jääskeläinen, P. Huerta, C. S. de La Lama, J. Takala, and J. Ignacio Martinez. Customized exposed datapath soft-core design flow with compiler support. In *Proceedings of the 2010 International Conference on Field Programmable Logic and Applications*, pp. 217–222, August 2010.

Index

9 781138 034013